Electronic Waste Management

Electronic Waste Management

Policies, Processes, Technologies, and Impact

Edited By

Sunil Kumar
Waste Re-Processing Division
CSIR-National Environmental Engineering Research Institute
(CSIR-NEERI), Maharashtra, India

Vineet Kumar
Department of Microbiology, School of Life Sciences
Central University of Rajasthan, Rajasthan, India

WILEY

This edition first published 2024

Registered Office
John Wiley & Sons, Inc., 111 River Street, Hoboken, NJ 07030, USA

For details of our global editorial offices, customer services, and more information about Wiley products visit us at www.wiley.com.

Wiley also publishes its books in a variety of electronic formats and by print-on-demand. Some content that appears in standard print versions of this book may not be available in other formats.

Library of Congress Cataloging-in-Publication Data
Names: Kumar, Sunil (Environmental engineer), editor. | Kumar, Vineet (Vineet Kumar Rudra), editor. | John Wiley & Sons, publisher.
Title: Electronic waste management : policies, processes, technologies, and impact / edited by Sunil Kumar, Vineet Kumar.
Description: Hoboken, NJ : JW-Wiley, 2024. | Includes bibliographical references and index.
Identifiers: LCCN 2023041026 (print) | LCCN 2023041027 (ebook) | ISBN 9781119891512 (hardback) | ISBN 9781119891529 (pdf) | ISBN 9781119891536 (epub) | ISBN 9781119891543 (ebook)
Subjects: LCSH: Electronic waste.
Classification: LCC TD799.85 .E36 2024 (print) | LCC TD799.85 (ebook) | DDC 621.3815028/6--dc23/eng/20231018
LC record available at https://lccn.loc.gov/2023041026
LC ebook record available at https://lccn.loc.gov/2023041027

Cover Design: Wiley
Cover Image: © gopixa/Shutterstock

Set in 9.5/12.5pt STIXTwoText by Integra Software Services Pvt. Ltd, Pondicherry, India

Printed and bound by CPI Group (UK) Ltd, Croydon, CR0 4YY

C9781119891512_100124

Contents

Preface

Electronic waste (e-waste) is a growing problem around the world. Every year, millions of tons of electronic devices are discarded, and much of this waste ends up in landfills. This can have a number of negative impacts, including environmental pollution, health risks, and economic losses. The environmental impacts and health risks of e-waste are significant. When e-waste is dumped in landfills, it can leach harmful chemicals into the soil and water. These chemicals can contaminate drinking water, harm wildlife, and pollute the air. E-waste also contains hazardous materials, such as lead, mercury, and cadmium, which can pose a health risk to people who are exposed to them. The economic impacts of e-waste are also significant. When e-waste is not adequately managed, it can cost millions of dollars to clean up. In addition, the loss of valuable resources, such as metals and plastics, can cost the global economy billions of dollars each year. The unprecedented growth in electronic consumption has brought about a significant challenge—the mounting problem of e-waste. Recognizing the urgency to address this global issue, we are pleased to present *Electronic Waste Management: Policies, Processes, Technologies, and Impact*.

The primary goal of this book is to foster a deeper understanding of the complex issues surrounding electronic waste and to explore innovative solutions that can help mitigate its adverse effects on the environment and health of humans and animals. We have invited together a diverse group of experts, including researchers, policymakers, and industry professionals, who generously shared their knowledge and experiences in the field to tackling this global issue. Their diverse perspectives and in-depth research have made this book a valuable resource for policymakers, researchers, industry professionals, and anyone interested in understanding and tackling the challenges posed by e-waste disposal.

The chapters of this book encompass a wide range of topics, covering both the theoretical foundations and practical applications of electronic waste management. The journey begins with an examination of the environmental consequences resulting from the presence of electronic waste in our ecosystems. Through rigorous scientific analysis and data-driven insights, we explored the toxicological implications, pollution pathways, and ecological disruptions caused by the mishandling and improper disposal of electronic waste.

This book is intended for a wide audience, including policymakers, businesses, and individuals. It is a valuable resource for anyone who wants to learn more about e-waste management and how to reduce the environmental and social impacts of this growing problem.

We hope that this book will help to raise awareness of the e-waste problem and promote the development of its sustainable management practices around the globe.

Sunil Kumar
Vineet Kumar

Acknowledgment

Dr. Vineet Kumar expresses his sincere thanks to Dr. Pradeep Verma, Professor in the Department of Microbiology, School of Life Sciences at the Central University of Rajasthan, Rajasthan, India, for his kind and moral support and for providing a fantastic facility in his laboratory in the last phase of compilation of the authors manuscript and to complete this task. Dr. Vineet Kumar gratefully acknowledges the Science and Engineering Research Board, Government of India for providing a Postdoctoral fellowship (F. No.: PDF/2022/000038).

Sunil Kumar
Maharashtra, India

Vineet Kumar
Rajasthan, India

About the Editors

Sunil Kumar is a well-rounded researcher with more than 22 years of experience in leading, supervising, and undertaking research in the broader field of Environmental Engineering and Science with a focus on Solid and Hazardous Waste Management. Dr. Kumar is a graduate in Environmental Engineering and Management from the Indian Institute of Technology, Kharagpur, India. He completed his Ph.D. in Environmental Engineering from Jadavpur University, Kolkata, India. His primary area of expertise is solid waste management (Municipal Solid Waste, Electronic waste etc.) over a wide range of environmental topics including contaminated sites, EIA, and wastewater treatment. His contributions in these fields led to a citation of approx. 15300, h-index of 62, and i10-index of 252 (Google scholar). His contributions since inception at CSIR-National Environmental Engineering Research Institute (NEERI), India in 2000 include approx. 300 refereed publications, 05 books, and 40 book chapters, 10 edited volumes, and numerous project reports to various governmental bodies and private, local, and international academic/research bodies. He is the Associate Editor of peer-reviewed journals of international repute i.e., Environmental Chemistry Letter, International Journal of Environmental Science & Technology, ASCE Journal of Hazardous, Toxic and Radioactive Waste. He also served as Editorial Board Member of Bioresource Technology, Elsevier during 2017-2023. He has completed more than 22 research projects as PI with 15 (11 awarded) Ph.D. and 20 M. Phil/M. Tech thesis/dissertations. Dr. Kumar was awarded the most prestigious award Alexander von Humboldt-Stiftung Jean-Paul-Str.12 D-53173 Bonn, Germany as a Senior Researcher for developing a Global Network and Excellence for more advanced research and technology innovation.

Vineet Kumar is presently working as a National Postdoctoral Fellow in the Department of Microbiology, School of Life Sciences at Central University of Rajasthan, Rajasthan, India. He earned his Ph.D. (2018) in Environmental Microbiology from Babasaheb Bhimrao Ambedkar (A Central) University, Lucknow, India. Dr. Kumar's research work mainly focuses on the development of integrated and sustainable treatment techniques that can help in minimizing or eliminating hazardous waste in the environment. He has published more than 46 articles in reputed international peer-reviewed Journals with citations of more than 2300, and *h*-index of 28. In addition, he is the author/co-author of 4 Proceeding Articles, 52 Book Chapters, and 4 Scientific Magazine articles. Moreover, he has published 2 authored and 22 edited books on different aspects of Phytoremediation, Bioremediation, Wastewater Treatment, Waste Management, Omics, Genomics, and Metagenomics from reputed international publishers. Dr. Kumar has been serving as a guest editor and reviewer in many prestigious International Journals. He has served the editorial board of various reputed journals. He has presented several papers relevant to his research areas at national and international conferences. He is an active member of numerous scientific societies including the Microbiology Society (UK), the Indian Science Congress Association (India), the Association of Microbiologists of India (India), etc. He is the founder of the Society for Green Environment, India (website: www.sgeindia.org). He can be reached at vineet.way18@gmail.com; drvineet.micro@gmail.com.

List of Contributors

Solórzano Acosta
Dirección de Supervisión y Monitoreo en las Estaciones Experimentales Agrarias
Instituto Nacional de Innovación Agraria (INIA)
Lima
Peru

Oyetayo Olaoluwa Adefiranye
Department of Microbiology
Faculty of Science
University of Lagos
Lagos
Nigeria

Joshua Adeniyi Adeniran
Centre for Nanoengineering and Tribocorrosion (CNT)
University of Johannesburg
Johannesburg
South Africa

Ismaila Adejare Adesiyan
Renewable Energy and Biomass Research Group
Department of Chemical Engineering
University of Johannesburg
Johannesburg
South Africa

Richard Andi
Faculty of Pharmacy and Biochemistry, Biotechnology and Omics
Life Sciences Research Group
Universidad Nacional Mayor de San Marcos
Lima
Perú

Andrea Appolloni
Faculty of Economics
University of Rome, Tor Vergata
Rome
Italy

Michael Osei Asibey
Department of Planning, College of Art and Built Environment
Kwame Nkrumah University of Science and Technology
Kumasi
Ghana

Olusola Olaitan Ayeleru
Centre for Nanoengineering and Tribocorrosion (CNT)
University of Johannesburg
Johannesburg
South Africa

Gonzalo Martínez Barrera
Laboratorio de Investigación y Desarrollo de Materiales Avanzados (LIDMA)
Facultad de Química
Universidad Autónoma del Estado de México
San Cayetano
México

Carlos Eduardo Barrera Díaz
Facultad de Química
Universidad Autónoma del Estado de México
San Cayetano
México

Ruchi Bharti
Department of Chemistry
University Institute of Sciences
Chandigarh University
Mohali, Punjab
India

Pritha Bhattacharjee
Department of Environmental Science
University of Calcutta
Kolkata, West Bengal
India

Barnali Bhui
Department of Chemical Engineering
Indian Institute of Technology Guwahati
Assam
India

Ana Laura De la Colina Martinez
Facultad de Química
Universidad Autónoma del Estado de México
Toluca
México

Liliana Ivette Ávila Córdoba
Facultad de Ingeniería
Universidad Autónoma del Estado de México
Toluca
México

Hema Diwan
National Institute of Industrial Engineering
Powai
India

Laurence Dyer
Department of Mining Engineering and Metallurgical Engineering
Western Australian School of Mines
Curtin University
Kalgoorlie
Australia

Humberto Estay
Advanced Mining Technology Center
Universidad de Chile
Santiago
Chile

Bimbo Lolade Fafowora
School of Journalism and Media Studies
Rhodes University
Makhanda
South Africa

Department of Mechanical Engineering
Oduduwa University
Ipetumodu, Osun State
Nigeria

Flavio Augusto de Freitas
Centro de Biotecnologia da Amazônia–CBA
Manaus, Amazonas
Brasil

Universidade Federal do Amazonas
Programa de Pós-Graduação em Química–PPGQ
Manaus, Amazonas
Brasil

María Magdalena García Fabila
Laboratorio de Análisis Instrumental
Facultad de Química
Universidad Autónoma del Estado de México
Toluca
México

Fernando López Gayarre
Departamento de Construcción e Ingeniería de Fabricación
Universidad de Oviedo
Campus de Gijón
Gijón, Asturias
Spain

Gopikrishnan T
Department of Civil Engineering
National Institute of Technology Patna
Patna, Bihar
India

David Joaquín Delgado Hernández
Facultad de Ingeniería
Universidad Autónoma del Estado de México
Toluca
México

Edwin Hualpa-Cutipa
School of Environmental Engineering
Universidad Continental
Lima
Perú

Christian Ihle
Department of Mining Engineering
Universidad de Chile
Santiago
Chile

Advanced Mining Technology Center
Universidad de Chile
Santiago
Chile

Karpagaraj A
Department of Mechanical Engineering
National Institute of Technology Puducherry
Karaikal, Puducherry
India

Anmol Kaur
Satish Chandar Dhawan Government College
Ludhiana, Punjab
India

Muammer Kaya
Mining Engineering Department
Eskisehir Osmangazi University
Eskisehir
Turkey

Daniela Landa-Acuña
Environmental Engineering
Faculty of Engineering
Private University of the North
Lima–Los Olivos Campus
Peru

Facultad de Farmacia y Bioquímica
Departamento Académico de Farmacología, Bromatología y Toxicología
Centro Latinoamericano de Enseñanza e Investigación en Bacteriología Alimentaria (CLEIBA)
Research Group Biotechnology and Omics in Life Sciences
Universidad Nacional Mayor de San Marcos
Lima
Perú

Juan Mandy
National Institute of Industrial Engineering
Powai
India

Janelle Mendoza León
Instituto Nacional de Salud–Centro Nacional de Control de Calidad
Lima
Perú

Facultad de Farmacia y Bioquímica
Universidad Nacional Mayor de San Marcos
Lima
Perú

Boris Miguel López Rebollar
Instituto Interamericano de Tecnología y Ciencias del Agua
Universidad Autónoma del Estado de México
San Cayetano
México

Helen Uchenna Modekwe
Renewable Energy and Biomass Research Group
Department of Chemical Engineering
University of Johannesburg
Johannesburg
South Africa

Daniel Nascimento Motta
Centro de Biotecnologia da Amazônia–CBA/SUFRAMA
Manaus, Amazonas
Brasil

Universidade Federal do Amazonas
Programa de Pós-Graduação em Biotecnologia–PPGBIOTEC
Manaus, Amazonas
Brasil

Sayan Mukherjee
Department of Environmental Science
University of Calcutta
Kolkata, West Bengal
India

Aniruddha Mukhopadhyay
Department of Environmental Science
University of Calcutta
Kolkata, West Bengal
India

Md Shah Newaz
Faculty of Economics
University of Rome
Tor Vergata
Rome
Italy

Nitha T.S.
National Institute of Industrial Engineering
Powai
India

Fernando Ureña Núñez
Instituto Nacional de Investigaciones Nucleares
La Marquesa Ocoyoacac
México

Tarhemba Tobias Nyam
Centre for Nanoengineering and Tribocorrosion (CNT)
University of Johannesburg
Johannesburg
South Africa

Renewable Energy and Biomass Research Group
Department of Chemical Engineering
University of Johannesburg
Johannesburg
South Africa

Olawale Charles Ogunnigbo
School of Journalism and Media Studies
Rhodes University
Makhanda
South Africa

Peter Apata Olubambi
Centre for Nanoengineering and Tribocorrosion (CNT)
University of Johannesburg
Johannesburg
South Africa

Renewable Energy and Biomass Research Group
Department of Chemical Engineering
University of Johannesburg
Johannesburg
South Africa

Matthew Adah Onu
Centre for Nanoengineering and Tribocorrosion (CNT)
University of Johannesburg
Johannesburg
South Africa

Renewable Energy and Biomass Research Group
Department of Chemical Engineering
University of Johannesburg
Johannesburg
South Africa

Annu Pandey
Department of Chemistry
University Institute of Sciences
Chandigarh University
Mohali, Punjab
India

Shekhar Jyoti Pathak
Department of Chemical Engineering
Indian Institute of Technology Guwahati
Assam
India

Carlos Perea
Department of Mining Engineering
Universidad de Chile
Santiago
Chile

Advanced Mining Technology Center
Universidad de Chile
Santiago
Chile

William Pinheiro
Centro de Biotecnologia da Amazônia–CBA
Manaus, Amazonas
Brazil

Chemistry Graduate Program–PPGQ
Federal University of Amazonas–UFAM
Manaus, Amazonas
Brazil

Prabu Vairakannu
Department of Chemical Engineering
Indian Institute of Technology Guwahati
Assam
India

Vanessa Leal de Queiroz Hermino
Centro de Biotecnologia da Amazônia–CBA
Manaus, Amazonas
Brasil

Universidade Federal do Amazonas
Programa de Pós-Graduação em Biotecnologia–PPGBIOTEC
Manaus, Amazonas
Brasil

Ginarajadaça Ferreira dos Santos Oliveira
Centro de Biotecnologia da Amazônia–CBA
Manaus, Amazonas
Brasil

Renu Sharma
Department of Chemistry
University Institute of Sciences
Chandigarh University
Mohali, Punjab
India

Vidya Shetty K
Department of Chemical Engineering
National Institute of Technology Karnataka, Surathkal
Srinivasnagar
Surathkal, Mangalore
India

Edson Pablo da Silva
Centro de Biotecnologia da Amazônia–CBA
Manaus, Amazonas
Brasil

Instituto de Tecnologia e Educação Galileo da Amazônia
Manaus, Amazonas
Brasil

Universidade Federal do Amazonas
Programa de Pós-Graduação em Biotecnologia–PPGBIOTEC
Manaus, Amazonas
Brasil

Rajwinder Singh
Department of Civil Engineering
DR BR Ambedkar National Institute of Technology
Jalandhar, Punjab
India

Shri Krishna Singh
Department of Civil Engineering
National Institute of Technology Patna
Patna, Bihar
India

Karanvir Singh Sohal
Geo Media Engineering and Consultancy
Services Pvt. Ltd.
Bathinda, Punjab
India

Sedevino Sophia
Department of Chemical Engineering
National Institute of Technology Karnataka, Surathkal
Srinivasnagar
Surathkal, Mangalore
India

Ajay Thakur
Department of Chemistry
University Institute of Sciences
Chandigarh University
Mohali, Punjab
India

Arti Thanki
UniStar Environment & Research Labs Pvt. Ltd.
Dahej, Gujarat
India

Angela Manka Tita
Mining Engineering Department
Eskisehir Osmangazi University
Eskisehir
Turkey

Monika Verma
Department of Chemistry
University Institute of Sciences
Chandigarh University
Mohali, Punjab
India

Akhilesh Kumar Yadav
Department of Mining Engineering
Indian Institute of Technology Bharti Hindu University
Varanasi
Uttar Pradesh
India

1

Electronic Waste Management in Developing Countries—The Sub-Saharan Africa Experience

Helen Uchenna Modekwe[1], *Olusola Olaitan Ayeleru*[2,*], *Joshua Adeniyi Adeniran*[3], *Bimbo Lolade Fafowora*[4], *Matthew Adah Onu*[2], *Tarhemba Tobias Nyam*[2], *and Peter Apata Olubambi*[2]

[1] *Renewable Energy and Biomass Research Group, Department of Chemical Engineering, University of Johannesburg, Johannesburg, 2028, South Africa*
[2] *Centre for Nanoengineering and Advanced Material, University of Johannesburg, Doornfontein Campus, Johannesburg, South Africa, 2028, South Africa*
[3] *Hydrogen South Africa (HySA) Systems Competence Centre, South African Institute for Advanced Materials Chemistry (SAIAMC), University of the Western Cape, South Africa*
[4] *School of Journalism and Media Studies, Rhodes University, South Africa*
[*] *Corresponding author*

1.1 Introduction

The twenty-first century has witnessed a continued increase in the rise of a technology-driven economy. Such is the rise in dominance of the fourth industrial revolution (Industry 4.0 or 4IR), which is predominantly driven by electrical and electronic gadgets in this era (Ssewanyana and Busler 2007). The increase in demand for the use of electronic equipment during this era of the fourth industrial revolution has led to the proliferation of the market, especially in Sub-Saharan Africa (SSA) and Asia, with cheaper electronic gadgets, many of which are of inferior quality and thus have a shorter life span compared to premium quality electronic gadgets (Kumar and Dixit 2018). The use of inferior quality or sub-standard electronic appliances has led to an increase in electronic waste (e-waste) generation in many parts of the world, particularly in developing nations where the appetite for the consumption of substandard or secondhand electronic appliances is high (Maphosa and Maphosa 2020). Indiscriminate disposal of e-waste is a worrisome trend that needs to be curtailed in SSA. Heacock et al. (2016) reported that ~42 million tons of e-waste were generated worldwide in 2014, and this number will continue to increase if the situation is left unchecked (Heacock et al. 2016). E-waste is the remains of spoiled or scrapped electronic or computer gadgets such as laptops, mobile phones, televisions, electronic tablets, etc. Often, especially in the SSA region and other developing nations of the world, these items are dumped in landfill sites (LSs) as waste with no proper regulation (Asante et al. 2019). E-waste poses environmental and health hazards to the human population if it is not properly checked. For example, mobile phones and computer batteries contain harmful chemicals that are poisonous to people and the environment if they are not properly disposed of. Some of these noxious chemicals include arsenic, lead, and mercury, which are potentially carcinogenic in nature (Tetteh and Lengel 2017).

A proper framework for disposing of e-waste and its enforcement is lacking in many SSA countries (Asante et al. 2019). Only a few countries in SSA have policies on the handling of e-waste. South Africa, Rwanda, and Uganda are reported to be among the nations in SSA to have enacted laws or formulated policies on the handling of e-waste (Baldé et al. 2017). In many parts of the developing world, collection, sorting, and management of e-waste are done in an unorganized and elementary manner that poses a risk to the health of the untrained men and women who work in the e-waste recycling sector (Bakhiyi et al. 2018). Most of these workers handle e-waste without wearing protective personal equipment (PPE). This development puts the region at risk of economical, health, and environmental catastrophes if the situation is not put under control. Another worrisome trend that has been reported in the management of e-waste in SSA is the prevalence of use of minors (young boys and girls) for the collection of e-waste (Huang et al. 2014; Oteng-Ababio 2010). These are children who should be in school but who due to the endemic poverty in the region are forced to engage in the horrible task of e-waste collection and handling in the informal and unregulated terrain to make ends meet (World Health Organization 2021).

Consequently, this study provides an overview of the regulations needed to curb the worrisome trend of the proliferation of substandard electronic appliances in Sub-Saharan

Electronic Waste Management: Policies, Processes, Technologies, and Impact, First Edition. Edited by Sunil Kumar and Vineet Kumar.

African markets, the standards and certification systems for secondhand electrical and electronic devices, policy instruments, and recycling facilities required for proper handling/effective e-waste recycling, and the roles of informal sectors and wastes pickers in e-waste recycling.

1.2 Description of the Study Area

Sub-Saharan Africa (SSA) is one of the most populated regions of the world. It is estimated that the region had a population of ~1.7 billion in 2021 (World Bank Group 2022). Most of the population is reported to be young people, with ~60% being under 30 years of age (United Nations 2022). The population of the youth has been predicted to increase in the SSA region (Garcia and Fares 2008). The relatively high population of SSA and the consequent higher proportion of the youth in the region make the region susceptible to the purchase of electronic equipment, especially with the potential affinity of young people to the use of electronic gadgets.

Poverty is one of the major challenges that are endemic in SSA. In most of the countries in the SSA region, the per capita income is low, thus adding to poverty. With the low financial purchasing power of many people in the SSA, the attraction for low-quality electronic gadgets becomes alluring due to their affordable prices. This development makes the SSA region an environment with the potential for an increase in e-waste generation if proper policies or guidelines are not put in place to ameliorate the menace of e-waste generation. The United Nation's Sustainable Development Goal (SDG) (3) emphasizes the need for the attainment of good health and well-being (Salvo et al. 2021). To attain this goal in the SSA by the year 2030, serious efforts must be made to address the proliferation of substandard new electronic facilities or the moribund used electronic gadgets that have infiltrated the SSA region. The expired and substandard electronic equipment has the capacity to remain in the dumping sites for many years since it is usually non-biodegradable. This, in addition to the challenge of improper sanitation in the majority of countries in the SSA region, particularly in the low income areas/suburbs that coincidentally include the major percentage of the population, can lead to serious health and environmental challenges.

1.3 Global Regulatory Perspectives and Origin of Substandard Products in the Developing Countries

The European Union passed the Waste Electrical and Electronic Equipment (WEEE) directive in 2003. The legislation was designed to prevent the generation of electrical and electronic waste and promote recovery through reuse, refurbishing, recycling, etc. in order to reduce the amount of waste generated to a minimum, while improving the environmental performance of economic operators involved in waste treatment (Umesi and Onyia 2008). Again, the Restriction of Hazardous Substances (RoHS) directive sets limits on the use of various toxic components and materials used in electronic devices, hence, requiring the need for substitution with environmentally friendly options (Umesi and Onyia 2008). Several studies have reported the illegal trans-border movement of used EEE and e-waste from developed countries to developing countries including the Sub-Saharan African countries such as Nigeria, Ghana, South Africa, etc., some of which are masked as "donations" mislabeled, smuggled, and undeclared as commercial goods, have been reported (Maphosa and Maphosa 2020; Odeyingbo et al. 2019; Okorhi et al. 2019; Oteng-Ababio et al. 2020). Developing countries (SSA inclusive) have become the receiver of both standard and substandard electrical and electronic goods because of the lack of institutionalized quality assurance mechanisms and weak safety rules and regulations, unlike in the developed countries (Omobowale 2012). Figure 1.1 shows the list of top ten countries globally based on the quantity of e-waste generated in 2016, and Figure 1.2 shows guesstimates of e-waste generated for every grouping in 2016.

During the time of standards and labeling in energy-efficient appliances in Europe in the 1970s, European countries started to dispose of their used, outdated, and energy-inefficient electrical/electronic appliances (Agyarko et al. 2020). Waste management companies and individuals in Europe began to ship all discarded energy-inefficient appliances to developing countries including the Sub-Saharan African countries (Agyarko et al. 2020). The SSA market saw a huge influx of used and secondhand electrical and electronics equipment such as air-conditioners, refrigerators, electric irons, washing machines, heaters, microwave ovens, etc. from Europe and other western countries (Agyarko et al. 2020). This market grew and boomed in Ghana until the appliance performance regulation was initiated in mid-2000s, and by June 2013, a ban on the importation of used refrigerators and air-conditioners was imposed (Agyarko et al. 2020; Oteng-Ababio et al. 2020). It is estimated that between 2010 and 2014, about 157 000 tons of new and used/waste EEE were imported into Ghana (Oteng-Ababio et al. 2020). However, the implementation and compliance of this regulatory ban on such importation at the port of entry by the enforcement officials were challenged due to some manipulations and interference from politicians, importers, and influential people in society whose businesses seemed disrupted by the ban (Agyarko et al. 2020). Ghana's Appliance Labelling

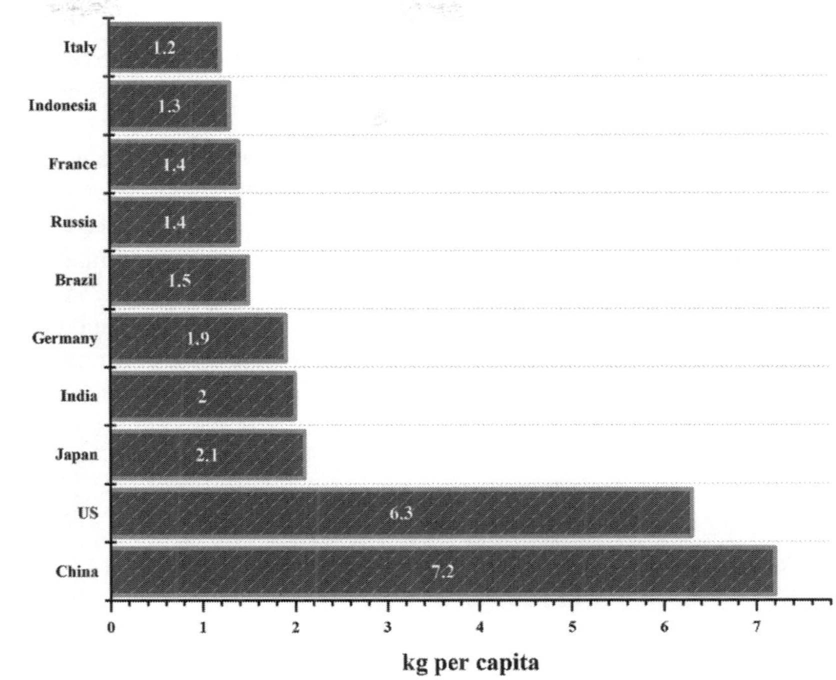

Figure 1.1 Ten top nations in the world by the quantity of e-waste generated. Adapted from https://cdn.statcdn.com/Infographic/images/normal/2283.jpeg.

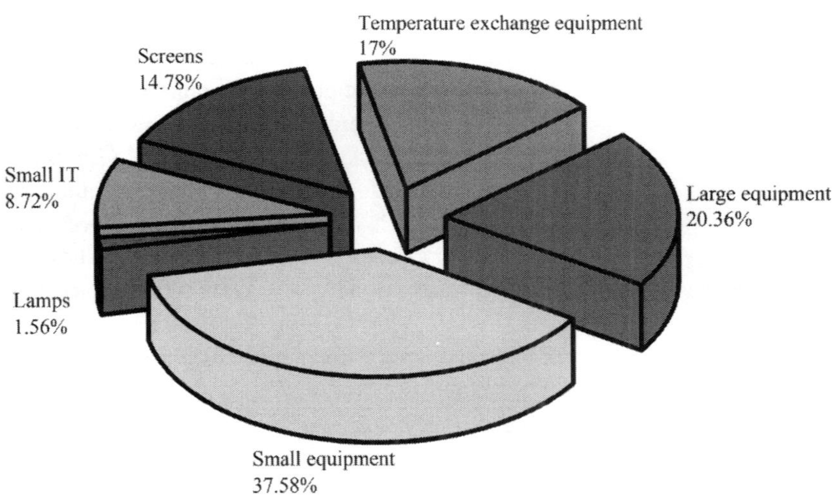

Figure 1.2 Guesstimates of e-waste generated for every class in 2016. Adapted from Lenz et al., 2019.

and Standards Program (GEALSP) was launched as part of the plan for the control of substandard appliances and the promotion of appliances' energy efficiency in Ghana. The Minimum Energy Performance Standards (MEPS) were also established for refrigerators, air conditioners, and lighting systems (Agyarko et al. 2020). Strict regulations and control for high energy standards on these appliances were due to the influx of used and junk energy-inefficient appliances in addition to some faulty components such as thermostats, compressors, etc., which escalated the appliances' electricity consumption, and the energy supply and utilization in Ghana were badly affected at that time (Agyarko et al. 2020). Other measures such as the use of the phone app "Certified Appliance App" are being developed to enable customers to verify the energy performance labels of air conditioners and refrigerators before purchasing (Agyarko et al. 2020).

In the Nigerian context, the inflow of substandard electronics and electrical goods started during the Structural Adjustment Programme (SAP), which was introduced in the mid 1980s by the Nigerian government. The program resulted in the economy dwindling, which affected the purchasing power of Nigerians, with most of the populace living in abject poverty unable to afford the purchase of luxury products (Omobowale 2012). Then, came the onset of several counterfeit and substandard goods popularly referred to as "Taiwan" imported from Asian countries (Omobowale 2012). The Taiwan are new substandard electronics gadgets that replicate the popular electronic brands but are, however, poor in quality and break down after a very short time (Omobowale 2012). Shortly after, the Nigerian electronics market was infiltrated with yet another substandard product from China named and referred to as "Chinco". According to Omobowale (2012), these substandard products are imported by "Nigerian importers who specify very low standards to be produced for them by the Chinese manufacturers to make the products more affordable to potential buyers and as well make a substantial profit." Hence, the Chinco replaced the Taiwan substandard products (Omobowale 2012). Taiwan and Chinco substandard gadgets continued in the Nigerian market until the 1990s, when the secondhand gadgets called "Tokunbo" slowly started flooding into the market from the developed countries of Europe, the United States, Japan, etc. (Omobowale 2012). Consumers prefer secondhand electronic gadgets mostly because of their affordability and durability compared to substandard new products. Thus, the increased demand and consumption of used Tokunbo electronic gadgets has made Nigeria a destination of used and secondhand products, which eventually translates into e-waste because the products are nearly at the end of their life cycle (Omobowale 2012). In 1989, Nigeria became a signatory to the Basel Convention on controlling trans-boundary movement and disposal of hazardous waste. The Basel Convention agreement was adopted in March 1989 and was enforced in May 1992 (Umesi and Onyia 2008). It was aimed at taking all necessary practical steps in regulating the movement of hazardous waste including EEE waste between countries in such a way that human health and the environment are protected against the adverse effects of such waste (Nnorom and Osibanjo 2008b; Umesi and Onyia 2008).

The Nigerian government, like other Sub-Saharan African nations, failed to properly enforce or implement the regulatory provision of the Basel Convention and the Basel Action Network (BAN), which stipulates adequate testing and labeling of imported used electronic gadgets to certify their functionality and quality to ensure that they do not result in hazardous waste or constitute a nuisance to the environment (Nnorom and Osibanjo 2008b). Also, in Nigeria, like most SSA nations, there is no specific legislation dealing with e-waste, and no integrated framework as regards the monitoring and management of hazardous and toxic waste as well as checking the illicit trade in secondhand electronic gadgets, and there is lax enforcement of the existing laws dealing with general waste management (Nnorom and Osibanjo 2008b). Several regulations and standards have been published, which include the Federal Environment Protection Agency Act No. 58 of 1988 as amended by Act 59 of 1992 and further amended by Act 14 of 1999 (Umesi and Onyia 2008). Under this act, the Federal Ministry of Environment was established and entrusted with the responsibility of protecting the environment. The grassroots approach to the management, monitoring, and pollution control was initiated by states through the Environmental Protection Agencies (EPAs) (Nnorom and Osibanjo 2008b). Regulations under the National Environmental Protection strategies are as follows:

- The provisions of Pollution Abatement in Industries and Facilities Generating Wastes Regulation S.1.9 of 1999 enforce restrictions on the release of toxic substances and specify rules and guidelines for pollution monitoring.
- The Waste Management Regulations S.1.15 of 1991 stipulates rules and guidelines for the collection, treatment, and disposal of solid and hazardous wastes, providing a broad list of chemical wastes and chemicals categorized according to their toxicity (Umesi and Onyia 2008).

Other regulations to control/limit the introduction of substandard electrical and electronic appliances and components into Nigeria were the Harmful Waste (Special Criminal Provisions) Act Cap HI LFN of 2004, the National Environmental (Sanitation and Waste Control) Regulations S.1. 28 of 2009, and the National Environmental (Electrical/Electronics Sector) Regulations S.1. No. 23 of 2011 (Basel Convention) (Okorhi et al. 2019). It should be noted that the Harmful Waste Act in Nigeria described the banning of materials based on their effects rather than on their composition (Sthiannopkao and Wong 2013). One of the problems in regulating substandard EEE or issues associated with e-waste management in Nigeria is lax enforcement by regulatory agencies such as the Standard Organization of Nigeria (SON), Nigeria Customs Services (NCS), National Environmental Standards and Regulations Enforcement Agency (NESREA), etc. (Nnorom and Odeyingbo 2020).

According to Maphosa and Maphosa (2020), e-waste policies in South Africa rely mainly on various legislation covering hazardous waste management, hence no specific policies/regulations on e-waste have been legislated (Maphosa and Maphosa 2020). Generally, the implementation of laws and regulations on e-waste in SSA is very challenging, and it arises from inter-regional trade negligence, and corrupt agency officials and market inspectors that allow the black market, smuggling, and substandard products and components to persist in the SSA market (Oteng-Ababio et al. 2020).

1.4 Standards and Certification System for Secondhand Electrical and Electronic Devices

Trade in secondhand devices has enhanced resource recovery; however, it has also made effective disposal of electrical and electronic devices more complex due to their short life expectancy (Lee and Lee 2013). In most countries' export regulations, secondhand electrical and electronic devices are exported with no secondhand mark (code), and on several occasions they are treated as e-scrap (Nnorom and Osibanjo 2008a). On the other hand, imported secondhand/used electrical and electronic goods usually undergo quality approval procedures in importing countries, where they are handled the same way as imported new devices (Lee and Lee 2013). Therefore, data on exportation and importation declarations are unreliable since the term "used" is frequently omitted on exportation and importation affirmations, and waste is recurrently declared as used goods (Lee and Lee 2013). This makes it impossible to regulate the tracking of secondhand goods and collection of the needed data to manage the inflow of exported and imported devices for quality assurance or a sound recycling system (Lee and Lee 2013). Each country has its own standard regulatory requirements for electrical and electronic devices that are expected to be met before the importation of the product into the country. These local laws and product regulations are put in place to ascertain that imported goods are harmless and sound, of good quality, and suitable for the intended needs of the consumers (United Nations Conference on Trade and Development 2016).

1.5 Policy Instruments and Recycling Facilities for Proper Handling/Effective E-waste Recycling

In today's world, the dynamics in electronic products are catalysts for social and economic development (Murthy and Ramakrishna 2022; Zlamparet et al. 2017). This has led to the enormous number of out-of-date products culminating in e-waste globally, and if not properly managed, it will constitute a large environmental and health hazard in the future. The problem of e-waste is of special concern in developing nations, especially in Asia and Africa, since a large quantity of these products are exported as secondhand goods into these continents (Abalansa et al. 2021; Dhingra and Maheshwari 2018; Sovacool 2019). Most of the nations in Asia and Africa often do not have the policies, legal frameworks, rules, expertise, and infrastructure required to safely managed e-waste in an environmentally friendly and healthy way (Abalansa et al. 2021; Maphosa and Maphosa 2020). Also, there is usually a lack of understanding of the adverse effects of inadequate e-waste management, hence the continuous proliferation of e-waste in the regions. So, it is crucial for every country to create legal, regulatory, and policy frameworks that will enable the environmentally responsible handling of e-waste. These frameworks should cover both the design and organization of the e-waste management system and its implementation in accordance with strict minimum standards (International Telecommunication Union (ITU) 2018).

1.5.1 Policies on E-waste for Proper Management

The increase in technological evolution and demand for electronic devices has brought about the drastic generation of e-waste (Borthakur 2020). The management of e-waste is crucial to the environment and to public health. The policy of recycling, reusing, and reducing consumption has been postulated by a group of researchers (Alblooshi et al. 2022). Feng et al. (2019), in their studies, suggested that remanufacturing policy incorporated with subsidy policy and carbon tax policy play a very important regulatory role in accelerating the recycling industry and carbon emission control (Feng et al. 2019). The use of subsidy instruments to provide incentives and motivation is adopted by the government to encourage producers to remanufacture goods. Extended producer responsibility (EPR) is another approach to promoting the reduce, reuse, and recycle model in e-waste management (Bimir 2020). With this strategy, producers become involved in the provision of solutions to the environmental and health impacts of e-waste. EPR places the responsibility on both producers and consumers for pollution caused by their activities rather than the local or national government being responsible for the costs of disposal (Bimir 2020; OECD 2019). The e-waste and End-of-Life Vehicle directives from Europe provide a way for original equipment manufacturers to manage the end-of-life of their products by identifying practical ways to lessen waste and environmental problems brought about by the products (implementation of the take-back recovery system) (Zlamparet et al. 2017). This approach enhances the use of remanufacturing, which enables firms and industries to grow their economies and create new jobs.

Currently, policies and regulations on e-waste are being developed across the globe by different countries; about 78 nations have put policies, laws, or regulations in place to deal with the problem of e-waste (Forti et al. 2020;

Maes and Preston-Whyte 2022). The Bamako Convention in Africa, international labor standards, international laws, and conventions (the Basel Convention and the Rotterdam Convention) and national laws that promote the stop of exportation of electronic materials of environmental concerns to developing nations are some of steps being taken to address the issues associated with e-waste management (Abalansa et al. 2021). Indeed, different nations have policies, rules, and regulations that cover actions to control the spread of e-waste at the national, regional, and international levels. Nevertheless, the problem still exists because of inadequate implementation and enforcement, coupled with legal loopholes. Despite the 1991 Bamako Convention, a regional treaty in the African continent that prohibited the importation of dangerous wastes into Africa and regulated the transboundary movement of this waste within Africa (Daum et al. 2017; Maes and Preston-Whyte 2022), the problem of importation of secondhand goods is still prevalent within the African region (Sovacool 2019). In Latin America, the Agreement on Environmental Cooperation of North America and the Mercosur Policy Agreement of 2006 to control the management of e-waste are still in force (Forti et al. 2020). India has also introduced specific policies and guidelines to deal with e-waste. For example, the Guidelines for Environmentally Sound Management of E-Waste were approved by the Ministry of Environment and Forest in 2008 (Borthakur 2020). Narrowing down the legislative framework on e-waste in Africa, there is a lack of commitment to formulating guidelines and policies for e-waste management. The continent ratified the Basel Convention, which most of the countries in the region are yet to formally pass into law (Baldé et al. 2017). South Africa, Rwanda, and Uganda have been reported as the only countries in Sub-Saharan Africa that have taken some measures in enacting policies and laws toward the management of e-waste (Maphosa and Maphosa 2020). According to Bimir (2020), there are various laws and regulations on environmental standards that existed in most African countries before the advent of e-waste (Bimir 2020); however, these standards are not up to date. Hence, there is a need for a comprehensive review of some of the environmental laws to adequately incorporate e-waste management and its enforcement structure.

1.5.2 Recycling Facilities and Proper Handling

Recycling and reusing e-waste are integral parts of e-waste management that encompass economic, environmental, and health benefits (Shahabuddin et al. 2023). Recycling e-waste could be done in a formal or an informal setting. Formal e-waste recycling involves the dismantling of the devices, extricating and classifying the materials for mechanical shredding, which are sorted with advanced separation machinery (Cho 2018). It is expedient for companies to follow health and safety guidelines and apply contamination control tools that minimize the health and environmental exposures of e-waste. In the informal approach, e-waste materials are collected, assembled, and literally burned to melt away non-valuable materials and manually disassemble the items by hand to extract valuable materials (Cho 2018; Maes and Preston-Whyte 2022). This practice is highly hazardous, as in most cases the workers do not make use of protective equipment and lack awareness of the danger of their practices (Annamalai 2015). Figure 1.3 depicts informal e-waste handling in Guiyu in China. Informal e-waste recycling is serving as a source of income for some impoverished urban people and the sector is contributing ~25% of e-waste recycling in South Africa with ~9000 workers (Maes and Preston-Whyte 2022). In Ghana, e-waste activities generated an average of US$187 million, creating employment of ~200 000 persons in the country in 2014 (Oteng-Ababio and Amankwaa 2014). Despite the financial benefits among the poor, the basic processing by the informal sectors creates elevated environmental and health consequences (Maes and Preston-Whyte 2022; Maphosa and Maphosa 2020; Yang et al. 2018).

Figure 1.3 Informal e-waste handling in Guiyu in China. (Photo credit: Basel actionnetwork. This work is shared under a Creative Commons Attribution–Non-Commercial–No Derivative Works License, and copyright has been granted by the authors). (*Source:* baselactionnetwork / Flickr / CC BY-ND 2.0.)

1.6 Roles of Informal Sectors and Wastes Pickers in E-waste Recycling

Informal waste pickers play a principal role at the initial stages of e-waste management as the wastes generated from homes and businesses are be collected and sorted by them (Shahabuddin et al. 2023). The collection of e-waste could be done by formal public or private and informal waste pickers. The informal collectors play an important role in providing a pick-up service and collecting e-waste that could remain stored or be improperly disposed of (International Labour Organization 2014). According to the International Labour Organization, informal sector waste workers are individual or small groups of enterprises that intervene in waste management without being registered and without being formally levied for such services. Gupta (2012) stated that without the involvement of the informal sector (waste pickers, scrap collectors, traders, and recyclers), the waste management systems of many developing nations would be difficult to achieve (Gupta 2012). In the formal sector in developed countries, government-funded agencies are responsible for most waste collection and treatment. With the lack of resources in many developing nations, the formal sector cannot keep pace with rapid waste production, and this has made the activities of the informal waste pickers and scavengers thrive (Yang et al. 2018). The informal waste pickers move around the streets, open spaces, and even go to dumpsites to salvage recyclable and reusable materials from the mixed waste stream and itemize them for the next operational stage (Samson et al. 2020). Even though the informal sector tends not to be recognized officially in the waste management stream, the members provide substantial contributions to the waste management services of municipalities by collecting, organizing, processing, and trading waste materials in the recycling processes (Gupta 2012). A study in the Republic of South Africa has recently proposed integration guidelines for the incorporation of informal waste pickers into the mainstream of waste management as they are key actors in the extraction of recyclables from the waste stream (Barnes et al. 2022; Samson et al. 2020). The roles of informal players in waste management can never be underestimated in developing nations, hence, adequate support and encouragement from government, nongovernmental organizations (NGOs), and voluntary donors are to be provided to enhance their health and safety as they currently serve as the backbone of the system that diverts recyclables away from landfills to recyclers in most developing nations. In general, policies that will encourage the informal sectors' integration should be put in place, which will lead to a rise in the rate of material recovery (Gupta 2012). Yang et al. (2018) stated that the effective integration of the formal and informal sectors could enhance recycling, livelihoods, and occupational and environmental health, in addition to reduction in the cost of waste management (Yang et al. 2018).

1.7 Legislation to Separate Residential Areas from Electronic Markets

A great amount of e-waste has been produced recently, and the unregulated disposal and recycling are particularly hazardous to the environment (Ghosh et al. 2022). Due to the threat of pollution, it is expedient to provide a clear separation of residential areas from electronic markets to avoid exposure of people to pollutants. This required urban planning and implementation so that residential areas can be demarcated from both the electronic market and e-waste dumpsites. People living around e-waste recycling facilities are susceptible to health and safety threats, such as radiation exposure and inhalation of harmful compounds (Maphosa and Maphosa 2020). The dismantled e-waste materials that may require disposal often contain traces of harmful substances, which could pollute the underground water, thereby endangering the life of people sourcing water from shallow aquifers (Kowsar et al. 2010). A healthy environment is the right of everyone and, as such, governments of nations of the world need to put modalities in place that guarantee the safety of residents in the provision of social amenities including citing the electronic market. There is a need for an institutional framework to oversee and assess the human and administrative capacity of enforcing the compliance of developmental planning (Amin et al. 2021). Nigeria like most other countries in SSA lacks legislation on electronic facilities versus residential areas. This has brought a high level of exposure of residents in Nigeria to e-waste contamination. Therefore, the federal government of Nigeria (FGN) needs to legislate the separation of electronics markets from residential areas to prevent continuous exposure to e-waste pollution (Shamim et al. 2015). There is a growing need for governments of countries in SSA to design national legislation and strategy for electronics stewardship that will drastically reduce the exposure to e-waste hazards (Fischer et al. 2020; Lebbie et al. 2021; Mutsau et al. 2015). The environmental and health concerns emanating from e-waste require initiatives toward responsible e-waste management (Borthakur 2020; Madanhire et al. 2020). For instance, the South African Constitution establishes fundamental environmental rights, such as the right to a healthy environment, fair administrative actions, and information access. This serves as the foundation for the nation's environmental and waste management laws (Lawhon 2013; Nel 2016).

1.8 Personal Protective Equipment (PPE) for Workers in the Electronic Markets

Personal protective equipment (PPE) shields those who wear it in opposition to any form of hazard at the workplace (National Audit Office 2020). PPE (Figure 1.4) consists of aprons, gloves, face masks, goggles, faceshields, etc. (World Health Organization [WHO] 2020). Since e-waste management and recycling in the developing nations is predominantly run by the informal sectors, it is the responsibility of the government and nongovernmental organizations to train workers at the e-waste recycling sites on when PPE is required, what PPE is required, how long to use PPE, the drawbacks of PPE, and the preservation of PPE (Occupational Safety and Health Administration 2004). Due to its limitations, PPE is the minimum efficient approach to control risk in the workplace. Although it offers a way to shield workers from being exposed to possible dangers, since it is left up to the workers to choose which PPE to wear, when to wear it, and how to take care of it, the possibility of being exposed to risk becomes very high (UW Environmental Health & Safety Department 2022). Therefore, to reduce risks at the e-waste recycling facility, the choice of the safest action becomes very crucial, and this is via engineering or administrative control. The selection of each control is done via a "hierarchy of controls" (Figure 1.5) that ranks controls based on their effectiveness (UW Environmental Health & Safety Department 2022).

1.9 Hindrances to the Effective Management of E-waste

For effective e-waste management to be achieved, there must be a good awareness campaign in place, good knowledge of the hazardous nature of e-waste, cooperation among stakeholders (including the public, informal sector, formal sector, municipalities, research institutions, and nongovernmental organizations). In most of the developing economies, there is hardly a synergy among the stakeholders, which makes it difficult for proper e-waste recycling to be realized (Yuan et al. 2019). Also, there are complications with regard to the cataloging of consignments since they are often wrongly labeled and instead of proper assessment by the regulatory body and enforcement officials, bribes from importers allow the passage of substandard electronic gadgets into the SSA region (Milovantseva and Fitzpatrick 2015). Figure 1.6 depicts some of the hindrances to effective

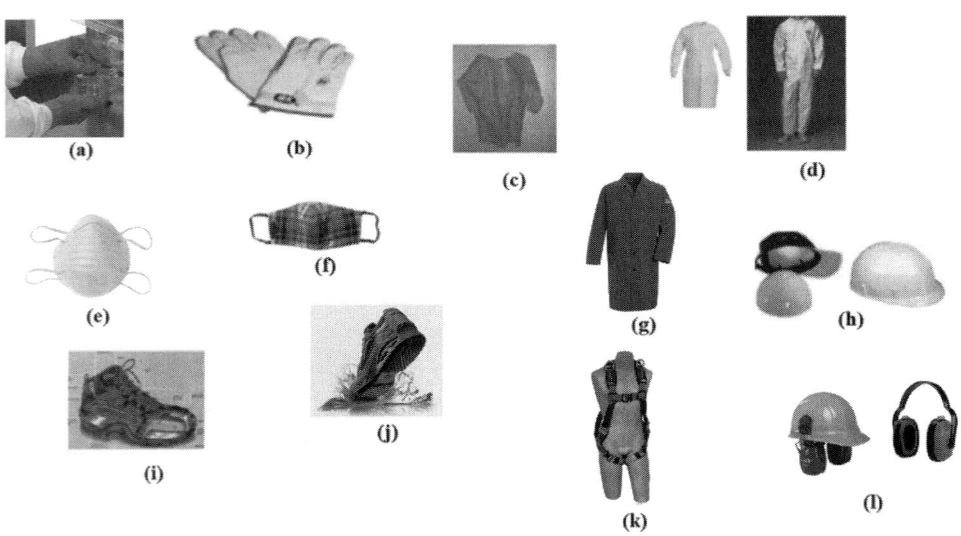

Figure 1.4 Personal protective equipment, consisting of: (a) disposable gloves, (b) leather gloves, (c) disposable gowns, (d) Tyvek gown/coveralls, (e) dust mask, (f) cloth masks, (g) lab coat, (h) bump cap, (i) safety shoes, (j) slip-resistant shoes, (k) full body harness, and (l) ear muffs. Courtesy of The Bulwark, '@Exofit.

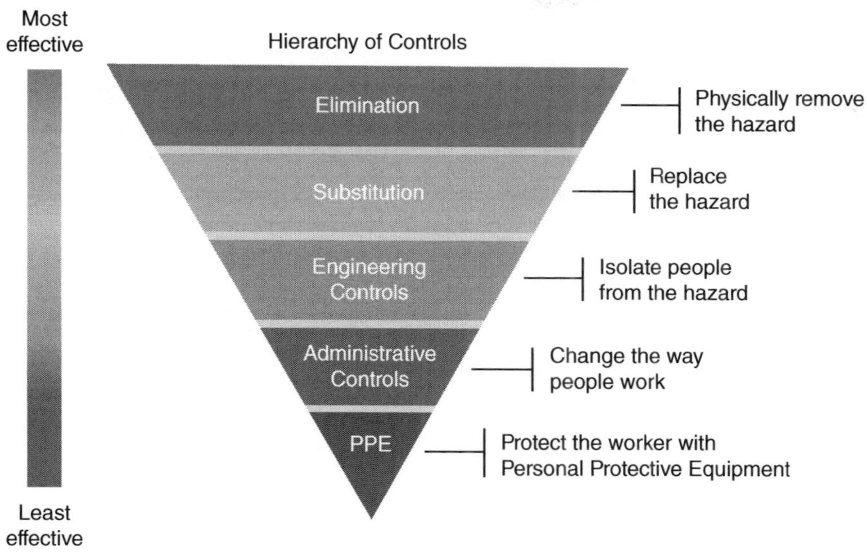

Figure 1.5 Hazard control (administrative and engineering methods). (*Source:* UW Environmental Health & Safety Department 2022.)

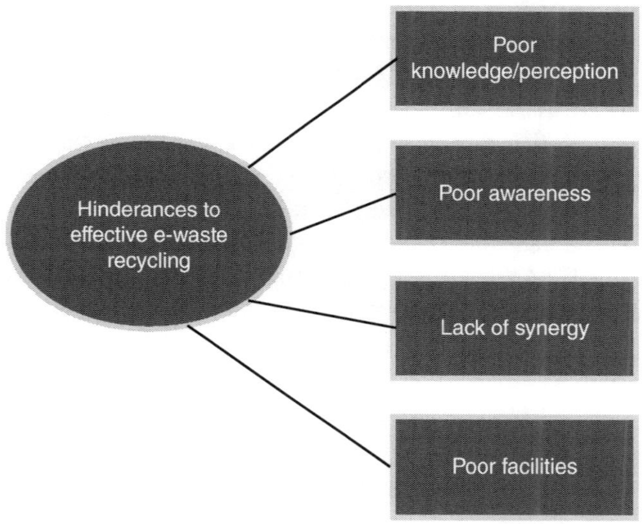

Figure 1.6 Hindrances to effective e-waste recycling and management. Adapted from Yuan et al., 2019.

e-waste recycling and management in the developing nations. Other hindrances to effective e-waste management include poor legislation, lack of workable policies to prevent used/secondhand electronic gadgets in the market, unreliable data/lack of data, illegal importation, poor operating standards, and corruption on the part of law enforcement agencies, etc. (Avis 2021; Naik and Eswari 2022).

1.10 Conclusion

This study presents an overview of regulations required to control the introduction of substandard electronic appliances into Sub-Saharan African markets. It was observed that standards and certification systems needed for secondhand electrical and electronic devices, policy instruments, and recycling facilities for proper handling/effective e-wastes recycling are almost nonexistent in many nations in the SSA region. Informal sectors and waste pickers play major roles in e-waste management and recycling. Legislation to isolate residential areas from electronic markets is lacking in the developing nations. Most of the workers at the e-waste recycling facilities do not apply PPE, and many of them are unaware of the hazardous nature of e-waste. The situation was attributed to a poor or lacking awareness campaign, poor synergy among stakeholders in the waste management sector, and weak legislation for the protection of workers or integration of informal workers into the waste management sector.

Acknowledgment

This work is based on the research supported in part by the National Research Foundation of South Africa (Ref Numbers: SASPD22091357540).

References

Abalansa, S., El Mahrad, B., Icely, J., and Newton, A. (2021). Electronic waste, an environmental problem exported to developing countries: the GOOD, the BAD and the UGLY. *Sustainability* 13 (9): 5302.

Agyarko, K.A., Opoku, R., and Van Buskirk, R. (2020). Removing barriers and promoting demand-side energy efficiency in households in Sub-Saharan Africa: a case study in Ghana. *Energy Policy* 137: 111149.

Alblooshi, B.G.K.M., Ahmad, S.Z., Hussain, M., and Singh, S.K. (2022). Sustainable management of electronic waste: empirical evidences from a stakeholders' perspective. *Business Strategy and the Environment* 31 (4): 1856–1874.

Amin, A., Lewis-Lettington, R., and Njuguna, S. (2021). Effectiveness of planning law in Sub-Saharan Africa. In: *Land Issues for Urban Governance in Sub-Saharan Africa* (ed. R. Home), 103–120. Nairobi: Springer.

Annamalai, J. (2015). Occupational health hazards related to informal recycling of E-waste in India: an overview. *Indian Journal of Occupational and Environmental Medicine* 19 (1): 61.

Asante, K.A., Amoyaw-Osei, Y., and Agusa, T. (2019). E-waste recycling in Africa: risks and opportunities. *Current Opinion in Green and Sustainable Chemistry* 18: 109–117.

Avis, W. (2021). *Drivers, barriers and opportunities of e-waste management in Africa* (K4D Helpdesk Report No. 1074). Retrieved from https://opendocs.ids.ac.uk/opendocs/handle/20.500.12413/17152 (accessed 30 August 2023).

Bakhiyi, B., Gravel, S., Ceballos, D. et al. (2018). Has the question of e-waste opened a Pandora's box? An overview of unpredictable issues and challenges. *Environment International* 110: 173–192.

Baldé, C.P., Forti, V., Gray, V. et al. (2017). *The Global E-Waste Monitor 2017: Quantities, Flows and Resources*. Bonn/Geneva/Vienna: United Nations University, International Telecommunication Union, UNU-VIE SCYCLE, and International Solid Waste Association.

Barnes, K., Blaauw, D., Schenck, R., and Pretorius, A. (2022). Buyback centres in Cape Town: the key integration point between formal and informal sectors in the waste economy of the Western Cape. *GeoJournal* 87 (3): 2051–2065.

Bimir, M.N. (2020). Revisiting e-waste management practices in selected African countries. *Journal of the Air & Waste Management Association* 70 (7): 659–669.

Borthakur, A. (2020). Policy approaches on E-waste in the emerging economies: a review of the existing governance with special reference to India and South Africa. *Journal of Cleaner Production* 252: 119885.

Cho, R. (2018). What can we do about the growing E-waste problem. *General Earth Institute, August*, 2018. Retrieved from https://news.climate.columbia.edu/2018/08/27/growing-e-waste-problem (accessed 20 July 2022).

Daum, K., Stoler, J., and Grant, R.J. (2017). Toward a more sustainable trajectory for e-waste policy: a review of a decade of e-waste research in Accra, Ghana. *International Journal of Environmental Research and Public Health* 14 (2): 135.

Dhingra, S. and Maheshwari, N.S. (2018). E-waste management awareness and intentions among youth consumers: the mediating role of motivation. *International Journal of Indian Culture and Business Management* 16 (4): 359–370.

Feng, D., Ma, L., Ding, Y. et al. (2019). Decisions of the dual-channel supply chain under double policy considering remanufacturing. *International Journal of Environmental Research and Public Health* 16 (3): 465.

Fischer, D., Seidu, F., Yang, J. et al. (2020). Health consequences for e-waste workers and bystanders—a comparative cross-sectional study. *International Journal of Environmental Research and Public Health* 17 (5): 1534.

Forti, V., Balde, C.P., Kuehr, R., and Bel, G. (2020). *The Global E-Waste Monitor 2020: Quantities, Flows and the Circular Economy Potential*. Bonn, Geneva and Rotterdam: United Nations University/United Nations Institute for Training and Research, International Telecommunication Union, and International Solid Waste Association.

Garcia, M. and Fares, J. (eds.) (2008). Why is it important for Africa to invest in its youth. *Youth in Africa's Labor Market* 3–14.

Ghosh, B.K., Mekhilef, S., Ahmad, S., and Ghosh, S.K. (2022). A review on global emissions by E-products based waste: technical management for reduced effects and achieving sustainable development goals. *Sustainability* 14 (7): 4036.

Gupta, S.K. (2012). Integrating the informal sector for improved waste management. *Private Sector and Development* 15: 12–17.

Heacock, M., Kelly, C.B., Asante, K.A. et al. (2016). E-waste and harm to vulnerable populations: a growing global problem. *Environmental Health Perspectives* 124 (5): 550–555.

Huang, J., Nkrumah, P.N., Anim, D.O., and Mensah, E. (2014). E-waste disposal effects on the aquatic environment: Accra, Ghana. *Reviews of Environmental Contamination and Toxicology* 229: 19–34.

International Labour Organization (2014). *Tackling Informality in E-Waste Management: The Potential of Cooperative Enterprises*. Geneva: ILO.

International Telecommunication Union (ITU) (2018). *Developing an e-waste national policy and regulatory framework for Malawi*. Retrieved from Malawi: https://

www.itu.int/en/ITU-D/Climate-Change/Documents/2018/Developing_e-waste_national_policy_and_regulatory_framework_Malawi.pdf (accessed 20 August 2022).

Kowsar, R., Hashia, H., and Khan, F. (2010). E-waste and its health impacts. *Global Journal of Human Social Science* 10: 88.

Kumar, A. and Dixit, G. (2018). Evaluating critical barriers to implementation of WEEE management using DEMATEL approach. *Resources, Conservation and Recycling* 131: 101–121.

Lawhon, M. (2013). Dumping ground or country-in-transition? Discourses of e-waste in South Africa. *Environment and Planning C: Government and Policy* 31 (4): 700–715.

Lebbie, T.S., Moyebi, O.D., Asante, K.A. et al. (2021). E-waste in Africa: a serious threat to the health of children. *International Journal of Environmental Research and Public Health* 18 (16): 8488.

Lee, E.C. and Lee, K. (2013). A study of cross-border trade in second-hand goods. *Journal of Distribution Science* 11 (11): 89–99. https://doi.org/10.13106/jds.2013. vol11.no11.89.

Lenz, K., Afoblikame, R., Karcher, S.Y. et al. (2019). *E-waste training manual*. Retrieved from Austria: https://www.giz.de/de/downloads/giz2019-e-waste-management.pdf (accessed 14 July 2022).

Madanhire, I., Mugwindiri, K., and Mbohwa, C. (2020). Evaluating the feasibility of effective E waste management: case study. Paper presented at the Proceedings of the International Conference on Industrial Engineering and Operations Management, Dubai, UAE, March 10–12, 2020.

Maes, T. and Preston-Whyte, F. (2022). E-waste it wisely: lessons from Africa. *Springer Nature Applied Sciences* 4 (3): 1–12.

Maphosa, V. and Maphosa, M. (2020). E-waste management in Sub-Saharan Africa: a systematic literature review. *Cogent Business & Management* 7 (1): 1814503.

Milovantseva, N. and Fitzpatrick, C. (2015). Barriers to electronics reuse of transboundary e-waste shipment regulations: an evaluation based on industry experiences. *Resources, Conservation and Recycling* 102: 170–177.

Murthy, V. and Ramakrishna, S. (2022). A review on global E-waste management: urban mining towards a sustainable future and circular economy. *Sustainability* 14 (2): 647.

Mutsau, S., Billiat, E., and Musingafi, C.C.E. (2015). Electronic waste management in Zimbabwe: a slow onset public health disaster. *Civil and Environmental Research* 7 (10): 84–88.

Naik, S. and Eswari, J.S. (2022). Electrical waste management: recent advances challenges and future outlook. *Total Environment Research Themes* 1–2: 100002.

National Audit Office (2020). The supply of personal protective equipment (PPE) during the COVID-19 pandemic. Retrieved from London: https://www.nao.org.uk/wp-content/uploads/2020/11/The-supply-of-personal-protective-equipment-PPE-during-the-COVID-19-pandemic.pdf (accessed 20 August 2022).

Nel, V. (2016). SPLUMA, zoning and effective land use management in South Africa. Paper presented at the Urban Forum.

Nnorom, I.C. and Odeyingbo, O.A. (2020). Electronic waste management practices in Nigeria. In: *Handbook of Electronic Waste Management* (eds. M.N.V. Prasad, M. Vithanage, and A. Borthakur), 323–354. Elsevier.

Nnorom, I.C. and Osibanjo, O. (2008a). Electronic waste (e-waste): material flows and management practices in Nigeria. *Waste Management* 28 (8): 1472–1479.

Nnorom, I.C. and Osibanjo, O. (2008b). Overview of electronic waste (e-waste) management practices and legislations, and their poor applications in the developing countries. *Resources, Conservation and Recycling* 52 (6): 843–858.

Occupational Safety and Health Administration (2004). *Personal Protective Equipment*. Retrieved from https://www.osha.gov/sites/default/files/publications/osha3151.pdf (accessed 24 August 2022).

Odeyingbo, A.O., Nnorom, I.C., and Deubzer, O.K. (2019). Used and waste electronics flows into Nigeria: assessment of the quantities, types, sources, and functionality status. *Science of the Total Environment* 666: 103–113.

OECD (2019). *Waste Management and the Circular Economy in Selected OECD Countries: Evidence from Environmental Performance Reviews*. OECD.

Okorhi, J.O., Amadi-Echendu, J.E., Aderemi, H.O. et al. (2019). Disconnect between policy and practice in developing countries: evidence of managing e-waste from Nigeria. *African Journal of Science, Technology, Innovation and Development* 11 (4): 523–531.

Omobowale, A.O. (2012). Global e-waste management and second-hand consumption in the third world: substandard context and Tokunbo phenomenon in Nigeria. *World* 10 (1).

Oteng-Ababio, M. (2010). E-waste: an emerging challenge to solid waste management in Ghana. *International Development Planning Review* 32 (2).

Oteng-Ababio, M.M. and Amankwaa, E.F. (2014). The e-waste conundrum: balancing evidence from the North and on-the-ground developing countries' realities for improved management. *African Review of Economics and Finance* 6 (1): 181–204.

Oteng-Ababio, M.M., van der Velden, M., and Taylor, M.B. (2020). Building policy coherence for sound waste electrical and electronic equipment management in a developing country. *The Journal of Environment & Development* 29 (3): 306–328.

Salvo, D., Garcia, L., Reis, R.S. et al. (2021). Physical activity promotion and the United Nations Sustainable Development Goals: building synergies to maximize impact. *Journal of Physical Activity and Health* 18 (10): 1163–1180.

Samson, M., Timm, S., Chidzungu, T. et al. (2020). *Lessons from waste picker integration initiatives: development of evidence based guidelines to integrate waste pickers into South African municipal waste management systems.* Retrieved from http://wasteroadmap.co.za/wp-content/uploads/2020/03/1-Wits_Final_Report_Johannesburg-Technical-Report.pdf (accessed 29 July 2022).

Shahabuddin, M., Uddin, M.N., Chowdhury, J.I. et al. (2023). A review of the recent development, challenges, and opportunities of electronic waste (e-waste). *International Journal of Environmental Science and Technology* 20 (4): 4513–4520.

Shamim, A., Mursheda, A.K., and Rafiq, I. (2015). E-waste trading impact on public health and ecosystem services in developing countries. *International Journal of Waste Resources* 5 (4): 1–18.

Sovacool, B.K. (2019). Toxic transitions in the lifecycle externalities of a digital society: the complex afterlives of electronic waste in Ghana. *Resources Policy* 64: 101459.

Ssewanyana, J. and Busler, M. (2007). Adoption and usage of ICT in developing countries: case of Ugandan firms. *International Journal of Education and Development Using ICT* 3 (3): 49–59.

Sthiannopkao, S. and Wong, M.H. (2013). Handling e-waste in developed and developing countries: Initiatives, practices, and consequences. *Science of the Total Environment* 463: 1147–1153.

Tetteh, D. and Lengel, L. (2017). The urgent need for health impact assessment: proposing a transdisciplinary approach to the e-waste crisis in sub-Saharan Africa. *Global Health Promotion* 24 (2): 35–42.

Umesi, N.O. and Onyia, S. (2008). Disposal of e-wastes in Nigeria: an appraisal of regulations and current practices. *The International Journal of Sustainable Development & World Ecology* 15 (6): 565–573.

United Nations (2022). *Young People's Potential, the Key to Africa's Sustainable Development.* Retrieved from https://scholar.google.com/scholar?hl=en&as_sdt=0%2C5&q=UNITED+NATIONS&btnG=#d=gs_cit&t=1660830870356&u=%2Fscholar%3Fq%3Dinfo%3A2R3ZptOBWkcJ%3Ascholar.google.com%2F%26output%3Dcite%26scirp%3D1%26hl%3Den (accessed 22 August 2022).

United Nations Conference on Trade and Development (2016). *United Nations Guidelines for Consumer Protection.* Retrieved from Geneva: https://unctad.org/system/files/official-document/ditccplpmisc2016d1_en.pdf (accessed 2 September 2022).

UW Environmental Health & Safety Department (2022). *Guidelines for Personal Protective Equipment (PPE).* Retrieved from https://www.ehs.washington.edu/system/files/resources/ppeguidelines.pdf (accessed 22 July 2022).

World Bank Group (2022). Retrieved from https://data.worldbank.org/indicator/SP.POP.TOTL?locations=ZG.

World Health Organization (2021). *Children and digital dumpsites: e-waste exposure and child health.* Retrieved from Geneva: file:///C:/Users/user/Downloads/9789240023901-eng%20(3).pdf.

World Health Organization (WHO) (2020). *Rational use of personal protective equipment for coronavirus disease (COVID-19) and considerations during severe shortages* (WHO/2019-nCov/IPC_PPE_use/2020.3). Retrieved from https://apps.who.int/iris/bitstream/handle/10665/331695/WHO-2019-nCov-IPC_PPE_use-2020.3-eng.pdf (accessed 24 July 2022).

Yang, H., Ma, M., Thompson, J.R., and Flower, R.J. (2018). Waste management, informal recycling, environmental pollution and public health. *Journal of Epidemiology and Community Health* 72 (3): 237–243.

Yuan, P.W.S., Mun, N.K., and Rajendran, S.D. (2019). A study on barriers to implement E-waste management. Paper Presented at the E3S Web of Conferences.

Zlamparet, G.I., Ijomah, W., Miao, Y. et al. (2017). Remanufacturing strategies: a solution for WEEE problem. *Journal of Cleaner Production* 149: 126–136.

2

Contextualizing Electronic Waste for Effective Management in Ghanaian Cities

Local Perspectives and Experiences

Michael Osei Asibey

Department of Planning, College of Art and Built Environment, Kwame Nkrumah University of Science and Technology, Kumasi, Ghana

2.1 Introduction

The increasing number of consumers for electrical and electronic products is noted to have crossed the "digital divide" in many countries across the globe (Nartey et al. 2016; Reddy 2015; Samson 2015). Discarded electronic and electrical products have become one of the fastest growing waste streams in the world (International Labour Organisation [ILO] 2012, 2014). It is, however, certain that the global e-waste stream generation has been growing annually. Estimates reveal that the amount of e-waste generated increased by 5.9% between 2010 and 2011 from 37.3 million tons to 39.4 million tons. The Solving the E-waste Problem (StEP) Initiative (2015) reported that about 45.6 million tons of e-waste were generated in 2012. The waste stream of old EEE (electrical and electronic equipment) was further estimated to be growing with an estimated 41.8 million tons in the year 2014. Baldé et al. (2015) further project the global e-waste generation to reach 55 million tons (49.8 Mt) in 2030. The waste is largely generated in the industrialized countries, where China, USA, and Europe lead the amounts generated with 7.9, 9.7, and 6.5 million metric tons, respectively.

Debates on the concept of electronic waste (e-waste) have evolved over time and have been widely contested among scholars. This has to do with inconsistencies in how e-waste is conceptualized or perceived. The term e-waste is often interchanged with electrical waste, electrical and electronic waste, or waste electrical and electronic equipment (WEEE). It should, however, be stated scientifically and technologically that there is an important distinction in the manipulation of energy between electrical and electronic devices, although it may appear subtle (AspenCore 2016). As such, e-waste has often been conceptualized in terms of composition of such equipment in various jurisdictions.

Global discourse on e-waste began in the mid-1970s, when the waste was either dumped into landfills or incinerated (Widmer et al. 2005). E-waste before 1976 was noted to mainly comprise discarded television sets and computers (StEP Initiative 2015). This suggests that any other electronic equipment was not recognized as e-waste even when it was of no more use and was discarded (AspenCore 2016). This practice occurred until 1976 when it was realized that unsafe handling could have detrimental effects on the environment and human health. However, until then, there was no globally agreed definition of the term e-waste and what constitutes such waste. Having realized the enormous implications of e-waste, there has been increased attention toward conceptualizing the term to safeguard its management in the last decade. The inconsistencies in what qualifies to be e-waste have been widely contested, specifically as to: (i) it being a foreign concept based on its categorization, (ii) at which point it becomes waste (to the user or disposer), or (iii) agreeing to accept a global definition, taking cognisance of adding up to the categorization of to-be-manufactured electrical and electronic equipment (EEE). The complexities surrounding the focus of e-waste over the years paved the way for some indicators, focusing on the multidimensionality of what constitutes e-waste (EU WEEE Directive 2012).

Due to increasing industrialization and evolution of technology, new models of several EEE were manufactured and used. Widmer et al. (2005) hence defined e-waste to include "electrical products" that are of no more value to owners of the products or devices. This definition, contrary to that of StEP (2013), only describes electrical and not electronic

Electronic Waste Management: Policies, Processes, Technologies, and Impact, First Edition. Edited by Sunil Kumar and Vineet Kumar.

products. Oswald and Armin (2011) also conceptualize e-waste to denote "EEE that are broken and discarded because of obsolescence and partial or complete inoperability." According to the StEP Initiative (2013), e-waste is the reverse supply chain that collects products no longer desired by a given consumer and "refurbishes for other consumers, recyclers, or otherwise processes waste." E-waste is further broadened to describe all types and forms of "electronic equipment" including their parts that are of no more use or have been discarded as waste with no intention of reuse (StEP Initiative 2015). These, according to Khurrum et al. (2011), can "include anything from large household appliances, such as air conditioners and refrigerators, to consumer electronics, including televisions, radios, cell phones, and computers."

The European Union's WEEE Directive (2012, p. 13) further describes e-waste to include "all the components, consumables and subassemblies which form part of any electronic and/or electrical product during periods of discarding." The US EPA (2016) also describes e-waste based on the type of electronic devices to imply discarded or waste electronic devices that include among others, televisions, computers, consoles, video games, monitors, and any other device that uses a program of printed wiring boards. Finally, Ghana's Hazardous and Electronic Waste Control and Management Act, Act 917 of 2016, defines e-waste to include all the components, consumables, and subassemblies that form part of any electronic and/or electrical product during periods of discarding. The foregoing indicates that the meaning and composition of e-waste depend on the needs of the owner and/or the country of origin or destination. The EU WEEE Directive (2012), due to the definitional and compositional inconsistencies, categorized e-waste into 10 groups, which appears to have become the globally accepted standard in conceptualizing e-waste (Table 2.1).

Table 2.1 Electrical and electronic equipment categorization.

Category1	Abbreviation
Large household appliances	Large HH
Small household appliances	Small HH
IT and telecommunications equipment	ICT
Consumer equipment and photovoltaic panels	CE
Lighting equipment	Lighting
Electrical and electronic tools (except large-scale stationary industrial tools)	E&E
Toys, leisure, and sports equipment	Toy
Medical devices (with the exception of all implemented and infected products)	Medical
Monitoring and control instruments	M&C
Automatic dispensers	Dispensers

Adapted from EU WEEE Directive (Directive 2012/19/EU).

Although the concept of e-waste has evolved over time, there appears to be little semblance of definitional consensus among stakeholders. This can pose a challenge to local authorities in coming up with effective policies to guide the management of such waste, considering the call for local initiatives to manage this waste stream (Organisation for Economic Co-operation and Development [OECD] 2014; Agyei-Mensah and Oteng-Ababio 2012). As a result of this, significant academic interest has been roused, with scholars actively engaged in the challenging task of adequately defining the concept of e-waste. This burgeoning interest and proliferation of e-waste interpretations, coupled with the lack of a universally agreed upon definition, may have resulted in a great deal of confusion regarding the "true meaning" of the concept, and consequently affecting its management.

Ghana has one of the largest informal e-waste dumpsites in Africa (Daum et al. 2017; Oteng-Ababio et al. 2014; Prakash et al. 2010). About 215 000 tons of e-waste are imported from European countries into Ghana annually (Amoyaw-Osei et al. 2011). The e-waste sector employing between 109 620 and 181 440 people is an important source of livelihood to some urban poor in Ghana (Prakash et al. 2010). Taking Ghana as a case region, the objective of this study is to analyze the lack of uniform conceptualization of e-waste accounting for the increased risks and poor management and handling exemplified by two dumpsites in Ghana's largest cities, Kumasi and Accra. Specifically, the study examines (i) local agency officials' and e-waste workers' understanding and interpretation of e-waste, its impact and its composition (i.e. the observed and most handled e-waste items at the recycling sites) and (ii) if knowledge of e-waste has influence on management practices from the perspectives of agencies and e-waste recyclers in Ghana. Finally, the aim is to propose the most appropriate definition of e-waste for managing e-waste sites in Ghana.

2.2 Research Approach

Two informal e-waste recycling sites in Ghana were selected as study areas: Agbogbloshie E-waste Processing Site (AEPS) in Accra and Dagomba Line E-waste Processing Site (DEPS) in Kumasi. The study sites were hence selected due to their (i) increasing urban population (Ghana Statistical Service [GSS] 2013, 2014), (ii) evidence of an intense e-waste impact with respect to social, economic, health, and environmental implications (Amoyaw-Osei et al. 2011; Nartey et al. 2016), (iii) presence of relevant agencies to promote environmental management and development (Oteng-Ababio et al. 2014), (iv) the number of years the activity has been going on (about five years or more). AEPS has been in existence for over two decades in contrast to DEPS, which is an emerging site (in existence for less than a decade; approximately eight years) (Asibey et al. in press).

E-waste is more pronounced in AEPS (specifically, the Agbogbloshie community) (Daum et al. 2017), which is an old slum neighborhood in the city of Accra (Grant and Oteng-Ababio 2012). The number of residents in Agbogbloshie has increased steadily from 7055 people in 2000 to 19 727 people in 2010 (GSS 2013). The DEPS is located in the Asokore Mampong Municipality in Kumasi, Ghana's second largest city (GSS 2014). Its economic endowment and position as a transitional transport hub have influenced its rapid urbanization, attracting many immigrants not only from Ghana but from other countries as well. There has thus been the emergence of informal economic activities such as scrap dealing, e-waste recycling, waste picking, among others.

Ghana's institutional framework for controlling and managing e-waste is provided for by and detailed in Section 10 of the Environmental Protection Agency (EPA) Act, 1994 (Act 490). The Act sets up a multi-stakeholder Hazardous Chemical Committee tasked with the responsibility of managing, destroying, or disposing of unwanted or obsolete hazardous waste including WEEE. The specific functions of the Hazardous Chemical Committee according to the EPA Act, 1994 (Act 490), IV-7 are: (i) to monitor the use of hazardous chemicals by collecting information on the importation, exportation, manufacture, distribution, sale, use, and disposal of those chemicals; (ii) to perform any other functions relating to chemicals that the Board or the executive director may direct; and (iii) to advise the Board and the executive director on the regulation and management of hazardous chemicals. The Hazardous Chemical Committee, chaired by the EPA and relevant to this paper, comprises among others the Environmental Protection Agency (Chair); Ministry of Environment, Science, Technology and Innovation (MESTI); Ministry of Health or Ghana Health Services; and the Ministry of Local Government and Rural Development.

Premised on the above, semi-structured interviews were held with six e-waste management-related Ministries, Departments, and Agencies. These agencies included three decentralised departments of the Accra and Asokore Mampong Municipal Assemblies; specifically, the Development Planning Unit in charge of making plans for socio-economic development, Waste Management Department responsible for general waste management, Environmental Health Directorate and Metropolitan Health Directorate concerned with tackling general health issues related to the unsafe environmental practices, MESTI responsible for creating wealth and sound environmental governance, and the EPA mandated to improve, conserve, and promote Ghana's environment in a sustainable manner.

Additionally, two focus group discussions with e-waste workers and leaders of relevant associations each at the AEPS and DEPS were also conducted. The group discussions were facilitated by two interviewers; one facilitated the discussions while the other took field notes. Recordings were made using digital recorders after seeking the consent of interviewees. The recordings were later transcribed under appropriate themes. The semi-structured interviews with agency officials and e-waste workers covered: (i) their knowledge and interpretation of e-waste; (ii) the composition of e-waste, i.e., what qualifies as e-waste; (iii) the economic, human health, and environmental impact of e-waste; (iv) their perception on whether knowledge of e-waste has influence on management practices.

The content analysis was used to analyze interview transcripts and secondary data through an inductive and deductive coding process. The content analyses of relevant datasets for this paper focused on assessing institutional and workers' perspectives on the interpretation of e-waste, its composition, and the economic, human health, and environmental impact. The process was facilitated by the use of the NVIVO 12 software package for coding the data and developing categories. Codes were inductively developed from the research participants' perspectives on the meaning of e-waste, its composition and impact. Categories such as how knowledge of e-waste can influence management efforts were deductively created based on the findings from the review of relevant documents. Guided by the inconsistencies that might arise in the analysis process, however (Bryman 2004), relevant relationships were established by reconciling and refining codes and categories into more conceptual categories based on common relationships. This helped increase understanding of the whole dataset, as the research participants' perspectives on e-waste, its composition and impact in Ghana were reflected in the final categories. The analysis was also done based on the context, consistency, intensity, and specificity of the responses to the questions.

2.3 Results

2.3.1 Local Conceptualization of E-waste, Composition and Impact

Responses from most agency officials on their knowledge of e-waste, its composition and impact demonstrated a good understanding of the phenomenon. Responses on the meaning of e-waste varied among the agency officials (Table 2.2). In spite of the variations, the major theme extensively discussed by officials was "all electrical or

Table 2.2 Local people and planning agencies' perception of e-waste, its composition and impact.

Respondent	Summary of responses about e-waste benchmarking with Act 914
Environmental Health Directorate at both areas	Unaware of what e-waste is and its composition. The EHD largely handles general health issues related to the environment.
Metropolitan Health Directorate at both areas	E-waste is any waste that comprises electronic or electrical items such as TVs, fridges, printers, computers, and household appliances. It is, however, not so much the focus of the Health Directorate since the mandate is to respond to general health issues reported at various centers.
Development Planning Unit of the KMA	E-waste is electronic equipment that is of no use. It is, however, not so much of a focus, because it is an emerging activity with minimal impact in the city. There are ongoing efforts to manage it.
Development Planning Unit of the AMA	E-waste is electrical and electronic equipment, which the user has no more use of and has discarded. Examples include TVs, computers, mobile phones, etc.
Waste Management Department of the KMA	E-waste is all types of EEE not of use to the user and as such discarded. It is, however, not of much interest because the focus is on municipal SWM.
Waste Management Department of the AMA	E-waste is all electronic equipment that is no longer needed and discarded by the user.
Environmental Protection Agency–Kumasi	E-waste refers to all EEE not of use to the user and as such disposed of. Examples include computers, refrigerators, phones, etc and all types of EEE.
Environmental Protection Agency–Accra	E-waste is the discarded electrical and electronic equipment inclusive of all components, subassemblies, and consumables that are part of the product at the time of discarding.
Ministry of Environment, Science, Technology, and Innovation	E-waste is all discarded electrical and electronic including all components, subassemblies, and consumables that are part of the product at the time of discarding.
Informal Workers at Dagomba Line E-waste Processing Site	E-waste is the things we do not need and have dumped in the open environment.
Informal Workers at Agbogbloshie E-waste Processing Site	Nothing is waste as long as it is functional or repairable. We do not know what e-waste or its composition is.

electronic products that are discarded by consumers and are of no more use to them" (see Table 2.2).

On definitional issues, the Waste Management Department at both study areas considered e-waste to be electronic products that are of no more use to producers and consumers and are discarded. For instance, the Waste Management Department official of the Asokore Mampong Municipal Assembly defined e-waste as:

> All types of electrical and electronic equipment (as the name suggests) which are of no more use to the user or consumer and are either sold or discarded ... However, every waste is waste and can be potentially harmful and so the need to pay attention to all ...

Similarly, the Development Planning Unit of the Asokore Mampong Municipal Assembly defined e-waste as any electronic equipment that is of no more use to the user. The official, however, stated that e-waste is not so much a focus of city authorities because it is an emerging activity with minimal impact in the city, compared to general municipal solid waste. In spite of this, the official stated that there were ongoing efforts (in the form of collaborations with relevant NGOs and the EPA) to manage the sector and minimize any potential adverse impact. The Development Planning Unit of the Accra Metropolitan Assembly similarly defined e-waste as all electrical and electronic equipment that the user has no more use of and discards.

Discussions with officials of the Environmental Health Directorate at both study sites revealed general unawareness and limited understanding of the concept, despite admitting that improper handling of general solid waste has adverse implications on the health of humans and the environment. To them, the Environmental Health Directorate is mainly concerned with tackling general health issues related to the unsafe environmental practices. The Environmental Health Directorate of the Accra Metropolitan Assembly respectively remarked that

> Waste is generally hazardous depending on their content—heavy metals and toxic materials

The Environmental Health Directorate of the Asokore Mampong Municipal Assembly similarly stated:

> What is e-waste? ... Cannot tell you exactly what it is or constitute, although have heard about it, especially, in Accra (Agbogbloshie). Your research might probably tell us more about the issues in Kumasi ...

Further, it was observed that although officials of the Metropolitan Health Directorate at both study areas had a good understanding of e-waste as any waste that comprises electronic or electrical items which users have no more need of, they stated that e-waste was, however, not so much a focus of theirs since their core mandate is to respond to general health issues reported at various health centers.

The EPA and MESTI also respectively described e-waste as follows:

> E-waste are all to all EEE not of use to the user and as such disposed of and no longer satisfy the needs of the initial purchaser ... [EPA Official at Kumasi].
>
> E-waste refers to the discarded electrical and electronic equipment inclusive of all components, subassemblies and consumables which are part of the product at the time of discarding ... [EPA Official in Accra].
>
> E-waste are all types of electronic and electrical equipment (small, medium or large) that have reached its end-of-life and of no more use to consumers ... [MESTI Official].

Interviews with local workers at both sites, however, showed variance in the understanding of the term "e-waste" as they regarded discarded EEE (which they handled at the sites) as not waste, but rather useful secondhand items and a source of affordable EEE to households and individuals. To the local e-waste recyclers, e-waste (as generally defined by officials as unwanted and of no more use) is an important source of income and livelihood and a means to acquiring affordable and advanced EEE. A worker during a focus group discussion at AEPS remarked that

> The EEE you see here are not waste. They are valuable to us and afford us the opportunity to own some electronic products we otherwise would not have access to, hence, we (workers) treating them here ... They only become waste after we think we have no need of them and dump them ...

Local workers at the DEPS similarly expressed the same concern as that shared by those at the AEPS. The following is one of such responses:

> Nothing is waste here so long as it is repairable ... The only waste you see are the things we don't need and have dumped in the open environment ... They are electronic equipment, most of which are functional. More of secondhand goods, not necessarily waste ... They do contain valuable assets such as brass, copper, gold, and zinc, among others, which generate income for us.

The foregoing shows that there appears to be some similarities in the interpretation of the concept and little consensus among respondents as well as inconsistency in how e-waste is conceptualized. Respondents (specifically agency officials) who understood the term generally interpreted e-waste to mean either electrical, electronic, or both EEE that are of no more use to the consumer and are thus discarded. Key issues of concern about the interpretation of the concept were: (i) whether e-waste comprises electronic or electrical products or both, considering that technically, there appears to be a difference between the two; (ii) the definitions do not indicate whether functional EEE that are of no more use to the user but may be useful to another person are still considered waste; (iii) whether discarded functional EEE products are "waste to the workers at the e-waste site." The differences in conceptualizing the product scope, according to the EPA and MESTI, may have partly affected their (sound) management of the sector. To better understand and clarify the term, respondents were asked to give the composition of what items qualify as e-waste.

The findings on e-waste composition generally showed that all respondents, irrespective of their knowledge about the definition of e-waste, did give responses on what items are deemed e-waste. For instance, the Metropolitan Health Directorate at both study areas further stated that e-waste comprised electronic or electrical items such as television sets, fridges, printers, computers, and household appliances. The Development Planning Unit of the Accra Metropolitan Assembly also stated that e-waste comprised television sets, computers, mobile phones, and printers, among others. The MESTI similarly noted that e-waste comprises among others television sets, washing machines, computers, mobile phones, refrigerators, and keyboards. Interview findings with the EPA (which acts as the chair of the Hazardous Chemical Committee tasked with the responsibility of managing, destroying, or disposing of unwanted or obsolete hazardous waste) showed that the most common e-waste items imported into the country are personal computers and accessories (central processing unit, mouse, screen, and keyboard), laptop computers, printers, cellular phones, television sets, video cameras and recorders, refrigerators, freezers, air conditioners, and washing machines. According to the EPA official, the importation of cathode-ray tubes (CRTs) and CRT glass are prohibited under the Hazardous and Electronic Waste Control and Management Act, Act 917 (2016). Field observations and interviews with local workers showed that CRTs are among the e-waste items treated at the recycling sites.

Discussions with e-waste recyclers or workers at the study sites and direct observations revealed that the most common EEE handled at the sites are television sets, computers and accessories, batteries, electric irons, washing machines, and refrigerators (Figures 2.1 and 2.2). According to the workers, the two most common EEE to the interest of the workers are televisions and computers, mainly because of their increased demand by households in the cities. The findings from all respondents generally revealed similarities in the composition of items considered as EEE, although workers did not consider them as waste.

It should, however, be stated that officials in Kumasi generally did not consider e-waste as a priority in development interventions as they did not see it as a category of special waste that requires attention, compared to the officials in Accra. The Waste Management Department of the Asokore Mampong Municipal Assembly hence stated that e-waste was not of much priority to city officials, considering that informal e-waste recycling was only one of the many emerging activities that had not received much attention in local, national, and global discourse. The official thus stated that:

> It appears that the issue of this type of waste (e-waste) is important reading around and listening to debates on it, especially, the site in Accra. I should however state that the Waste Management Department is largely involved in general municipal waste as it constitutes the chunk of waste we have to deal with in the city and, because we probably do not see anything special about e-waste and [it] has even not been a major concern in the city compared to the situation in Agbogbloshie, Accra … .

On the impact of e-waste, interviews with the research participants showed that the e-waste sector has both positive and adverse implications. First, interview findings with all agency officials showed that the sector provides livelihood to thousands of urban poor in Ghana. All agency officials and local workers interviewed at both sites confirmed this by indicating that the sector has provided employment to several youths who are either engaged in the collection, manual dismantling, or repair and sale of e-waste items in the respective cities and other parts of the country. It was observed that the e-waste items were processed in small workshops where workers employed rudimentary tools and methods, largely manual disassembling and open burning. The e-waste was stripped of its valuable components and metals, and processed directly into reusable parts in various refining and conditioning processes. These were later either exported to other countries

Figure 2.1 Some electrical and electrical items at the study sites: computer cases, refrigerators, circuit boards, printer cables, and electrical wires, among others. Credit: Michael Osei Asibey. 2021 / Reproduced With permission from Elsevier.

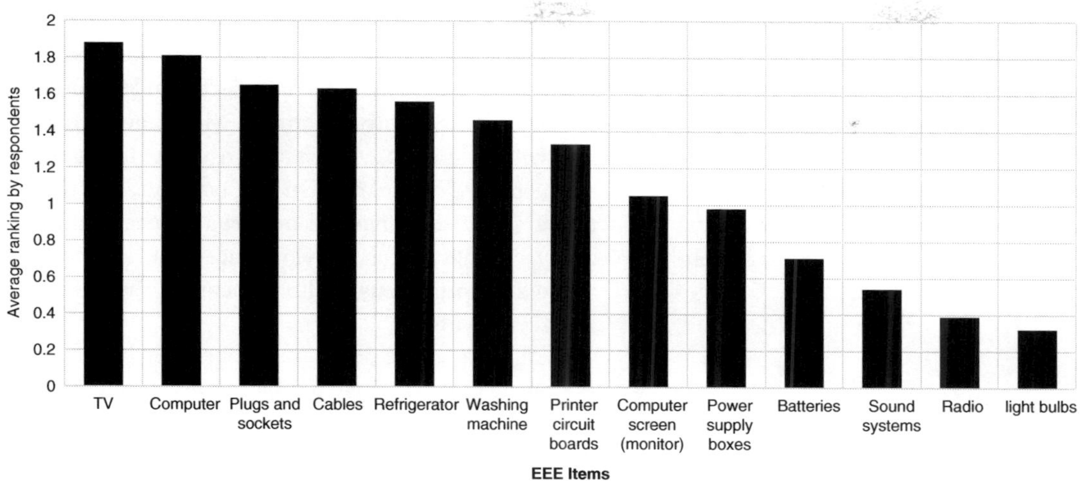

Figure 2.2 Ranking of abundance of electrical and electronic equipment handled at the study sites by e-waste recyclers. Ranking scale: 0 (not at all); 1 (a bit/rarely); 2 (often).

(specifically, Asia and Europe) or purchased by companies and households in the country. These different activities provided jobs to several individuals. Hence, most local workers were involved in informal e-waste recycling activities because of it being a major source of livelihood. This was confirmed by a local worker at DEPS during a focus group discussion who remarked that

> The e-waste sector is very lucrative and has provided jobs to several youths in this community and surrounding communities.

Additionally, the MESTI confirming responses of the EPA and Development Planning Unit of the Accra Metropolitan Assembly indicated that the e-waste sector, largely informal, has provided a source of livelihood to more than 100 000 people in Ghana. Specifically, according to the EPA official, the estimated total numbers of collectors, recyclers, and dependents on e-waste activities range between 121 800 and 201 600 (representing about 1.04% and 1.72%) of the urban population and 0.50% to 0.82% of Ghana's total population. Additionally, the estimated indirect contribution of the informal e-waste sector to the national economy ranges between US$105 and US$268 million, depicting the lucrative nature of the sector.

In spite of the positive impact, e-waste was reported to have adverse health and environmental implications. The unsafe and environmentally unfriendly e-waste recycling practices (such as open burning of the EEE items, open disposal and dumping of the items as reported extensively in other research works) were reported by workers and most agency officials as contributory factors to adverse human health and environmental impact. For instance, all agency officials attributed the open burning and dumping or disposal of e-waste items to contributing to environmental pollution: air, soil, and water. The EPA official of Accra indicated that

> The poor e-waste recycling practices such as open burning of plastic cases, wires and unwanted parts as well as open dumping into the environment, have severe detrimental effects on the health of workers themselves and the environment ...

The Environmental Health Directorate and EPA officials at both study areas further indicated that the unsafe handling of e-waste to human health risks is critical and requires alternative safer management efforts. Health issues such as cancer, cuts, burns, headaches, skin irritations, and anaemia, among others, were reported as some of the short- to long-term adverse health impacts associated with the unsafe practices among e-waste workers. This finding lends credence to arguments that the improper handling of e-waste has had adverse health effects as a result of its contribution to environmental pollution.

Interviews with agency officials and field observations further showed that e-waste recycling activities cause large-scale environmental pollution. Expectedly, interview results point to the polluted site at Agbogbloshie over the past two decades due to the pollution and contamination of the environment from e-waste items. The foregoing suggest that e-waste has both positive (in terms of job creation and income generation) and adverse impacts (specifically, environmental pollution and exposure to health risks); hence, the need to understand its impact, and put in place the necessary efforts so as not to exacerbate the risks at both the individual and community levels. The findings, among others, show that although informal recycling activities could be a catalyst for economic development and well-being, unsafe recycling practices present numerous health and environmental risks that adversely affect well-being. Considering that the differences in conceptualizing e-waste and its associated impact could affect the sound management of the sector, the next section presents results on the perspective of respondents on how local conceptualization of e-waste may influence management efforts in addressing the associated impact.

2.3.2 Influence of Local Conceptualization of E-waste on Management Efforts

All the respondents (agency officials and informal e-waste workers) interviewed affirmed that knowledge or understanding of e-waste, its composition and impact, can have significant influence on its management toward addressing associated risks and maximizing the benefits of the e-waste sector. To the respondents, knowledge of the concept can directly determine what should be done with such waste, the forms of collaborations required to improve waste handling practices, and the needed interventions to effectively address the risks associated with the enterprise and maximize and sustain the benefits brought about by e-waste recycling activities. For instance, the EPA official remarked that

> E-waste recycling activities are potential pathways of heavy metal pollution and associated risks such as effects on the kidneys, gastrointestinal tract, joints and reproductive system, and acute or chronic damage to the nervous system, liver damage, and damage to circulatory and nerve tissue, among others. They do contain heavy metals which are any metallic chemical element that has a relatively high density and is toxic or poisonous at low concentrations. Understanding what e-waste is and its components would help put in place the necessary measures to control its treatment and disposal.

The MESTI shared a similar view and stated that

> E-waste contains heavy metals which bio-accumulate, that is, increase in the concentration of a chemical in a biological organism over time, compared to the chemical's concentration in the environment. Heavy metals can enter a water supply by industrial and consumer waste, or even from acidic rain breaking down soils and releasing heavy metals into streams, lakes, rivers, and groundwater. If one does not know what e-waste is, its likely impact, composition and hazardous fraction, it would be difficult to put in place the necessary measures to address the risks...

Discussion participants (local workers) during group discussions at the study sites also reported that

> Knowing what e-waste is as well as what it comprises influences decisions on what to look out for when targeting such items (in terms of importation and sorting). It would consequently determine who to even collaborate with to understand what such waste are made up of, the hazard or threats of such substances, and who to talk to sensitize individuals on how to process and generally, handle them. These would go a long way to address the numerous health and environmental issues associated with it.

Although the Environmental Health Directorate in both study areas indicated its mandate is to oversee the general health conditions of individuals, and not necessarily on waste management, they were of the view that knowledge and understanding of a phenomenon (in this case, e-waste) can contribute greatly to finding solutions to problems surrounding such issues. The Directorate official of the Accra Metropolitan Assembly, expressing a similar view of the official of Asokore Mampong Municipal Assembly, reported that

> Even in the health and environmental sectors, if you do not know about an environmental or health issues or problems, you can find solutions to them. Therefore, understanding what e-waste is as well as its impact on the environment and health of individuals is crucial to finding prudent measures to effectively manage, maximize the benefits from the sector and address the adverse impact associated with it ...

None of the respondents indicated otherwise that knowledge of e-waste, its composition, and impacts can have influence on efforts in addressing associated risks. The findings show that not having an understanding of the concept of e-waste could pose a challenge to coming up with effective policies to guide the management of such waste, considering the call for local initiatives to manage this waste stream. The finding, among others, shows that although informal recycling activities could be a catalyst for economic development and well-being, unsafe recycling practices present numerous health and environmental risks that adversely affect well-being. An important entry point of this study is that the knowledge of workers about associated risks is critical to developing effective risk-handling practices.

2.4 Discussion

The subject of e-waste is argued to be an urban phenomenon and a contentious one (ILO 2012). This is partly due to the lack of consistency on how such waste is conceptualized.

There are arguments that the phenomenon of e-waste is largely a foreign concept with a global outlook, which in several instances, may affect local management efforts. Localizing e-waste issues, which can have a direct influence on developing effective measures, would thus be an important pathway toward attaining set international targets, such as the Sustainable Development Goals (SDGs). Hence, this paper examines the various interpretations and understanding of e-waste and its impact and as well as respondents' perception on whether knowledge of the term can influence its management.

Since its recognition and dominance in global discourses, various researchers and international organizations have provided different interpretations of the term. Generally, respondents' conceptualization of e-waste, in terms of its definition and composition, was found to be similar with previous studies (e.g. Atiemo et al. 2012; EU WEEE Directive 2012; Puckett et al. 2002), which generally regard e-waste as the end of the life of electrical and electronic products including computers, printers, photocopy machines, television sets, mobile phones, and toys, which are made of sophisticated blends of plastics and metals, among other materials. In most e-waste destination countries, including Ghana, large and small EEE are the most common items found at recycling workshops and dumpsites (Asibey et al. in press; Grant and Oteng-Ababio 2012; Nartey 2016). Ongondo et al. (2011) further assert that due to stricter regulations and higher costs associated with disposal and recycling of e-waste in the industrialized countries, obsolete and discarded EEE (e-waste) are transported to developing economies as "secondhand, charity, donation, or slightly used equipment." A greater proportion are old with very little or no satisfaction value and consigned as waste equipment for disposal within a short time (Nartey 2016). Most of these end up at the informal recycling sites, where mainly open burning of electronic gadgets, manual dismantling, and unsafe disposal methods are practiced, which expose the environment and communities to pollution and contamination.

The most common e-waste items found at the study sites are television sets, computers, cell phones, refrigerators, and washing machines, among others. Studies (e.g., Reddy 2015; Samson 2015; StEP Initiative 2013, 2015; Khurrum et al. 2011) indicate that large e-waste in recipient countries is mostly found in the first, third, and fourth categories of the EU WEEE (2012) categorization and comprises equipment such as television sets, refrigerators and freezers, personal computers, washing machines, among others. Due to their sizes and weights, they are mostly gathered by scavengers from households and various companies. The smaller e-waste such as laptops, music players (DVD and MP3), radio sets, mobile phones, and video cameras, among other portable devices, find their way into household bins and are mostly disposed of with other municipal waste. However, due to the toxic nature and content of most e-waste (Bridgen et al. 2008; Nartey 2016; Atiemo et al. 2012; Atiemo et al. 2016; Yates 2017), it is unsafe to human health and the environment to be disposed of without proper treatment.

The findings showed that officials of the Environmental Health Directorates and Metropolitan Health Directorates were generally unaware of the concept of e-waste, despite admitting that improper handling of general solid waste has adverse implications on the health of humans and the environment. This, however, raises a major issue of concern, considering that they form an important part of the entire existing institutional framework and stakeholders mandated to ensure proper management of the e-waste sector (Government of Ghana 1994; Nartey 2016; Oteng-Ababio et al. 2014). Local e-waste workers regarded the EEE as assets that generated enough revenue, and as such do not necessarily regard them as waste. This notwithstanding, all respondents asserted that knowledge on e-waste, its composition, and impact can influence efforts in effectively managing the sector to minimize the adverse implications. Health issues such as cancer, cuts, burns, headaches, skin irritations, and anaemia, among others, are commonly reported as some of the short- to long-term adverse health impacts associated with current unsafe informal e-waste practices (Bridgen et al. 2008; Burns et al. 2019; Lines et al. 2016; Yu et al. 2016). This finding lends credence to several studies which argue that the improper handling of e-waste has had adverse effects as a result of its contribution to environmental pollution (Asante et al. 2012; Caravanos et al. 2011; Nartey et al. 2016; Oteng-Ababio 2012).

Evidently, several studies show that e-waste recycling activities have large-scale environmental pollution evolving autonomously with considerable adverse implications on workers and communities (see Bansode et al. 2015; Atiemo et al. 2016; Caravanos et al. 2011). This reiterates earlier assertions that not having an understanding of the concept of e-waste could pose a challenge to local authorities in coming up with effective policies to guide the management of such waste, considering the call for local initiatives to manage this waste stream (Agyei-Mensah and Oteng-Ababio 2012; OECD 2014). Premised on the foregoing, this paper defines e-waste to include all the components, consumables, and subassemblies that form part of any electronic and/or electrical product during periods of discarding that are of no more use to the consumer or producer.

2.5 Conclusion

This study sought to contextualize and examine agencies and workers' understanding of e-waste, its composition, and impact, as well as how knowledge of the phenomenon influences management of the sector in Ghana. Relevant agencies that play an important role in Ghana's e-waste sector as well as local e-waste workers at Agbogbloshie (Accra) and Dagomba-Line (Kumasi) were interviewed for relevant data. The premise of this research is that established assumptions about e-waste underpin planning systems and the attitudes of many planners that have been transferred from the "global north" to the "global south" through colonial and postcolonial processes and intertwined with the institutionalization of "local" ideologies.

The study showed that although the concept of e-waste has evolved over time, there appears to be little semblance of definitional consensus amongst respondents. The findings show that not having an understanding of the concept of e-waste could pose a challenge to coming up with effective interventions to guide the management of such waste. Generally, the paper has revealed that responses from agency officials on knowledge of e-waste, its composition, and impact demonstrated a good understanding of the phenomenon that is similar to other definitions from international organisations. In spite of this, the findings have shown that there still appears to be inconsistencies in terms of whether e-waste should comprise only electronic or electrical products or both and at which point they become "waste." This, according to respondents, have (in) direct influence on the management of the waste stream. Workers' knowledge of the e-waste appears to have a positive connotation where they regard e-waste as not waste, but rather as valuable and affordable electrical and electronic items. The study further revealed that the most common e-waste items imported into the country and found at most recycling sites in Ghana are television sets, computers, cell phones, refrigerators, and washing machines, among others, where workers engaged in the collection, manual dismantling, repair, and resale of functional items, and open burning and disposal of unwanted items. The study has revealed that the e-waste enterprise has both positive and adverse implications: revenue generation and a major source of livelihood as well as sources of environmental pollution and human health risk exposure.

Based on the above, the study suggests training and education of all relevant agencies at the national, regional, and local levels as well as of workers on this growing phenomenon. As earlier discussed, officials interviewed have an understanding of e-waste and its impact, yet find it difficult to better contextualize its composition and the point at which it does become waste. This study advocates that relevant agencies and workers be educated via workshops, seminars, and training sessions on this phenomenon and their respective roles in its management, considering attempts by the Hazardous and Electronic Waste Control and Management Act, Act 917 (2016) to better contextualize e-waste and help shape effective management of the sector.

Next, to advance the education process, standardization of e-waste, in terms of the set of attributes, is deemed important for urban planning practice. As earlier indicated, literature on e-waste conceptualization suggests various interpretations but there is a set of attributes that cuts across. In order to characterize and integrate e-waste issues in management efforts, there is the need for standardization––standardization (as appears to be done by the Act 917) to determine the scope of e-waste, composition, and impact on management efforts so as to characterize, better understand, and incorporate such knowledge in both local and national e-waste and environmental planning contexts. This recommendation will ensure that relevant stakeholders determine a set of criteria, or attributes or standards for policy and practice, and set a minimum threshold for understanding and incorporating e-waste into relevant policies and practice. The standardization process will guarantee consistency in the application of e-waste planning and management in Ghanaian environmental and sanitation policy and practice.

References

Agyei-Mensah, S. and Oteng-Ababio, M. (2012). Perceptions of health and environmental impacts of e-waste management in Ghana. *International Journal of Environmental Health Research* 22 (6): 500–517.

Amoyaw-Osei, Y., Agyekum, O.O., Pwamang, J.A. et al. (2011). Ghana e-waste country assessment, SBC e-waste Africa project. http://www.basel.int/Portals/4/Basel%20Convention/docs/eWaste/EwasteAssessmentGhana.pdf (accessed 18 January 2019).

Asante, K.A., Agusa, T., Biney, C.A. et al. (2012). Multi-trace element levels and arsenic speciation in urine of e-waste recycling workers from Agbogbloshie, Accra in Ghana. *Science of the Total Environment* 424: 63–73.

Asibey, M.O., Anne Mette, L., and King, R.S. (in press). Understanding the factors for increased informal electronic waste recycling in Kumasi, Ghana. *International Journal of Environmental Health Research*. https://doi.org/10.1080/09603123.2020.1755016.

AspenCore (2016). Electrical devices vs electronic devices—what's the difference? https://circuitglobe.com/difference-between-electrical-and-electronic-devices.html (accessed 11 June 2019).

Atiemo, S., Faabeluon, L., Manhart, A. et al. (2016). Baseline assessment on e-waste management in Ghana. Sustainable Recycling Industries (SRI) Baseline Assessment. Accra, June. http://www.sustainable-recycling.org/sampson_2016_sri-ghana (accessed 13 January 2020).

Atiemo, S.M., Ofosu, F.G., Aboh, I.J.K., and Kuranchie-Mensah, H. (2012). Assessing the heavy metals contamination of surface dust from waste electrical and electronic equipment (e-waste) recycling site in Accra, Ghana. *Research Journal of Environmental and Earth Sciences* 4: 605–611.

Baldé, C.P., Kuehr, R., Blumenthal, K. et al. (2015). *E-waste Statistics: Guidelines on Classifications, Reporting and Indicators*. Bonn, Germany: United Nations University, IAS–SCYCLE.

Bansode, R., Kumar, S., Malav, C.L. and Malav, M.K. (2015). *Xenobiotic Compounds and Their Types*. http://www.biotecharticles.com/Others-Article/Xenobiotic-Compounds-and-Their-Types-3378.html (accessed 30 November 2018).

Bridgen, K., Labunska, I., Santillo, D., and Johnston, P. (2008). Chemical contamination at e-waste recycling and disposal sites in Accra and Koforidua, Ghana. *Greenpeace Research Laboratories Technical Note* 1–23. https://www.greenpeace.to/publications/chemical-contamination-at-e-wa.pdf (accessed 10 August 2023).

Bryman, A. (2004). *Social Research Methods*, 2e. Oxford: Oxford University Press.

Burns, K.N.L., Sayler, S.K., and Neitzel, R.L. (2019). Stress, health, noise exposures, and injuries among electronic waste recycling workers in Ghana. *Journal of Occupational Medicine and Toxicology* 14: 1.

Caravanos, J., Clark, F.R., and Lambertson, C. (2011). Assessing worker and environmental chemical exposure risks at an e-waste recycling and disposal site in Accra, Ghana. *Journal of Health Pollution* 1: 16–25.

Daum, K., Stoler, J., and Grant, R.J. (2017). Toward a more sustainable trajectory for e-waste policy: a review of a decade of e-waste research in Accra, Ghana. *International Journal of Environmental Research and Public Health* 14: 1–18.

EU WEEE Directive (2012). 2012/19/EU of the European Parliament and of the Council of 4 July 2012 on waste electrical and electronic equipment (WEEE). *Official Journal of the European Union*, L197/38-L197/71. https://eur-lex.europa.eu/LexUriServ/LexUriServ.do?uri=OJ:L:2012:197:0038:0071:EN:PDF (accessed 23 August 2018).

Ghana Statistical Service (2013). 2010 Population & Housing Census. In: *Regional Analytical Report, Greater Accra Region*, 30. Accra: Ghana Statistical Service (GSS).

Ghana Statistical Service (2014). 2010 Population & Housing Census. In: *Analytical Report of Kumasi Metropolitan Area*, 5–21. Accra: (GSS) Ghana Statistical Service.

Government of Ghana (GoG) (1994). *Environmental Protection Agency Act 1994 (Act 490)*. Accra: Environmental Protection Agency (EPA).

Grant, R. and Oteng-Ababio, M. (2012). Mapping the invisible and real African economy: urban e-waste circuitry. *Urban Geography* 33: 1–21.

ILO (2012). The global impact of e-waste: addressing the challenge. Geneva: SAGE Publications.

ILO (2014). *Tackling Informality in E-waste Management: The Potential of Cooperative Enterprises. International Labour Organization, Sectoral Activities Department*. Geneva: Cooperatives Unit.

Khurrum, M.B.S., Omar, A., and Yang, X. (2011). Electronic waste: a growing concern in today's environment. *Economics Research International* https://www.hindawi.com/journals/ecri/2011/474230.

Lines, K., Garside, B., Sinha, S., and Fedorenko, I. (2016). *Clean and Inclusive? Recycling E-waste in China and India*. London: International Institute for Environment and Development (IIED).

Nartey, K.V. (2016). Environmental and health impacts of informal e-waste recycling in Agbogbloshie, Accra, Ghana: recommendations for sustainable management. http://hss.ulb.uni-bonn.de/2016/4325/4325.pdf (accessed 11 January 2019).

Nartey, K.V., Greve, K., and Atiemo, S.M. (2016). Spatial assessment of soil contamination by heavy metals from informal electronic waste recycling in Agbogbloshie, Ghana. *Environmental Health and Toxicology* 31: 1–10.

Ongondo, F.O., Williams, I.D., and Cherrett, T.J. (2011). How are WEEE doing? A global review of the management of electrical and electronic wastes. *Waste Management* 31: 714–730.

Organisation for Economic Co-operation and Development (OECD) (2014). The cost of air pollution. Health impacts of road transport. http://www.keepeek.com/Digital-Asset-Management/oecd/environment/the-cost-of-air-pollution_9789264210448 (accessed 17 August 2018).

Oswald, I. and Armin, R. (2011). E-waste: a story of trashing, trading, and valuable resources. *Gaia* 20 (1): 42.

Oteng-Ababio, M. (2012). Electronic waste management in Ghana-issues and practices. http://cdn.intechopen.com/pdfs/38097/InTech-Electronic_waste_management_in_ghana_issues_and_practices.pdf (accessed 17 January 2020).

Oteng-Ababio, M., Amankwaa, E.F., and Chama, M.A. (2014). The local contours of scavenging for e-waste and higher-valued constituent parts in Accra, Ghana. *Habitat International* 43: 163–171.

Prakash, S., Manhart, A., Amoyaw-Osei, Y., and Agyekum, O.O. (2010). Socio-economic assessment and feasibility study on sustainable e-waste management in Ghana. http://www.resourcefever.org/publications/articles/CARE_INNOVATIONS_11_2010_Prakash_text.pdf (accessed 23 January 2019).

Puckett, J., Byster, L., Westervelt, S. et al. (2002). Exporting harm: the high-tech trashing of Asia. Seattle: The basel action network and silicon valley toxic coalition. http://ban.org/EE-waste/technotrashfinalcom.pdf (accessed 15 January 2020).

Reddy, R.N. (2015). Producing abjection: e-waste improvement schemes and informal recyclers of Bangalore. *Geoforum* 62: 166–174.

Samson, M. (2015). Accumulation by dispossession and the informal economy––Struggles over knowledge, being and waste at a Soweto garbage dump. *Environmental and Planning D: Society and Space* 33: 813–830.

Solving the E-waste Problem Initiative (StEP) (2015). *2013/2014 Annual Global Report on Electrical Electronic Waste*. Bonn: United Nations University.

StEP (2013). *StEP Launches Interactive World E-Waste Map*–United Nations University. http://unu.edu/media-relations/releases/step-launches-interactiveworld-e-waste-map.html#info (accessed 27 January 2018).

US EPA (2016). *Electronic Products Generation and Recycling Methodology Review*. U.S. Environmental Protection Agency.

Widmer, R., Oswald-Krapf, H., Sinha-Khetriwal, D. et al. (2005). Global perspectives on e-waste environmental impact. *Assessment Review* 25 (5): 436–458.

Yates, R. (2017). Week in a toxic waste dump: the Insider, Series 2. https://inclusivevcc.wordpress.com/publications (accessed 18 January 2020).

Yu, E.A., Akormedi, M., Asampong, E.L. et al. (2016). Informal processing of electronic waste at Agbogbloshie, Ghana: workers' knowledge about associated health hazards and alternative livelihoods. *Global Health Promotion* 24 (4): 90–98.

3

Multiomics Approaches on Extremophiles and Their Application in the Biological Management of E-waste

Edwin Hualpa-Cutipa[1,*], *Richard Andi*[2], *Solórzano Acosta*[3], *Daniela Landa-Acuña*[4,5], *and Janelle Mendoza León*[6,7]

[1] *School of Environmental Engineering, Universidad Continental, Lima-Perú*
[2] *Faculty of Pharmacy and Biochemistry, Biotechnology and Omics in Life Sciences Research Group, Universidad Nacional Mayor de San Marcos, Lima-Perú*
[3] *Dirección de Supervisión y Monitoreo en las Estaciones Experimentales Agrarias, Instituto Nacional de Innovación Agraria (INIA), Lima, Peru*
[4] *Professional career in Environmental Engineering, Faculty of Engineering, Private University of the North, Lima–Los Olivos Campus*
[5] *Facultad de Farmacia y Bioquímica, Departamento Académico de Farmacología, Bromatología y Toxicología, Centro Latinoamericano de Enseñanza e Investigación en Bacteriología Alimentaria (CLEIBA), Research Group Biotechnology and Omics in Life Sciences, Universidad Nacional Mayor de San Marcos, Lima, Perú*
[6] *Instituto Nacional de Salud–Centro Nacional de Control de Calidad, Lima, Perú*
[7] *Facultad de Farmacia y Bioquímica, Universidad Nacional Mayor de San Marcos, Lima, Perú*
* *Corresponding author*

3.1 Introduction

3.1.1 The E-waste Problem and Future Potential

Electronic and electrical waste has increased exponentially in recent years due to technological advancement (Işıldar et al. 2019) and has become a very important potential source of environmental pollution (Arya and Kumar 2020). High concentrations of heavy metals (Zn, Hg, Cd, Cr, and Cu) are used for the manufacture of electronic devices, which generates high accumulation and biomagnification in organisms associated with the food chain (Cheng et al. 2014). Persistent organic pollutants (POPs), including PCBs that are used as a refrigerant reagent in dielectric fluids, lubricants, and as oil in transformers, and other persistent chemicals impose serious risks to the environment and human health (Khalid et al. 2021), because even the recycling and disposal of these wastes produces contamination (Song and Li 2014). In addition, waste electrical and electronic equipment (EEE) represents a source of valuable raw materials of great interest, and in the case of metals, e-waste could become a precious alternative source (Bindschedler et al. 2017).

3.1.2 Microorganisms that Biodegrade E-waste

Various indigenous soil microorganisms, including bacteria, fungi, actinomycetes, and archaea (Song et al. 2018), have been explored for e-waste treatment through an enrichment culture approach followed by screening, identification, and their bioformulation for in situ investigation (Debbarma et al. 2018). Rhizospheric effects on soil microbial community structure in e-waste recycling sites have also been studied (Song et al. 2018). Even the most well-known biodegradation pathways have been elucidated as in the case of PCBs such as anaerobic oxidation and dechlorination (Huang et al. 2019; Passatore et al. 2014), which indicates that even these unconventional wastes are amenable to efficient remediation through sustainable and case-specific strategies.

3.1.3 Omics to Research, Improve, and Treat E-waste

The microorganisms used in the remediation of e-waste are quite complex beings with a defined metabolic machinery and physiology as a result of their genotypic constitution and the action of environmental conditions in which they develop and in which e-waste is treated. Therefore, technological platforms are needed that can clarify these conditions and create models that can be transformed from natural evidence and scientific curiosity into new-generation engineering tools that can solve the consumption of electronic devices and their management without detriment to the environment in which we live. These technological platforms are the omics (genetics, genomics, metabolomics, and proteomics) that through

Electronic Waste Management: Policies, Processes, Technologies, and Impact, First Edition. Edited by Sunil Kumar and Vineet Kumar.

bioinformatics management are allowing us to understand the interactions between genes, proteins, and molecules with the environment in which the treatment of electronic waste is performed. Even the possibility of strain manipulation at the chemical or genetic level to improve the storage capacity and efficiency of contaminants is open (Baniasadi et al. 2019; Liu et al. 2018).

The aim of this chapter is to provide information on microorganisms that degrade e-waste and their study under a multiomics perspective, related to strategies, species, and waste types.

3.2 Main Biological Strategies for E-waste Management

Digital devices are now widespread in modern life and are used in almost every aspect of human activity (Mensah 2019). Despite the ease and value that smart products have provided, the insatiable desire for newer and more attractive models has also led to a growing ecological issue known as e-waste or e-waste management (Cruz-Sotelo et al. 2017). There is no sign of a reduction in the demand placed on e-waste disposal systems as the lifespan of electronic gadgets gets shorter and shorter (Tan et al. 2016).

E-waste management is an approach to all actions taken to safeguard environmental resources from the dangers of electronic waste components and other types of waste in order to lessen their negative impacts on human health and the environment and with the potential to recover raw materials from them (Heacock et al. 2016; Ogbuene et al. 2013). In that sense, the technology conventionally employed is bioremediation for the removal of environmental contaminants associated with e-waste due to the extensive metabolic machinery of microorganisms (Khalid et al. 2021). More specifically, we refer to microremediation, which is the process of removing, containing, or transforming pollutants in the environment into less- or non-hazardous forms through the life cycle of microorganisms (Mulligan et al. 2001). Information on microbes that degrade contaminants associated with e-waste is summarized in Table 3.1.

Among the most studied wastes are PCBs (Wu et al. 2018; Xia et al. 2018) containing polychlorinated biphenyls (PCBs), which are complex to degrade due to their chemical stability (Bourzama et al. 2021; Esmaeili et al. 2022; Narayanasamy et al. 2021). Furthermore, it is claimed, there are several microbes that can biotransform or mineralize PCBs aerobically or anaerobically (Adhapure et al. 2013; Liang et al. 2013) in which the enzyme biphenyl dioxygenase stands out as the main enzyme used for aerobic degradation of PCBs in the metabolic pathway of these microorganisms (Khalid et al. 2021).

In the microremediation of hazardous metals, there are six primary mechanisms, which include: (1) bioleaching, (2) biosorption, (3) bioaccumulation, (4) biotransformation, (5) biomineralization, and (6) microbial enhanced metal chemisorption (Patel and Kasture 2014; Bourzama et al. 2021; Esmaeili et al. 2022; Narayanasamy et al. 2021).

Regarding the most studied microorganisms or those reported to have greater activity in e-waste degradation, those belonging to the genera *Bacillus* stand out (Arshadi and Mousavi 2015; Faramarzi et al. 2014; Guan et al. 2013; Liang et al. 2019; Mohan et al. 2016; Wang et al. 2008), also *Pseudomonas* (Faramarzi et al. 2014; Ruan et al. 2014; Xin et al. 2014) and *Penicillium* (Esmaeili et al. 2022; Xia et al. 2018).

The purposes of the use of microorganisms in the treatment of e-waste can be grouped into two large blocks: the remediation of toxic components and the recovery of metallic elements. In this sense, the contaminants that are most often remediated are PCBs and the most recovered elements are Ni, Cu, Zn, and Au (Table 3.1). It is equally important to note that the remediation or recovery efficiency is very often above 90%, which makes these techniques potential technologies for e-waste management.

3.3 Integration of Omics (Multiomics) for the Study of Extremophilic Microorganisms with E-waste Biodegradation Capacity

Microorganisms play an important role in ecosystems and their relationship with humans, due to the important role they play in maintaining the basic functions of living organisms (Nikolaki and Tsiamis 2013), both as complete organisms and in their metabolites. This has allowed scientists to study the behavior of microorganisms within communities, as well as their ecology, evolution, and metabolism, with tools applicable to genomic, metabolomic, proteomic, and transcriptomic studies—omics tools that provide important advances in the knowledge and understanding of biological aspects (Baker and Dick 2013). Thus, the development and technological progress has allowed the use of the capabilities of microorganisms in various fields to produce a wide range of antibiotics and drugs for clinical use, as markers for the detection of diseases, in the food industry, and for the remediation of contaminated soil and water (National Research Council [US] 2007).

Table 3.1 Biodegradation strategies with microorganisms for electronic waste.

E-waste	Mechanism or strategy	Microorganism	Element recovered or remediated	Efficiency recovery or degradation	Reference
Computer printed circuit board	Bioleaching	*Penicillium simplicissimum*	Cu, Al y Ni	100% (Cu), 70% Ni, 98% (Al)	Esmaeili et al. 2022
Printed circuit board	Bioleaching	*Frankia* sp.	Hg y Pb	98% (Hg), 89% (Pb)	Narayanasamy et al. 2021
Batteries and circuit boards	Keratinolytic enzymes	*Geotrichum candidum*	None (weight degradation)	23% (batteries), 71% (circuit boards)	Bourzama et al. 2021
Batteries and circuit boards	Keratinolytic enzymes	*Rhizopus stolonifer*	None (weight degradation)	7% (batteries), 60% (circuit boards)	Bourzama et al. 2021
Low-power electronic sewage	Specific microbe immobilization cell (SMIC)	Various	Tetramethylammonium hydroxide	higher than biological activated carbon (BAC)	Chung et al. 2020
Electronic waste water	Enzymes	*Bacillus* sp.	2,4,6-Tribromophenol (TBP)	98.80%	Liang et al. 2019
Electronic waste water	Biosorption	*Phomopsis* sp.	Gold and toxic dyes	80% (gold)	Xu et al. 2019
Printed circuit boards (PCB)	Magnetotactism	*Magnetospirillum* sp.	Cd, Pb, Ni	97% (Cd), 100% (Pb), 99% (Ni)	Sannigrahi et al. 2019
Panels LCD	Bioleaching	*Acidithiobacillus* sp.	Indio	100% sulfur-mediated	Xie et al. 2019
Recycled electronic waste	Bromination, hydroxylation, and ring-opening processes	*Stenotrophomonas* sp.	Polybrominated diphenyl ethers (PBDE)	55.15% (BDE209)	Wu et al. 2018
Printed circuit boards (PCB)	Bioleaching	*Penicillium chrysogenum*	Cu	47%	Xia et al. 2018
Electronic waste	Hidroxilación	*Burkholderia cepacia*	éter 2,2′,4,4′-tetrabromodiphenyl (BDE-47)	82.40%	Huang et al. 2017
Electronic plastics	Depolymerization	*Enterobacter* sp., *Citrobacter sedlakii, Alcaligenes* sp. y *Brevundimonas diminuta*	High-impact polystyrene (HIPS)	12.40%	Sekhar et al. 2016
Electronic waste	–	*Enterococcus casseliflavus*	BDE-209	–	Tang et al. 2016
Recycled electronic waste	Debromination	*Rhodococcus* sp.	BDE-209	65.10%	Liu et al. 2016
Electronic plastics	Biomass degradation	*Bacillus* spp.	High-impact polystyrene (HIPS)	23% (peso)	Mohan et al. 2016
Computer printed circuit board	Bioleaching	*Bacillus megaterium*	Au	63.8% (Au in pretreated sample)	Arshadi y Mousavi 2015
electronic waste	–	*Pseudomonas putida*	BDE-47	49.96%	Xin et al. 2014
Printed circuit boards (PCB)	Bioleaching	*Pseudomonas chlororaphis*	Au, Ag, Cu	8.2% (Au), 12.1% (Ag), 52.3% (Cu)	Ruan et al. 2014
Electronic waste	Clearing	*Bacillus cereus*	Decabrominated diphenyl ethers (BDE209)	14.16%	Wang et al. 2008

(Continued)

Table 3.1 (Continued)

E-waste	Mechanism or strategy	Microorganism	Element recovered or remediated	Efficiency recovery or degradation	Reference
Printed circuit boards (PCB)	Bioleaching	*Chromobacterium violaceum, Pseudomonas fluorescens, Bacillus megaterium*	Ni	–	Faramarzi et al. 2014
Electronic waste	Bioleaching	*Chromobacterium violaceum*	Au	11%	Natarajan y Ting, 2014
Electronic waste	–	*Bacillus* sp.	2,2′,4,4′-tetrabromodifenil éter (BDE-47)	50%	Guan et al. 2013
Printed circuit boards (PCB)	Bioleaching	*Acidithiobacillus thiooxidans y Acidithiobacillus ferrooxidans*	Cu	92.6% (Cu)	Liang et al. 2013
Printed circuit boards (PCB)	Bioleaching	*Consorcio aislado de bauxita y pirita*	Cu, Zn, Ni	96.93% (Cu), 93.33% (Zn), 10.26% (Ni)	Adhapure et al. 2013
Electronic waste	Debromination	*Ochrobactrum* sp.	Tetrabromobisphenol-A (TBBPA)	91.80%	An et al. 2011
Printed circuit boards (PCB)	Bioleaching	*Consorcio bacterias acidófilas*	Cu, Al, Zn	96.8% (Cu), 88.2% (Al), 91.6% (Zn)	Zhu et al. 2011

In this sense, the use of omics has gained importance in the generation of biotechnological tools for the bioremediation of natural sources contaminated by materials of diverse origin. Such is the case for electronic waste, whose remediation is complicated because it represents a risk for living organisms and their environment (Mudila et al. 2021). The permanence of e-waste in ecosystems is encouraged by the demand and production of electronic products, which when left unused end up in landfills that cause a growing permanence of electronic waste and electrical appliances (e-waste), while the regulations and laws focused on the treatment of these wastes, recycling, and immobilization vary from one country to another (United Nations Environment Programme 2005). It is because of these variations in the laws of each country that it is possible to observe crude and rudimentary methods applied in the recovery of valuable compounds from e-waste, generating the release of various organic pollutants at high levels such as polycyclic aromatic hydrocarbons (PAH), polychlorinated dibenzo-p-dioxins and furans (PCDD/F), polybrominated diphenyl ethers (PBDE), polychlorinated biphenyls (PCB), and toxic heavy metals such as Pb, Cd, Cr, and Cu (Gullett et al. 2007). The monitoring of environments requires the use of indicators that are minimally invasive, easy to transport, store, analyze, and with little risk to health—a clear example of which is human hair—as they can reflect the level of exposure over long periods of time, from weeks to years (Wang et al. 2009).

3.4 The Role of Omics in the Bioremediation of E-waste by Extremophilic Microorganisms

With respect to omics sciences, their application has led to great advances in the recognition of electronic waste (e-waste) bioremediation processes. Molecular biology in the field of biomining has provided valuable information on metal–microbial interaction and the function of microbes in the recovering and extraction of metals (Vera et al. 2013). Microorganisms such as *Acidithiobacillus ferrooxidans*, *Acidithiobacillus thiooxidans*, and *Leptospirillum ferrooxidans* have been reported as iron and sulfur oxidizers, in addition to other metal-mobilizing extremophilic microbial species found in various genera such as *Acidithiobacillus*, *Acidiphilum*, *Acidiferrobacter*, *Sulfolobus*, *Acidianus*, *Metallosphaera*, *Sulfurisphaera*, *Ferroplasma*, and *Ferro vum*, among others. Proteobacteria, Nitrospirae, Firmicutes, and Actinobacteria are the four phyla in which they are spread (Dave et al. 2018). It is worth mentioning that, under genetic engineering, bacteria of the genus *Acidithiobacillus* have been modified to improve bioleaching performance, as is the case of *A. ferrooxidans*, whose expression of arsenic-tolerant genes resulted in strains with greater resistance to arsenic, compared to *A. ferrooxidans* whose expression of the mer operon generated variants with sensitivity to mercury ions (Kusano et al. 1992; Peng et al. 1994).

Biological tools have allowed us not only to discover new capabilities in relation to e-waste biodegradation, but it has also been possible to improve and enhance the catalytic capabilities of enzymes, substrate specificity, and tolerance to contaminants (Capeness and Horsfall 2020; De Lorenzo, 2018). *Pseudomonas putida* 06909 was genetically improved to produce a peptide with cadmium-binding affinity in addition to alleviating cell toxicity caused by cadmium; on the other hand, co-assemblies (biotic-abiotic) with the ability to effectively reduce cadmium and lead by a heavy metal capture mechanism on the cell surface and magnetic nanoparticles have been designed in synthetic bacterial cells (Zhu et al. 2020).

As is known, one of the main interests in e-waste is the recovery and recycling of certain materials, such as gold. *Chromobacterium violaceum* is a cyanogenic bacterium with the ability to biodissolve gold. Its capacity was increased with the introduction of exogenous promoters pBAD and pTAC to decouple quorum control; this in principle because the operon (hcnABC) that produces the enzyme involved in cyanide leachate production limits its production to 20 mg/L at the beginning of the stationary phase, which improved cyanide production by 70% into modified strains and more than doubled the gold recovery that could be obtained (Natarajan and Ting 2014; Tay, 2013). Although promising, the use of biologically modified microorganisms for e-waste bioremediation remains in its early stages and requires a great deal more study and development before the technique can be used commercially (Han et al. 2022).

Analysis techniques focused on FTIR, TG-DTG-DTA, and SEM have allowed confirmation of the potential of bacteria to biodegrade on electronic waste, where from enriched soil, TG shows that the biodegradation process is more effective at lower temperatures than in the control sample, while FTIR spectra show that the microbial communities use the carbon contained in the chemical structure of the e-waste as the main source of carbon. On the other hand, the images obtained with SEM showed the decomposition of the samples on their surface (Debbarma et al. 2018). Xu et al. (2015) mention that the functional mechanisms of macroorganisms may favor the stimulation of genes and microbial diversity associated with the presence of organic contaminants in e-waste such as POPs. Meanwhile, genes such as chrR (chromium detoxification), metC (mercury detoxification), merB (mercury detoxification), and the silafin gene (silicon biosynthesis) have been identified in microorganisms involved in the homeostasis of heavy metals, as well as genes involved in the biodegradation of polycyclic aromatic compounds, including xylL, benD, and nhh (Yin et al. 2015).

It is worth mentioning that there is still a wide diversity of non-culturable microorganisms that could have e-waste bioremediation potential, whose study and new methods for their investigation are under development (Salam and Varma 2019). Other biphenyl- and PCB-degrading extremophilic microorganisms have been identified in e-waste contaminated soils by the independent DNA stable isotope probe culture method, including the genera *Ralstonia* and *Cupriavidus*, which were enriched in the 13C DNA heavy fraction, confirming PCB-degrading action (Jiang et al. 2018). *Stenotrophomonas* sp. is another extremophilic bacterium isolated from recyclable e-waste with the ability to degrade decabromodiphenyl ether (BDE 209) under aerobic conditions, pH 5, 25°C, 0.5% salinity, 150 ml minimum average salt volume, conditions determined by orthogonal tests, making possible the use of three metabolic pathways for the degradation of BDE 209, four degradation genes (one hydrolase, one dioxygenase, and two dehalogenases) to degrade between 5.15% of 65 g/L of BDE 209 (debromination, hydroxylation, and ring-opening activities) (Wu et al. 2018).

References

Adhapure, N.N., Waghmare, S.S., Hamde, V.S., and Deshmukh, A.M. (2013). Metal solubilization from powdered printed circuit boards by microbial consortium from bauxite and pyrite ores. *Prikladnaia Biokhimiia I Mikrobiologiia* 49 (3): 279–284. https://doi.org/10.7868/S0555109913030033.

An, T., Zu, L., Li, G. et al. (2011). One-step process for debromination and aerobic mineralization of tetrabromobisphenol-A by a novel *Ochrobactrum* sp. T isolated from an e-waste recycling site. *Bioresource Technology* 102 (19): 9148–9154. https://doi.org/10.1016/j.biortech.2011.06.080.

Arshadi, M. and Mousavi, S.M. (2015). Enhancement of simultaneous gold and copper extraction from computer printed circuit boards using *Bacillus megaterium*. *Bioresource Technology* 175: 315–324. https://doi.org/10.1016/j.biortech.2014.10.083.

Arya, S. and Kumar, S. (2020). E-waste in India at a glance: current trends, regulations, challenges and management strategies. *Journal of Cleaner Production* [online] 271: 122707. https://doi.org/10.1016/j.jclepro.2020.122707.

Baker, B.J. and Dick, G.J. (2013). Omic approaches in microbial ecology: charting the unknown. *Microbe* 8 (9): 353–359. https://doi.org/10.1128/MICROBE.8.353.1.

Baniasadi, M., Vakilchap, F., Bahaloo-Horeh, N. et al. (2019). Advances in bioleaching as a sustainable method for metal recovery from e-waste: a review. *Journal of Industrial and Engineering Chemistry* 76: 75–90.

Bindschedler, S., Bouquet, T.Q.T.V., Job, D. et al. (2017). Fungal biorecovery of gold from e-waste. *Advances in Applied Microbiology* 99: 53–81.

Bourzama, G., Iratni, N., Ennaghra, N. et al. (2021). In vitro removal of electronic and electrical wastes by fungi isolated from soil at Annaba area in northeast of Algeria. *Environment and Natural Resources Journal* 19 (4): 302–309. https://doi.org/10.32526/ennrj/19/2020294.

Capeness, M.J. and Horsfall, L.E. (2020). Synthetic biology approaches towards the recycling of metals from the environment. *Biochemical Society Transactions* 48 (4): 1367–1378.

Cheng, Z., Mo, W.Y., Man, Y.B. et al. (2014). Replacing fish meal by food waste in feed pellets to culture lower trophic level fish containing acceptable levels of organochlorine pesticides: health risk assessments. *Environment International* 73: 22–27.

Chung, J., Choi, J., and Chung, S. (2020). Pilot study of specific microbe immobilization cells (SMICs) technology in removing tetramethyl ammonium hydroxide for reuse of low-strength electronics wastewater. *Journal of Hazardous Materials* 384. https://doi.org/10.1016/j.jhazmat.2019.120829. contamination. Sci. Rep., https://doi.org/10.1038/srep14266.

Cruz-Sotelo, S.E., Ojeda-Benítez, S., Jáuregui Sesma, J. et al. (2017). E-waste supply chain in Mexico: challenges and opportunities for sustainable management. *Sustainability* 9 (4): 503.

Dave, S.R., Sodha, A.B., and Tipre, D.R. (2018). Microbial technology for metal recovery from e-waste printed circuit boards. *Journal of Bacteriology and Mycology* 6 (4): 241–247.

De Lorenzo, V., Prather, K.L., Chen, G.Q. et al. (2018). The power of synthetic biology for bioproduction, remediation and pollution control: the UN's sustainable development goals will inevitably require the application of molecular biology and biotechnology on a global scale. *EMBO Reports* 19 (4): e45658.

Debbarma, P., Zaidi, M.G.H., Kumar, S. et al. (2018). Selection of potential bacterial strains to develop bacterial consortia for the remediation of e-waste and its in situ implications. *Waste Management* 79: 526–536.

Esmaeili, A., Arshadi, M., and Yaghmaei, P.O.C.A.P.E.D.S. (2022). Simultaneous leaching of Cu, Al, and Ni from computer printed circuit boards using *Penicillium simplicissimum*. *Resources, Conservation and Recycling* 177: https://doi.org/10.1016/j.resconrec.2021.105976.

Faramarzi, M.A., Stagars, M., Pensini, E. et al. (2014). Metal solubilization from metal-containing solid materials by cyanogenic *Chromobacterium violaceum*. *Journal of Biotechnology* 113 (1–3): 321–326. https://doi.org/10.1016/j.jbiotec.2004.03.031.

Guan, J.F., Wang, J.H., Liu, X. et al. (2013). Biodegradation of 2,2′,4,4′-tetrabromodiphenyl ether in an aerobic environment by a novel strain of *Bacillus* sp. doi: 10.4028/www.scientific.net/AMR.771.45. Retrieved from www.scopus.com.

Gullett, B.K., Linak, W.P., Touati, A. et al. (2007). Characterization of air emissions and residual ash from open burning of electronic wastes during simulated rudimentary recycling operations. *The Journal of Material Cycles and Waste Management* 9: 69–79.

Han, P., Teo, W.Z., and Yew, W.S. (2022). Biologically engineered microbes for bioremediation of electronic waste: wayposts, challenges and future directions. *Engineering Biology* 6 (1): 23–34. https://doi.org/10.1049/enb2.12020.

Heacock, M., Kelly, C.B., Asante, K.A. et al. (2016). E-waste and harm to vulnerable populations: a growing global problem. *Environmental Health Perspectives* 124 (5): 550–555.

Huang, C., Zeng, Y., Luo, X. et al. (2019). In situ microbial degradation of PBDEs in sediments from an e-waste site as revealed by positive matrix factorization and compound-specific stable carbon isotope analysis. *Environmental Science & Technology* 53 (4): 1928–1936.

Huang, T., Duan, X., Tao, X. et al. (2017). Identification of an aerobic strain efficiently degrading 2,2′,4,4′-tetrabromodiphenyl ether and its biodegradation pathway. [2,2′,4,4′-四溴联苯醚高效好氧降解菌的鉴定及其降解路径].

Işıldar, A., van Hullebusch, E.D., Lenz, M. et al. (2019). Biotechnological strategies for the recovery of valuable and critical raw materials from waste electrical and electronic equipment (WEEE)––a review. *Journal of Hazardous Materials* [online] 362: 467–481. https://doi.org/10.1016/j.jhazmat.2018.08.050.

Jiang, L., Luo, C., Zhang, D. et al. (2018). Biphenyl-metabolizing microbial community and a functional operon revealed in e-waste-contaminated soil. *Environmental Science & Technology* 52 (15): 8558–8567.

Khalid, F., Hashmi, M.Z., Jamil, N. et al. (2021). Microbial and enzymatic degradation of PCBs from e-waste-contaminated sites: a review. *Environmental Science and Pollution Research* 28 (9): 10474–10487.

Kusano, T., Sugawara, K., Inoue, C. et al. (1992). Electrotransformation of *Thiobacillus ferrooxidans* with plasmids containing a mer determinant. *Journal of Bacteriology* 174 (20): 6617–6623.

Liang, G., Tang, J., Liu, W., and Zhou, Q. (2013). Optimizing mixed culture of two acidophiles to improve copper recovery from printed circuit boards (PCBs). *Journal of Hazardous Materials* 250–251: 238–245. https://doi.org/10.1016/j.jhazmat.2013.01.077.

Liang, Z., Li, G., Mai, B., and An, T. (2019). Biodegradation of typical BFRs 2,4,6-tribromophenol by an indigenous strain *Bacillus* sp. GZT isolated from e-waste dismantling area through functional heterologous expression. *Science of the Total Environment* 697: https://doi.org/10.1016/j.scitotenv.2019.134159.

Liu, J., Chen, X., Shu, H.Y. et al. (2018). Microbial community structure and function in sediments from e-waste contaminated rivers at Guiyu area of China. *Environmental Pollution* 235: 171–179.

Liu, L., Zhang, Y., Liu, R. et al. (2016). Aerobic debromination of BDE-209 by *Rhodococcus* sp. coupled with zerovalent iron/activated carbon. *Environmental Science and Pollution Research* 23 (4): 3925–3933. https://doi.org/10.1007/s11356-015-5663-4.

Mensah, J. (2019). Sustainable development: meaning, history, principles, pillars, and implications for human action: literature review. *Cogent Social Sciences* 5 (1): 1653531.

Mohan, A.J., Sekhar, V.C., Bhaskar, T., and Nampoothiri, K.M. (2016). Microbial assisted high impact polystyrene (HIPS) degradation. *Bioresource Technology* 213: 204–207. https://doi.org/10.1016/j.biortech.2016.03.021.

Mudila, H., Prasher, P., Kumar, A. et al. (2021). E-waste y su gestión de peligros mediante procesos específicos de biorremediación microbiana. In: *Microbial Rejuvenation of Polluted Environment. Microorganisms for Sustainability*, vol. 26. (eds. D.G. Panpatte, Y.K. Jhala, and C.O. Adetunji). Singapore: Springer. https://doi.org/10.1007/978-981-15-7455-9_6.

Mulligan, C.N., Yong, R.N., and Gibbs, B.F. (2001). Remediation technologies for metal-contaminated soils and groundwater: an evaluation. *Engineering Geology* 60 (1): 193–207.

Narayanasamy, M., Dhanasekaran, D., and Thajuddin, N. (2021). Bioremediation of noxious metals from e-waste printed circuit boards by *Frankia*. *Microbiological Research* 245. https://doi.org/10.1016/j.micres.2021.126707.

Natarajan, G. and Ting, Y. (2014). Pretreatment of e-waste and mutation of alkali-tolerant cyanogenic bacteria promote gold biorecovery. *Bioresource Technology* 152: 80–85. https://doi.org/10.1016/j.biortech.2013.10.108.

National Research Council (US) (2007). *Committee on Metagenomics: Challenges and Functional Applications, the New Science of Metagenomics: Revealing the Secrets of Our Microbial Planet.* Washington, DC, USA: National Academies Press.

Nikolaki, S. and Tsiamis, G. (2013). Microbial diversity in the era of omic technologies. *BioMed Research International*. https://doi.org/10.1155/2013/958719.

Ogbuene, E.B., Igwebuike, E.H., and Agusiegbe, U.M. (2013). The impact of open solid waste dump sites on soil quality: a case study of UgwuajiIn Enugu. *British Journal of Advanced Academic Research* 2 (1): 43 53.

Patel, S. and Kasture, A. (2014). Review Article E (Electronic) Waste management using biological systems—overview. *International Journal of Current Microbiology and Applied Sciences* 3 (7): 495–504. https://www.ijcmas.com/vol-3-7/Shuchi%20Patel%20and%20Avani%20Kasture.pdf.

Passatore, L., Rossetti, S., Juwarkar, A.A., and Massacci, A. (2014). Phytoremediation and bioremediation of polychlorinated biphenyls (PCBs): state of knowledge and research perspectives. *Journal of Hazardous Materials* 278: 189–202.

Peng, J.B., Yan, W.M., and Bao, X.Z. (1994). Expression of heterogenous arsenic resistance genes in the obligately autotrophic biomining bacterium *Thiobacillus ferrooxidans*. *Applied and Environmental Microbiology* 60 (7): 2653–2656.

Ruan, J., Zhu, X., Qian, Y., and Hu, J. (2014). A new strain for recovering precious metals from waste printed circuit boards. *Waste Management* 34 (5): 901–907. https://doi.org/10.1016/j.wasman.2014.02.014.

Salam, M.D. and Varma, A. (2019). A review on impact of e-waste on soil microbial community and ecosystem function. *Pollution* 5 (4): 761–774.

Sannigrahi, S. and Suthindhiran, K. (2019). Metal recovery from printed circuit boards by magnetotactic bacteria. *Hydrometallurgy* 187: 113–124. https://doi.org/10.1016/j.hydromet.2019.05.007.

Sekhar, V.C., Nampoothiri, K.M., Mohan, A.J. et al. (2016). Microbial degradation of high impact polystyrene (HIPS), an e-plastic with decabromodiphenyl oxide and antimony trioxide. *Journal of Hazardous Materials* 318: 347–354. https://doi.org/10.1016/j.jhazmat.2016.07.008.

Song, M., Cheng, Z., Luo, C. et al. (2018). Rhizospheric effects on the microbial community of e-waste-contaminated soils using phospholipid fatty acid and isoprenoid glycerol dialkyl glycerol tetraether analyses. *Environmental Science and Pollution Research* 25 (10): 9904–9914.

Song, Q. and Li, J. (2014). Environmental effects of heavy metals derived from the e-waste recycling activities in China: a systematic review. *Journal of Waste Management* 34: 2587–2594.

Tan, M.J., Owh, C., Chee, P.L. et al. (2016). Biodegradable electronics: cornerstone for sustainable electronics and transient applications. *Journal of Materials Chemistry C* 4 (24): 5531–5558.

Tang, S., Yin, H., Chen, S. et al. (2016). Aerobic degradation of BDE-209 by *Enterococcus casseliflavus*: isolation, identification and cell changes during degradation process. *Journal of Hazardous Materials* 308: 335–342. https://doi.org/10.1016/j.jhazmat.2016.01.062.

Tay, S.B., Natarajan, G., Rahim, M.N.B.A. et al. (2013). Enhancing gold recovery from electronic waste via lixiviant metabolic engineering in *Chromobacterium violaceum*. *Scientific Reports* 3 (1): 1–7.

United Nations Environment Programme (2005). E-waste, the hidden side of IT equipment's manufacturing and use.

Vera, M., Schippers, A., and Sand, W. (2013). Progress in bioleaching: fundamentals and mechanisms of bacterial metal sulfide oxidation–Part A. *Applied Microbiology and Biotechnology* 97 (17): 7529–7541.

Wang, T., Fu, J., Wang, Y. et al. (2009). Use of scalp hair as indicator of human exposure to heavy metals in an electronic waste recycling area. *Environmental Pollution* 157 (8–9): 0–2451. doi: https://doi.org/10.1016/j.envpol.2009.03.010.

Wang, T., Yin, H., Peng, H. et al. (2008). Effects of low concentration heavy metals on biodegradation of BDE209 by *Bacillus cereus*. *Huanjing Kexue/Environmental Science* 29 (7): 1967–1972. Retrieved from www.scopus.com.

Wu, Z., Xie, M., Li, Y. et al. (2018). Biodegradation of decabromodiphenyl ether (BDE 209) by a newly isolated bacterium from an e-waste recycling area. *AMB Express* 8 (1): https://doi.org/10.1186/s13568-018-0560-0.

Xia, M., Bao, P., Liu, A. et al. (2018). Isolation and identification of *Penicillium chrysogenum* strain Y5 and its copper extraction characterization from waste printed circuit boards. *Journal of Bioscience and Bioengineering* 126 (1): 78–87. https://doi.org/10.1016/j.jbiosc.2018.02.001.

Xie, Y., Wang, S., Tian, X. et al. (2019). Leaching of indium from end-of-life LCD panels via catalysis by synergistic microbial communities. *Science of the Total Environment* 655: 781–786. https://doi.org/10.1016/j.scitotenv.2018.11.141.

Xin, J., Liu, X., Liu, W., and Zheng, X.L. (2014). Aerobic transformation of BDE-47 by a *Pseudomonas putida* sp. strain TZ-1 isolated from PBDEs-contaminated sediment. *Bulletin of Environmental Contamination and Toxicology* 93 (4): 483–488. https://doi.org/10.1007/s00128-014-1306-6.

Xu, M., Jiang, Y., Pradhan, S., and Yadavalli, V.K. (2019). Use of silk proteins to form organic, flexible, degradable biosensors for metabolite monitoring. *Frontiers in Materials* 6: https://doi.org/10.3389/fmats.2019.00331.

Xu, X., Zeng, X., Boezen, H.M., and Huo, X. (2015). E-waste environmental contamination and harm to public health in China. *Frontiers of Medicine* 9: 220–228.

Yin, H., Niu, J., Ren, Y. et al. (2015). An integrated insight into the response of sedimentary microbial communities to heavy metal. *Scientific Reports* 5.

Zhu, N., Xiang, Y., Zhang, T. et al. (2011). Bioleaching of metal concentrates of waste printed circuit boards by mixed culture of acidophilic bacteria. *Journal of Hazardous Materials* 192 (2): 614–619. https://doi.org/10.1016/j.jhazmat.2011.05.062.

Zhu, N., Zhang, B., and Yu, Q. (2020). Genetic engineering-facilitated coassembly of synthetic bacterial cells and magnetic nanoparticles for efficient heavy metal removal. *ACS Applied Materials & Interfaces* 12 (20): 22948–22957.

4

Worldwide E-waste Management Models

Delving into Pros and Cons and the Way Forward

*Md Shah Newaz and Andrea Appolloni**

Faculty of Economics, University of Rome Tor Vergata, Rome, Italy
** Corresponding author*

4.1 Introduction

The definition of electronic waste, also known as waste electrical and electronic equipment (WEEE), keeps evolving. However, traditionally it is compartmentalized as monitors, printers, tablets, computers, servers, and cell phones. Nowadays, the term e-waste encompasses almost every business and household electronic product or device that uses some sort of power or battery supply. Eventually, when this electrical and electronic equipment (EEE) reaches its end-of-life (EOL) or its useful life it becomes electronic waste or WEEE, which these days is known as e-waste (Read 2021). Therefore, the most generic and simplistic way to define e-waste would be the definition agreed upon by StEP (Solving the E-waste Problem): E-waste is a term used to cover items of all types of electrical and electronic equipment (EEE) and its parts that have been discarded by the owner as waste without the intention of reuse.

Interestingly enough, e-waste may account for 3% (at the most) of household solid waste, but it is considered one of the most unsafe types of waste to manage (Kumar et al. 2017). For instance, around 70% of cadmium and mercury that are found in the landfills of the USA comes from electronic waste streams (Tanskanen 2013). Annually, approximately 50 million tons of global WEEE end up in the environment (UN 2009), while the volume grows by 3–5% per year (Yu et al. 2010). Around 80% of e-waste is not even documented.

Meanwhile, the developing nations turn out to be among the biggest contributors to the WEEE due to their prevalent informal system. The deficiencies of their current laws and policies on the management of WEEE have let the informal recycling sector develop (Borthakur and Govind 2017), where the most valuable materials are extracted. In the meantime, the rest of the hazardous elements are tossed away into the environment, bringing harm to human health and obviously the environment (Balde et al. 2017). Thus, these mentioned deficiencies in WEEE treatment ultimately contribute to the expansion of informal e-waste sectors in the countries of Asia, South America, Oceania, and many other African and poor countries. Furthermore, even with the presence of a formal electronic waste collection/recycling/management system in place, the informal system could still be operational and significant (Shevchenko et al. 2021).

In this dire situation, to address the looming challenge of e-waste management this chapter looks at the problem from a circular economy and business model perspective. The economic opportunities stemming from e-waste are elaborated on and connected to the role of the Circular Economy Business Model (CEBM) in this paradigm (Sections 4.3 and 4.4). These sections also explain how the business model canvas (BMC) can be used as a guiding principle or a frame to explain the better management of e-waste. Then, Sections 4.5 through 4.17 illustrate some generic CE business models under the lens of BMC and sustainability. Each of the four cases focuses on a specific sector (e.g. lithium-ion battery, consumer and professional EEE sectors) that fits either into the EU or developing economic context. The final sections show the path forward for the future of e-waste management in developing economies and wraps up with some realistic suggestions for the implementation of such business models.

Electronic Waste Management: Policies, Processes, Technologies, and Impact, First Edition. Edited by Sunil Kumar and Vineet Kumar.

4.2 Economic Opportunities Underneath the Garbage

Fortunately, e-waste is not necessarily always hazardous or harmful for humankind as it includes a wide range of electronic (e.g. consumer, digital, industrial) products encompassing both hazardous yet valuable substances. On one hand, electric wastes contain zillions of harmful substances contributing to the massive pollution and health hazard if not treated properly. On the flip side, these WEEEs are also bountiful in terms of metal resources, and rare earth materials essential for the emerging and manufacturing industries to survive. From 2020 to 2030, the generation of e-waste has been anticipated to soar from 15.5 million tons to 28.4 million tons, which include but are not limited to air conditioners, washing machines, computers, and refrigerators as the top products to be turned into e-waste.

By 2030, the value of these huge amounts of secondary materials—waiting to be extracted from e-waste—will be roughly 73.4 billion US dollars. This immense economic opportunity ultimately contributes to promoting e-waste closed-loop recycling and the development of WEEE recycling business models (Islam and Huda 2019). Furthermore, the demand for secondhand materials/resources and EEE products drives earning opportunities for the locals in the emerging and third world countries, which further yields opportunities for refurbishment and dismantling through informal e-waste management channels (Wang et al. 2012).

As precious metals are abundantly found in the e-waste, recovering the metals comes with economic opportunity through the conservation of expensive and scarce materials for the manufacturing of electronics and many other devices. However, this opportunity to make an economic profit from the secondhand materials/resources contained in the WEEE is further distinguished as one of the significant challenges to achieving sustainability (Isernia et al. 2019). Thus, numerous countries have taken different steps to make e-waste management—particularly, recycling—better. This includes measures such as undertaking pilot projects, materializing funds, infrastructure development, enhancing consumers' understanding, and regulatory improvements. However, the effects are yet to bring expected outcomes. For instance, in 2016, only 20% of the global electronic waste is documented as either recycled or collected. This illustrates the fact that traditional WEEE collection and recycling processes face enormous challenges to begin with. Therefore, building national or regional-level closed-loop recycling systems has become a necessity. In this vein, it is crucial to explore innovative and alternative business models for the WEEE collection, reuse, and recycling that can also offer learning opportunities for other countries (Zuo et al. 2020).

To delve into new business models for e-waste management, manufacturers are recently paying great attention to the waste prevention and environmental outcomes of their production processes by reusing, recycling, and reducing the final dumping of the end-of-life (EOL) products via circular economy (CE) approaches, since this delivers the opportunity to enhance their competitive advantage. By establishing the economic value and underlying business opportunities stemming from the electronic waste, the next section of this chapter elaborates on the circular economic approach toward managing e-waste that eventually takes the business model to the next level, which is the circular economy business model (CEBM) or sustainable business model (SBM).

4.3 E-waste and the Circular Economy Business Model (CEBM)

The conventional linear economy uses the "take-make-dispose" model, which is both energy-consuming and resource-intensive, and which eventually brings adverse environmental and social impacts. On the contrary, by design, the concept of a circular economy (CE) is restorative and regenerative and is usually suggested as a viable business model to achieve resource efficiency. Hence, the benefits of the CE promote the organizations to transition from the traditional linear to a circular economic model (Mboli et al. 2022).

The circular economy is regarded as a new business model with the ability to uphold sustainable development by securing as many resources, materials, products, and as much energy as possible. This goal can be attained through recycling, refurbishing, reusing, and remanufacturing materials, products, and waste. Adoption of the CE approach has the capability to optimize the management of WEEE that can foster social, environmental, and economic benefits or—to use another word—sustainability (Isernia et al. 2019). Thus, this attribute of CE can turn a business model into a sustainable business model (SMB) by guaranteeing minimal environmental pollution, sustainable economic growth, and maximum social welfare.

In this context, the circular economy business model (CEBM) refers to the way an organization captures and proposes values while enhancing resource efficiency through (a) closing material loops and (b) stretching the life cycle of the products, materials, and components (Pollard et al. 2021). CEBMs are predominantly regarded as a part or subset of sustainable business models (SBM) (Rosa et al. 2019) that prioritize economic and environmental values (Whalen 2019).

Unlike the traditional linear models, a circular model (CEBM) can reap the benefits that incorporate maximizing profits, enhancing resource efficiency, optimizing cost benefits, and as a result move toward a safer environment (Mboli et al. 2022). In fact, when supported by government regulations, dissemination of the CE processes of reuse and recycling of EOL products can generate new entrepreneurial activities, economic markets, and business models/strategies that are expanding in numerous developing and developed economies (Ghisellini et al. 2016).

However, lack of scaling up is one of the common complaints of CE, where technological innovation can act as the enabler to scale up a CE business model (Mboli et al. 2022). In this calling, organizations are faced with the difficulty to determine which business model can help them overcome these challenges while concentrating on capturing the value propositions. However, before choosing the appropriate business models for better e-waste management, it is paramount to set up a theoretical framework that can verify these new business models will succeed.

In general, either the ReSOLVE framework of The Ellen MacArthur Foundation (EMF) or the Business Model Canvas (BMC) methodology by Osterwalder and Pigneur (2010) can be used to verify a new and prospective business model when it comes to e-waste management. In this chapter, the Business Model Canvas (BMC) is used (in the following sections) as the methodological framework to assess several alternative e-waste business models. Hence, the next section takes a dive into the business model canvas to conceptualize the CEBMs.

4.4 E-waste Management Under the Lens of the Business Model Canvas

The majority of the EEE industry is built upon a linear economic model in which products are manufactured, utilized, and finally dumped into the waste streams at either technical or functional obsolescence. This, along with the shortening of both product lifespans and their innovation cycles, is ushering forward the rapid growth of e-waste. To solve the growing WEEE concern, the CE strives to sustain the economic growth from resource consumption by conserving the value of materials all through the life cycle.

Due to recent regulations and adherence to the UNSDG 2030 goals, a strong level of agreement among the EEE stakeholders has been observed to seize the opportunities that exist for the adoption of CEBMs, notably economic opportunities to maximize value creation and capture the consumer segments (Pollard et al. 2021).

Unfortunately, in the EEE sector, a paucity of exploration and limited understanding persists in the application of both circular economy business models (CEBM) and circular economy business model innovation (CEMBI) processes. Moreover, this sector is still far from undertaking the required changes in the business model (BM) level in order to truly transition to CE. Therefore, development of business model innovation at the CE level is vital for the EEE sector in order to address the needs and consequently support the EEE manufacturers in their quest to develop CEBMs (Pollard et al. 2021), which this chapter particularly aims to explore by explaining several prospective CEBMs using the lens of the business model canvas (BMC).

4.4.1 How Does the Sustainable Business Model Canvas Work?

A business model guides the companies to form and deliver their value propositions (Osterwalder and Pigneur 2010); therefore, it acts as a toolkit to conceptualize the operational building blocks and figure out the core activities to deliver value in order to make a profit. Thus, a BM helps to conceptualize and capture a holistic view of the organizational activities (Schneider and Spieth 2013).

To explain a business model, the Business Model Canvas (BMC) helps the user to pinpoint the smaller components of a model and equips the user with a visual illustration to document such details with a step-by-step procedure. Upon completion of the steps, the BMC provides a one-page graphical illustration or a dashboard view of *what* the business does to attain its goals. From an ontological perspective, four operational areas of a business—namely, customer interface, product, infrastructure management, and financial aspects—are used as a framework to create nine smaller blocks to facilitate in-depth analysis during the business modeling process (see Figure 4.1).

The Infrastructure Management section is situated on the left-hand side of the canvas; comprised of three building blocks: Key Activities, Key Resources, and Key Partners. These three building blocks usually contribute to the cost structure of the business. The product component is represented by the Value Proposition building block, and it is placed in the middle of the canvas. The Customer Interface quadrant is placed all the way to the right side of the canvas. Customer Segments, Customer Relationships, and Channels represent the customer interface component, which shows how the business segments its customers and how it interacts with them to propose values. Together, these three building blocks basically help the company generate profit. Finally, the last two building blocks—the Revenue Stream and Cost Structure—are placed at the bottom of the page, and they represent the financial aspects of the business (Carter and Carter 2020).

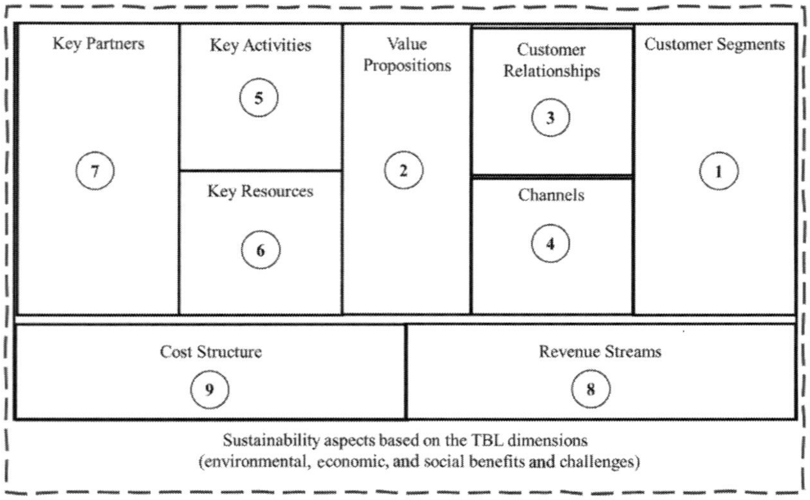

Figure 4.1 Sustainable Business Model Canvas. (*Source:* Adapted from Osterwalder and Pigneur 2010.)

These nine building blocks work in an orderly fashion. The order of the blocks is as follows: (1) customer segments, (2) value proposition, (3) customer relationships, (4) channel, (5) key activities, (6) key resources, (7) key partners, (8) revenue streams, (9) cost structure. It is important to deal with each block following the order since they are connected to each other.

The business model canvas as a proven tool can be used to support and embed sustainability (TBL, the triple bottom line concept) into the toolkit. BMC as an instrument can be extremely effective for the users and stakeholders of e-waste to comprehend an organization's business model. Expanding the BMC via adding another layer—triple bottom line (TBL) or sustainability—modifies the BMC into a sustainable business model canvas (SBMC) that integrates the sustainability aspects into the model, considering social, environmental, and economic outcomes from the circulation of materials perspective. Adding the TBL perspective as another layer to the BMC, the organizations can assess and formally account for the sustainability impacts since the TBL dimensions are a widely established and well-understood philosophy that can be easily integrated into devising business models to support more sustainable actions (Joyce and Paquin 2016). This eventually translates circularity into the value stream of the BM, which makes the sustainable business model a key driver of the transition toward a CE (Calvo Porral and Lévy Mangin 2020).

In the following sections, based on the BMC or better to state the extended version, SBMC, this chapter further elaborates on and breaks down four prospective and innovative CE business models as generic e-waste management business models focusing on four key areas: namely, (1) electronic vehicles' lithium-ion batteries, (2) CEMBI for the EEE manufacturers, (3) CEBM for the professional EEE sector powered by IoT and blockchain, and (4) a smart reverse e-waste collection system for consumer electronics. Next, each of these four generic models is represented as an individual case.

4.5 Case 1: Electric Vehicles' Lithium-Ion Batteries in the European Context

When it comes to greenhouse gas emissions in Europe, the road transportation sector is responsible for approximately 75% of the emissions. Thus, by 2030 EU is supposed to cut about 40% of its greenhouse gas emissions. As the way forward, electrification of the transportation sector has been deemed as one of the feasible strategies to decarbonize transport systems in the EU (Alonso Raposo et al. 2019). This transformation toward electrification is paying off; as a result, the number of EVs is also rapidly increasing in the EU.

On the flip side, this electrification of transportation has a dark side filed with environmental and practical challenges stemming from the enhanced production of lithium-ion (LIB) batteries, which are both energy-intensive and environmentally degrading (Emilsson and Dahllöf 2019) and necessitate the use of rare materials (Alonso Raposo et al. 2019), which links numerous social impacts stemming from the material extraction (Meyer et al. 2018). On top of that, managing the EOL of LIBs is expensive and demands technological innovations, which altogether limits the rate of material recovery.

The rate of LIBs entering waste streams now may be relatively low; however, they are rapidly adding to the e-waste stream. It is forecasted that more than six million battery packs (potentially with remaining storage capacity) would engulf the LIB marketplace by 2030 (Jiao 2020), which has to be managed in an environment-friendly way. Thus, when reaching their EOL the LIBs must be managed in a way that can reduce environmental impacts, by lowering the wastage cost and efficiently managing the supply of critical raw materials (Idjis and Costa 2017). Fortunately, to counterbalance the situation to manage the EOL of LIBs, substantial business interest and opportunities prevail for capitalizing the ingrained value of LIBs at their EOL cycle. Reflecting upon this, the European Commission (EC) has earmarked LIB as one of the critical products to be upscaled via repurposing, reusing, and recycling activities. Therefore, there is a need to thoroughly investigate the variety of CE strategies that are implemented/adopted, followed by an in-depth examination of the corresponding business models.

Hence, this case study illustrates which CEBMs for LIBs have been implemented by the Original Equipment Manufacturers (OEMs) in the EU context under the lens of BMC. Examining these activities through the BMC framework presents a comprehensive insight on the building blocks of the implemented business models in terms of repair, refurbishing, repurposing, and recycling strategies focusing on the take-back system.

4.6 Take-Back System

A vast majority of the closed-loop and EOL CEBMs are based on the retrieval of LIBs through a take-back system. In Europe, most companies manage the take-back of LIBs (before reaching their EOL) through dealer networks. In terms of LIB diagnostics, differences prevail within the setups of the take-back systems. For example, Volkswagen has its own trained dealers capable of conducting thorough diagnostic analyses on the LIBs, while most of the other companies' dealers just move the LIBs to the OEM's central hubs. Renault, VDL, and Daimler are some of the companies that are working with central hubs, focusing on the optimization of logistics. Currently, the increasing number of retrieved LIBs has been predicted to cause an upsurge in the number of central hubs, consequently, diminishing the logistics demand. This take-back system is quite prevalent in the EU jurisdiction. Hence, how the LIBs-CBM based on the business canvas framework can look is explained by focusing on the core of the model, which is the take-back-system, to acquire a comprehensive understanding in a real-world context.

4.6.1 Repair Strategies

Numerous enterprises deliver LIB repair services although detailed information is not publicly available. Plus, these repair services vary from company to company. For instance, the majority of Renault's returned LIBs are repaired but not necessarily sent back to the initial users. On the other hand, Volkswagen's LIB repair services include exchanging each battery module by returning it back to the original owner. Therefore, different OEMs work out their repair strategy in a different manner, although it is common that the design for repair is usually considered early in the development phase.

In terms of the nine points of BMC, most of these OEMs keep a close relationship with their professionals or private customers by proposing value to hybrid and electric passenger vehicles with extensive repair services through the dealer networks. As for the key activities, they design for repair and do repair service via highly skilled technicians and replacement modules. Therefore, voluntary return by the dealer network acts as the core of the take-back system of their model. To execute these key activities, the costs incur in terms of LIB development, human labor, logistics, and spare parts. Finally, the OEMs make revenue through selling vehicles, sale of one-time repair services, or a monthly service fee.

4.6.2 Refurbishing Strategy

When it comes to refurbished LIBs, the customer segment is quite price-sensitive; hence, not necessarily every customer looks for the maximum capacity of a LIB. Usually, refurbished LIBs are in high demand due to the low price, particularly when the residual value of the vehicle is also low, justifying not going for a new LIB. Therefore, the OEMs usually provide value to the price-sensitive customer segment with refurbished LIBs that are lower capacity and lower priced than new ones. The same is true for the repair strategy: here, they keep close ties with the customers using the dealer network, where voluntary return through this network is required in the take-back system. To cater to this customer segment with the proposed value, the OEMs usually end up designing for disassembly, running diagnostics, logistics, and refurbishing the LIBs. However, these key activities can differ greatly from company to company. For instance, Nissan disassembles their battery packs completely, analyzes the modules, and reassembles them based on their state of health, while Renault exchanges up to three faulty modules. Moreover, the whole refurbishing process might also be outsourced as part of their strategy (e.g. Ford and Nissan). To execute these key activities the key partners involved are usually the dealers, spare parts suppliers, and refurbishing partners who use highly skilled mechanics and replacement modules at their disposal as key resources to generate revenue from monthly payments or leasing contracts for the refurbished LIBs. The cost structure incorporates spare parts, labor, and logistics support.

4.6.3 Repurposing Strategy

Willingness to pay for second-life LIB is considerably lower than for the new ones; usually, second-life LIBs are traded if their price is 30% to 50% lower than new ones. This makes the customer segment quite a niche, mainly including environmentally conscious customers (both B2C and B2B), EV-charging networks, and the large energy consumers. This further makes sense of why only a handful of full-scale repurposing CEBMs exist among automotive OEMs. In the repurposing sector, most of the projects are still in the experimentation or piloting stage. In terms of mature initiatives, Volkswagen offers mobile charging as a value proposition by collaborating with other OEMs such as Renault or Nissan along with other energy storage providers such as Eaton and Connected Energy. Thus, regarding value proposition, several energy storage solutions are prominent with multiple services by keeping a close transactional relationship with the customers. As the relationship is more transactional, building trust is a key factor and is often attained via establishing extensive direct connection between the manufacturer and customer.

As the repurposing LIB projects are at the piloting stage, utilizing the direct contacts or the partner companies' distribution network is the best channel strategy. Now, as far as the key activities go, design for disassembly, technological development, logistics, diagnostics, repurposing process development for prospective reassembly and disassembly, collaboration with energy utilities, and customer services are the main concerns for utilizing expertise (e.g. engineers), staffs for customer service, warehousing and repurposing facilities, financial capital, and new parts as the key resources. As key partners, energy trading and energy utilities companies and repurposing partners are the best possible options to employ in the repurposing LIB model.

Finally, revenue streams come from the sales of storage solutions, energy sales, services provided, and leasing of LIB packs, while the cost structure involves development costs, labor, and logistics costs.

4.6.4 Recycling Strategy

OEMs usually engage one or more recyclers who comply with at least 50% of the legal recycling target and are able to sell the recycled secondary components. These secondary materials can be recycled using either a closed-loop or an open/linear model. For the closed-loop recycling model, a handful of examples show the involvement of the OEMs in the EU. For instance, BMW has established a technology consortium with Umicore and Northvolt with the aim to set up a closed-loop life cycle, whereas Audi headed for strategic research collaboration for setting up a closed-loop recycling system in order to recover 90% of the nickel and cobalt from LIB modules.

Lack of interest from the OEMs makes evaluating the recycling strategies by the OEM through the BMC quite simple. To start with, for the customer segment the target is simply the recycling market, where OEMs can have conventional recycling options that are compliant with legislation. The same as the repurposing strategy, the recycling strategy also only involves a transactional (but not too close) relationship with the customers. To cater to the customer segment recyclers, OEMs can organize reintegration by establishing contracts and direct contacts between enterprises or on a commodity marketplace to establish the channel strategy.

When it comes to recycling LIBs, the OEMs are usually not much attracted to the key activities beyond paying a fee to a recycler (key partner), as the revenues made by the recycler can barely cover the costs. This makes sense as the costs stem from logistics and treatment processes such as human labor, energy, etc., while the revenue source comes only from sales of recycled raw materials in the secondary market. In this sector, and with the help of process optimization targeting, a particular LIB might be able to achieve a higher recycling rate and profits from secondary raw material sales, which is still a long shot.

All nine points of the BMC: customer relationship, customer segments, value proposition, key resources, key partners, key activities, channels, cost structure, and revenue streams, are summarized and graphically presented in the one-page version of the BMC shown in Figure 4.2, which illustrates how the repair, refurbishing, repurposing, and recycling strategies looks for the European automotive OEMs' LIB business model.

4.7 Sustainability Issues for the LIB in the EU

Uncertainty turns out to be a pivotal issue shared by different sustainability drivers while establishing the LIB sustainable business model. Thus, based on this main theme some sustainability aspects of the LIB business model are elaborated on next.

4.7.1 Economic Sustainability Aspects

- Pricing and the cost structure of new LIBs, along with economic value and demand of energy storage factors, are highly influential in making the decision regarding which CE strategies to pursue.

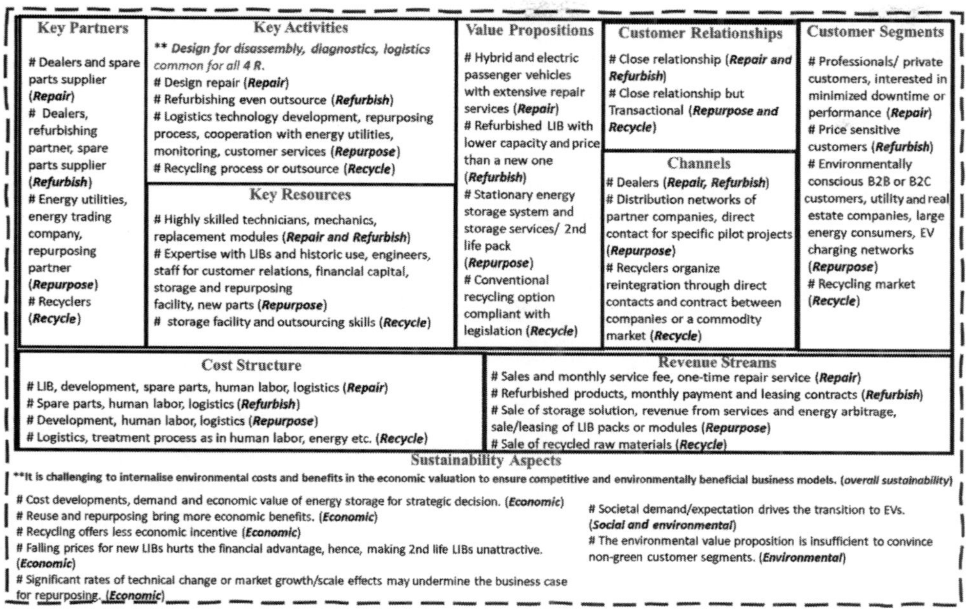

Figure 4.2 Sustainable business model canvas for EVs LIB in the EU. (*Source:* Adapted from Albertsen et al. 2021.)

- Compared to recycling, repurposing and reusing LIBs turn out to be more economically profitable. However, advancements in recycling techniques and advanced technologies can change the situation.
- For recycling to be economical, cell development can impact the demand for high-value components in battery cells.
- The price of new LIBs is falling, which may diminish the financial benefit or competitive edge of second-life batteries.

4.7.2 Social and Environmental Aspects

- Transition to EVs has usually been upscaled via a consistent social demand for environment-friendly progress in which sustainable management of LIB plays a crucial role. Thus, societal expectation works as a motivating factor for the implementation of CE strategies.
- Substantial economic gains emerge from second-life LIBs, specifically in power storage applications. However, several conditions may vandalize the prospects for repurposing, which stem from uncertainty raised by either disruptive technological transformation or market growth/scale effects.
- The price instability of cobalt, coupled with the underlying environmental and social risks related to the global supply chains, is regarded as a catalyst behind measures to downsize the use cobalt for LIBs. For instance, Tesla strives to produce LIBs without cobalt as soon as possible; this trend certainly would have a significant planetary impact in the foreseeable future.

Overall, when it comes to sustainable development of a business model the second-life features typically attract the green customer segment; however, sustainability as a value proposition appears to be inadequate for persuading other larger consumer segments because the operational and economic value propositions also must be competitive. Therefore, internalization of environmental costs and benefits in the economic valuation persists to be the ultimate challenge while ensuring business models that entail both competitive and sustainability components. For a holistic view of the sustainability on the BMC, see Figure 4.2.

Finally, to wrap up the case of electronic vehicles' LIBs under the jurisdiction of the EU, the main limitation toward developing a sound business model is the lack of public information. The EV or automobile OEMs share an insignificant amount of public information on the use intensification, refurbishing, remanufacturing, and repair of LIBs, which makes the development of a proper framework cumbersome. However, this defensive scenario is understandable since most of the project related to repair, repurposing, refurbishing, or recycling is still at a nascent level or pilot stage, which makes verifying the technical and economic feasibility difficult. On one hand, these limitations do hinder new companies in developing a successful business; on the other hand, this also shows them the untapped possibility in this paradigm.

4.8 Case 2: CEBM for the Electrical and Electronic Equipment Sector in the EU

The EEE sector is built upon the bedrock of the linear economy with products manufactured, utilized, and tossed into the waste streams at the time of their technical or functional obsolescence (Baldé et al. 2017); this, along with the miniaturization effect and shortening of innovation and product life cycles, is causing the hyper-growth of WEEE (Forti et al. 2020). Faced with this increasing quantity of e-waste, along with a scarcity of materials and a lower recovery rate, the EEE sector has become more relevant than ever in driving the transition toward the CE.

In order to spearhead the transition toward CEBM, the EEE sector faces numerous pressures to change. Social pressures, such as fast-improving technology and increasing consumer demand and changing behavior for EEE, are the key factors behind the disposal of still-functioning products. Economic pressures emanate from the increasing material prices, cost of WEEE management, and loss of valuable materials. Also, the supply scarcity is propelling the shift into the EEE industry's linear production and consumption (Charles et al. 2020). Altogether, the EEE sector, as well as the policymakers, are constantly encountering pressures concerning the health issues and environmental pollution that are associated with the growth of hazardous materials from e-waste (Salhofer et al. 2016).

With the pressure to change built on, the adoption of the CE depends on the organizations to integrate CE principles into the business models (Lewandowski 2016). The need is for a model that can help them to create, deliver, and capture value (Osterwalder and Pigneur 2010), and serve as a conceptual frame to provide a holistic or dashboard view of the core activities (Schneider and Spieth 2013). Unfortunately, the EEE sector is still far from undertaking the required changes at the business model level to fully transition to CE. The EEE sector lacks knowledge and understanding of the CEBM applications. Hence, CEBM processes need more development in this sector, to support the EEE manufacturers in developing their business model. Therefore, the goal of this case is to assist the EEE manufacturers with the design and implementation of a detailed CEBM that can be customized using the BMC as the framework.

4.9 Business Model Canvas as a Framework to Explain the E-waste Model in the EEE Sector

Components of the CEBM canvas present the EEE manufacturers with the opportunity to contemplate the potential of integrating circularity principles in their sales strategy, delivery, and value propositions. The BMC as a framework is used to encourage the EEE manufacturer to detect circular actions to address each building block of the BMC and verify their applicability within their business. Several examples of circular actions are presented next that could be implemented or adopted by an EEE industry following the nine building blocks of the BMC.

4.9.1 Customer Segments

Adopting circular economy activities to fit the B2C or B2B customer segments can be a good strategy to capture a wide market. For example, to capture this segment coming up with more refurbished products is desirable to the environmentally conscious customer segment. Also identifying and taking advantage of certain customer segments is a viable strategy for the B2B market; for instance, targeting the early adopters, niche markets, and the eco-friendly segments.

4.9.2 Value Propositions

The above customer segments can be served in several ways to generate value for them. The first is by enhancing options of products as services or, in other words, selling functionality rather than ownership. Second is by introducing leasing or rental or shared product options. Another is introducing products with improved sustainable consumption patterns that provide intangible benefits (e.g. green impact, energy efficiency). Finally, and most important, is introducing and enhancing the circular EOL options for the products by including the take-back system where feasible.

4.9.3 Customer Relationships

To make sure to reach out to the mentioned customer segments with the proposed values it is vital to launch circular customer relationship strategies. To embed circularity into the business model, customer relationship initiatives can include but are not limited to using social media platforms, changing traditional relationships, and offering after-sales services. For instance, offering circular solutions to the dedicated customer pool, enabling CE feedback on the company website would be helpful, modifying the existing or new contracts to offer extra support through the feedback system, and providing a warranty for the circular offerings are some examples to enhance the CE into the customer relationship.

4.9.4 Channels

A circular channel or distribution channel can be achieved by ICT platform-based customer communication (showcasing CE offerings), branding the company image (including CE activities in the events), participating in the eco-labelling certification and standards, and enhancing the data security issues for returned and refurbished equipment. Most important, embedding the CE messages into the marketing strategies is also vital as a channel strategy.

4.9.5 Key Activities

In order to reach the mentioned customer segments with proposed values, several key activities are essential for the firms to perform.

- *Diversifying the circular activities* by expanding the product recovery and refurbishment program.
- *Embracing the eco-design* by improving the durability of the components.
- *Strategizing the circular production* by constantly auditing the supplier and production process against the CE indicators.
- *Improving reverse logistics* for greater product take-back to induce circularity in the logistics and distribution system.
- *Enhancing EOL circularity* by providing recyclers with a material declaration to aid recycling.
- *Tracking materials* and components using RFID or QR for better recycling.

4.9.6 Key Resources

To make sure the above key circular activities are performed properly, three key resources need to be in place. First, a competitive financing model needs to be in place to save costs via the reduction of components and obtaining reused/recycled components. Second, the development of a CE enhancement skills and training program is necessary. Third is expanding the use of ICT to foster CE into the system.

4.9.7 Key Partners

Circularity in the partnership can be established by going into new strategic alliances with better suppliers who have circularity elements embedded into their system. Furthermore, forming partnerships with the public and private procurement (PPP) bodies is helpful to expand the customer base. Finally, partnering with the media to achieve an active media campaign on refurbished products is a good strategy too.

4.9.8 Revenue Streams

In terms of generating revenue from a CE business model, three sources are most dominant and found viable.

- *Recurring revenues* can be generated from bundle sales and services, rental services, leasing options, and remanufacturing through the take-back system. A take-back system can be put into the revenue stream by incentivizing the customer to return the EOL products for high-end or premium product lines.
- *Rental options for B2B customers* can be explored. Adapting the financial administration to enable CEBM to focus on rental or leasing for the B2B segment is a great source of revenue.
- *Obtain value from the waste.* Reusing EOL parts and components in refurbished products is another good option to generate revenue. For instance, a take-back system for remanufacturing, part exchange option for products, or preparation for recycle and reuse are some valid sources of revenue streams able to generate value from the e-waste.

4.9.9 Cost Structure

When it comes to cost structure, the legislative fees, cost of design, cost of the take-back system, and several hidden costs of e-waste management are the main sources to consider. Cost structure strategy is the last vital strategy to develop when preparing the CEBM based on the BMC framework. For instance, reducing the legislative compliance fees for the e-waste management through the collection program can help to mitigate some additional costs. Also, with the manufacturing and sales process development, the cost of design for recycling can be achieved. Finally, improving the tracing and reverse logistic system can mitigate the cost of the take-back system.

Now the one-page BMC or the sustainable BM canvas combining all these nine components and strategies is presented in Figure 4.3.

4.10 Sustainability Opportunities and Challenges for the EEE Sector

Developing or transitioning to the CEBM comes with several opportunities and challenges from a sustainability perspective. Following are some key economic and social opportunities and challenges that prevail to implement CEBM in the EEE sector that can significantly influence the growth of the WEEE.

4.10.1 Economic Opportunities and Challenges

Gaining economic efficiencies turns out to be the greatest opportunity for the implementation of CEBM. Apart from the efficiency, capturing wider and new markets, reduction of production and raw material and waste disposal costs are some other economic gains from the CEBM in this sector. However, on the flip side, the potential costs related to recycling, refurbishment, and repairs act as a financial challenge.

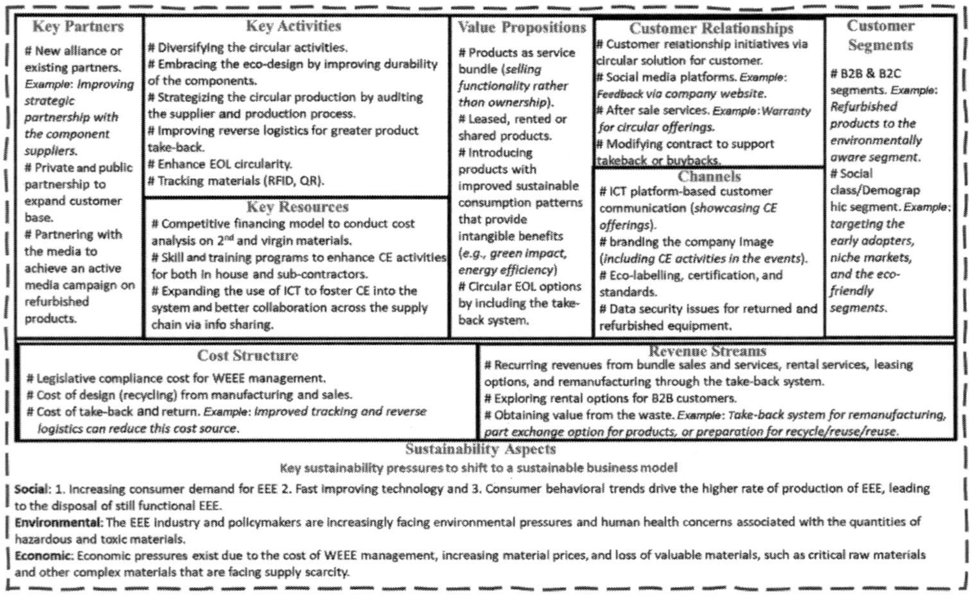

Figure 4.3 Sustainable business model canvas for the EEE sector in the EU. (*Source:* Adapted from Pollard et al. 2021.)

Thus, high prices for secondary raw materials and low prices for primary raw materials are two major economic challenges to overcome. Also, limited leasing services and reverse logistics further hinder the economic gains.

4.10.2 Social Opportunities and Challenges

A lack of knowledge or awareness of the CE and circular products along with very limited social acceptance of second-life (recycled, refurbished, and reused) products act as the key social challenges to overcome. Adding to the challenge, the trend of replacing instead of repairing and the negative perception of recycled components in the product and limited renting options hinder the CEBM adoption. To counter these social challenges, enhancing the CE knowledge and skills is one way to address the issues related to awareness. On the other hand, product certifications, warranties, and guarantees to promote the purchase of circular products can be used as key social opportunities to deal with the challenge of limited social acceptance for the second-life products.

4.11 Case 3: Digital Technology (Blockchain) Based Circular Management of E-waste

A circular economy requires a breakthrough in the domain of material economics for both economic and environmental purposes. Plus, the circular economic strategies that help to avert waste can also contribute to reduce the emission of GHG and corresponding environmental damage. Unfortunately, the world economy is only 8.6% circular (Circle-Economy 2021).

Furthermore, the high innovation frequency along with the constant decreasing prices of EEE devices, as well as the planned obsolescence, are skyrocketing the WEEE growth with almost 2 million metric tons of global e-waste per year (Kristoffersen et al. 2019). Hence, in this context, transitioning the EEE sector to a CE-based model would have significant environmental and social impacts.

As more and more innovation through industry 4.0 technologies is taking place in the EEE sector, expectations are also growing exponentially for digitally-enabled circular economy (Antikainen et al. 2018; Bressanelli et al. 2018b) solutions, since these disruptive technologies have been recognized as the enablers for the CEBMs (Rosa et al. 2020). However, not many case studies have linked the CE and industry 4.0 or digital technologies (Bressanelli et al. 2018b, 2018a; Kristoffersen et al. 2019, 2020) in the domain of e-waste or WEEE.

To this calling, a blockchain-based digital CEBM can offer a more transparent, secure, trustable, reliable, and independent solution for e-waste management, which in turn helps to capture full control over the value chain. Hence, this case elaborates the implementation of blockchain-based CEBM by exploring an innovative approach toward a digital tracking BM for professional EEE and e-waste management. However, although such a model is not yet possible for an informal economy, in the EU it is possible due to established WEEE directives and some sort of formal network.

4.12 A Blockchain-Based Solution for the Professional EEE Sector (WEEE) with Consumer Involvement

4.12.1 Assumptions of the Model

To begin with, each professional EEE device needs to be equipped with sensors (for instance, GPS) and an identifier (e.g. a tag), which would enable the key stakeholder to extract and key in information about the condition and location of the product throughout the product's life cycle. Here, the providers and producers act as the main actors, as well as the repair centers who would utilize a blockchain platform to accumulate data concerning the lifetime of products or components, maintenance, use phase, geo-localization of EEE, and the repair cycles. The desired outcome is substituting the virgin materials and new products to save money for the EEE manufacturers, while the role of the sensors would be to help decide if remanufacturing or reuse is a possibility. Hence, for the EEE users/consumers, a potential expansion of the life cycle would be a possibility of leasing toward saving money. Finally, the end consumers while visiting the areas where EEE is located can send a photo of the EEE itself. The photo (validated by an image processing software) would be dispatched to the BC system, with other information gathered by the sensors. To visualize the model the process diagram of this proposed BC-based model is shown in Figure 4.4.

4.12.2 Benefits of Data Storage and Transfer

This business model allows several benefits to generate from the data collection, storage, and transfer capability of the BC technology. Some benefits are elaborated on next.

- Product characteristics related data can help optimizing the logistics.
- The lifetime-related data of the products and parts, along with use phase, maintenance, and repair cycles data can come in handy for enhancing both the product design and demand forecasting of the components.
- Data related to traceability, such as data on condition, location, and availability, can be useful to facilitate the maintenance and repair of the current EEEs.
- The geo-localization of electric equipment and its corresponding e-waste ensures compliance with regulations for WEEE management in Europe.
- Level and usage of chemical substances-related data can be further passed throughout the supply chain to the interested parties working on e-waste management.

Finally, the picture of EEE or WEE captured and sent by the customer is another source for the information repository to be used in later phases to analyze and research this field.

4.12.3 Sustainability Benefits of the Environmental Model

Inducing traceability in the professional EEE supply chain could help retain valuable materials and extend the product life cycle, therefore, moving toward a more resource-efficient WEEE regime, which in turn helps to reduce the health hazard and tremendous pollution arising from the WEEE.

4.12.4 Social Benefits

The professional EEE sector can boost its CSR performance by raising consumer awareness and contributing to the shift toward sustainable development. The blockchain can assure the stakeholders that no unethical practice such as child labor was involved, and no human rights were revoked in the manufacturing and supply chain process (Upadhyay et al. 2021).

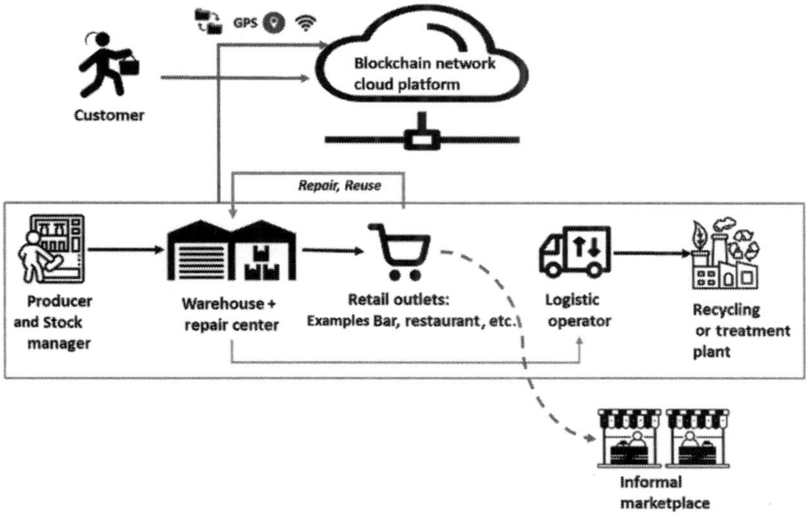

Figure 4.4 Blockchain-based circular model of e-waste management. (*Source:* Adapted from Magrini et al. 2021.)

4.12.5 Legal Benefits

Overcoming limitations related to copyright violations in the management of the disposal of e-waste and recycling, remanufacturing, and reuse is also possible with the BC-based model (Chen and Ogunseitan 2021).

4.12.6 Overall Challenges of the Model

To sum up, several challenges also persist that may hinder the implementation of the explained model. To start with, the implementation of blockchain-based solutions is still at its infancy or piloting stage. Thus, the implementation still suffers from infrastructural-level challenges encompassing failures of interoperability, stability, and technological security-related issues. Also, challenges linger in the scalability of the solutions due to a lack of funding, insufficient demand, and inadequate regulatory framework (Hedberg et al. 2020). Finally, a full integration of the supply chain would be essential too. Just establishing a closed-loop supply chain is evidently not adequate; it is paramount for the organizations to specify their value proposition with such supply chains (Lüdeke-Freund et al. 2019). The one-page adaptation of the business model is shown in Figure 4.5 using the business model canvas as the framework.

4.13 Case 4: Smart E-waste Management System of EOL Consumer Electronics for Developing Economies

Efficiently managing WEEE is one of the crucial strategies to conserve materials (e.g. valuable, and rare metals and essential minerals) with delimited global reserves. Recently, issues related to e-waste collection have attained growing attention, particularly in the developing nations, due to the presence of an informal system and low collection rates. In the informal system, it is critical to develop adequate infrastructures to maintain the timely return of EoL household EEEs by the consumers that eventually cater to economic incentives. Furthermore, for the developing economies with an informal channel or system, a smart reverse system for WEEE management is suitable as it induces recycling behavior for the consumers within a short period with minimal transport and infrastructural costs for collection.

Therefore, in this calling, this case aims to elaborate on a smart reverse system for WEEE collection from EOL electronics holders (households) to the recycling facilities by involving IT tools (such as interactive online maps for collection and tracking requests) and local delivery services to collect e-waste.

4.14 Explanation of the E-waste Collection Business Model

The IoT or smart technologies can help transforming the WEEE practices to facilitate the shift toward the CEBMs. Hence, the idea is to use a smart IoT-based WEEE collection scheme that is focused on optimizing the collection routes of e-waste from households, which would lower the transport cost and material sorting cost.

First, for the collection of e-waste, local delivery service enterprises are considered, since in the developing economies available transport potential for WEEE collection exists in the urban setting. Thus, vehicles from any local delivery service companies can be taken as movable collection points capable

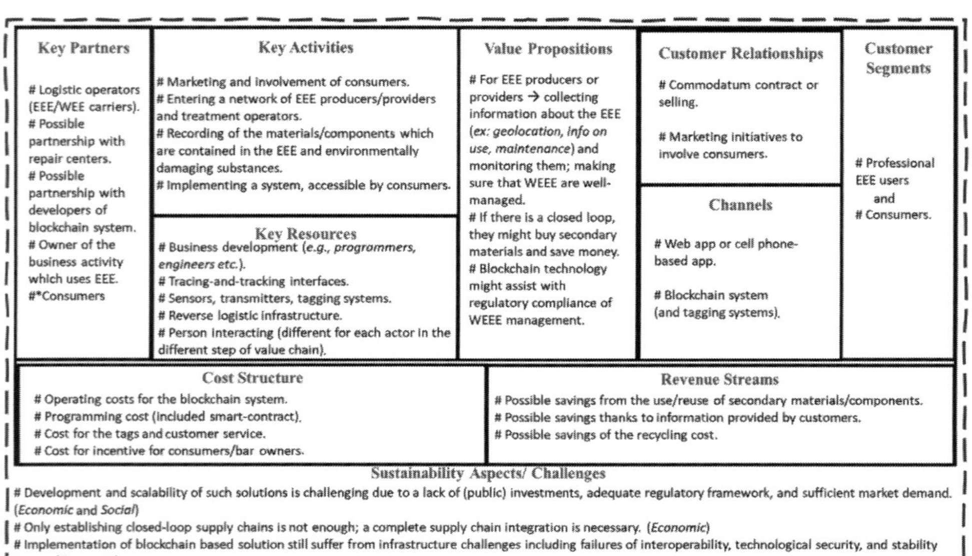

Figure 4.5 Sustainable business model canvas for the BC-based professional EEE sector in the EU. (*Source:* Adapted from Magrini et al. 2021.)

of collecting and delivering the obsolete or end-of-life EEEs to the specialized e-waste processing center. Furthermore, incorporating delivery service companies for WEEE collection is possible when economic incentives are offered by the local authorities, which makes the model both economically and environmentally viable as this would minimize transport and infrastructure costs while reducing CO2.

Second, interactive online maps as a support tool will be used for e-waste collection, which would allow the involvement of any local delivery service. The online maps are a suitable mechanism to track the EEE holders' requests for e-waste collection.

Third, from the user's perspective, the end-user will make a request for e-waste collection by filling in the interactive online map with the necessary information. Here, a local operator will play a key role in the system of reversing EoL or obsolete EEE by connecting the link between:

1) End-user (WEEE holders) and the delivery service vehicle as potential e-waste collectors,
2) The local delivery service enterprises (responsible for collection) and the local authority, and
3) The specialized WEEE processing company and the local delivery service companies.

Finally, the delivery service vehicle would act as the e-waste transport operator and mobile collection point. To visualize the whole mechanism the flow diagram of the IoT-based collection model is shown in Figure 4.6.

4.15 Benefits and Challenges of the Model

4.15.1 Benefits

The online maps application for e-waste collection encourages consumers to dispose of their EOL electronics properly since this IT tool offers marginal transaction costs in terms of effort and time, while using shared platforms for both collection and delivery services would further decrease the mileage. This eventually leads to energy savings that in turn save cost and reduce CO2 emissions.

Overall, this model would specifically work for developing nations as there is a lack of specialized and specific infrastructure for WEEE collection. Moreover, the proposed model allows for saving costs in terms of infrastructure development and maintenance. This comes with a huge economic opportunity to significantly lower the infrastructure, consumer transaction costs, and transportation costs.

4.15.2 Challenges

The proposed model does not discard involving non-specialized transportation from the delivery companies. Thus, following safety conditions while transporting sensitive WEEEs must be taken into consideration. For instance, damaged LIBs need to be moved into special mini boxes for safety purposes.

In order to ensure higher involvement from the consumer in the collection process, adequate economic incentives and a convenient collection infrastructure are paramount. Without these key factors, the system may fall apart.

Without a good environmental regulatory framework and formal channels in place, it is quite cumbersome to develop the proposed WEEE collection system.

4.16 Issues to Be Resolved or Questions to Be Answered Before Implementation

1) How would the financing of the system be resolved by sharing the WEEE collection costs among stakeholders (manufacturers, distributors, and municipalities)?

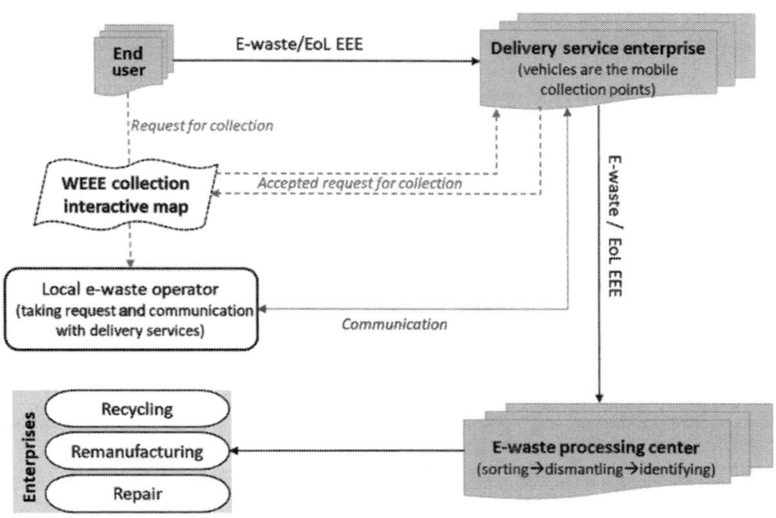

Figure 4.6 Logical flow diagram of the smart e-waste collection model. (*Source:* Adapted from Shevchenko et al. 2021.)

2) How can costs for economic incentives (to promote recycling intention) be disseminated within the interested parties?
3) What are the financial conditions for the delivery services?
4) How will the costs of the delivery services be distributed among the stakeholders?

4.17 Business Model Canvas to Verify the E-waste System

The BMC framework can economically justify the smart e-waste management model. To start with, all the EEE users, particularly the households, represent the consumer segment. Here, the specialized WEEE processing organization proposes the value for the users, while for the partners it creates a positive image via sustainable operation or impact on the SDG. When it comes to the channel strategy: (a) the local operator and consumers interact with each other via city delivery services, and (b) the specialized recycling center prepares the agreements for the requirement of transport services. Thus, to provide these proposed values the key activities of the WEEE operator and the specialized organization are: collecting e-waste; estimating the value of products and components for remanufacturing, disassembling, and identification of components; and repairing and refurbishing secondary raw materials for recycling. To execute these key activities, trading companies with delivery services, enterprises providing delivery services, and finally, repair and recycling companies act as the main partners. The key resources needed are technical support for interactive online maps, a support team to process consumer requests, website, software, and other communication systems, and finally, machineries to dismantle and identify components.

All these above activities generate costs in terms of cost of equipment (e.g. sorting, dismantling, and identifying parts), salaries, and material-and-technical equipment. Finally, the revenue can be generated by selling secondary raw materials to the recycling companies, sale of EoL EEEs to manufacturing or repairing companies, and selling operational parts to repair or remanufacturing companies. The visual representation of the one-page BMC is shown in Figure 4.7.

4.18 Future of E-waste Management in the Developing Economies

The search for better e-waste management schemes keeps soaring, and numerous nations/governments have taken numerous measures to improve WEEE recycling, which include but are not limited to the establishment and enhancement of infrastructures, executing pilot projects, materializing funds, improving waste dumping, and discarding laws, improving consumers' awareness, and so forth (Sousa et al. 2018; Tanskanen 2013; Zeng et al. 2017). Unfortunately, the results are not yet up to the mark (Sastre et al. 2018). Adding to this context, the traditional offline collection models are still operational in full swing, which is not only labor-intensive but also capital- and resource-intensive and unable to satisfy the advancement of WEEE (Wang et al. 2018).

This represents the intricate situation of global WEEE recycling that demands innovative solutions. In this vein,

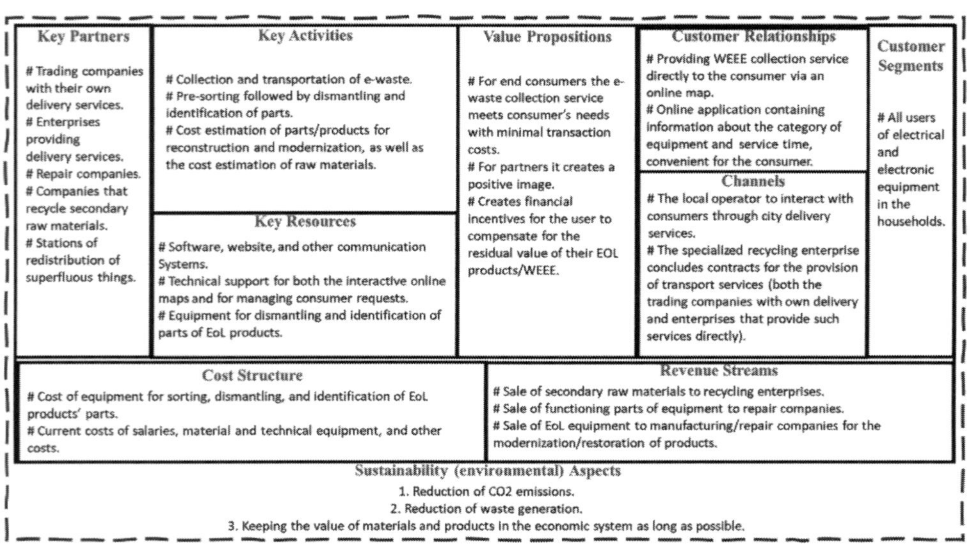

Figure 4.7 Sustainable business model canvas for smart e-waste collection system for informal channels. (*Source:* Adapted from Shevchenko et al. 2021.)

the advancement of the internet opens new gates to WEEE collection or recycling solutions. Therefore, grasping these technological opportunities encourages the promotion and integration of internet and WEEE collection/recycling, which is probably one of the most convenient strategies or ways forward for the informal channel/system-based developing and underdeveloped economies.

Recently in China, many internet-based platforms have arisen, such as Taolv, Wangyou-used net, Aihuishou, Huishouge, Yizaisheng, Sinnorrc, and Zaishenghuo (Sun et al. 2018; Wang et al. 2018), which are also known as Internet + or Online to Offline (O2O) models. This O2O (integrating offline collection with online platform) model has also been adopted by many companies in the developing economies, China being the pioneer (Zuo et al. 2020).

The O2O e-waste collection model has further become popular in both the capital market and among customers due to its capability to reduce transaction costs, expand the collection scale and scope, and contribute to efficient WEEE recycling, which particularly fits with the informal e-waste structure of the developing countries (Zuo et al. 2020). Before the year 2000, in China, e-waste collection was led by the informal sector and even now in a lot of developing economies such as Pakistan, India, and Bangladesh the informal sector rules strongly. Therefore, the O2O model can help the developing nations to transition from informal to formal channels.

4.19 How Does the O2O Model Work?

The O2O WEEE collection model is comprised of two main components, the offline collection network and the online information platform. The development of the O2O collection model can take different routes. However different the routes may be, three components—namely, material flow, information flow, and fund flow—always have to be there and connected effectively to make the model work. To make sure of the fluency of these three components, the O2O e-waste model establishes information support platforms with user-friendly interfaces, precise pricing systems, and several distribution channels. Figure 4.8 shows an example of a real-life O2O e-waste recycling/collection model from China that works for the collection of consumer electronics such as mobile phones, laptops, etc., illustrated with an explanation.

Figure 4.8 shows how an O2O model starts as a dot-com and vertically integrates its business into WEEE collection, dismantling, and reprocessing. First, the consumer posts his/her inquiry by describing the appearance and performance of the product he/she wants to recycle or give away. Second, the O2O platform (company) sends the information to its partner collectors to obtain the offers and presents an offer to the consumer. Third, upon acceptance of the offer through the O2O platform, the consumer delivers the waste products either to (a) the offline store of the O2O company or to (b) the third-party logistics center for delivery to the O2O company. Finally, upon collection of the e-waste, the O2O company either separates the usable or repairable e-waste to sell as second-hand products or separates the bad condition e-waste to be disassembled and processed as scraps by professional enterprises.

4.20 Benefits of the O2O Model

- The O2O e-waste collection model can contribute to the traceability for manufacturers, accessibility to the customers, hybridity in the collection, transparency for the public, profitability for recyclers, and friendliness to the environment (Gu et al. 2019; Tong et al. 2018), which is extremely essential for the sustainable development of the WEEE recycling industry.
- O2O platforms help to balance the information asymmetry and break the business impediments by improving the interplay between the demanders and suppliers. They can further lower the cost of seeking for information and provide a platform to create and exchange value.
- Most importantly, the e-wastes accumulated by employing the O2O platforms can finally enter the formal disposal and dismantling channels, which further diminishes the environmental pollution. Hence, this platform offers

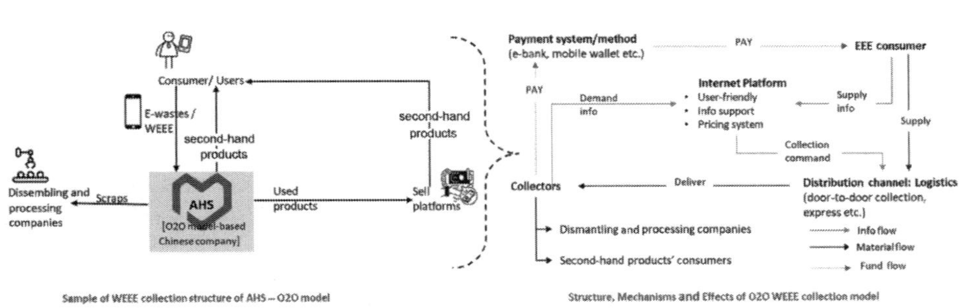

Figure 4.8 O2O e-waste collection generic model. (*Source:* Adapted from Zuo et al. 2020.)

the opportunity to transcend from the informal to a more formal channel for e-waste management.

4.21 Critical Factors to Be Considered for Better Implementation

4.21.1 Structure Characteristics

Material flow, information flow, and fund flow are the major components to promote the advancement of the model. O2O collection enterprises need to develop accurate systems and channels in order to ensure the fluency and sustenance of these three flows.

4.21.2 User-Friendly Interface

Website and mobile apps are valuable intermedia for the O2O e-waste model. Plus, customers' willingness and satisfaction are directly affected by the friendliness of these interfaces.

4.21.3 Information Support Platform

Improving information transparency is a critical success factor of the O2O model. For instance, setting up individual shops, company databases, and blogs are paramount to sharing information with members to develop better communication opportunities and can significantly reduce the information asymmetry.

4.21.4 Distribution Channels

The channel strategy is also vital, as the e-wastes for the consumer electronics are usually scattered, which requires solving 'the last kilometer' problem. This can be solved by establishing door-to-door collection teams and offline stores in the core areas, and third-party logistics are suggested to use outside the core areas.

4.21.5 Pricing System and Convenient Payments

This is the final key success factor to make this model work. Legal collection at the highest market price based on an accurate pricing system is the most attractive competitive advantage one can have. This is possible through building a detailed WEEE evaluation system from the aspects of the brand, model, appearance, and function. Moreover, convenient, quick, and safe payments such as E-banks and mobile wallets are best suited for the O2O collection platforms.

Overall, this kind of model can be the future of e-waste management, particularly for the developing countries, as it suits even the medium to small startups, which usually do not have the resources and capabilities to cover the full industrial chain of WEEE recycling. Therefore, the small startups can develop their core business and competitiveness according to the strategic goals and choose proper partners to complement the weak links. However, the biggest drawback is the initial funding, because, at the early stage, massive investment is needed for the construction of internet platforms, offline collection sites, and teams before a profit can be made.

4.22 Concluding Remarks

The bitter truth of the unnatural growth of electronic waste and its impact on social and environmental degradation is well established, as also the economic opportunities that lie beneath the pile of this electro-junk. Along with the regulation and policies, there is a heightened urge for easier and compatible ways to establish sustainable circular business models to manage the e-waste problem. The policies, regulations, or penalties can only take us so far. In particular, the informal channels in developing and underdeveloped economies are facing the biggest challenge when it comes to e-waste management. Unlike the developed economies, where good infrastructure and rules are established and the formal channel of waste management works, as the dumping yard of the developed nations the developing countries are struggling even to come up with regulations.

In this context, the plausible solution for the developing nations is to motivate, promote, and subsidize the startups (SMEs) and existing electronic manufacturers and the manufacturing industry to come up with innovative circular business models to deal with the ever-encompassing issue. Sustainable circular business models are the fastest way to resolve this issue and make a win-win situation for the stakeholders by achieving both economic and social gains, hence, impacting the sustainable development goals.

In this vein, this chapter explored and represented several generic sustainable CEBM cases that can be followed and implemented by EEE industries, startups, and recycling firms by taking both the end consumers and professional users of EEE into consideration. The business models are explained from different EEE sector perspectives using the BMC as the generic and well-established framework to assess the validity of each model. Also, the benefits and implementation challenges are explained in-depth. This should help the industry leaders and decision-makers to visualize and enhance their business models based on their consumer base, value proposition, key activities and partners, and competitive situation. Some of the models can even directly fit into a particular scenario if the BMC building blocks match perfectly within a given context.

Most importantly, the path forward in this paradigm is surely through the industry 4.0 or digital technologies and their integration with the internet, hence, the internet+ model development with the combination of both online and offline channels and strategies to make the change happen. In this calling, the developing nations have to seize the opportunity and come up with more innovative business models. The manufacturing industries are at the cynosure of the problem since they are utilizing and importing tons of machines yearly, hence, contributing to the e-waste generation heavily. Therefore, they should use the internet+ model and industry 4.0 technologies at their disposal to make their machine or WEEE supply chain better with closed-loop supply chain models or a circular economy business model. Finally, in the developing economies the consumer awareness regarding e-waste is still very low, and it is the responsibility of the stakeholders and governments to enhance pro-environmental behavior, otherwise, no business model can capture the consumer electronics' WEEE market to make the situation better.

Glossary

BC Blockchain
BM Business model
BMC Business model canvas
CBM Circular business model
CE Circular economy
CEBM Circular economy business model
EEE Electrical and electronic equipment
EOL End-of-life
EU European Union
EV Electronic vehicle
LIB Lithium-ion battery
O2O Online to offline models
OEM Original equipment manufacturers
SBM Sustainable business model
SBMC Sustainable business model canvas
TBL Triple bottom line
UNSDG United Nation's sustainable development goals
WEEE Waste electrical and electronic equipment

Acknowldgment

This study was funded by the European Union – *NextGenerationEU, in the framework of the GRINS -Growing Resilient, INclusive and Sustainable project* (GRINS PE00000018 – CUP E83C22004690001). The views and opinions expressed are solely those of the authors and do not necessarily reflect those of the European Union, nor can the European Union be held responsible for them.

References

Albertsen, L., Richter, J.L., Peck, P. et al. (2021). Circular business models for electric vehicle lithium-ion batteries: an analysis of current practices of vehicle manufacturers and policies in the EU. *Resources, Conservation and Recycling* 172: 105658.

Alonso Raposo, M., Ciuffo, B., Ardente, F. et al. (2019). *The Future of Road Transport*. Joint Research Centre (Seville site).

Antikainen, M., Uusitalo, T., and Kivikytö-Reponen, P. (2018). Digitalisation as an enabler of circular economy. *Procedia CIRP* 73: 45–49.

Balde, C., Forti, V., Gray, V. et al. (2017). *The Global E-waste Monitor–2017*. Bonn/Geneva/Vienna: United Nations University, International Telecommunication Union & International Solid Waste Association.

Borthakur, A. and Govind, M. (2017). Emerging trends in consumers' E-waste disposal behaviour and awareness: a worldwide overview with special focus on India. *Resources, Conservation and Recycling* 117: 102–113.

Bressanelli, G., Adrodegari, F., Perona, M., and Saccani, N. (2018a). Exploring how usage-focused business models enable circular economy through digital technologies. *Sustainability* 10: 639.

Bressanelli, G., Adrodegari, F., Perona, M., and Saccani, N. (2018b). The role of digital technologies to overcome circular economy challenges in PSS business models: an exploratory case study. *Procedia CIRP* 73: 216–221.

Calvo Porral, C. and Lévy Mangin, J.-P. (2020). The circular economy business model: examining consumers' acceptance of recycled goods. *Administrative Sciences* 10: 28.

Carter, M. and Carter, C. (2020). The creative business model canvas. *Social Enterprise Journal* 16: 141–158.

Charles, R.G., Douglas, P., Dowling, M. et al. (2020). Towards increased recovery of critical raw materials from WEEE—evaluation of CRMs at a component level and pre-processing methods for interface optimisation with recovery processes. *Resources, Conservation and Recycling* 161: 104923.

Chen, M. and Ogunseitan, O.A. (2021). Zero E-waste: regulatory impediments and blockchain imperatives. *Frontiers of Environmental Science Engineering* 15: 1–10.

Circle-Economy (2021). The circularity gap report. Amsterdam, The Netherlands.

Emilsson, E. and Dahllöf, L. (2019). *Lithium-Ion Vehicle Battery Production-Status 2019 on Energy Use, CO2*

Emissions, Use of Metals, Products Environmental Footprint, and Recycling. IVL Svenska Miljöinstitutet.

Forti, V., Baldé, C.P., Kuehr, R., and Bel, G. (2020). *The Global E-waste Monitor 2020*. United Nations University.

Ghisellini, P., Cialani, C., and Ulgiati, S. (2016). A review on circular economy: the expected transition to a balanced interplay of environmental and economic systems. *Journal of Cleaner Production* 114: 11–32.

Gu, F., Zhang, W., Guo, J., and Hall, P. (2019). Exploring "internet+ recycling": mass balance and life cycle assessment of a waste management system associated with a mobile application. *Science of the Total Environment* 649: 172–185.

Hedberg, A., Šipka, S., and Bjerkem, J. (2020). *The Circular Economy: Going Digital*. European Policy Centre.

Idjis, H. and Costa, P.D. (2017). *Is Electric Vehicles Battery Recovery a Source of Cost or Profit? The Automobile Revolution*. Springer.

Isernia, R., Passaro, R., Quinto, I., and Thomas, A. (2019). The reverse supply chain of the e-waste management processes in a circular economy framework: evidence from Italy. *Sustainability* 11: 2430.

Islam, M.T. and Huda, N. (2019). Material flow analysis (MFA) as a strategic tool in E-waste management: applications, trends and future directions. *Journal of Environmental Management* 244: 344–361.

Jiao, N. (2020). Second-life electric vehicle batteries 2020–2030. *IDTechEx*. https://www.mckinsey.com/uk/~/media/McKinsey/Industries/Automotive%20and%20Assembly/Our%20Insights/Second%20life%20EV%20batteries%20The%20newest%20value%20pool%20in%20energy%20storage/Second-life-EV-batteries-The-newest-value-pool-in-energy-storage.pdf.

Joyce, A. and Paquin, R.L. (2016). The triple layered business model canvas: a tool to design more sustainable business models. *Journal of Cleaner Production* 135: 1474–1486.

Kristoffersen, E., Aremu, O.O., Blomsma, F. et al. (2019). Exploring the relationship between data science and circular economy: an enhanced CRISP-DM process model. *Conference on e-Business, e-Services and e-Society*. Springer, 177–189.

Kristoffersen, E., Blomsma, F., Mikalef, P., and Li, J. (2020). The smart circular economy: a digital-enabled circular strategies framework for manufacturing companies. *Journal of Business Research* 120: 241–261.

Kumar, A., Holuszko, M., and Espinosa, D.C.R. (2017). E-waste: an overview on generation, collection, legislation and recycling practices. *Resources, Conservation and Recycling* 122: 32–42.

Lewandowski, M. (2016). Designing the business models for circular economy—towards the conceptual framework. *Sustainability* 8: 43.

Lüdeke-Freund, F., Gold, S., and Bocken, N.M. (2019). A review and typology of circular economy business model patterns. *Journal of Industrial Ecology* 23: 36–61.

Magrini, C., Nicolas, J., Berg, H. et al. (2021). Using internet of things and distributed ledger technology for digital circular economy enablement: the case of electronic equipment. *Sustainability* 13: 4982.

Mboli, J.S., Thakker, D., and Mishra, J.L. (2022). An internet of things-enabled decision support system for circular economy business model. *Software: Practice and Experience* 52: 772–787.

Meyer, K., Buchert, M., Degreif, S. et al. (2018). Ensuring a sustainable supply of raw materials for electric vehicles: a synthesis paper on raw material needs for batteries and fuel cells. *Öko-Institut*.

Osterwalder, A. and Pigneur, Y. (2010). *Business Model Generation: A Handbook for Visionaries, Game Changers, and Challengers*. John Wiley & Sons.

Pollard, J., Osmani, M., Cole, C. et al. (2021). A circular economy business model innovation process for the electrical and electronic equipment sector. *Journal of Cleaner Production* 305: 127211.

Read, L. (2021). E-waste: turning a challenge into business advantage. *ERM*. ERM Foundation. https://www.erm.com/insights/e-waste-turning-a-challenge-into-business-advantage.

Rosa, P., Sassanelli, C., and Terzi, S. (2019). Towards circular business models: a systematic literature review on classification frameworks and archetypes. *Journal of Cleaner Production* 236: 117696.

Rosa, P., Sassanelli, C., Urbinati, A. et al. (2020). Assessing relations between circular economy and industry 4.0: a systematic literature review. *International Journal of Production Research* 58: 1662–1687.

Salhofer, S., Steuer, B., Ramusch, R., and Beigl, P. (2016). WEEE management in Europe and China—a comparison. *Waste Management* 57: 27–35.

Sastre, S., Llopart, J., and Ventosa, I.P. (2018). Mind the gap: a model for the EU recycling target applied to the Spanish regions. *Waste Management* 79: 415–427.

Schneider, S. and Spieth, P. (2013). Business model innovation: towards an integrated future research agenda. *International Journal of Innovation Management* 17: 1340001.

Shevchenko, T., Saidani, M., Danko, Y. et al. (2021). Towards a smart E-waste system utilizing supply chain participants and interactive online maps. *Recycling* 6: 8.

Sousa, R., Agante, E., Cerejeira, J., and Portela, M. (2018). EEE fees and the WEEE system—a model of efficiency and income in European countries. *Waste Management* 79: 770–780.

Sun, Q., Wang, C., Zuo, L.-S., and Lu, F.-H. (2018). Digital empowerment in a WEEE collection business ecosystem: a comparative study of two typical cases in China. *Journal of Cleaner Production* 184: 414–422.

Tanskanen, P. (2013). Management and recycling of electronic waste. *Acta Materialia* 61: 1001–1011.

Tong, X., Tao, D., and Lifset, R. (2018). Varieties of business models for post-consumer recycling in China. *Journal of Cleaner Production* 170: 665–673.

UN (2009). United Nations environmental program: recycling e from E-waste to resources.

Upadhyay, A., Mukhuty, S., Kumar, V., and Kazancoglu, Y. (2021). Blockchain technology and the circular economy: implications for sustainability and social responsibility. *Journal of Cleaner Production* 293: 126130.

Wang, F., Huisman, J., Meskers, C.E. et al. (2012). The best-of-2-worlds philosophy: developing local dismantling and global infrastructure network for sustainable e-waste treatment in emerging economies. *Waste Management* 32: 2134–2146.

Wang, H., Han, H., Liu, T. et al. (2018). "Internet+" recyclable resources: a new recycling mode in China. *Resources, Conservation and Recycling* 134: 44–47.

Whalen, K.A. (2019). Three circular business models that extend product value and their contribution to resource efficiency. *Journal of Cleaner Production* 226: 1128–1137.

Yu, J., Williams, E., Ju, M., and Shao, C. (2010). Managing e-waste in China: policies, pilot projects and alternative approaches. *Resources, Conservation and Recycling* 54: 991–999.

Zeng, X., Duan, H., Wang, F., and Li, J. (2017). Examining environmental management of e-waste: China's experience and lessons. *Renewable and Sustainable Energy Reviews* 72: 1076–1082.

Zuo, L., Wang, C., and Sun, Q. (2020). Sustaining WEEE collection business in China: the case of online to offline (O2O) development strategies. *Waste Management* 101: 222–230.

5

Electronic Waste Management Strategy in a Circular/Control Economy

Helen Uchenna Modekwe[1], Olusola Olaitan Ayeleru[2,], Olawale Charles Ogunnigbo[3], Bimbo Lolade Fafowora[4], Tarhemba Tobias Nyam[2], Matthew Adah Onu[2], and Peter Apata Olubambi[2]*

[1] *Renewable Energy and Biomass Research Group, Department of Chemical Engineering, University of Johannesburg, Johannesburg, 2028, South Africa*
[2] *Centre for Nanoengineering and Advanced Material, University of Johannesburg, Doornfontein Campus, Johannesburg, South Africa, 2028, South Africa*
[3] *Department of Mechanical Engineering, Oduduwa University, Ipetumodu, Osun State, Nigeria*
[4] *School of Journalism and Media Studies, Rhodes University, South Africa*
* *Corresponding author*

5.1 Introduction

Electrical and electronic products such as cellphones, personal computers, game gadgets, photocopiers and printers, televisions, lighting equipment, home appliances (washing machines, refrigerators, air conditioners, etc.), and power storage batteries, to mention a few, are waste when they have reached the end of their useful life, are outdated and old-fashioned, or are nonfunctional. The rapid increase in information and communication technology (ICT) and increasing usage of electrical and electronic equipment in everyday life result in the generation of an enormous volume of electronic waste (E-waste) all around the world. Globally, about 54 million tons of E-waste were generated in 2019, which was an average of 7.3 kg per capita, of which only around 18% of the generated E-waste was properly recycled; this vast figure has been projected to double by 2045 (Chakrabarty and Nandi 2021; Parajuly et al. 2020; Sharma et al. 2021). E-waste is now recognized as the quickest growing municipal solid waste stream with over 4% annual growth (Srivastava and Pathak 2020; Yong et al. 2019). E-waste contains a mixture of valuable resources and hazardous components that require careful and proper handling or management to recover the valuable materials (metals and non-metals) and to mitigate against the negative impact of the toxic materials present in E-waste (Srivastava and Pathak 2020; Yong et al. 2019). Mercury, lead, brominated flame retardants (BFRs), and chlorofluorocarbon (CFC) compounds, such as polybrominated diphenyl ethers (PBDEs), polychlorinated dibenzo-p-dioxins (PCDDs), polychlorinated dibenzofurans (PCDFs), and polychlorinated biphenyl), are some of the hazardous and carcinogenic materials present in electronic waste that are harmful to human health and to environmental sustainability (Chakrabarty and Nandi 2021; Srivastava and Pathak 2020). Effective E-waste management practice involves a combination of actions and steps taken in the management and handling of E-waste, including collection, transportation, treatment/preprocessing, and disposal (Nnorom and Odeyingbo 2019). The success of each handling step has great influence on the entire E-waste management output (Nnorom and Odeyingbo 2019). For example, the efficiency and effectiveness of the E-waste collection system influence the overall amount of waste materials available for recovery, the amount directed toward recycling, and the amount lost in the process of storage and disposal (Nnorom and Odeyingbo 2019). Several studies have emphasized the importance of digitalization and technology transfer in tracking material flow, dissemination of best practices, and planning in E-waste management (Awasthi et al. 2019; Shevchenko et al. 2019, 2021). The strategy for taking-back and restoring E-waste to new productive cycles is encompassed in the principle of the circular economy (CE) (Xavier et al. 2021). This approach, unlike the linear consumption economic model of take-make-use-dispose, encourages a regenerative concept in the production and consumption of resources (Sharma et al. 2021). The CE paradigm inculcates a slow-, narrow-, and closed-loop system where resources are reused and kept in a cycle of continuous usage, and by so doing, resources can create more value for a longer period of time (Sharma et al. 2021). According to Sharma et al. (2021), "the CE system is a regenerative approach that

Electronic Waste Management: Policies, Processes, Technologies, and Impact, First Edition. Edited by Sunil Kumar and Vineet Kumar.

attempts to incorporate the three pillars of sustainable development (social, economic, and environmental) through an interdependent approach for energy and material recovery from waste, and designs durable products with an extended useful life" (Sharma et al. 2021).

Considerable quantities of valuable materials are present in E-waste, which drives interest in formal and informal E-waste recycling businesses around the globe (Yong et al. 2019). In the CE, E-waste is regarded as an "urban mine" based on the potential of recovering various materials from it via recycling, which could be reutilized as substitutes for the primary raw materials in production (Sharma et al. 2021; Xavier et al. 2021). Up to 69 elements could be found in electrical and electronic equipment, which consists of critical (magnesium, antimony, bismuth, cerium, beryllium, lead, etc.), non-critical (aluminum, etc.), and precious metals (silver, gold, copper, etc.) (Chakrabarty and Nandi 2021). For instance, every computer and monitor contain an average of 2–4 kg of lead, rounding up to a total of about 20% lead in each computer and monitor by weight (Nnorom and Osibanjo 2008b, 2008a). Also, characteristic metal components in printed circuit boards (PCBs) are about 30%, while the residual fractions are made up of non-metals like plastics, ceramics, and resins (Yong et al. 2019).

Material recovery through reuse or recycling from E-waste conserves resources and saves energy, provides feedstock for continuous usage in production, and also averts air, water, and land pollution, as well as greenhouse gas emissions linked to material production and manufacturing (Nnorom and Osibanjo 2008a). Hence, the estimated global worth of raw materials generated from E-waste in 2019 was about US$58 billion (Chakrabarty and Nandi 2021; Sharma et al. 2021). Also, around 98 million metric tons of carbon dioxide equivalent ($CO_{2\text{-eq}}$) were emitted worldwide from poor handling of E-waste obtained from refrigerators and air conditioners (Sharma et al. 2021).

The aim of this study is to explore the various robust circularity strategies toward a sustainable E-waste management and the perceived roles of various stakeholders as a collective behavioral dynamic approach to address electronic waste management. In this chapter, E-waste management approaches in a circular and control economy are discussed with elaborate discussions on the stakeholders' involvement for an effective implementation of these strategies as a means of reaching sustainable E-waste management. An overview of hazardous materials in E-waste, and negative impacts of poor handling methods on the environment and the associated health issues, are highlighted. E-waste scenarios in Nigeria, with a critical assessment of the current techniques of E-waste handling as well as effective mitigative measures to curb these challenges, are also detailed.

5.2 E-waste Management Strategies in a Circular Economy

The concept of CE is to manage resources efficiently through the reverse logistics/take-back system for innovative, redesigned, and collaborative ecosystems (Sharma et al. 2020; Xavier et al. 2021). This concept is impactful in improving both the economic performance of the electronic and electrical equipment (EEE) industry and maintaining a sustainable environment in order to preserve the ecosystem for future generations (Sharma et al. 2020). The framework for E-waste management strategies in a circular economy has been proposed and is based on incorporating the ideas of sustainable and green practices such as cleaner technologies, eco-design, and green packaging into the management of E-waste systems while enhancing the product life span through usage in diverse forms (Chakrabarty and Nandi 2021; Sharma et al. 2020). The CE requires a smarter approach to material flow to avoid loss of resources, slow the resource loop, and reduce waste generation (Sharma et al. 2021). Managing E-waste is quite challenging due to advanced technological upgrading resulting in increased consumption of electrical and electronic products and the complexity of their treatment. Adopting E-waste management in the CE is generally influenced by economic growth, environmental adaptability, social inclusiveness and acceptability, technological performance, and policy formulation (Chakrabarty and Nandi 2021).

The introduction of the 3R concept—Refuse, Rethink, and Reduce—to encourage smarter product manufacture and smarter product use is the first strategy in managing electronic waste (E-waste) and transitioning from the traditional linear material flow to a CE (Sharma et al. 2021). "Refuse" implies making needless the use of a product by abandoning its function or offering a similar purpose with a different product. "Rethink" refers to intensifying a product's use through product sharing. "Reduce" implies using fewer raw materials or resources in the manufacture of a product (Sharma et al. 2021). The second E-waste management approach in a CE is the 5R principle, which is "Reuse, Repair, Refurbish, Remanufacture, and Repurpose" (Sharma et al. 2021). The 5R protocol is the outcome of a product after its usable lifetime (Chakrabarty and Nandi 2021). This approach ensures that manufactured electrical and electronic devices along with their parts have extended lifespans. The concept of reuse implies reusing a product after it has been discarded by another user when the product can still accomplish its original function. Repair denotes maintaining a malfunctioning product so that it can be used for its original purpose. Refurbish signifies the remodeling of an old product to bring back its functionalities. Remanufacturing deals with using parts of a discarded product in a new product for the same function. Repurposing refers to the use of a discarded product or its parts to create new products with entirely different functions (Sharma et al. 2021). Figure 5.1 shows the 10Rs strategies for E-waste management in a circular economy.

The third CE strategy in managing E-waste focuses on increasing the useful application of the electronic waste materials and relies heavily on the principles of "recovery and recycling" (Chakrabarty and Nandi 2021; Sharma et al. 2021). Presently, most circularity policies and targets focus more on these strategies in material recycling and recovery, and over time, the combination is considered one of the most practicable E-waste management practices (Sharma et al. 2021).

Therefore, a framework based on the 10Rs (i.e., recover, recycle, repurpose, remanufacture, refurbish, repair, reuse, reduce, rethink, and refuse) is commonly studied as an effective and efficient E-waste management strategy in transitioning to a CE (Chakrabarty and Nandi 2021; Sharma et al. 2021). The 10Rs strategies in E-waste management in a CE shown in Figure 5.1 are organized according to the useful application of materials, the extended lifecycle of products and their parts, smarter product manufacturing, and smarter product consumption (Sharma et al. 2021).

5.3 Stakeholders' Involvement in E-waste Management

Electronics have become indispensable tools in today's world with their rapidly advancing technologies that quickly render previous technologies obsolete. These products therefore resurfaced and have raised the issues of accumulated mountains of waste electrical and electronic products (E-waste). The attention E-waste has received is not only due to its volume but also to its toxicity and hazards. However, beneath the risks of E-waste, there are economic and social opportunities that recycling E-waste can tap into. Due to the incessant emphasis on the recycling issue, the circular economy (CE) is acknowledged as a strategic approach for reducing the stress on the environment and an effort to balance the economies (Sharma et al. 2020) for sustainable development. Sustainable development in the electronic industry is "adopting strategies and activities that meet the needs of the enterprise and its stakeholders today while protecting, sustaining, and enhancing the human and natural resources that will be needed in the future" (Goosey et al. 2009).

The scene in Sub-Saharan Africa (SSA), like most developing economies, questions the sustainability of its development because it is one of the locations with growing E-waste streams due to factors such as: increasing importation of E-waste and growing internally generated E-waste; a low level of awareness of the toxicity associated with E-waste and the effects of improper handling on the part of some informal recyclers; permeable or non-existence of a specific legal framework for E-waste management; absence of technical and infrastructural capacity in the value chain (collection, sorting, reuse, dismantling, final disposal) to handle, treat, and dispose of generated E-waste; poor data on generation and recycling of E-waste; non-existing incentives for consumers or enterprises for return of end-of-life or obsolete electrical and electronic equipment, etc.

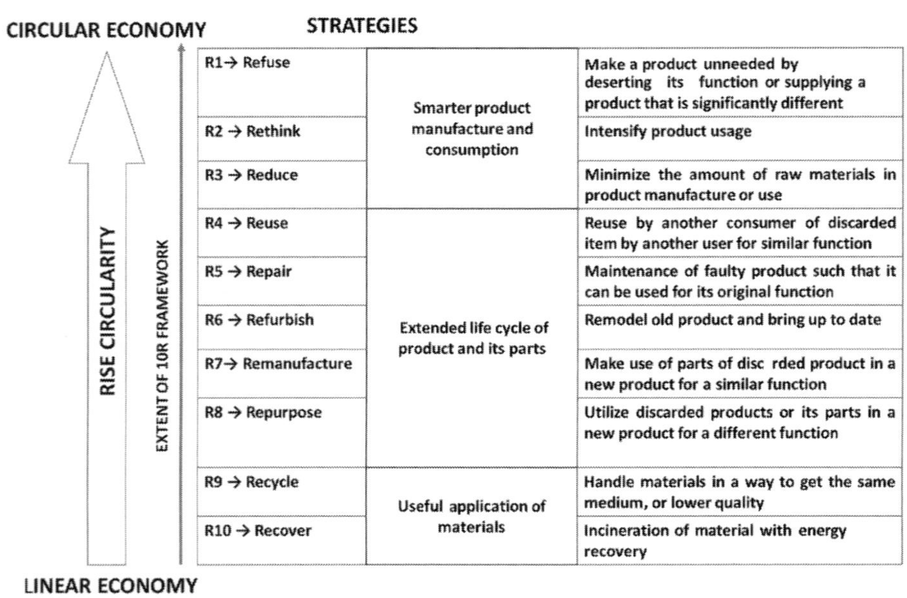

Figure 5.1 The 10R strategies of E-waste management in a circular economy. (*Source:* Adapted from Chakrabarty and Nandi 2021 and Sharma et al. 2021 with some modifications.)

The global growth rate of E-waste is three times that of municipal waste (EACO 2017) and at a growth rate of 4%, which is posing a major recycling challenge in developing countries (DCs) (Woggsborg and Schröder 2018). Attitudes to E-waste in the global north and south differ. In the global south, E-waste management, collection, and recycling are mostly dominated by the informal sector (Heeks et al. 2015), whose treatment and disposal techniques are inappropriate (EACO 2017). According to Heeks et al. (2015), only 5% of E-waste is recycled in India and this is handled entirely by the informal sector, a situation that is giving rise to a high growth rate of E-waste (Heeks et al. 2015). This is not far from what is obtainable in SSA and other developing economies, where the recycling rate is less than 7%. The global south and its key actors are plagued by a knowledge gap that is manifest in the weak formal recycling systems, a weak or lack of legislation and policy, inappropriate handling, and lack of collaboration.

The stakeholders identified for the waste electrical and electronic sector include the following (EACO 2017):

i) Development partners: UNIDO (United Nations Industrial Development Organization), ITU (International Telecommunication Union), UNEP (United Nations Environment Programme), World Bank, etc.
ii) Private sector
iii) States
iv) Agencies in states
v) E-waste scheme: extended producer responsibility, recyclers, collectors
vi) Public
vii) Formal and informal sectors: recyclers, collectors
viii) Producers, importers, and distributors
ix) Non-governmental organizations: environmental, humanitarian, health, education
x) Tertiary and research institutions
xi) Companies and international corporations

A CE entails minimizing waste by attaching a resource value to what would have been termed or handled as waste. To this end, waste handling falls within the purview of the formal and informal sectors. There is a compelling need to focus on E-waste management owing to the opportunities therein, which include E-waste as a source of valuable resources, a source of business and employment opportunities, and the reduction of possible negative environmental impact.

5.3.1 Manufacturers

Manufacturing of electronics in developed economies has an attendant high cost of not only establishing infrastructure as required by legislation for the development of new products, maintenance, and end-of-life treatment of those products, but it extends to the high cost of environmental and health degradation as well. This cost has prompted 'well-financed and strategic marketing' of locating production of components, suppliers, and assemblers around the globe as cost reduction measures (Christian 2012). Christian (2012) went further by stating that the trans-boundary market created is not only for new goods but also for used electrical electronics equipment, which is shipped to low-income regions with weak community voice as well as weak occupational safety and health enforcement. This can be seen as an extended "international division of labor," as practiced by original equipment manufacturers (OEM), that has reconstructed exclusion, economic insecurity, and marginalization.

Manufacturers have engaged in competitiveness and profiteering through product design that avoids longevity but promotes obsolescence and disposability (Christian 2012; Parajuly 2017). This penchant has implied more raw material sourcing (mining of metals and fossil fuel refining) to produce electrical and electronic equipment, and more disposal at the end-of-life with little reuse and recycling. As a result of the raw material quest and waste electrical and electronic disposal, environmental degradation occurs; the sustainability of the "manufacture, use, and dispose of" model is called into question.

A control economy speaks of the government's role in E-waste recycling, even as manufacturers have a major role to play in the attainment of a circular economy. In Singapore and Hong Kong, control economies are practiced, where the government is fully involved or takes over the control of industries and decides what to produce and in what quantity it is to be sent to the market. In Singapore, whatever is not explicitly authorized is outlawed, and in Hong Kong, whatever is not prohibited is authorized. Both Singapore and Hong Kong are known for their economic intervention and free-market approaches, respectively, and these approaches have resulted in the nations enjoying continuous economic booms (Forti et al. 2020; Lam 2000; Sthiannopkao and Wong 2013).

When the CE model is efficiently managed, there will be provisions for the redesign of products, redesign for remanufacturing, reuse, recycling, and safety for the environment. The attention garnered by E-waste is mainly for the inherent toxicity to the environment from leakage of plastics and heavy metal components used that can emanate from poor handling or disposal; a proper design for environmental preservation must seek to reduce toxicity. Models for design for reuse should be predicated on: (i) provisions for an upgrade, removal and replacement, and (iii) refurbishing of the electrical and electronic equipment in the country of origin. Remanufacture, a possible way of reducing E-waste, is in its nascent stage and hence is not widespread (Linder and Williander 2017).

Promotion of a redesign must be driven by economic, health, social, and environmental considerations and include investment in research and development of alternative components to replace toxic ones for sustainable development (Sharma et al. 2020). Manufacturing legislations like restriction of hazardous substances (RoHS) and registration, evaluation, authorization, and restriction of chemicals (REACH) by the EU are commendable and can be emulated, but call for balance in the promotion of innovation, cost reduction, and improved foreign trade (Christian 2012). The increasing demand for electrical electronics amid depleting resources for production has made necessary the improved redesign, recycling, and reuse for the added advantage of using fewer virgin resources and energy and causing less pollution. Legislation requiring original equipment manufacturers (OEMs) to source recyclable components from recyclers of their products, as well as fiscal and financial incentives for OEMs to promote recycling and provide technical information such as material composition for recyclers, can go a long way to bring needed behavioral change in OEMs.

5.3.2 Traders/ Distributors

Traders have been identified as critical stakeholders in developing economies into which used electrical electronics are imported. In Sub-Saharan Africa (SSA), this trade brings in goods for sale, repair, resale, and dumping, whereas in Asia, the imports are mostly for recycling. The imports, which are a mix of waste and used electrical and electronic products, are brought into developing economies situated in Africa and Asia (Ignatuschtschenko 2017; Sharma et al. 2020; Woggsborg and Schröder 2018), where there is a dearth of competent technical manpower and infrastructure to manage the imports. The ensuing sorting, repair/ refurbishing, and sale of wares gives rise to informal sectors made up of poor people engaged in scavenging, recycling, and refurbishing (Ignatuschtschenko 2017; Nnorom and Osibanjo 2008b; Sharma et al. 2020). The operators in this informal sector are poorly educated on the inherent dangers of mishandling waste electrical electronics and non-use of personal protective equipment while working on or disposing of the waste electrical and electronic product (Nyam et al. 2015; Woggsborg and Schröder 2018).

5.3.3 Companies and International Corporations

It has been established that very large organizations (VLOs) deploy ICT from a large proportion of internal E-waste generators and demonstrate a proactive attitude toward E-waste management because they see to its evacuation as being in conformity to existing regulatory requirements. This is in complete contrast to the indifference being displayed by small and medium-sized enterprises (Heeks et al. 2015). These attitudes toward E-waste are responses to external and internal factors such as corporate reputation, peer pressure, client and government requirements, organizational leadership, organizational culture, and financial impact. The downside, however, is that even the VLOs use informal recyclers largely because of the almost complete absence of formal recyclers in most developing countries (Ignatuschtschenko 2017; Sharma et al. 2020; Woggsborg and Schröder 2018).

5.3.4 Formal and Informal Sector Recyclers

In the developing economies, the workers in these groups lack sufficient know-how in the handling and disposal of end-of-life electrical and electronic products, a fact that manifested itself in the monumental environmental degradation noticed in Ghana, India, China, and Nigeria from the activities of these recyclers (Andeobu et al. 2021). The number of workers and facilities in the informal sector (mostly poor illiterates) outweighs those in the formal sector, so the quantum of recycled materials is skewed in favor of the former, and they work from shacks where they dismantle, break, brush, carry out chemical stripping in an open pit, by river beds, and open-air burn to harvest metals. There poor monitoring and enforcement of legal provisions are major banes (Dutta and Goel 2021; Ignatuschtschenko 2017). As equipment manufacturers brace up for circular models that promote E-waste management systems for a skilled workforce specific to E-waste recycling, such training has to be green for the adoption of environmentally cleaner technologies to protect workers and the ecosystem (Khoshand et al. 2019). Both the formal and informal sectors have proven that incentives rather than punitive measures could be motivating factors in enforcing standards (Ignatuschtschenko 2017).

5.3.5 Public Involvement in the Implementation of E-waste Management Strategies

Recycling is key to the success or failure of strategies deployed to promote circularity in the treatment of waste. The public, or users of electronic products, occupy the place of import in the chain as the primary source of the waste generated; hence, first collectors, the subsequent collection and recycling rates are immensely dependent on these users. Users have over time come to adopt "home stockpiling" of this waste from e-products for lack of urgency stemming from the perceived 'clean and valuable' nature of the waste (Casey et al. 2019). In Europe, home stockpiling resorts to a behavioral response by the users to "convenience and lack of flexibility of formal collection systems" (Nowakowski 2016). Similar reasons may be adduced in other climes; in the developing countries, especially in

Sub-Saharan Africa, where there is a weak or no formal sector system for E-waste, users are predisposed to turning in their waste to informal collectors (Nnorom and Osibanjo 2008a; Nyam et al. 2015). Another reason users subscribe to informal collectors is that they pay more for returned wares than the formal collectors (Ignatuschtschenko 2017). On one hand, consumers' awareness of the toxic nature of some parts of E-waste is questioned, and on the other, there exist differences between their claims of participation in sorting, reusing, or leasing of electrical electronic items and their actual actions (Parajuly 2017).

Consumers or the public play a threefold role in the life of a product: purchase, use, and end-of-life management; this principle applies to electrical and electronic goods as well (Parajuly 2017). The recycling business is worth ~50 billion euros and covers buyers of secondhand EEE, scavengers, informal recycling, and recyclers—recycling generally. The place of consumer behavior is at the heart of the circular economy and the end-of-life of a product (Parajuly 2017; Sharma et al. 2020); however, Parajuly et al. (2020) decried the neglect of the consumer's role in sustainable production and consumption of electrical and electronic products. As emphasis is placed on improving manufacturing, strategies for a circular economy aim to optimize public use of "artifacts," particularly when viewed as resources. Acceptance and adaptation to investing in more durable products with circular business models without fear of obsolescence, the option to repair and reuse broken or malfunctioned products, and the appropriate time to disposal of product end-of-life are critical and require behavioral change dynamics to address (Parajuly 2017).

Economic packages, such as subsidies and paybacks for users who return E-waste to formal recyclers, a deposit refund scheme paid to customers as an incentive for return of obsolete products to formal recyclers, and payments to recyclers for recovery of rare earth metals, can be put in place as end-of-life incentives for users. Full public participation is required for the success of E-waste management strategies in the control economy, which can be achieved by behavioral change aided by social enticements (Parajuly 2017). Sharma et al. (2020), in advocating a control economy for E-waste, listed enhanced community participation, strict E-waste-specific legislation, the implementation of environmental management systems, and the manufacture of green products as 'enablers' in India (Sharma et al. 2020). It was strongly recommended that the identification of the enablers in other climes factor should be closely tied to:

i) Environmental management systems: policies supporting manufacturers for setting environmental goals, planning, and assuming responsibility for monitoring end-of-life products through activities such as recycling, resource recovery, and environmentally fitting disposal.
ii) Reduction of hazardous and toxic materials.
iii) Green image.
iv) Reduction in landfill practices.
v) Recovery of precious metals.

5.3.6 Government Involvement

East Africa Communication Organisation (EACO) (2017) identified political will and regional political instability as threats to achieving control economy in the E-waste sector, which led to policy discontinuation especially in Sub-Saharan Africa (SSA); this led to discontinuation of policies or non-implementation of ratified agreements, and sometimes reversal of policies (EACO 2017). Government loses in economic terms from the improper handling of E-waste because precious metals from primitive extraction methods deployed by informal operators are not properly situated in the economy as a source of revenue to the government (Ignatuschtschenko 2017). Governments at the national and subnational level in Asia, and to a large extent in SSA, have not prioritized E-waste management until recently, as evidenced by some of their non-E-waste-specific policies and regulations (Nowakowski 2016; Nyam et al. 2015). This neglect has affected the implementation of policies and regulations as well as the provisions of formal recycling facilities, sector-specific data collation systems, and the monitoring and evaluation of the sector's key players and indicators.

Governments' policies and legal frameworks for embracing circularity have to envisage an intentional proactive improvement of product design—a design that promotes desired end-of-life realities—or else they will fail (Parajuly 2017). These regulations must be far-reaching to various key actors and processes, including electrical and electronic production, post-consumption waste, the inclusion of very large organizations and small and medium enterprises (SMEs), an annual environmental compliance report, etc. (Heeks et al. 2015). Policymakers should emphasize the incorporation of key enablers during the strategy formulation process for developing an efficient E-waste management system (Sharma et al. 2020). Governments' strategies for achieving a circular economy in E-waste management should be guided by the inclusion and participation of all identifiable key stakeholders in the waste management industry (electrical and electronics system), envisioning a collaborative framework for all stakeholders, aligning with national, regional, and global development agendas, and advancing fair and equitable distribution of initiative among states (EACO 2017;

Ignatuschtschenko 2017). Governments should also consider the following:

- Reset minimum recycling goals, which may drastically reduce home E-waste stockpiling.
- Enact legislation to encourage recycling and prevent waste generation by encouraging reuse.
- Use cultural and techno-economic variation in the design of behavioral interventions for circular economy implementation.
- The business model and infrastructure will facilitate regulatory tools, particularly closed-loop recycling systems.
- Include end-users in electrical electronics products' eco-design.

The green dot policy and other approaches implemented in Germany have made inroads for the success of solid waste management (SWM), and incentives can have a positive impact on E-waste management. Incentives (monetary reward) and legal requirements can be relied on to influence consumers' behavior, which is to either sell or buy secondhand E-waste, explore reuse channels, return E-waste to an appropriate collector for recycling, or discard it wrongly—decisions that can affect the collection and recovery of resources at the end-of-life of electrical and electronic products, hence the failure or success of a circular system for the products (de Vicente Bittar 2018). There is a need to factor in a re-direction of consumer behavior in the control economy for E-waste by creating awareness for customers to prefer leasing of E-products and to increase demand for remanufactured and refurbished goods (de Vicente Bittar 2018).

5.3.7 Non-governmental Organizations (NGOs)

The perception of waste e-products as resources or waste is a function of the awareness and actions of the users. There exists a variance between the awareness of "green products," sorting, and recycling and the actual action of purchasing green, leasing, repairing of malfunctioned equipment for reuse, and collection and disposing of E-waste at recycling centers (Parajuly 2017). There is a need for awareness campaigns to identify this knowledge–action gap in consumption in the control economy. Collaboration has been reportedly lacking among stakeholders in the waste electronic sector (Ignatuschtschenko 2017; Parajuly 2017). The established factors responsible for the low level of action taken despite knowledge of circular strategies to curb the accumulation of E-waste by users must be factored into awareness campaign strategies to achieve the necessary consumer behavior change. The trans-boundary nature of E-waste assumed over the last decades, where weak countries have fallen victims, necessitates legal empowerment in the form of policies and frameworks aimed at mitigating the trans-boundary harm caused by E-waste movement (Christian 2012). This problem can be brought to the forefront by development partners and NGOs. NGOs could yet be involved in fostering collaboration among stakeholders and the development of certificates of reuse for issuance on products where the circular economy model entrenches the designation of electrical and electronics into used electrical and electronic equipment (UEEE) and waste electrical and electronic equipment (WEEE). Isomorphic pressure from professional associations in the form of discussions on E-waste issues and awards of excellence on environmental matters can influence behavior change among traders and companies that deal in electrical electronics (Heeks et al. 2015).

5.3.8 Extended Producer Responsibility

Extended producer responsibility (EPR) is a policy tool that involves equipment manufacturers taking responsibility (physically or financially) to manage their used products' end-of-life (Esenduran et al. 2019; Leclerc and Badami 2020; Woggsborg and Schröder 2018). EPR is a tool that is consistent with the "polluter pay principle."

The primary objectives of EPR are to a large extent in tandem with those of a circular economy. Leclerc and Badami (2020) listed the objectives of EPR to include abatement of waste generated and disposed of, improved design for the environment, abatement in the use of virgin materials, and reduction of hazardous components of the waste stream and pollution in the product use phase. EPR is implemented as previously stated, either through infrastructural means or as a physical burden upon the OEM, wherein a product take-back arrangement is put in place with recycling rate targets; the take-back may be mandatory or voluntary (Leclerc and Badami 2020). On the other hand, the implementation of the EPR could be a financial burden upon the OEM, where payments are made for the end-of-life treatment of their products under various tags such as the deposit refund system (DRS), recycling fees combined with recycling subsidies, upstream tax, or an advanced recycling fee. The implementation of EPR as made popular by the OECD has resulted in economic earners worth billions of dollars for countries. African countries have not been left out despite the late establishment of the program. In Nigeria, for instance, the program was only established in 2016 (Woggsborg and Schröder 2018).

The EPR tool is contextualized by the operating country to take cognizance of each country's peculiarities; despite this provision, studies have shown that operating EPR are not without hiccups. For instance, the absence of or weak legislation that should enhance enforcement, poor knowledge of material flow in an E-waste management system, poor public awareness of the reality and benefits of E-waste recycling, lack of upgrade of the E-waste recycling infrastructure, and the presence of strong informal sectors are among some of the factors that have hindered the realization of the objectives of establishing the EPR program. Sadly, these challenges are not typical to developing countries alone (Leclerc and Badami 2020).

To expand a circular economy in the used or waste electrical and electronics sector, operating strategies must address the issues of participants' competencies for all rungs in the E-waste recycling chain as part of EPR. The effect will manifest in the conservation of resources and an improved environment characterized by more reuse and recycling with less waste generation and disposal as outcomes. Effective regulation places product design at the heart of E-waste generation and management, as this can curb obsolescence (Christian 2012). Design is a factor in making products that are less toxic, more serviceable, and more reusable after repairs. Overtly, intervention strategies for the control/circular economy in the E-waste subsector are strengthened by a far-reaching policy and regulatory framework that encapsulates some of the provisos earlier mentioned, such as proper and effective coordination and institutional alignment, enforcement of extended producer responsibilities (EPR) and ARF, resource mobilization in terms of E-waste funds (fees collection and disbursement for E-waste management), public-private partnerships (in the collection, transportation, and treatment) and engagement of partners (e.g. ITU, UNEP, UNIDO, etc.), capacity building for both the formal and informal sector operators, provision of infrastructure for E-waste management, and communication and knowledge or experience sharing on E-waste.

5.3.9 Research and Academic Institutions

The shift toward a control economy can be enhanced by the quest for components and processes to be used in the manufacture of electrical and electronic appliances as replacements for more toxic components. Research and academic institutions have been exploring ways of recycling E-waste in an environmentally friendly and cost-effective way, and an integrated process still eludes the E-waste recycling industry; this calls for more work. If research can optimize precious and rare earth metals recovery from E-waste, recycling has the potential to become the primary source of these metals for reuse without relying on virgin metals from mine fields (Cordova-Pizarro et al. 2019).

5.3.10 Media Organizations

The type of mass media penetration varies with communities, and this is critical in choosing the media channel to deploy for awareness creation. The internet and social media appeal to urban dwellers, while the rural folks are reached more effectively by traditional broadcast media like radio, television, and print. These outlets have been shown to be sources through which awareness of E-waste-related issues has been disseminated to the public in many countries (Bhat and Savale 2020). Therefore, strategies that involve behavioral change for attaining a circular economy in E-waste must situate the media type for each community, taking cognizance of the community's cultural, economic, and social peculiarities (Oyero et al. 2018). It is necessary to raise awareness among users and recyclers about the inherent dangers and toxic nature of E-waste on the environment. However, access to E-waste-related media content has been described as limited (Ben-Enukora et al. 2017). Even more limited is the media's focus on the menace of E-waste. In most developing economies, low awareness has led to huge illegal trans-boundary importation of E-waste for economic gains by traders, accumulation of internally generated waste and home stockpiling by users, bizarre ways of transporting and dismantling E-waste for recovery of parts by recyclers, and economic losses to governments as well. The low amount of media content has been attributed to a number of factors, which include: non-prioritization of E-waste management issues generally by media outfits as compared to gender, education, or health issues; low budgetary provision for E-waste issues; fear of government retributory action for any media exposure of shortcomings in E-waste management practices; and a preference to cover more profitable issues (Koser 2017).

The media have been a critical tool in setting the agenda on various issues; they are a veritable tool to be deployed to improve E-waste management systems that will engender sustainable practices in the management of E-waste. The creation of opportunity from the challenges of E-waste is at the top of the agenda setting that the media can champion (Torres et al. 2016). The gamut of the waste chain (from generation to disposal) is influenced by the behavior of producers, users, regulators, and workers. The media are capable of deliberate orchestration for the promotion of green design of products by original equipment manufacturers for durability, upgradability, reuse, repair, and recycling. Likewise is true for the promotion of green purchases and practices on the part of consumers to prefer more repair and reuse to less dumping of old products for new ones. Though incentivization has previously been shown to impact consumer participation in recycling models, a sustainability model should work toward encouraging more consumers to return their wares to recycling ports or agents on a free basis. Most developing economies have not achieved free or voluntary return of end-of-life products to collection points (Vishwakarma et al. 2022); the media and NGOs can create deeper awareness and engagement campaigns to influence consumers' behavior. The electrical and electronics sector is one that is prone to frequent upgrades of products and devices. This scenario poses a challenge to the technical manpower available in developing economies for repairs and recycling. There have been attempts at closing the gaps through initiatives like the Mobile Phone Partnership Initiative (MPPI) and Partnership for Action on Computing Equipment (PACE); such initiative was coordinated by the World Intellectual Property Organization (WIPO) (Torres et al. 2016; Vishwakarma et al. 2022). However, there are gaps in the system that the media, non-governmental organizations, and research institutes can collaboratively work together to fill.

According to Oyero et al. (2018), the communicated message must include the following: the meaning of E-waste, its types, and evidence; the environmental and health consequences of waste accumulation and improper handling on the environment and its people; and behavioral practices that people, including recyclers, should imbibe and practice. In their study, they also stated that the language in use in communication is as important as the channels used to communicate the message, for which they advocated a mixed-media approach—the folk media, interpersonal channels, and the mass media (Oyero et al. 2018).

There are numerous international initiatives and policies on E-waste whose ratification and domestication are dependent on political willpower (EACO 2017), which is lacking in most developing economies. Similarly, the implementation and enforcement of regulations that are relevant to, and in some cases, non-specific to E-waste call for attention. By torchlight, the media, as society's watch dog, has a role in monitoring policy and regulations enforcement for compliance and identifying malpractices and flaws. This will go a long way in curbing unhealthy practices in the sector.

According to Vishwakarma et al. (2022), various malpractices exist in the E-waste sector, with producers, formal operators, and informal operators all having a piece of the pie. Only 17% of the world's E-waste generation has been documented to be collected and recycled. These malpractices question the mutilation of sales and actual collection data on the part of producers with relation to the execution of the extended producer responsibility policies. Furthermore, the formal and informal sectors have roles in the creation and perpetuation of illegal dumps and landfills for E-waste and the use of illegal channels for trans-boundary transportation of the waste (Vishwakarma et al. 2022).

5.4 Overview of Hazardous Materials in E-waste

Typical E-waste is a heterogeneity of appliances, each made up of a mixture of glass, plastics, ceramics, and metals. Notably, the proportion of these components varies according to the appliances. The thermal, electrical, and mechanical properties of these components are determined by the intended use to which the original product will be put, operating conditions, and aesthetics. The mix of metals is shared between 48% non-hazardous metals, for example, steel and iron; and 13% non-ferrous metals made up of precious metals (platinum, palladium, silver, and gold), rare earth metals such as indium, neodymium, praseodymium, and tantalum, and hazardous metals (e.g. arsenic, cadmium, cobalt, chromium, copper, mercury, lithium, and lead) (Vishwakarma et al. 2022). The components are also made of inorganic and organic chemicals. Plastics make up about 21% of the components in electrical and electronic devices.

Pre-treatment of E-waste is done in three ways: (i) mechanical dismantling, which uses size reduction methods of shredding and crushing to separate metals from non-metals; (ii) manual dismantling, which separates non-hazardous and hazardous materials like casings, batteries, monitors, and printed circuit boards; and (iii) a combination of mechanical and manual methods by use of semi-automatic processes, (Shahabuddin et al. 2022). Open-air burning and acid (acid bath and stripping) treatment operations are also employed.

The emergent possibilities from the poor, primitive, and crude handling of E-waste, especially on the part of the informal operators, are evidenced by the release into the environment of the hazardous and toxic substances contained in the E-waste. These substances, which include cadmium, polycyclic aromatic hydrocarbons (PAH), lead, chromium, etc., are in the form of gases and vapor that go into the atmosphere and water bodies, leachates into groundwater supplies, particulate matter in the form of fly ash into soil, and open discharge of solvent liquid into the ground or water bodies.

As indicated in Table 5.1, informal workshops or factories for artisanal recovery operations such as shredding and incineration are open spaces, dumpsites, markets, and riverbanks, and the workers do not use the correct personal protective equipment. They incur lacerations occur from sharp objects, inhale fouled air, and are exposed to the deposition of contaminants in washing and drinking water. A study has shown higher concentrations of toxic heavy metals such as antimony, zinc, lead, and copper at workshops and surrounding areas (soil, river sediments, and air), as well as in the homes of E-waste workers (Ignatuschtschenko 2017). There are reports of higher lead levels in children associated with E-waste treatment and plant contamination (Torres

Table 5.1 E-waste treatment methods in the informal sector.

E-waste component	Treatment technique
Various components	Physical dismantling by using bare hands and tools such as hammers, chisels, screwdrivers to separate different materials
Plastics	Shredding and melting without proper ventilation
Printed circuit boards (PCBs)	Open burning over coal-fired grills
Wires and cables	Open burning for recovery of copper
Chips and gold-plated components	Metals chemically stripped in open-pit acid baths along riverbanks
Residues	Uncontrolled disposal of residues in fields and riverbanks
Toner cartridges	Toner recovery by use of paintbrushes
Printed wire boards (PWBs)	Acid leaching

et al. 2016) and deposits of toxic materials on the human body (workers at E-waste recycling hubs), and their effects have been established (Kishore 2010).

In 2019, a total of 50 metric tons (Mt) of mercury and 71 000 Mt of brominated flame retardant (BFR) plastics were discovered in undocumented flows of E-waste, and 98 Mt of CO_2-equivalents, including chlorofluorocarbons (CFC) or hydrochlorofluorocarbons (HCFC), were released from discarded refrigerators and air conditioners, most of which were released into the environment and impacted the health of the exposed workers (Sharma et al. 2021). A similar report in 2014 stated that the E-waste supply of toxins was made up of 2.2 Mt of lead glass, 0.3 Mt of batteries, and 4000 Mt of ozone-depleting substances (CFC) (Baldé et al. 2015). In open burning of materials, heavy metals and other toxic materials contained in fly ash are discharged, resulting in inhalation of these toxic materials and contamination of food, soil, and surface water after deposition. For example, the recovery of copper wires through the burning of polyvinyl chloride (PVC) and PBDE-protected cables can release toxic chlorinated and brominated dioxins (PCDD/PBDD) and furans (PCDF/PBDF). The open incineration of computer plastic casings and circuit boards stripped of metal parts can result in toxic ashes containing polycyclic aromatic hydrocarbons (PAHs) being released into the environment (Borthakur and Sinha 2013).

5.5 Environmental and Health Issues Associated with Poor E-waste Handling

There is a growing concern over E-waste exposure to the environment and public health, particularly for the children living very close to the areas where E-waste is treated. Three major sectors contribute to the exposure of E-waste: environmental, formal recycling, and informal recycling. Environmental exposure occurs as a result of hazardous E-waste compounds released into the environment. Formal electronic waste recycling facilities specifically utilize designed technology to safely remove salvageable materials from obsolete ones in order to protect workers from the adverse health effects of electronics exposure (Forti et al. 2020; Sthiannopkao and Wong 2013). However, these centers are costly to set up and run and are not easily adopted in less developed countries. Hence, different national safety standards imply that workers at these recycling centers still risk being exposed to hazardous compounds at a low rate (Tsydenova and Bengtsson 2011). Informal electronic waste recycling is the use of primitive techniques to retrieve valuable elements from electronics that have reached their end-of-life value, without or with very little technology to reduce exposure or equipment protection, giving room for harmful chemicals to be emitted in the process (Parvez et al. 2021). Laboratory research has revealed that leachates obtained from electronic waste significantly contain high amounts of metals, including beryllium (Be), vanadium (V), aluminum (Al), lead (Pb), cadmium (Cd), copper (Cu), chromium (Cr), cobalt (Co), nickel (Ni), and barium (Ba), compared to the generated leachates from other waste streams (Kiddee et al. 2013, 2014).

5.6 E-waste Scenarios in Nigeria

The population of Nigeria is approximately 200 million according to the Worldometer's elaboration in 2022, and by estimation, the implication is that a huge amount of E-waste is generated. For instance, the commercial capital of Nigeria, Lagos, is arguably among the largest dumps of E-waste from DCs. Piles of e-waste in Nigeria originate both at home and abroad. The country generated 300 000 metric tons of electronic waste in 2017 (a 170% increase) as against 2009 (Odeyingbo et al. 2019). Meanwhile, Nigeria remains a major recipient of used electronics from abroad. While the true amount of overseas-generated waste landing in Nigeria is difficult to quantify, United Nations University research has revealed that more than 60 000 metric tons of used electrical and electronic equipment are shipped into the country annually via the Lagos ports alone, with an unknown amount imported over land routes from neighboring countries (Odeyingbo et al. 2019). More than 25% of these materials are dead on arrival, heading straight to dumpsites or dismantled (Odeyingbo et al. 2019). According to the report by International Labour Organization (ILO), up to 100 000 people work in the informal E-waste recycling sector in Nigeria, collecting and dismantling electronics by hand to reclaim the saleable components. Informal workers are directly exposed to hazardous chemicals and commonly suffer respiratory and dermatological problems, eye infections, and a lower-than-average life expectancy (Orisakwe et al. 2020). The waste with no economic value is often dumped or burned, releasing pollutants, including heavy metals and toxic chemicals (including dioxins, furans, and flame retardants) into the air, water, and soil (Orisakwe et al. 2020). Over 52 000 metric tons of brominated (flame retardant) plastics, 4000 metric tons of lead, 80 metric tons of cadmium, and 0.3 metric tons of mercury are burned or dumped in Nigeria every year (Nnorom and Odeyingbo 2019; Odeyingbo et al. 2019).

According to a survey carried out by the Basel Centre in 2010, it was revealed that at least 500 containers of used computers and other electrical appliances are imported into the country through Lagos port every month (Odeyingbo et al. 2019). In 2011, the National Environmental Standards and Regulations Enforcement Agency (NESREA) barred six cargos loaded with toxic electronic waste from offloading. The population of Lagos is currently around 15 million, generating around 9000 metric tons of waste daily. Based on the 2006 census, Ibadan has an estimated population of 3 million, Kaduna has 6 million, Kano has 9 million, Abuja has 8 million, Katsina has 6 million, and Aba has 8.3 million. Very likely, the population of these cities has increased far more than these figures, resulting in a corresponding increase in the volume of generated E-waste. Also, Nigeria became mobile phone compliant a few decades ago, and as of December 2009, it had recorded 73 million active mobile subscribers, which made it the fastest-growing mobile phone market and electronic waste volume in Africa (Andeobu et al. 2021; Odeyingbo et al. 2019). Because of its natural disposition, both the good, the bad, and the ugly easily come into Lagos through the seaports. Every day, many containers are unloaded with a variety of goods, some of which are also transported to Tinker Island in Lagos. This indicates that E-waste is a serious problem whose impact will be felt most acutely by the most vulnerable groups of people in developing countries, and which has the potential to worsen in the coming years if not properly managed (Andeobu et al. 2021). It is worth noting that Nigeria's population and land mass have easily made it a major stakeholder in Africa's regions, particularly with regard to E-waste.

5.6.1 Critical Assessment of the Current Methods of E-waste Handling Techniques in Nigeria

Unfortunately, the country lacks official, effective management policies and a recycling program. Electronic waste material is quickly disposed of with municipal solid waste, informally dumped or left in homes, warehouses, and landfills (Andeobu et al. 2021). Large volumes are crudely recycled and burned outdoors, thereby polluting the environment and endangering the health of workers on site and residents nearby. Also, because Nigeria does not have a formal recycling sector for safe management of E-waste, 500 000 tons of electronic and electrical equipment are dumped in open spaces, workshops, landfills, and water sources, where more than half is near the end of its useful life or completely damaged (Odeyingbo et al. 2019).

5.7 Effective Mitigative Measures

To meet up with the challenges posed by the E-waste handling system in Nigeria, the following mitigative measures are encouraged to curb the effects of exposure to health and ultimately the environment.

a) Recycling
   E-waste can be a valuable source of income when broken down into other components through recycling. Several people have realized that money could be made from trash without endangering the environment. Lack of implementing the EPR creates a huge challenge to E-waste management in Nigeria; therefore, it is important that producers/major retailers of EEE products are charged with the responsibility of managing the collection, disposal, and recycling of their end-of-life products (Nnorom and Odeyingbo 2019).
b) Informal sector upgrade
   While formal E-waste handlers might have adequate waste handling and processing, the informal E-waste approach poses a great threat to the handlers, public health, and the environment in general. In Nigeria, the main process employed by the informal handlers is burning, and by openly burning E-waste, the air is infiltrated, the lungs are affected, and public health is jeopardized due to the presence of toxins inherent in E-waste. The informal handlers' activities need to be put in check and upgraded by the National Environmental Standards Regulations Enforcement Agency (NESREA). The informal E-waste sector could be formalized through training programs in order to protect the

environment and workers would be well-informed of the implications of their operations (Nnorom and Odeyingbo 2019). It is important for the government to act, by making and enforcing policies that would ensure effective E-waste handling techniques in Nigeria under the highest environmental and industrial standards. Strict measures must be put in place to ensure that valuable extractions and E-waste recycling follow all regulations charged with the responsibility of ensuring both the safety of those handling the materials and the public.

c) Regulatory bodies synergy
The Nigerian government needs to strike a synergy between the regulatory bodies that oversee environmental issues within the country. This goes beyond intergovernmental agency interactions; it stretches into the cooperation that exists between the handlers and government agencies, as well as the public in general. If the government works frictionlessly with the professional handlers in the country, it will ease the current challenges of electronic wastes. For example, legislation should be enacted to encourage corporate entities to use responsible handlers for the disposal of their E-waste.

d) Sensitizing the public on the dangers of E-waste
As shocking as it may sound in this age and time, many Nigerians know little or nothing about the dangers of E-waste in the country, across the continent, and globally. There are millions who are ignorant of the dangers inherent in inadequate disposal of E-waste and on how they can be properly mitigated; hence, the need for enlightenment. There should be appropriate planning and execution of E-waste awareness campaigns in various communities across the country. These campaigns can be executed by the government, through the regulatory bodies, or by the available pool of professional handlers. It would be even more interesting if the government worked with corporate organizations and waste handlers in the country to organize E-waste awareness campaigns. If there is adequate awareness of the dangers posed by E-waste by the people and they are informed of their solutions, the public would work in synergy with the appropriate authorities to see to it that their E-waste is disposed of in the most professional manner, thereby minimizing the magnitude of E-waste that goes into landfills and invariably solving E-waste challenges.

e) Optimum combination of E-waste management tools
There are ongoing studies on E-waste management that aim to mitigate the challenges that could come up at both the national and international levels. There has been tremendous advancement in the development of various tools and applications for the management of E-waste such as life cycle assessment, which is used to design environmentally friendly electronic devices and to minimize E-waste challenges (McAllister 2015); material flow analysis (MFA), which is used to study the route of material (E-waste) flowing into recycling sites or disposal areas and stocks of materials in time and space. MFA links the sources, pathways, intermediates, and final destinations of the material (McAllister 2015). Multi-criteria analysis is also a decision-making tool developed for considering strategic decisions and solving complex multi-criteria problems that include qualitative and quantitative aspects of the problem (Garfi et al. 2009). Extended producer responsibility (EPR) is also an environmental policy technique that attributes responsibility to manufacturers in retrieving their products after use based on 'polluter pays' principles (OECD 2001; Widmer et al. 2005). The management of E-waste in developed countries has taken a proactive step forward with the release of a Waste Electric and Electronic Equipment (WEEE) directive (Directive 2002/96/EC), which is believed to minimize waste disposal and improve environmental quality. Although Nigeria has yet to explore these tools for mitigating E-waste, they can be useful in mitigating E-waste if adopted.

5.8 Conclusion

In this study, the joint efforts of all stakeholders, which include the public, manufacturers, governments, policymakers, and media organizations, have been highlighted for sustainability to be achieved in the E-waste management sector. Media organizations have a crucial role to play by reporting the real scenarios in E-waste management and by creating awareness among the public about the dangers associated with exposure to E-waste. E-waste management strategy in a circular economy and control economy were also discussed, in which manufacturers take responsibility for the products at the end of their useful lives. Government involvement in industries is a means of influencing what kind of products are produced and the quantities that are permitted to be sent to market in order to reduce the negative impact of improper E-waste management on public health and the environment. A critical assessment of the current methods of E-waste handling techniques in Nigeria has shown that the country has a very long way to go in the handling of E-waste for environmental safety due to the fact that the majority of its E-waste is recycled or salvaged by the informal handlers/recyclers employing crude treatment methods.

Acknowledgement

This work is based on the research supported in part by the National Research Foundation of South Africa (Ref Numbers: SASPD22091357540).

References

Andeobu, L., Wibowo, S., and Grandhi, S. (2021). An assessment of e-waste generation and environmental management of selected countries in Africa, Europe and North America: a systematic review. *Science of the Total Environment* 792: 148078.

Awasthi, A.K., Li, J., Koh, L., and Ogunseitan, O.A. (2019). Circular economy and electronic waste. *Nature Electronics* 2 (3): 86–89.

Baldé, C., Wang, F., Kuehr, R., and Huisman, J. (2015). *The Global E-waste Monitor–2014*. Bonn, Germany: United Nations University, IAS–SCYCLE. *ISBN Print*, 978–992.

Ben-Enukora, C., Okorie, N., Oresanya, T., and Ekanem, T. (2017). Awareness and perception of media campaign on E-waste effects among residents of Ado Odo-Ota, Nigeria. *Covenant Journal of Communication* 4 (2): 18–32.

Bhat, V. and Savale, O. (2020). Media's role in E-waste awareness—a perspective of rural and urban population of Maharashtra. *International Journal of Modern Agriculture* 9 (3): 1176–1184.

Borthakur, A. and Sinha, K. (2013). Electronic waste management in India: a stakeholder's perspective. *Electronic Green Journal* 1 (36).

Casey, K., Lichrou, M., and Fitzpatrick, C. (2019). Treasured trash? A consumer perspective on small waste electrical and electronic equipment (WEEE) divestment in Ireland. *Resources, Conservation and Recycling* 145: 179–189.

Chakrabarty, A. and Nandi, S. (2021). Electronic waste vulnerability: circular economy as a strategic solution. *Clean Technologies and Environmental Policy* 23: 429–443.

Christian, G.E. (2012). End-of-life management of waste electrical and electronic equipment for sustainable development. Centre for International Sustainable Development Law. McGill University 3644.

Cordova-Pizarro, D., Aguilar-Barajas, I., Romero, D., and Rodriguez, C.A. (2019). Circular economy in the electronic products sector: material flow analysis and economic impact of cellphone e-waste in Mexico. *Sustainability* 11: 1361.

de Vicente Bittar, A. (2018). Selling remanufactured products: does consumer environmental consciousness matter? *Journal of Cleaner Production* 181: 527–536.

Dutta, D. and Goel, S. (2021). Understanding the gap between formal and informal e-waste recycling facilities in India. *Waste Management* 125: 163–171.

EACO (2017). *Regional E-waste management strategy draft*. Regional E-waste Management Steering Committee. Retrieved from https://www.eaco.int/admin/docs/publications/EACO (accessed 15 February 2023).

Esenduran, G., Atasu, A., and Van Wassenhove, L.N. (2019). Valuable e-waste: implications for extended producer responsibility. *IISE Transactions* 51 (4): 382–396.

Forti, V., Baldé, C.P., Kuehr, R., and Bel, G. (2020). *The Global E-waste Monitor 2020*. Bonn, Geneva, and Rotterdam: United Nations University/United Nations Institute for Training and Research, International Telecommunication Union, and International Solid Waste Association.

Garfi, M., Tondelli, S., and Bonoli, A. (2009). Multi-criteria decision analysis for waste management in Saharawi refugee camps. *Waste Management* 29 (10): 2729–2739.

Goosey, M., Stevens, G., and Herman, H.K. (2009). *Electronic Waste Management*, 27. Royal Society of Chemistry.

Heeks, R., Subramanian, L., and Jones, C. (2015). Understanding e-waste management in developing countries: strategies, determinants, and policy implications in the Indian ICT sector. *Information Technology for Development* 21 (4): 653–667.

Ignatuschtschenko, E. (2017). E-waste management in China: bridging the formal and informal sectors. *Journal of Chinese Governance* 2 (4): 385–410.

Khoshand, A., Rahimi, K., Ehteshami, M., and Gharaei, S. (2019). Fuzzy AHP approach for prioritizing electronic waste management options: a case study of Tehran, Iran. *Environmental Science and Pollution Research* 26 (10): 9649–9660.

Kiddee, P., Naidu, R., and Wong, M.H. (2013). Metals and polybrominated diphenyl ethers leaching from electronic waste in simulated landfills. *Journal of Hazardous Materials* 252: 243–249.

Kiddee, P., Naidu, R., Wong, M.H. et al. (2014). Field investigation of the quality of fresh and aged leachates from selected landfills receiving e-waste in an arid climate. *Waste Management* 34 (11): 2292–2304.

Kishore, J. (2010). E-waste management: as a challenge to public health in India. *Indian Journal of Community Medicine: Official Publication of Indian Association of Preventive & Social Medicine* 35 (3): 382.

Koser, R. (2017). *Role of Mass Media in Public Awareness about Household Waste Management: A Case Study in Lahore (Pakistan)*. Ås, Norway: Norwegian University of Life Sciences.

Lam, N.M.K. (2000). Government intervention in the economy: a comparative analysis of Singapore and Hong Kong. *Public Administration and Development: The International Journal of Management Research and Practice* 20 (5): 397–421.

Leclerc, S.H. and Badami, M.G. (2020). Extended producer responsibility for E-waste management: policy drivers and challenges. *Journal of Cleaner Production* 251: 119657.

Linder, M. and Williander, M. (2017). Circular business model innovation: inherent uncertainties. *Business Strategy and the Environment* 26 (2): 182–196.

McAllister, J. (2015). Factors influencing solid-waste management in the developing world. Master of Science thesis. Utah State University, Logan.

Nnorom, I.C. and Odeyingbo, O.A. (2020). Electronic waste management practices in Nigeria. In: *Handbook of Electronic Waste Management: International Best Practices and Case Studies* (eds. M.N.V. Prasad, M. Vithanage, and A. Borthakur), 323–354. Butterworth-Heinemann.

Nnorom, I.C. and Osibanjo, O. (2008a). Electronic waste (e-waste): material flows and management practices in Nigeria. *Waste Management* 28: 1472–1479.

Nnorom, I.C. and Osibanjo, O. (2008b). Overview of electronic waste (e-waste) management practices and legislations, and their poor applications in the developing countries. *Resources, Conservation and Recycling* 52 (6): 843–858.

Nowakowski, P. (2016). The influence of residents' behaviour on waste electrical and electronic equipment collection effectiveness. *Waste Management & Research* 34 (11): 1126–1135.

Nyam, T., Iliyasu, N., and Ohagoro, C.O. (2015). Electronic waste in Nigeria: prevalence and management in Kaduna Metropolis. *The Journal of Solid Waste Technology and Management* 41 (4): 403–411.

Odeyingbo, A.O., Nnorom, I.C., and Deubzer, O.K. (2019). Used and waste electronics flows into Nigeria: assessment of the quantities, types, sources, and functionality status. *Science of the Total Environment* 666: 103–113.

OECD (2001). *Extended Producer Responsibility: A Guidance Manual for Governments*. Paris: OECD.

Orisakwe, O.E., Frazzoli, C., Ilo, C.E., and Oritsemuelebi, B. (2020). Public health burden of E-waste in Africa. *Journal of Health & Pollution* 9 (22): 1–12.

Oyero, O., Oyesomi, K., Abioye, T. et al. (2018). Strategic communication for climate change awareness and behavioural change in Ota Local Government of Ogun State. *African Population Studies* 32 (1).

Parajuly, K. (2017). Circular economy in E-waste management: resource recovery and design for end-of-life. PhD thesis. University of Southern Denmark, Odense.

Parajuly, K., Fitzpatrick, C., Muldoon, O., and Kuehr, R. (2020). Behavioral change for the circular economy: a review with focus on electronic waste management in the EU. *Resources, Conservation and Recycling: X* 6: 100035.

Parvez, S.M., Jahan, F., Brune, M. et al. (2021). Health consequences of exposure to e-waste: an updated systematic review. *The Lancet Planetary Health* 5 (12): e905–e920.

Shahabuddin, M., Uddin, M.N., Chowdhury, J. et al. (2023). A review of the recent development, challenges, and opportunities of electronic waste (e-waste). *International Journal of Environmental Science and Technology* 20: 4513–4520.

Sharma, B.H., Vanapalli, K.R., Samal, B. et al. (2021). Circular economy approach in solid waste management system to achieve UN-SDGs: solutions for post-COVID recovery. *Science of the Total Environment* 800: 149605.

Sharma, M., Joshi, S., and Kumar, A. (2020). Assessing enablers of e-waste management in circular economy using DEMATEL method: an Indian perspective. *Environmental Science and Pollution Research* 27 (12): 13325–13338.

Shevchenko, T., Laitala, K., and Danko, Y. (2019). Understanding consumer E-waste recycling behavior: introducing a new economic incentive to increase the collection rates. *Sustainability* 11: 2656.

Shevchenko, T., Saidani, M., Danko, Y. et al. (2021). Towards a smart E-waste system utilizing supply chain participants and interactive online maps. *Recycling* 6: 8.

Srivastava, R.R. and Pathak, P. (2020). Policy issues for efficient management of E-waste in developing countries. In: *Handbook of Electronic Waste Management: International Best Practices and Case Studies* (eds. M.N.V. Prasad, M. Vithanage, and A. Borthakur), 81–99. Butterworth-Heinemann.

Sthiannopkao, S. and Wong, M.H. (2013). Handling e-waste in developed and developing countries: initiatives, practices, and consequences. *Science of the Total Environment* 463: 1147–1153.

Torres, D., Guzmán, S., Kuehr, R. et al. (2016). Sustainable management of waste electrical and electronic equipment in Latin America. Paper presented at the ITU, Secretariat of the Basel Convention, UNESCO, UNIDO, UNU, WIPO, BCRC-South America. ECLAC.

Tsydenova, O. and Bengtsson, M. (2011). Chemical hazards associated with treatment of waste electrical and electronic equipment. *Waste Management* 31 (1): 45–58.

Vishwakarma, S., Kumar, V., Arya, S. et al. (2022). E-waste in information and communication technology sector: existing scenario, management schemes and initiatives. *Environmental Technology & Innovation* 27: 102797.

Widmer, R., Oswald-Krapf, H., Sinha-Khetriwal, D. et al. (2005). Global perspectives on e-waste. *Environmental Impact Assessment Review* 25 (5): 436–458.

Woggsborg, A. and Schröder, P. (2018). Nigeria's E-waste management: extended producer responsibility and informal sector inclusion. *Journal of Waste Resources and Recycling* 1 (1): 1–9.

Xavier, L.H., Ottoni, M., and Lepawsky, J. (2021). Circular economy and e-waste management in the Americas: Brazilian and Canadian frameworks. *Journal of Cleaner Production* 297: 126570.

Yong, Y.S., Lim, Y.A., and Ilankoon, I.M.S.K. (2019). An analysis of electronic waste management strategies and recycling operations in Malaysia: challenges and future prospects. *Journal of Cleaner Production* 224: 151–166.

6

A Global Perspective on E-waste: From Cradle to Grave

*Sayan Mukherjee, Aniruddha Mukhopadhyay, and Pritha Bhattacharjee**

Department of Environmental Science, University of Calcutta, 35, Ballygunge Circular Road, Kolkata-700019, West Bengal, India
** Corresponding author*

6.1 Introduction to E-waste

E-waste comprises a varied range of electric circuit-based products that have a power supply. Electronic and electrical equipment (EEE) has become an integral part of the modern lifestyle. From domestic products like phones, laptops, and toys to health, transport, and security systems, every daily anthropogenic practice has been an important source of increasing contribution to global e-waste generation. Within a decade, global e-waste production has increased dramatically from 33.36 million metric tons (2010) to 57.4 million metric tons (2021). It is speculated further to reach 74.7 million metric tons globally, by 2030. Composed of various materials, e-waste when disposed of without proper management causes equal harm to environmental and human health.

The various types of EEE and e-waste can be grouped on the basis of various common factors like source, functions, components, etc. But for ease of analysis, they are classified into product-centric groups by the *E-waste Statistics Guidelines on Classification Reporting and Indicators–Second Edition* (Forti et al. 2018). These 54 categories are largely grouped into 6 categories for better management and handling. They are described in Table 6.1.

In this review, the main objective is to study the types of e-waste, their production, quantification, collection, and also recycling techniques. Herein, discussion is made on the impact on environment and biota, as a whole. These topics are discussed from a global perspective with an emphasis on the Indian scenario.

6.2 Production and Usage Trends

The extensive use of electronics has been increasing through the years. Irrespective of the place, whether households or offices and industries, e-waste has been piling up continuously. Different reports show various values in recorded e-waste production and their forecasts. However, all the reports emphasize the ever-increasing quantity of produced e-waste.

6.2.1 Sources and Disposal Practices

The sources of e-waste, its collection and disposal methods under e-waste management can be classified under four scenarios. Figure 6.1 explains the scenario and the outcomes of such practices.

6.2.1.1 Sources of E-waste

The main source of e-waste generation falls into two classes, domestic and industrial, although the types of e-waste generated from both sources are similar. They include:

1) Waste generated from data processing devices: viz. computer, monitor, speakers, keyboard.
2) Waste from devices for entertainment: viz. TV, DVD, CD, etc.
3) Device for communication: viz. phone, fax, etc.
4) Household equipment: viz. vacuum cleaner, microwave oven, washing machine, etc.
5) Audio visual components such as VCRs.

Electronic Waste Management: Policies, Processes, Technologies, and Impact, First Edition. Edited by Sunil Kumar and Vineet Kumar.

Table 6.1 Different categories of e-waste.

SL. No.	Category	Examples
1.	Temperature exchange equipment	Temperature regulating equipment: freezers, refrigerators, heat pumps, air conditioners, etc.
2.	Screens and monitors	TVs, laptops, computer screens, tablets
3.	Lamps	LED lamps, CFL lamps, tubelights, bulbs.
4.	Large equipment	Domestic and office use products: washing machines, dryers, electric stoves, copy and printing machines, etc.
5.	Small equipment	Household products: vacuum cleaners, microwave ovens, ventilation equipment, toasters, electric kettles, electric shavers, scales, calculators, radio sets, video cameras, electrical and electronic toys, small electrical and electronic tools, small medical devices, small monitoring, and control instruments.
6.	Small IT and telecommunication equipment	Mobile phones, Global Positioning System (GPS) devices, pocket calculators, routers, personal computers, printers, and telephones.

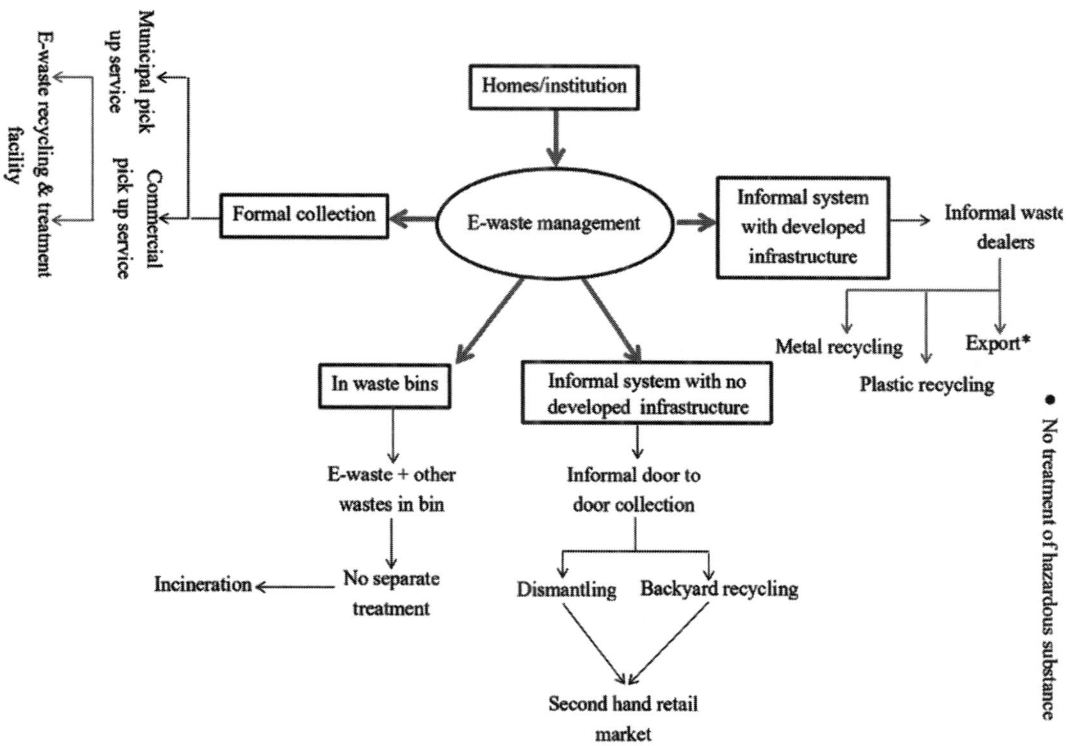

Figure 6.1 The outcomes of an e-waste disposal management system.

6.2.1.2 Disposal of E-waste

1) **Formal collection:** This process can be conducted by the designated organization, the government, and producers. Retailers, municipal collection points, and pick-up services are also involved in this process. After collection of the waste, it is disposed of in a specific site where the valuable materials got retrieved and hazardous wastes separated in a controlled way. Also, the residues will then go to controlled landfills and incineration.
2) **E-waste in waste bins:** In this process the user directly disposes of e-waste in a waste bin with other types of waste. As a result, the e-waste is treated as other household waste without retrieving valuable elements. Mainly incineration is used to treat the waste and then it is sent to landfill sites. This process is neither safe nor scientific, so this procedure is not recommended.
3) **E-waste collected outside of a formal system with a developed e-waste management infrastructure:** This process can occur in countries with developed laws of waste management. In this process individual dealers or companies collect the waste. Then the waste is traded and sent to its final destination. In this stage, metal recycling and plastic recycling are conducted. However, the waste is not depolluted and is not treated in a specialized recycling facility.

4) **E-waste collected outside of a formal system with no developed e-waste management infrastructure:** In most developing countries, self-employed people go door to door to buy the e-waste. Then they sell it to be repaired, refurbished, or dismantled. The waste is dismantled manually and converted to reusable parts. Sometimes recyclers burn, leach, and melt the e-waste to convert it to new materials. This process damages the environment severely.

6.2.2 Global Trends

According to the *Global E-waste Monitor 2020* report by United Nations University (UNU), the globally generated e-waste in 2019 was 53.6 metric tons (Mt), making the per capita production 7.3 kg. Only 17.4% of this waste was recycled worldwide. The contributions of different continents and countries vary due to the different management systems across the world, which can be seen in Table 6.2 and Figure 6.2.

Table 6.2 The scenario of e-waste generation (in weight) by different countries across all continents.

Sl no.	Region	Quantity of waste generated (kt)	Total (Mt)
	Africa		**2.9 Mt**
1.	Eastern Africa • Ethiopia • Kenya • Tanzania	55.2 kt 51.3 kt 50.2 kt	0.3 Mt
2.	Middle Africa • Angola • Cameroon • Congo	125.1 kt 26.4 kt 18.3 kt	0.2 Mt
3.	Northern Africa • Egypt • Algeria • Morocco	585.8 kt 308.6 kt 164.5 kt	1.3 Mt
4.	Southern Africa • South Africa • Botswana • Namibia	415.5 kt 18.8 kt 15.7 kt	0.5 Mt
5.	Western Africa • Nigeria • Ghana • Cote de Ivory	461.3 kt 52.9 kt 3. kt	0.6 Mt
	Americas		**13.1 Mt**
1.	Caribbean • Jamaica	18 kt	0.1 Mt
2.	North America • USA • Canada	6918 kt 757 kt	7.7 Mt
3.	Central America		1.5 Mt
4.	South America • Brazil • Argentina • Colombia	2143 kt 465 kt 318 kt	3.9 Mt
	Asia		**24.9 Mt**
1.	Western Asia • Turkey • Iraq	847 kt 278 kt	2.6 Mt
2.	Central Asia		0.2 Mt
3.	Southeastern Asia • Thailand • Philippines	621 kt 425 kt	3.5 Mt

(Continued)

Table 6.2 (Continued)

4.	Eastern Asia • China • Japan	10 129 kt 2569 kt	13.7 Mt
5.	Southern Asia • India • Pakistan	3230 kkt 433 kt	4.8 Mt
	Europe		**12 Mt**
1.	Eastern Europe • Russia • Ukraine	1613 kt 324 kt	3.2 Mt
2.	Northern Europe • UK • Sweden	1598 kt 208 kt	2.4 Mt
3.	Southern Europe • Italy • Spain	1063 kt 888 kt	2.5 Mt
4.	Western Europe • Germany • France	1607 kt 1362 kt	4 Mt
	Australia		**0.7 Mt**
1.	Australia and New Zealand • Australia • New Zealand	554 kt 96 kt	0.7 Mt
2.	Melanesia	–	0.02 Mt
3.	Micronesia	–	0.0005 Mt
4.	Polynesia	–	0.01 Mt

Source: Global E-waste Monitor, 2020.

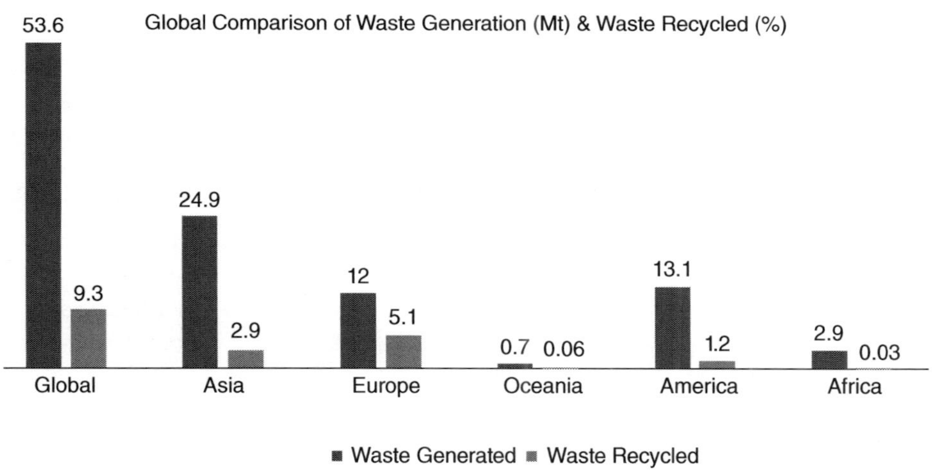

Figure 6.2 E-waste generation (Mt) and percentage of recycling by different continents.

6.2.2.1 Asia

In 2019, e-waste generation in Asia reached 24.9 Mt with a per capita value of 5.6 kg. Of this enormous amount, only 11.7% or 2.9 Mt of the generated waste was documented to have been collected and recycled properly. Only 17 countries of the entire continent have a national e-waste policy or regulation implemented. The region of Eastern Asia contributes the highest toward waste generation (around 13.7 Mt) followed by Southern Asia (4.8 Mt), Southeast Asia (3.5 Mt), Western Asia (2.6 Mt), and Central Asia (0.2 Mt).The e-waste management systems prevalent in different parts of Asia are quite diverse. Developed countries like Japan, South Korea, China, and Taiwan have implemented very advanced management systems. Informal activities mostly dominate in the larger part of Asian, unlike in China where it coexists with advanced management systems. The informal activities system involves collection, dismantling, and recycling of e-waste in the South Asian sub-region. China as the world's most populous country has a high demand for electrical and electronic equipment, and thus has a large EEE manufacturing industry. Hence the statistics show China to be the world's largest e-waste producer, generating 10.1 Mt of e-waste in 2019. China has focused on refurbishment, reuse, and recycling of e-waste, which has been fruitful as nearly 70 million items of e-waste are dismantled annually. The collection and recycling rate of all United Nations (UN)-classified 54 EEE products is 15%. However, due to the imposition of strict environmental laws there has been a decline in the informal sector. China has also banned the illegal importation of e-waste and has set targets of sourcing 20% of raw materials from recycling and recycling 50% of e-waste by 2025. Under Taiwan's Recycling Fund Management Board, the implemented 4-in-1 recycling system resulted in producing a recycling rate of 64% in 2018. Taiwan hosts around 20 e-waste recycling facilities with capacities larger than their required demand. Japan, through its advanced collection system and developed processing infrastructure under strong jurisdiction, has successfully collected 570.3 kilotons (Kt) officially in 2016.

The subregion of central Asia generated around 0.2 Mt of e-waste with a per capita generation of 7.1 kg. Kazakhstan dominates the region, producing 172 Kt of e-waste followed by Turkmenistan and Kyrgystan (39 and 10 Kt, respectively). Piling in landfills and illegal dumping sites are common in this region. The discarded EEE products from consumers are handed over to small companies responsible for dismantling and reuse of the components. In Kazakhstan, the setup of collection and recycling sites is inadequate to meet the demands. However, Uzbekistan, though having no regulatory measures regarding e-waste management, improved its municipal waste infrastructure in 2014–2016 and in 2017 launched a five-year program to improve its waste management system.

Western Asia generated 2.6 Mt of e-waste with 9.6 kg per capita, and Turkey, Saudi Arabia, and Iraq contributed around 847, 595, and 278 Kt, respectively, of e-waste in 2019 as the highest contributors. The management system is mostly informal. In the richer countries, the migrant workers play a major role in the reuse and repair sector.

6.2.2.2 Europe

In 2019, Europe generated 12 Mt of e-waste. The per capita waste generated in the continent was 16.2 kg. With around 37 countries having national legislation on policies, the continent has a high rate of documented recycling of 42.5% on 5.1 Mt of e-waste. The EU, having a well-developed infrastructure of environmentally sound e-waste collection and recycling by privately operated shops and municipalities, has shown a high rate of documented collection and recycling of e-waste in Northern and Western Europe (59% and 54%, respectively) in 2017. However, a tendency to export e-waste along with metal scraps has also been seen, consisting of EEE from laptops and other IT sectors sent to African countries. Countries like the Netherlands have reduced their share of e-waste from 11% to 9% since the 2010s (Balde et al. 2020).

Belarus is noted to have a system of advanced collection and recycling facilities, which led them to collect around 23 Kt of e-waste in 2019. Multiple municipal and private drop-off and collection points are key factors in this system. Ukraine and Russia generated 324 Kt and 1631 Kt of e-waste, respectively, in 2019, and also have implemented environment-friendly collection systems in this region.

6.2.2.3 Oceania

Oceania, comprised of Australia, New Zealand, Papua New Guinea, Fiji, and other island nations, collectively generated 0.7 Mt of e-waste with 16.1 kg per capita waste generation, of which 8.8% or 0.06 Mt of waste was recycled. Australia (554 Kt) and New Zealand (96 Kt) collectively generated around 650 Kt of e-waste. The established four co-regulatory arrangements manage the overall waste management in Australia. Under this management, as reported in 2019, around 291 Kt of TV and computer waste was collected and recycled. According to the 2017–2018 report 58 Kt of e-waste was recycled, with a recovery rate greater than 93%. Compared to Australia, New Zealand faces hurdles in waste management due to lack of an overall formalized system (Blake et al. 2019). More than 97 Kt of the generated e-waste is disposed of in the landfills, of which more than 98.2% of e-waste is household generated, since most of the waste does not go through mandatory recycling and treatment. The Pacific Island Region (PICTs) consists of 22 countries and territories. These islands have resorted to a coordinated system of e-waste management and pollution control strategy since due to its remoteness, expanse, and limited lands across the small islands and atolls, individual large-scale e-waste management facilities cannot be implemented.

6.2.2.4 North and South America

In 2019, the United States of America generated around 6918 Kt of e-waste, the highest across the entire continent of North America. The entire e-waste management system is upheld at the federal level, which certifies various independent recycling facilities under their two standard systems: Responsible Recycling (R_2) and Stewards Standards.

Latin American countries have focused more on the development of local recyclers. By 2019, under the R_2 certification, countries like Mexico have set up more than 15 companies. A similar rise in the steady collection has been observed in countries having established legal frameworks for collection, like Peru and Colombia.

Regulations and legal frameworks have resulted in better collection systems of diverse e-waste and subsequently have given rise to the growth of recyclers in these regions. However, the Latin American region needs to focus more on its R&D and to update the policies with time and bring awareness to consumers.

6.2.2.5 Africa

Africa's e-waste management system thrives mostly on the shoulders of the informal sectors where the government has very limited control. Lack of general awareness, government policies, and infrastructure along with poor financing can be observed all over the continent. A few countries like South Africa, Egypt, Morocco, Namibia, etc. have limited recycling facilities, which co-exist with the large informal sectors. Large nations like Nigeria annually import EEE from countries like Germany, Belgium, the UK, etc.

The overall scenario in Africa shows the lack of public awareness, government policies, legislation, and management systems. The informal system remains the dominant power in controlling the recycling system, which is often mismanaged, ill-equipped, and highly polluting.

6.2.3 Management System in India

The second most populous country in the world and the seventh-largest country, India, in 2019 generated 3230 Kt of e-waste, the highest in the Southern Asia sub-region and the second-largest generator in the Asian continent after China. India already has implemented proper legislation under which formal recycling facilities have been established. By 2019 the number of certified recyclers in India had reached around 312, having the capacity to recycle around 200 Kt of e-waste annually. However, most of the waste handling is conducted by the informal sector. Improper enforcement of legislation, inadequate infrastructure, and improper collection and disposal methods remain an enormous hurdle toward successful e-waste management.

6.3 Composition

In general, the most common constituents of e-waste are metals, metal oxides/glass fibers, and plastics. Part of these materials can be recovered in certain amounts and thereby can also be a part of the waste-to-wealth stream. However, the composition of e-waste does not stay the same for all but changes depending on its type, age, origin, and manufacturer. Metals in e-waste generally contain gold, iron, cadmium, lead, copper, aluminum, etc. Regarding the gold content in e-waste, it is said that one million mobile phones would generate 24 kg of gold, 250 kg of silver, 9 kg of palladium and 9000 kg of copper. According to the Solving the E-waste Problem (StEP) Annual Report (2015/2016), the total global gold content of Waste Electrical and Electronic Equipment (WEEE) inventory was said to be 300 tons in 2014, valued at 10.4 billion euros. On the other hand, the plastic content of e-waste contains polyesters and phenol formaldehyde (C–H–O polymers), halogenated polymers (polyvinyl chloride (PVC), polytetrafluoroethylene (PTFE), polybromo compounds), N-containing polymers (nylon, polyurethanes), etc. All the plastic contents are grouped into four major classes, i.e. acrylonitrile-butadiene styrene (ABS), high-impact polystyrene (HIP), polycarbonate (PC), and polypropylene (PP). Figure 6.3 sums up a general overview of the contents of e-waste along with their average weight compositions.

6.4 Consequences

Most of the components of e-waste are hazardous to the environment at large. It not only affects environmental health but also the health of the flora and fauna of the ecosystem in different ways (refer to Table 6.3).

1) **Effect on environment:** When e-waste starts to disintegrate, it forms other compounds. Most of them are hazardous heavy metals and other organic compounds. These toxins leach through the soil layer and contaminate groundwater. All over the world, soil is exposed to e-waste and the amount of waste is increasing day by day. According to Dharini et al. (2017), the presence of toxic material can induce the alkaline nature of soil. Assessment of the effects of Pb, Cu, Cr, Ni, and Cd were done in e-waste contaminated areas by Adesokan et al. (2016). Polycyclic Aromatic Hydrocarbons (PAH) were also found in the contaminated places. During the shredding, extracting metals, and dismantling process the heavy metal particles leached through the soil. These materials affect soil composition. In China, a study was conducted using samples from a paddy field, a desolate site, and an acid leaching site. The results revealed that the acid leaching site has a heavy load of Ni, Cu, Zn, Sn, Sb, Pb, and Cd. In the paddy field the main contamination was found at the surface and in mid-layer soil. On the other hand, the desolate site had most of the contaminated soil at the deep bottom (Quan et al. 2015).

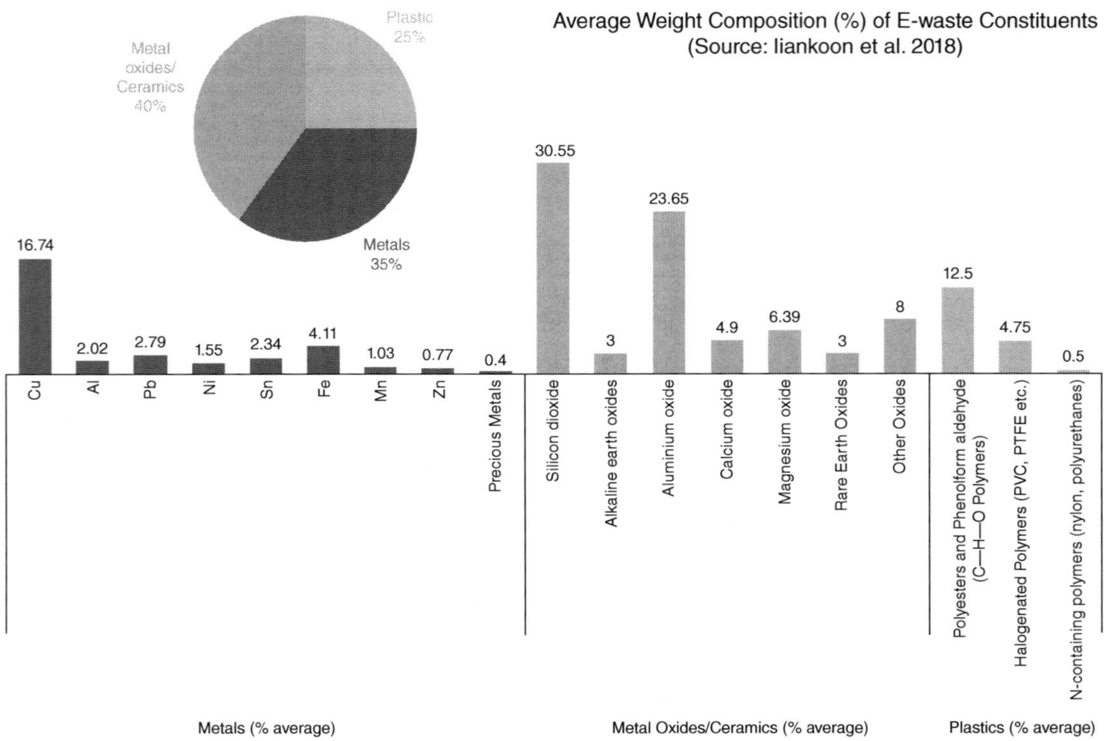

Figure 6.3 Average weight composition (%) of e-waste components.

Table 6.3 Impact on human health by different components of e-waste.

E-waste pollutants	E-waste sources		Health impact on humans
Heavy metals	Cd	Semiconductors, resistor chips, batteries, CRT screens, IR detectors, pigments	Bone disease, CNS damage, kidney and liver failure, teratogenic (affects embryo and fetus)
	Pb	Solder in circuit boards, gasket computer monitors, lead batteries, glass panels, CRT screens, solar, transistors	Renal failure,CNS and PNS damage, affects reproductive system, affects circulatory system, affects brain development in minors
	Hg	Hg wetted relays and switches, printed circuit boards, alkaline batteries, fluorescent lamps, thermostats, sensors, medical equipment, mobile phones	Respiratory and skin problem, brain damage, affects liver, kidney and fetus, CNS and PNS damage
	Cr	Untreated galvanized (corrosion protector) steel plate, floppy disks, data tapes, decorative hardener, dyes and pigments, solar cells	Bronchitis and asthma, complications in the liver, kidney, and gastrointestinal tract, carcinogenic, DNA destruction
	As	LED (as gallium arsenide), circuit boards, LCD displays and computer chips, solar cells, microwaves	Irritation in digestive system, carcinogenic (skin and liver)

(Continued)

Table 6.3 (Continued)

E-waste pollutants	E-waste sources		Health impact on humans
Other hazardous metals	Li	Mobile phones, lithium-ion batteries, photographic equipment	Present in breast milk and harm infants, cause irregular heart rhythms, kidney problems, and thyroid problems
	Ni	Ni-Cd batteries, CRT electron guns, alloys, relays, semiconductors, pigments	Dermatitis, eczema, asthma, chronic bronchitis, cancer of the lung and nasal sinus, allergic
	Cu	Electric wires and cables, printed circuit boards, pigments	Irritation, stomach cramps, nausea, liver damage (Wilson's disease), kidney damage
	Be	Motherboards, power supply boxes (containing Si controlled rectifiers), x-ray lens, finger clips	Respiratory disease, affects liver, kidneys, heart, nervous system, and the lymphatic system, skin sensitization
	Ag	Capacitors, switches, batteries, resistors	Argiria (skin pigmentation), skin irritation, drowsiness, kidney damage, eye damage, lung damage, liver damage, anemia, brain damage if taken in excessive amounts
	Se	Pigments, fax machines, photocopiers, photoelectric cells	Higher concentration (fivefold or more) leads to selenosis, deformation of nails and hairs, can impair lung tissues, and cause various commotion in regular functioning of body organs
	Zn	Alloys (brass and steel), luminous materials, disposed batteries and rechargeable batteries	Burning skin, lowers blood pressure in diabetic patients, poses problem with blood Fe at higher concentrations
	Co	Insulators	Thyroid and myocardial ailments, respiratory tract and skin inflammations, hyperplasia and polycythemia in bone marrow
Plastics			
Organic pollutants	VOCs	Paints, photocopiers, and printers	Detrimental for kidneys, liver, and CNS, carcinogenic, memory, and visual disorders
	PAHs	Contaminated soil and air near e-waste burning sites (incineration of polymers)	Irritation in eyes, throat, etc. as short-term effects; long-term effects involve cataracts, renal and liver mutilation
	PBDEs	Recycling e-waste components, PBDE treated electronic equipment, viz. computers, televisions, etc.	Neurobehavioral deviations, dam-age to reproductive and immune system, liver, and thyroid
	PCBs	Transformers (used as coolant), capacitors, and electrical gears, microscopes, fluorescent lighting	Cause liver, gallbladder, gastrointestinal, breast, brain cancer, etc., chloracne (severe skin acne)

Source: Panpatte et al. (2021).

2) **Effect on plants:** Toxins generated from the e-waste can percolate through soil and bio-accumulate in plants. According to Dharini et al. (2017), the increase of Ca in water and soil can affect the growth of plants by affecting cell wall structure. Also, the addition of lead (Pb) can decrease the concentration of available nitrogen, phosphorus, and potassium in soil. Thus, it is making the soil sterile and inhibiting plant growth.

   A study by Mudila et al. shows the concentration of As, Cu, and Cd to have a constant trend; on the other hand, Pb is observed to be negative. Heavy metals accumulated and translocated in plants due to the formation of chelates and organic compounds bind to the lipid in xylem or in the vacuoles of root cells (Mudila et al. 2021).
3) **Effect on humans:** Incineration of circuit boards, cables, etc. is an effective method for extracting precious metals, but it produces hazardous cancerous gases. On the other hand, dismantling of equipment can cause serious injury to workers. Toxins from e-waste show effects like ulcers, skin damage, etc. Children are more vulnerable to this substance than adults (Grant et al. 2013; Zeng et al. 2016). Many toxic materials derived from e-waste are entering the human digestive system with food and causing physiological, neurological, and behavioral changes.

Handling of e-waste is not so easy because of the complex method of disposal. It causes the accumulation of toxic substances in the ecosystem. In an e-waste recycling area, five heavy metals, Cd, Pb, Cu, Zn, and Ni, are found in the hair follicles of people living in that area. It is an indication of increasing risk and more stress toward the health of the particular community.

Children living near disposal areas are also exposed to these heavy metals, and a study shows the presence of heavy metals in their body. It was found that the concentration of seven antibodies, viz. diptheria-tetanus-pertussis (DPT), hepatitis B, Japanese encephalitis, measles, and polio are negatively correlated with the blood concentration of Cu, Zn, and Pb (Lin et al. 2017).

6.5 Legislation

To combat the rise of waste in every part of the world, different countries and their governments develop policies, laws, and acts and enforce them at a national, municipal, or local level. These policies and laws are generally non-binding; however, they do act as guidelines for regulatory bodies to enforce the legislation. These are meant to provide achievable targets and need proper enforcement and financial backing, something that we see a lack of in most countries.

According to the *Global E-waste Monitor 2020* report, the number of countries having policies or regulations at a national level has increased from 67 in 2017 to 78 in 2019. These policies include all the mandatory as well as non-binding policies. These policies cover 71% of the global population in 2019, 5% more than in the 2017 valuation (see Figure 6.4).

In the matters of e-waste legislation at an international level, the induction of the Basel Convention on the control of trans-boundary movements of hazardous wastes and their disposal, effective from 1992, is a landmark treaty that was targeted to control and suppress the exports of hazardous wastes mostly from the developed countries to developing countries for waste management. To date, 187 countries have signed this treaty. The check on the trans-boundary movement paved the way for alternative methods like a circular economy, rather than just recycling, treatment, and disposal. This also extends the life cycle of the EEE products and decreases the possibility of hazardous waste discharge (Forti et al. 2020).

The existing legislation of a few countries across the globe are discussed below as reported in the *Global E-waste Monitor 2020* by UNU:

6.5.1 Asia

- The **South Asian** region has begun to recognize the importance of proper e-waste management. India is the only country in Southern Asia with e-waste legislation, although several other countries are considering such legislation. In India, laws to manage e-waste have been in place since 2011, mandating that only authorized dismantlers and recyclers can collect e-waste. A manufacturer, dealer, refurbisher, and Producer Responsibility Organization (PRO) were brought under the ambit of the E-Waste (Management) Rules 2016. The National Resources Policy (still in draft at the time of publishing) also envisages a strong role for producers in the context of recovering secondary resources from e-waste.

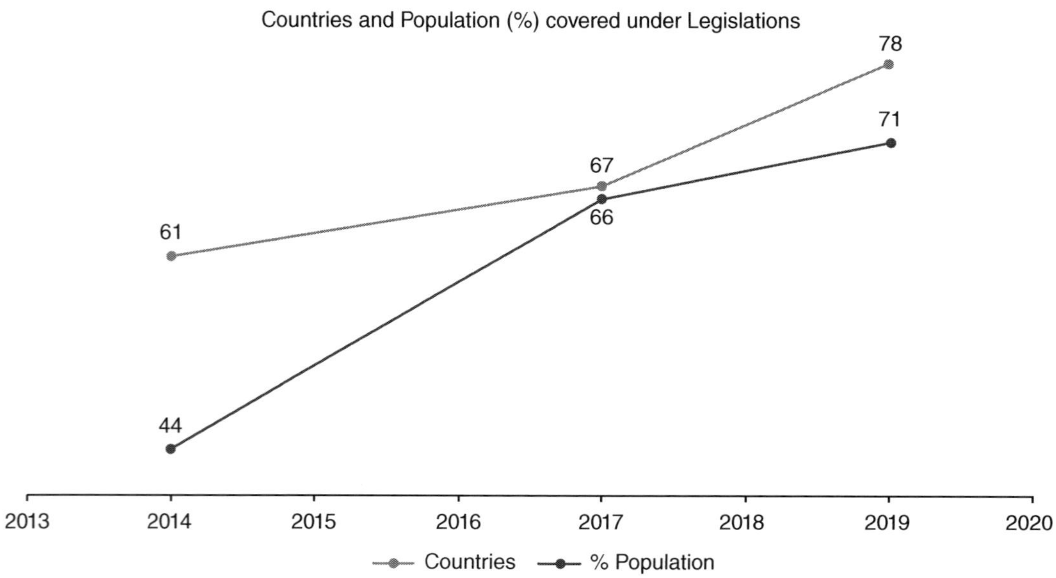

Figure 6.4 Number of countries and percentage of the global population covered by legislation across time.

- In **Southeast Asia**, some countries are more advanced. The Philippines does not have regulation specifically for e-waste management, but it does have a range of hazardous waste regulations that cover e-waste as it is considered hazardous waste. Cambodia has a specific law relating to e-waste management with the 2016 Sub-decree on Electrical and Electronic Equipment Waste Management (E-waste Management). Myanmar has recognized the importance of hazardous waste management and is currently working toward a Master Plan and guidelines for it. China has national legislation in force that regulates the collection and treatment of fourteen types of e-waste (i.e. five types, initially, and nine more were later added). Japan was one of the first countries globally to implement an EPR-based system for e-waste.
- In **Western and Central Asia**, e-waste legislation advances are still very poor. There is some formalized legislation regarding mercury-containing lamps. However, for types of e-waste, collection, legislation, and e-waste management infrastructure is mostly absent.

6.5.2 Europe

- In Europe, the majority of e-waste is regulated by the WEEE Directive. This regulation is in force in the European Union and in Norway. Other countries—including Iceland, Switzerland, and several Balkan countries, such as Serbia, Bosnia, and Herzegovina—have similar laws.
- Preparation for reuse is defined as the weight of whole appliances that have become waste and of components of e-waste that, following checking, cleaning, or repairing operations, can be reused without any further sorting or pre-processing. It also contains a decision on the registration of e-waste exports. Where e-waste is sent for treatment in another member state or exported for treatment in a third country in accordance with Article 10 of Directive 2012/19/EU, only the member state that has collected and sent or exported the e-waste for treatment may count it toward the minimum recovery targets referred to.
- In Ukraine, an EPR system based on the EU WEEE Directive is in development, by the association agreement between the EU and Ukraine.
- Recently, the two-year project "Implementation of Management System for Waste of Electric and Electronic Equipment and Batteries in Ukraine" has been completed. This project helped develop two laws: the Bill on Batteries and Accumulators and the Bill on Waste of Electric and Electronic Equipment, which is expected to pass Parliament in 2020.

6.5.3 Oceania

- The National Television and Computer Recycling Scheme (NTCRS) was implemented in Australia under the Australian Government's Product Stewardship Act 2011. This regulation provides Australian householders and small businesses with access to industry-funded collection and recycling services for televisions and computers.
- The plan provides approximately 98% of the Australian population with reasonable access to collection services. These services may include a permanent collection site at a local waste transfer station or retail outlet, or at one-off events. Through this, Australia plans to increase its recycling potential up to 80% by 2025–2026.

6.5.4 Americas

North America:

- **United States of America:**
  - The United States of America does not have national legislation on the management of e-waste, but 25 states and the District of Columbia have enacted some form of legislation.
  - Apart from California and Utah, all states that have implemented laws use an EPR approach.
- **Canada:**
  - Though there is no national legislation in effect on the management of e-waste, 12 provinces and territories have regulations in place with industry-managed programs.

South and Central America:

- E-waste regulations in Latin America are limited to a few countries. Apart from Mexico, Costa Rica, Colombia, and Peru are the leading forces in the region for environmentally sound e-waste management.
- Mexico is planning on reviewing the current regulation after its first five-year term and has been expanding several studies in order to redefine the responsibilities of involved stakeholders, establish clearly defined categories, and set mandatory collection targets, thereby increasing collected and formally recycled volumes.
- Costa Rica has been focusing on improving the implementation of the current regulation.

6.5.5 Africa

E-waste problems in African nations have been a serious issue in the past but some improvements have been observed in the legal, institutional, and infrastructural framework to tackle the problems. Some specific examples are illustrated here:

- **Ghana:** Technical Guidelines on Environmentally Sound E-waste Management for Collectors, Collection Centers, Transporters, Treatment Facilities, and Final Disposal are currently being enforced.
- **Nigeria:** A Producer Responsibility Organization (PRO) named E-waste Producer Responsibility Organisation of Nigeria (EPRON) was set up by electrical and electronic producers in March 2018.
- **Rwanda:** For the adoption of e-waste regulation, the strategy prioritizes (a) strengthening the policy, legal, and regulatory framework for sustainable resourcing of e-waste management, (b) putting in place the requisite e-waste management infrastructure, (c) establishing mechanisms for comprehensive and sustainable mobilization for e-waste management resources, (d) strengthening e-waste coordination structures at regional and national levels, and (e) promoting research and innovation in e-waste management.

6.5.6 Indian Scenario

E-waste management has been a serious issue in the Indian scenario. Before the enactment of E-Waste (Management and Handling) Rules, 2011 (effective from 1 May 2012), it was managed under Hazardous Waste Management (HWM) Rules. With such a scenario, the concept of Extended Producer Responsibility (EPR) became effective in India. After 2011, the act has been amended two times, first in 2016 and then in 2018. With the amendment of 2016 (effective from 1 October 2016) manufacturers, dealers, and the Producer Responsibility Organization (PRO) were also brought under the ambit of the HWM Rules. PRO is a professional organization taking responsibility for the collection and channelization of e-waste generated from their products. Furthermore, the applicability of the Rules was expanded to cover components, consumables, parts, and spares of EEE in addition to the equipment covered under the Rules.

The latest amendment of the act was in 2018, with the major aim of channelizing the e-waste generated in the country toward authorized dismantlers and recyclers in order to further formalize the e-waste recycling sector. As per the revised targets of e-waste collection, 10% of the quantity of waste generated was to be collected during 2017–2018. As per the 2018 amendment, the PROs have to apply to the Central Pollution Control Board (CPCB) for registration, which further strengthens the supervising power of CPCB.

For effective implementation of the E-Waste Rules, the producers have to take responsibility for their products. A synchronized endeavor by the industries, the pollution control boards, and local authorities is crucial to cope with the problem of e-waste in India. The relaxation in the EPR Plan by reduction of the e-waste collection targets and overture of realistic collection targets will result in increased observance of the E-Waste Rules by organizations. Also, the requirement of PROs to register with the Central Pollution Control Board (CPCB) would aid in the creation of an organized and legitimate industry and would pave the way for effective and improved e-waste management in India.

6.6 Treatment and Bioremediation

Treatment of waste by any means must be cost-effective and eco-friendly to prevent the discharge of hazardous elements in the environment, since improper handling of such waste can have a tremendous negative impact on human health (Murthy and Ramakrishna 2022). Various methods are used worldwide to treat e-waste before its disposal. Some of these are as follows:

1) **Physical method:** By employing various physical methods, essentials and valuable metals can be recovered from e-waste. This method is used as it is simple, easy, convenient, and with low processing cost. A few of the methods are discussed here.
   a) **Mechanical recycling:** In this method dismantling of instruments is involved and a magnet is used to separate the metallic parts. After that, the powdered sample is passed through a density separator (Zhang and Forssberg 1999).
   b) **Air classification:** This method is used to separate materials based on their weight. By this method, Al and Cu can be separated effectively (Laurmaa et al. 2011).
   c) **Magnetic separation:** By this method, magnetic and non-magnetic e-waste components can be separated. One of the major drawbacks of this process is some non-ferrous materials also get separated with ferrous materials (Sohaili et al. 2012).
   d) **Electrostatic separation:** This method is based on the electrical conductivity of materials. According to Guo et al. (2008), this method is capable of recovering 2.8% Al from the printed circuit boards.
2) **Chemical method:** The chemical method includes pyrolysis, gasification, hydrometallurgy, de-polymerization, etc. These are effective processes to remove useful components from e-waste.
   a) **Pyrolysis:** This is a process to heat a material in an anaerobic condition. It helps to recover purified metals from waste. In this process organic compounds are converted to lower molecular mass, but

materials with glass remain solid.

b) **Hydrometallurgy:** In this process e-waste is treated with acids to leach out the materials of interest. The next step of this treatment is separation and purification of the treated substance. The liquids generated in this process should be disposed of in a proper manner. Various liquid components are generated during this process, viz. H_2SO_4, HCl: HNO_3, HNO_3, and cyanides. To recover the metals, leach out to non-metallic substrate samples are prepared after crushing, pulverizing, and decontaminating waste materials. After that, they are treated with HNO_3. HNO_3 is used to recover Cu and Ag. To recover Au, the remains should be treated with aqua regia. Ammonium and sodium thiosulfate are used to recover gold.

3) **Bioremediation:** Microbes are capable of growing in diverse conditions including some adverse environments. For this property, they are used to degrade various types of pollutants, a process called bioremediation. The main purpose of this process is to convert the waste materials to biomass or energy. Bioremediation involves various processes, viz. degradation, sequestration, modification/alternation decontamination, or removal of pollutants. Bioremediators involve a number of enzymes, which include isomerase, hydrolase, transferase, ligase, etc. that allow the biological waste to nutrient for the growth of biomass (Abatenh et al. 2017; Singh 2014).

The effectiveness of bioremediation depends on several factors, viz. the nature of the organism, the nature of utilization, and the environmental scenario of the pollutant area. Bioremediation can be achieved using microorganisms. These organisms accumulate heavy metals and change them to harmless materials by a redox reaction (Jan et al. 2014). Bioremediation can be done using plants, known as phytoremediation. This process can be divided into in situ and ex situ. In situ remediation is done onsite by plant. On the other hand, for ex situ remediation, the process is to take contaminated soil out of its original site to a different location for treatment.

Microorganisms are very important components for the bioremediation of heavy metals as they have the capability to convert hazardous and toxic material to non-hazardous substances. According to a study conducted by Wang and Chen (2009), a bacterial mixture had greater efficiency of remediation than a single-strain culture. For microbial biodegradation, the following mechanisms are used by microbes:

1) Uptake of toxic metals by cell wall components or by metalloproteins and phytochelatins with other compounds, viz. catecholates.
2) Alternation of the biochemical pathway to block the uptake of metal.
3) Conversion of metals to non-toxic forms by enzymes.
4) Reduction of intracellular metal concentration by using efflux systems.

Bacteria produce iron-chelating substances to enhance the mobility and reduce bioavailability of metals. Sulfate-reducing bacteria convert sulfate to hydrogen sulfate, which then reacts with heavy metals to form metal sulfides (Chibuike and Obiora 2014).

The usage of microbes for different heavy metal removal is described in Table 6.4.

Table 6.4 List of microbes used for different heavy metal removal.

Types of microbes	Name of microbes	Heavy metal removed	References
Bacteria	*Bacillus cereus* strain XMCr-6	Cr (VI)	Dong et al. 2013
	Kocuria flava	Cu	Coelho et al. 2015
	Bacillus cereus	Cr (VI)	Coelho et al. 2015; Kanmani et al. 2012
	Sporosarcina ginsengisoli	As (III)	Coelho et al. 2015; Achal et al. 2012
	Pseudomonas veronii	Cd, Zn, Cu	Coelho et al. 2015; Vullo et al. 2008
	Pseudomonas putida	Cr (VI)	Balamurugan et al. 2014
	Enterobacter cloacae B2-DHA	Cr (VI)	Rahman et al. 2015
	Bacillus subtilis	Cr (VI)	Balamurugan et al. 2014
Fungi	*Aspergillus versicolor*	Ni, Cu	Coelho et al. 2015; Taştan et al. 2010
	Aspergillus fumigatus	Pb	Kumar Ramasamy et al. 2011
	Gloeophyllum sepiarium	Cr (VI)	Achal et al. 2011
	Rhizopus oryzae (MPRO)	Cr (VI)	Sukumar 2010
	Saccharomyces cerevisiae	Pb, Cd	Farhan and Khadom 2015; Livia and Benedito 2015

(Continued)

Table 6.4 (Continued)

Types of microbes	Name of microbes	Heavy metal removed	References
Algae	*Spirogyra* spp. and *Cladophora* spp.	Pb (II), Cu (II)	Coelho et al. 2015; Lee and Chang 2011
	Spirogyra spp. and *Spirullina* spp.	Cr Cu, Fe, Mn, Zn	Coelho et al. 2015; Mane and Bhosle 2012
	Hydrodictylon, Oedogonium, and Rhizoclonium spp.	As	Coelho et al. 2015

6.7 Circular Economy

Rising demands of industry-produced products require an enormous amount of different metals as raw materials. To meet such demands, mining of such elements is necessary, which happens even at the cost of depleting the Earth's natural resources. The cost of continuous production of virgin raw materials is also very expensive. This calls for alternative methods to meet the needs. The produced e-waste is known to contain around 69 elements, which includes precious metals like Au, Ag, Pt, Pd, Rh, etc., critical raw materials like Co, In, Ge, etc., and non-critical metals like Al and Fe. These heaps of e-waste generated can be perceived as mines for secondary raw materials, thus reducing the pressure on virgin materials.

The recycling process is often quite expensive since recovery of some metals is difficult due to their relatively low concentration or low demand in the market. Economically viable methods involve manual separation and recycling, obtaining material loss less than 5% (Deubzer et al. 2007). This process is suitable in the cases of precious metals with high concentration. For other Critical Raw Materials (CRMs), improved collection and treatment remains to be the only answer.

The 2019 global recycling rate, when taken into account, amounts to US$10 billion worth of raw materials recovered from e-waste along with 4 Mt of secondary raw materials for recycling. This also in a way potentially saved 15 Mt of CO_2-equivalent emissions in 2019. Metals like Al, Cu, and Fe constitute the majority of the elements in e-waste. In an ideal scenario where the entire globally generated e-waste is recycled, one would reproduce 25 Mt of these metals, 14 Mt short of the market demand for 2019. This demand gap is a result of the ever increasing demand for EEE. Hazardous elements like Hg, chlorofluorocarbons (CFCs), hydrochlorofluorocarbons (HCFCs), etc. are extremely harmful to the environment. Improper disposal in landfills and open dumpsites of Hg-laden EEE products often leads to leaching and groundwater contamination. This contamination enters the food chain, bioaccumulates, and damages the central nervous system (CNS), lungs, kidneys, and other vital organs (Balde et al. 2018).

Though the informal collection and noncompliant recycling system seems to be cheaper than the formal legally compliant system, the latter is much more eco-friendly in nature. This is due to the technical costs incurred in the different technical procedures. However, the result not only eradicates the issues of pollution but also helps turn waste into wealth.

6.8 Conclusion

The world in the current era heavily depends on the mining of natural resources to fulfill the needs for development. However, these demands have already started to put stress on the Earth's resource carrying capacity. With industrialisation and the growth of civilization, lifestyle choices have also changed drastically, which have also resulted in an increase in waste generation from households and industries. Dependence on EEE has been prominent and so is the rise of e-waste. The hazardous substances in e-waste are a threat to the environmental health and the health of the flora and fauna that come in contact with them. Most of the countries around the globe have resorted to some kind of policy or legislation to combat and manage the ever increasing generation of e-waste. However, practices like the export of e-waste have put a lot of developing and underdeveloped countries in an ill-fated position. International treaties and policies like the Basel Convention stand as guards to stop such mismanagement practices. The ideas of the 3Rs, remediation, and the circular economy are the only ways to mitigate this situation. Countries should take into account these ideas while creating policies, laws, and legislation. However, these will be successful and beneficial only if they are implemented properly.

References

Abatenh, E., Gizaw, B., Tsegaye, Z., and Wassie, M. (2017). The role of microorganisms in bioremediation—A review. *Open Journal of Environmental Biology* 2 (1): 038–046.

Achal, V., Kumari, D., and Pan, X. (2011). Bioremediation of chromium contaminated soil by a brown-rot fungus, *Gloeophyllum sepiarium. Research Journal of Microbiology* 6 (2): 166.

Achal, V., Pan, X., Fu, Q., and Zhang, D. (2012). Biomineralization based remediation of As (III) contaminated soil by *Sporosarcina ginsengisoli. Journal of Hazardous Materials* 201: 178–184.

Adesokan, M.D., Adie, G.U., and Osibanjo, O. (2016). Soil pollution by toxic metals near e-waste recycling operations in Ibadan, Nigeria. *Journal of Health and Pollution* 6 (11): 26–33.

Balamurugan, D., Udayasooriyan, C., and Kamaladevi, B. (2014). Chromium (VI) reduction by *Pseudomonas putida* and *Bacillus subtilis* isolated from contaminated soils. *International Journal of Environmental Sciences* 5 (3): 522–529.

Balde, C.P., van den Brink, S., Forti, V. et al. (2020). The Dutch WEEE Flows 2020: what happened between 2010 and 2018. United Nations Institute for Training and Research.

Baldé, K., D'Angelo, E., Forti, V. et al. (2018). Waste mercury perspective: 2010–2035 from global to regional. Vienna: United Nations University.

Blake, V., Farrelly, T., and Hannon, J. (2019). Is voluntary product stewardship for e-waste working in New Zealand? A Whangarei case study. *Sustainability* 11 (11): 3063.

Chibuike, G.U. and Obiora, S.C. (2014). Heavy metal polluted soils: effect on plants and bioremediation methods. *Applied and Environmental Soil Science* 2014.

Coelho, L.M., Rezende, H.C., Coelho, L.M. et al. (2015). Bioremediation of polluted waters using microorganisms. *Advances in Bioremediation of Wastewater and Polluted Soil* https://doi.org/10.5772/60770.

Deubzer, O., Griese, H., & Reich, H. (2007). Global impacts of the e-waste problem. Paper presented at International Symposium on Environmentally Conscious Design and Inverse Manufacturing (EcoDesign) 2007, Tokyo.

Dharini, K., Cynthia, J.B., Kamalambikai, B. et al. (2017 July). Hazardous E-waste and its impact on soil structure. *IOP Conference Series: Earth and Environmental Science* 80 (1), 012057. Bristol, UK: IOP Publishing.

Dong, G., Wang, Y., Gong, L. et al. (2013). Formation of soluble Cr (III) end-products and nanoparticles during Cr (VI) reduction by *Bacillus cereus* strain XMCr-6. *Biochemical Engineering Journal* 70: 166–172.

Farhan, S.N. and Khadom, A.A. (2015). Biosorption of heavy metals from aqueous solutions by *Saccharomyces cerevisiae. International Journal of Industrial Chemistry* 6: 119–130.

Forti, V., Balde, K., and Kuehr, R. (2018). *E-waste Statistics Guidelines on Classification Reporting and Indicators-Second Edition.* Bonn: United Nations University.

Forti, V., Balde, C.P., Kuehr, R., and Bel, G. (2020). *The Global E-waste Monitor 2020: Quantities, Flows and the Circular Economy Potential.* United Nations University.

Grant, K., Goldizen, F.C., Sly, P.D. et al. (2013). Health consequences of exposure to e-waste: a systematic review. *The Lancet Global Health* online 1 (6): e350–e361. https://doi.org/10.1016/s2214-109x(13)70101-3.

Guo, J., Cao, B., Guo, J., and Xu, Z. (2008). A plate produced by nonmetallic materials of pulverized waste printed circuit boards. *Environmental Science & Technology* 42 (14): 5267–5271.

Jan, A.T., Azam, M., Ali, A., and Haq, Q.M.R. (2014). Prospects for exploiting bacteria for bioremediation of metal pollution. *Critical Reviews in Environmental Science and Technology* 44 (5): 519–560.

Kanmani, P., Aravind, J., and Preston, D. (2012). Remediation of chromium contaminants using bacteria. *International Journal of Environmental Science and Technology* 9: 183–193.

Kumar Ramasamy, R., Congeevaram, S., and Thamaraiselvi, K. (2011). Evaluation of isolated fungal strain from e-waste recycling facility for effective sorption of toxic heavy metal Pb (II) ions and fungal protein molecular characterization—a mycoremediation approach. *Asian Journal of Experimental Biological Sciences* 2: 342–347.

Laurmaa, V., Kers, J., Tall, K. et al. (2011, February). Mechanical recycling of electronic wastes for materials recovery. In: *Recycling of Electronic Waste II: Proceedings of the Second Symposium*, pp. 1–10. Hoboken, NJ, USA: John Wiley & Sons, Inc.

Lee, Y.C. and Chang, S.P. (2011). The biosorption of heavy metals from aqueous solution by *Spirogyra* and *Cladophora* filamentous macroalgae. *Bioresource Technology* 102 (9): 5297–5304.

Lin, X., Xu, X., Zeng, X. et al. (2017). Decreased vaccine antibody titers following exposure to multiple metals and metalloids in e-waste-exposed preschool children. *Environmental Pollution* 220: 354–363.

Lívia de, C.F., Mario, H.B., and Benedito, C. (2015). Potential application of modified *Saccharomyces cerevisiae* for removing lead and cadmium. *Journal of Bioremediation & Biodegradation* 6 (2).

Mane, P.C. and Bhosle, A.B. (2012). Bioremoval of some metals by living algae *Spirogyra* sp. and *Spirullina* sp. from aqueous solution. *Proceedings of the International Academy of Ecology and Environmental Sciences* 4 (1): 1.

Mudila, H., Prasher, P., Kumar, A. et al. (2021). E-waste and its hazard management by specific microbial bioremediation processes. *Microbial Rejuvenation of Polluted Environment* 2: 139–166.

Murthy, V. and Ramakrishna, S. (2022). A review on global e-waste management: urban mining towards a sustainable future and circular economy. *Sustainability* 14 (2): 647.

Panpatte, D.G., Jhala, Y.K., and Adetunji, C.O. (eds.) (2021). *Microbial Rejuvenation of Polluted Environment*. Springer.

Quan, S.X., Yan, B., Yang, F. et al. (2015). Spatial distribution of heavy metal contamination in soils near a primitive e-waste recycling site. *Environmental Science and Pollution Research* 22: 1290–1298.

Rahman, A., Nahar, N., Nawani, N.N. et al. (2015). Bioremediation of hexavalent chromium (VI) by a soil-borne bacterium, *Enterobacter cloacae* B2-DHA. *Journal of Environmental Science and Health, Part A* 50 (11): 1136–1147.

Singh, R. (2014). Microorganism as a tool of bioremediation technology for cleaning environment: a review.

Sohaili, J., Muniyandi, S.K., and Mohamad, S.S. (2012). A review on printed circuit board recycling technology. *Journal of Emerging Trends in Engineering and Applied Sciences* 3 (1): 12–18.

Sukumar, M. (2010). Reduction of hexavalent chromium by *Rhizopus oryzae*. *African Journal of Environmental Science and Technology* 4 (7): 412–418.

Taştan, B.E., Ertuğrul, S., and Dönmez, G. (2010). Effective bioremoval of reactive dye and heavy metals by *Aspergillus versicolor*. *Bioresource Technology* 101 (3): 870–876.

Vullo, D.L., Ceretti, H.M., Daniel, M.A. et al. (2008). Cadmium, zinc and copper biosorption mediated by *Pseudomonas veronii* 2E. *Bioresource Technology* 99 (13): 5574–5581.

Wang, J. and Chen, C. (2009). Biosorbents for heavy metals removal and their future. *Biotechnology Advances* 27 (2): 195–226.

Zeng, X., Xu, X., Boezen, H.M., and Huo, X. (2016). Children with health impairments by heavy metals in an e-waste recycling area. *Chemosphere* 148: 408–415.

Zhang, S. and Forssberg, E. (1999). Intelligent liberation and classification of electronic scrap. *Powder Technology* 105 (1–3): 295–301.

7

The Impact of Electronic Waste and Its Implications for Soil, Air, and Water

Edson Pablo da Silva[1,2,3,*], *Daniel Nascimento Motta*[1,3], *Flavio Augusto de Freitas*[1,4], *Ginarajadaça Ferreira dos Santos Oliveira*[1], *and Vanessa Leal de Queiroz Hermino*[1,3]

[1] *Centro de Biotecnologia da Amazônia–CBA, Manaus, Amazonas, Brasil*
[2] *Instituto de Tecnologia e Educação Galileo da Amazônia, Manaus, Amazonas, Brasil*
[3] *Universidade Federal do Amazonas, Programa de Pós-Graduação em Biotecnologia–PPGBIOTEC, Manaus, Amazonas, Brasil*
[4] *Universidade Federal do Amazonas, Programa de Pós-Graduação em Química–PPGQ, Manaus, Amazonas, Brasil*
* *Corresponding author*

7.1 Introduction

The globalization process allied to the increasingly advanced technological development in several areas has made it possible for people across the globe to enjoy the benefits and convenience brought by these increasingly advanced technological development processes. However, these new technologies do not only bring benefits. With the ease of acquiring more and more modern equipment, the exchange of obsolete and/or outdated equipment has become a consumer trend, generating more and more electronic waste, which when improperly disposed of can have serious consequences for the environment. One of the greatest challenges of the century is the mitigation of the environmental impacts of these compounds in the air, soil, and water. According to Qianchi Ma et al. (2022), technological advances have led to a significant increase in electrical devices ranging from digital cameras, telephones, and computers, among other consumer electronics, collection, and parts replacement (Lv et al. 2021). E-waste is a term that expresses the waste of electrical and electronic equipment, considered the type of waste whose production grows the most in quantity today. According to Bukhari et al. (2022), the definition of e-waste (electronic waste) is very broad, covering six categories:

- Temperature exchange equipment: refrigerators, freezers, air-conditioning.
- Screens: monitors, laptops, televisions, and notebooks.
- Lamps: fluorescent lamps, high-intensity discharge lamps, and LED.
- Large equipment: tumble dryers, washing machines, large printers, electric stoves, photocopiers, panels.
- Small equipment: microwaves, toasters, vacuum cleaners, fans, shavers, electric kettles, scales, cameras, desktop calculators, radio sets, electric and electronic toys, small electric, small medical devices, and electronic tools.
- IT and telecommunications equipment: cell phones, telephones, GPS, routers, personal computers, pocket calculators, and printers.

As e-waste involves different types of equipment, chemical elements, organic and inorganic substances, causes of generation and for creating multiple impacts in different spectrums, its complexity becomes a key factor for a transdisciplinary economic and environmental analysis, as well as in the areas of health and work. Waste electrical and electronic equipment reached the impressive mark of 44.7 million tons generated worldwide in 2016, of which 1.5 million tons were produced in Brazil alone (Baldé et al. 2017). Being considered the fastest-growing type of waste in the world and having a generation projection for 2050 estimated at 120 million tons per year, the need to understand this phenomenon is evident, given its ambivalence as both dangerous and of economic value. As if its quantitative growth were not enough, there is also a qualitative increase in e-waste, due to its origin. Contamination of natural resources can be both natural and anthropogenic. According to Lv et al. (2021), the impact exerted by man is of two types: first, the consumption of natural resources at a faster pace than the one at which they can be renewed by the ecological system; second, the generation of waste products in amounts greater than can be integrated into the natural nutrient cycle.

The components of electrical and electronic equipment are constantly updated to satisfy the final consumer,

Electronic Waste Management: Policies, Processes, Technologies, and Impact, First Edition. Edited by Sunil Kumar and Vineet Kumar.

increasing consumption comfort and equipment efficiency. Specifically, e-waste contains a wide range of elements and chemical substances in its composition. Around 60 chemical elements can be found (practically half of the elements in the periodic table), among them: iron, copper, aluminum, gold, silver, palladium, rare-earth metals, in addition to economically usable polymers, such as high value; lead, cadmium, chromium, mercury, manganese, nickel, arsenic, zinc, and chrome; and brominated organic substances, elements, and substances known to be potentially or even already confirmed as dangerous (Heacock et al. 2016, p. 550). As estimated, global e-waste generation will reach 110 million tons by the end of 2050, as shown in Figure 7.1a (Li and E Achal 2020; Parajuly et al. 2019). As most electronic devices are powered by battery, global battery demand will also increase continuously and is believed to reach around 70 GWh by the end of 2030, as shown in Figure 7.1b (Fichtner et al. 2021).

This growing use of electronic devices causes serious problems to the environment by generating electronic waste (e-waste) (Bukhari et al. 2022). Concurrent with the advance of development and the advent of new, increasingly modern technologies, there is also an exponential increase in the generation of e-waste, making it difficult to estimate the production of electronic waste due to the economic and population development of each country (Akram et al. 2019). Figure 7.1 shows which countries generated the most e-waste in 2019 according to *The Global E-waste Monitor 2020*.

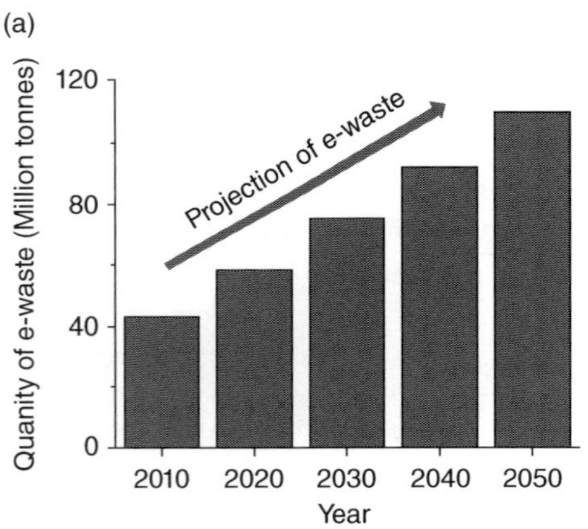

Figure 7.1a Global e-waste generation. Adapted from Fichtner et al., 2021.

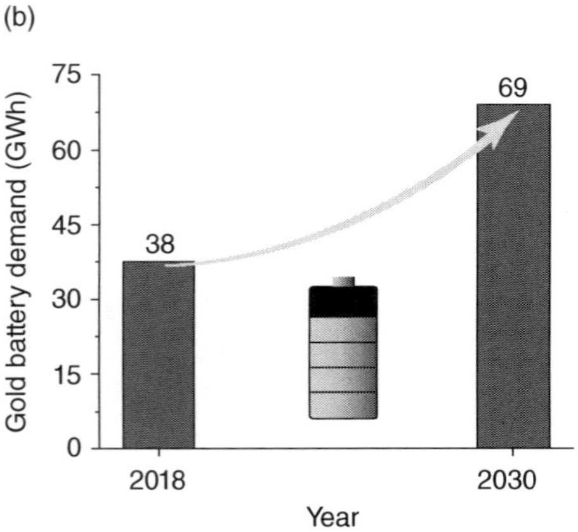

Figure 7.1b Global battery demand. Adapted from Fichtner et al., 2021.

In this way, the irregular disposal of waste electrical and electronic equipment generates problems on different scales, both locally, nationally, and globally. The Public Power does not indicate suitable places for the disposal of electronic waste and, therefore, people do not know the real importance of disposing of it in the right places and in the correct way. Incorrect disposal creates possibilities of contaminating the soil and groundwater and surface water, consequently affecting food production, the environment, and human beings themselves.

7.1.1 The Impact of Eletronic Waste and Its Implications for Soil

Soil can be defined as a body of unconsolidated material that covers the Earth's surface, composed of inorganic and organic matter. It is formed by the disintegration of rock particles that make up the Earth's mantle together with water, climate, relief, and the action of living beings. Soils present well-differentiated physical, chemical, and physico-chemical characteristics and properties. They can appear as clayey or sandy; red, yellow or whitish gray; rich or poor in organic matter; thick or shallow; homogeneous or structured in well-defined horizons depending on the factors that influence weathering. According to CETESB (2013), the soil acts as a filter, having the ability to purify and immobilize most of the impurities deposited in it. However, this capacity is limited, and soil quality may change due to the cumulative effect of the deposition of atmospheric pollutants, the application of pesticides and fertilizers, and the disposal of industrial and urban solid waste and toxic and radioactive materials. Improper disposal of waste in the soil can lead to the unfeasibility of its use, leading to its unusability, in extreme cases impacting the entire ecosystem. In recent years, the accelerated technological revolution has brought with it countless benefits with countless electronic products providing comfort, practicality, and leisure. On the other hand, these same products have been causing several disorders and changes in the environment and ecosystems due to inappropriate disposal. Chakraborty et al. (2022) report that soils are surfaces that can be leached by toxic substances contained in electronics and can cause several problems, especially when they contaminate agricultural land, impacting the food chain and causing intoxication at different levels in consumers. In view of this, the United Nations (UN), through the United Nations Environment Program (UNEP), has promoted international forums involving legal entities governed by public law, private law, and non-governmental organizations (NGOs) to find a solution to the issue of e-waste recycling and the accountability of all those involved in the problem of this waste.

This accelerated elevation in the production of electronic and electrical products has brought not only benefits but also consequences, in consumption as well as in the production of waste and in the final phase of disposal. This has been happening due to great and constant technological innovations that always generate new and more advanced products, which have short periods of use and soon become waste. Various electronic and electrical equipment such as computers, televisions, audio equipment, cell phones, batteries, among others, have had their life cycle shortened, generating technological waste that most of the time does not have an adequate destination. Many of these materials end up mixed with the common waste, since their recycling is still insufficient given the large proportion of their generation, thus contributing to environmental and public health impacts. These materials are composed of components and equipment that, due to breakage and/or end of life, are discarded in the trash.

According to Kapoor et al. (2021), when electronics are discarded in common waste, the chemical substances present, such as mercury, cadmium, arsenic, copper, lead, zinc, and aluminum, penetrate the soil and groundwater, contaminating plants and animals through the water. These substances from electronic waste can also cause contamination in plants, animals, and water that will later be used for various human activities either for consumption or irrigation.

Soils contaminated by heavy metals often come from urban solid waste, including electronics, which are often sent to inappropriate places such as dumps. Diagnostics carried out around the dump in the municipalities of Iguatu/Ceará showed the soil with slightly acidic organic matter and content ranging from 8 to 60 g kg^{-1}, due to the presence of metals such as Cu, Pb, Mn, Cr, Fe, and Zn. It was possible to verify through statistical data with Pearson's correlation that organic matter influences the retention of heavy metals Zn, Mn, and Fe in the soil. More metals such as Cr, Cu, and Pb show little interaction with organic matter, and may have greater mobility in the soil, thus reaching the water table. It was also observed that the concentrations found of the metals Cu, Zn, and Pb in the soil were above the limits established by CONAMA resolution 420/2009 (Chakraborty et al. 2022).

And the problems of contamination with heavy metals in the soil derived from e-waste are spread across the globe. In Latin America, Brazil leads the ranking of e-waste production, followed by Mexico, Colombia, Argentina, Peru, Venezuela, Chile, Ecuador, and Guatemala. In China, Han et al. (2019) observed that environmental heavy metal pollution caused by e-waste recycling is of growing concern; however, little information is available on metal emissions during e-waste processing activities. The results showed that the soils in the analyzed sites had significant levels of metals such as Cd, Sb, and Pb. Such results present a high risk to human health because the levels found were much higher than those found in acid leaching and/or burning sites. The results demonstrate that the management of raw electronic waste is extremely important, as it is a source of heavy metals for the soil, which can directly impact the health system from processes of exposure and human contamination, requiring correct adoption of e-waste management strategies.

Trace metals, when ingested by humans, are toxic, as they affect the mechanisms that govern biochemical activities and, when combined with the biomolecules of the human body, such as proteins and enzymes, form stable biotoxic compounds that are capable of mutilating. Their structures prevent them from reacting to their physiological. Qianchi Ma et al. (2022) report that in the city of Guiyu, China, in a region of high productive activity based on electronic waste, higher concentrations of trace elements were found in the human placenta compared to results from a controlled area, where no e-waste processing takes place. In Kunamoto, Japan, on October 10, 2013, the Minamata Convention was adopted by 140 countries at a diplomatic conference in Japan. The final text of this treaty had been agreed in January of the same year, when measures were established to control and reduce the use of mercury in a series of processes and products, including batteries and some types of fluorescent lamps.

The large amount of e-waste generated each year affects the human health, environment, society, and the privacy or security of data stored on electronic devices in the garbage (Figure 7.2a). Inadequate and questionable management of e-waste poses a danger in all these respects. Thus, the proper management of e-waste becomes essential today with the increase in discarded electronic devices and materials (Kapoor et al. 2021). Figure 7.2b illustrates the uncontrollable use of resources, environmental damage, and toxicity for all living beings (Parajuly et al. 2019).

It is worth mentioning that the contamination of the soil by e-waste, in addition to making the use of the soil unfeasible, can also start a process of contamination of the air in the elimination of particles beyond the waters through the contamination of the water tables. Finding measures that mitigate this process of inappropriate disposal is one of the great challenges to be overcome.

(a)

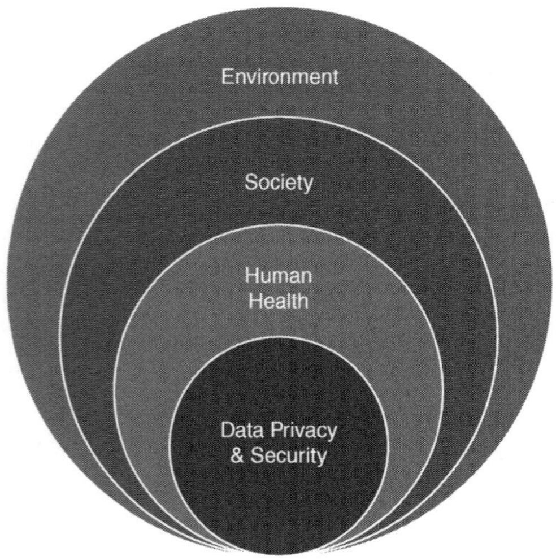

Figure 7.2a Impact of e-waste.

(b)

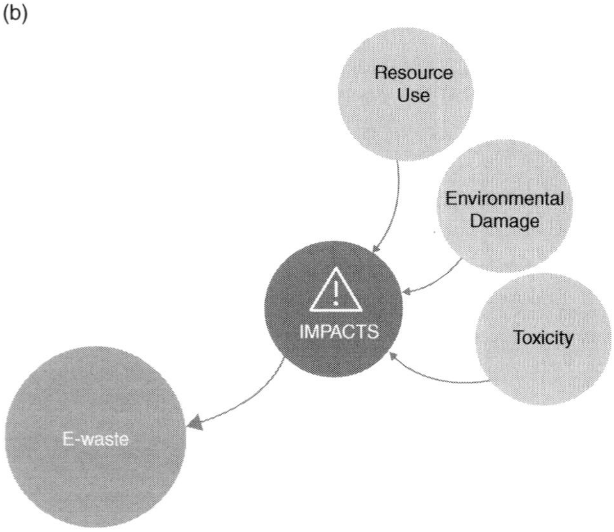

Figure 7.2b Impact of e-waste.

7.1.2 The Impact of Electronic E-waste and Its Implications for Air

With the advent of new technologies and processes in the electro-electronic industry, including semiconductor devices (SDs) (K. Li et al. 2021), electronic terminal products (ETPs), display devices (DDs), and printed circuit boards (PCBs) (Lv et al. 2021), a new problem also emerged––the waste generated––presenting a serious problem of contamination of the environment, mainly the air. This contamination is due to the use of volatile organic compounds (VOCs) in electronic cleaning, surface coating, and other processes used in the manufacturing of electronics (Feng et al. 2021).

Few studies have focused on the emission characteristics of the electronics industry on VOC source profiles. Studies include Wang et al. (2020), example thin-film transistor liquid crystal display (TFT-LCD), and acetone in the manufacture of printed circuit boards (PCBs), while those of integrated chips use isopropanol (56.4–98.3%) and acetone (30.8%). Some VOCs origin profiles in electrical terminal products (ETP) and polychlorinated biphenyl (PCB) organic pollutants, which lead to serious air contamination processes; the need for viable solutions is accelerating rather than slowing down at the moment. The binders, plastics and carbon materials, and electrolytes contained in the anode burn in an exothermic reaction generate toxic gases. These are some reasons for the need to reduce CO2 emissions to delay global warming (Birol 2018; Sommerville et al. 2021).

Organic pollutants such as polybrominated diphenyl ethers and polychlorinated biphenyls (PCBs) (Murthy and Ramakrishna 2022) are among many of the toxic chemicals that pose a danger to human health and to the environment. PCBs are emitted into the atmosphere by dielectrics from capacitors used in electronic devices (Kapoor et al. 2021). Particle/gas partitioning (P/G) governs the atmospheric transport and fate of semi-volatile compounds (SVOCs) from chlorinated biphenyls (PCBs) (Qiao et al. 2022).

With the rise of e-waste, informal commerce, and related crimes as well, the social impact of e-waste is linked even more to illegal activities. As far as data privacy and security are concerned, drives, USBs, memory cards, hard drives, etc. are where users tend to store important personal, confidential, institutional, and other information (Kapoor et al. 2021; Qiao et al. 2022).

7.1.2.1 The Health Effects of Particulates

The impacts caused by e-waste suspended in the air are diverse, with some reports being raised about potential damage to the respiratory system, DNA modification, and cancers, among other impacts. Kapoor et al. (2021) and Lou et al. (2021) report that nocive substances with the hexavalent chromium Cr(VI) can potentially damage DNA in individuals living near e-waste disposal sites, inducing significant instability of the rDNA array in human cells (Figure 7.3).

Exposure routes may vary depending on the substance and process (Table 7.1). Exposure to harmful health components found in e-waste generally occurs through dermal contact, inhalation, and ingestion (Grant et al. 2013).

In studies by Xiao et al. (2022), a magnetic sample analysis was done indicating that soft ferrimagnetic minerals of low coercivity were the main materials. That is, a mixture of single-domain and multi-domain magnetite dominated in magnetic parameters (PM) (Xia et al. 2014). The average annual PM 2.5 concentration was 60.17 $\mu g/m^3$,

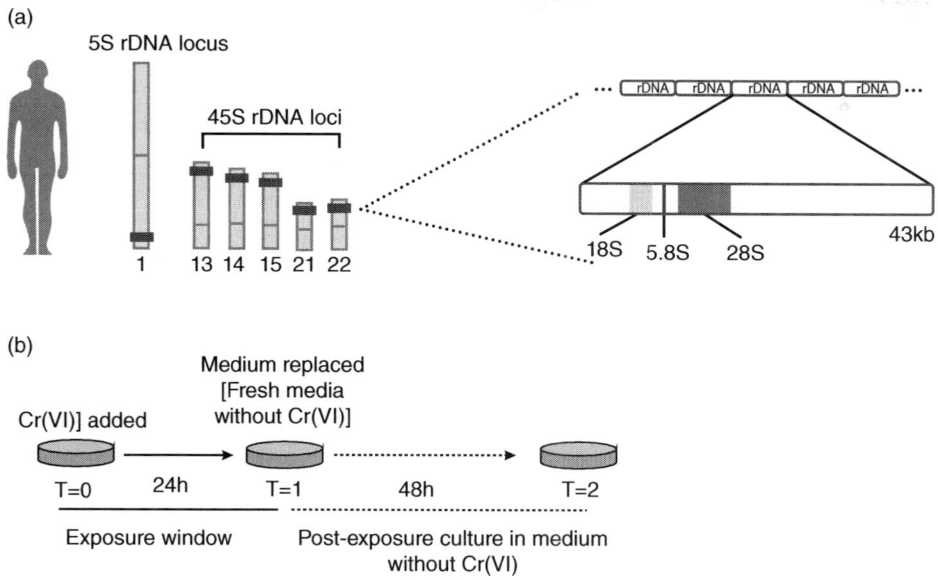

Figure 7.3 Ribosomal DNA (rDNA) structure and environmental exposure. (A) In the human genome, the 45S rDNA unit (encoding the 18S, 5.8S, and 28S rRNAs) is tandemly arrayed in nucleolar organizer regions (NORs) residing on five acrocentric chromosomes (13, 14, 15, 21, 22), while the tandemly repeated cluster encoding the 5S rRNAs is located on chromosome 1. (B) Paradigm for hexavalent chromium [Cr (VI)] exposure. (*Source:* Lou et al., 2021 / with permission of Elsevier.

and approximately 32.02% of the daily PM 2.5 concentrations exceeded the guideline value of the China National Ambient Air Quality Standard (Xiao et al. 2022). Values reflect health reactions causing a persistent elevation in one or more of the selected markers, potentially toxic by inhalation (Viegas et al. 2022). Concentrations for arsenic (As), cadmium (Cd), lead (Pb), and manganese (Mn) were better than for other metals with training and validation values of $R > 0.8$, while relatively poor for aluminum (Al), cobalt (Co), and chromium (Cr) in environmental magnetism analyzes for the evaluation of airborne heavy metals (Xiao et al. 2022).

Studies with cobalt (Co) by Viegas et al. (2022) that investigated the inhalation toxicity of these substances showed that cobalt induces pre-inflammatory biomarkers resulting in a group of substances that can cause inflammation. The in vivo activity was performed in animals exposed to cobalt for a period of 4 hours and after 14 days the animals were sacrificed (Figure 7.4). Eight compounds with cobalt in their composition were tested: cobalt powder (Co), acute tox. 1; cobalt dihydroxide [$Co(OH)_2$], acute tox. 1; cobalt monoxide (CoO), acute tox. 2; with lower contents for cobalt sulfate heptahydrate ($CoSO_{4.7}H_2O$), trichobalt tetroxide (Co_3O_4), lithium cobalt dioxide ($CoLiO_2$), and cobalt sulfide (CoS).

The main VOCs or their proportions studied by Lv et al. (2021) were 2-methylpentane, 3-methylpentane, and n-pentane, the main alkanes in SD, accounting for 58.0%, while cyclohexane, heptane, and methylcyclohexane were the main ones most abundant for ETP, accounting for 1.1%. Furthermore, the main alkene species of DD were isoprene (0.1%) and cis-2-pentene (0.1%), while those of ETP were acrylic (0.4%), n-butene (0.1%), and cis-2-butene (0.03%). The main halocarbon species of SD was ethylene trichlor (24.6%), while that of ETP was chloroform (0.02%). This is mainly because trichlorethylene is always used as a chlorine-containing cleaning agent, resulting in a higher contribution of total VOCs. The dangers of e-waste and its variety of components, which, if not handled properly, can harm human health and the environment and their consequences for individuals living near facilities that recycle or burn electronic waste, are recorded in Table 7.1. Some effects directly related to e-waste and health components are recorded in Table 7.2.

7.1.3 The Influence of Electronic E-waste and Its Implications for Water

The flow of water resources has increased, mainly, due to the rate of industrial and population growth, leading to an increase in the production and emission of pollutants in water bodies. Other factors also contributed to the degradation and influence of water, such as inadequate soil management for agriculture and livestock, the use of fertilizers and pesticides, the deforestation of the riparian forest, triggering ingestion, flooding, and depletion of underground water. Pollution by trace metals as it refers to water resources significantly impacts the habitat of the terrestrial or, because the surface of the planet consists of 29.3% of emerged land and 70.7% of water (Bowen Liang et al. 2022).

The global generation of electronic waste in 2016 was 44.7 Mt, and it was estimated that in 2018, 49.8 Mt (million

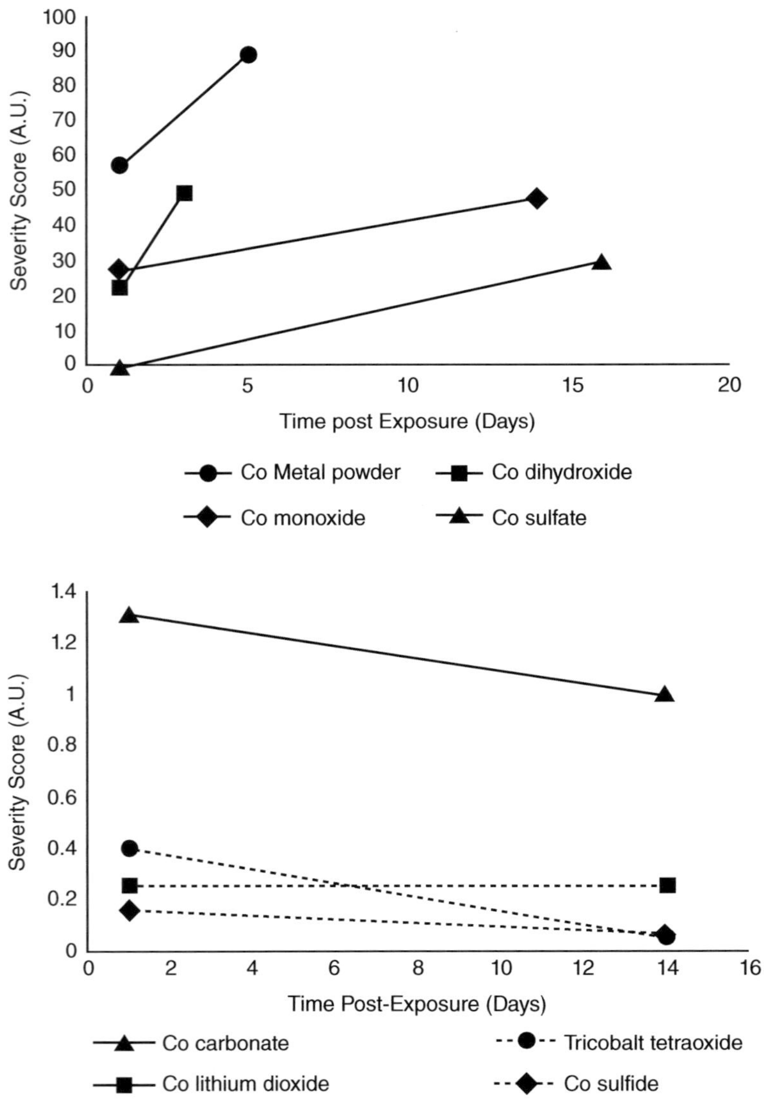

Figure 7.4 Two distinct trends are observed among reactive substances (Co metal powder, Co dihydroxide, Co monoxide, Co sulfate): there is an increase in severity from 1 day post-exposure to exposure days later in the observation period, up to 16 days. Based on the severity of the effect, animals exposed to cobalt dihydroxide and cobalt powder were sacrificed early, on days 3 and 5, respectively. The observation period for the others was 16 days and confirms the permanence of the less intense observed severity. (*Source:* Viegas et al., 2022 / with permission of Elsevier.

tons) were produced in the world (Baldé et al. 2017). Regarding the electronic waste generated in 2016, only 8.94 Mt were formally collected and processed, which corresponds to 20% of the annual global generation. This number is directly related to countries that have implemented internal regulations on e-waste management and treatment (Baldé et al. 2017). In recent decades, e-waste management has become a topic of global relevance. There have been great advances in implementing regulations for the management and disposal of electronic waste, but the vast majority of countries do not have regulations for this type of waste (Khanna et al. 2018).

Improper disposal of these electronic materials near water courses makes their use for consumption and/or use in other ways unfeasible, since the heavy metals present can cause serious damage to health. Nascimento et al. (2021) report that most consumers and large companies are not aware of the consequences, and many also do not have the necessary information about recycling electronic waste and its impact on the environment. The danger of throwing these products into the water is great, as these residues contain lead, cadmium, and mercury. These metals, which are present inside some cells and batteries, for example, can contaminate bodies of water such as lakes and rivers and reach human beings. This activity leads to deterioration of local drinking water, which can result in serious illness. Purushothaman et al. (2020) report in their work that a water sample from the

Table 7.1 Health hazards and environmental impacts.

E-waste component	Process	Potential health hazard	Impact
Lamps, ray tubes, televisions	Copper yoke breakage, copper yoke removal and open disposal in landfill.	Inhalation of phosphorus containing cadmium.	Toxic phosphorus release.
Printed circuit boards (PCBs)	Separating computer chips and desoldering.	Lead and tin inhalation. Inhalation of brominated beryllium, cadmium, dioxina, and mercury.	Probable emission of substances to the atmosphere.
Computer plastics and peripherals	Lower melting temperature.	Exposure to hydrocarbons and brominated dioxin.	Hydrocarbon emissions into the atmosphere.
Dioxins	Released in combustion.	Contamination by dibenzodioxin and polychlorinated dibenzofurans, biphenyls, Polychlorinated.	Inhalation and ingestion.
Polyaromatic hydrocarbons	Released as a by-product of combustion.	Acenaphthene, anthracene, phenanthrene, acenaphthylene, fluorene, chrysene, fluoranthene, and pyrene.	Inhalation and ingestion.
Chemical elements	Anti-corrosion coating. Connector components. Thermostats, sensors. Batteries. Fluorescent lamps.	Chrome Cadmium Mercury Nickel and lithium Barium	Inhalation, ingestion, and contact.

(*Source:* Murthy and Ramakrishna 2022; Grant et al. 2013.)

Table 7.2 Health effects of e-waste exposure.

Health effects of e-waste exposure			
E-waste	**Effect**	**Results**	**Source**
Polybrominated diphenyl ethers	Effect on the thyroid	Was significantly and positively correlated with T3 and T4; exposure can disrupt thyroid hormone homeostasis.	(Zhao et al. 2021)
Cadmium	Pulmonary function	Is associated with inflammatory cytokines in patients with chronic obstructive pulmonary disease (COPD).	(Jiang et al. 2022)
Exposure to chromium (Cr) and nickel (Ni)	Reproductive health	Showed evidence of neonatal exposure.	(Yüksel, Arıca e Söylemezoğlu 2021)
Polychlorinated biphenyls (PCBs)	Genotoxicity	Chromosomal instabilities induced in cells exposed to PCBs.	(J. K. Li et al. 2021)
Lead	Hematological toxicity	Exposure to Pb induces hematological toxicology.	(Wang et al. 2020)

Lianjiang River, near a Chinese "recycling village," had lead levels 2400 times higher than the World Health Organization's Drinking Water Guidelines, involving a serious health risk.

According to Sitaramaiah and Kusuma-Kumari (2016), children are especially vulnerable to the health risks that can result from exposure to e-waste and therefore need more protection. As they are still growing, children's intake of air, water, and food in proportion to their weight increases significantly compared to adults--and with it, the risk of dangerous chemical absorption. In addition, their bodies' functional systems, such as the central nervous, immune, reproductive, and digestive systems, are still developing, and exposure to toxic substances, by hampering development, can cause irreversible damage. Many children are exposed to chemicals derived from e-waste in their daily lives due to unsafe recycling activities that are often carried out in their homes, either by family members or by the children themselves. In addition, children can be exposed through dumps located close to their homes, schools, and play areas, which are most often close to water courses (Sitaramaiah and Kusuma-Kumari 2016).

It is evident from several studies that rudimentary recycling techniques coupled with the amounts of e-waste processed have already resulted in adverse impacts to the environment and human health, including contaminated soils and surface waters. Health problems have been reported in recent years, including diseases and problems related to the skin, stomach, respiratory tract, and other organs. Workers suffer high incidences of birth defects, infant mortality, tuberculosis, blood disorders, immune system abnormalities, kidney and respiratory system malfunction, lung cancer, brain underdevelopment in children, and damage to the nervous and blood systems (Ahirwar and Tripathi 2021).

7.2 Conclusion

The inadequate disposal of electronic waste has led to serious environmental and health problems worldwide, requiring the adoption of adequate measures for a better destination of the waste generated from discarded electronics. New forms of treatment that are increasingly advanced and more effective should be more accessible so that all populations have access, thus avoiding contamination by e-waste residues in water, air, or soil. The role of raising awareness of companies and the population about the possible risks caused by improper disposal should be constant, or treated as a public policy for social development.

Acknowledgments

Our thanks to the funding bodies that always help us in the development of projects: Capes, CNPq, Fapeam, and Inmetro.

References

Ahirwar, R. and Tripathi, A.K. (2021). E-waste management: a review of recycling process, environmental and occupational health hazards, and potential solutions. *Environmental Nanotechnology, Monitoring and Management* 15 (May 2020): 100409. https://doi.org/10.1016/j.enmm.2020.100409.

Akram, R., Natasha, F.S. et al. (2019). Trends of electronic waste pollution and its impact on the global environment and ecosystem. *Environmental Science and Pollution Research* 26 (17): 1–16. https://doi.org/10.1007/s11356-019-04998-2.

Baldé, C., Forti, V., Gray, V. et al. (2017). *The Global E-waste Monitor*. Bonn/Geneva/Vienna: United Nations University (UNU), International Telecommunication Union (ITU), and International Solid Waste Association (ISWA).

Birol, F. (2018). *World Energy Outlook 2018*. Paris: IEA.

Bukhari, M.U., Khan, A., Maqbool, K.Q. et al. (2022). Waste to energy: facile, low-cost and environment-friendly triboelectric nanogenerators using recycled plastic and electronic wastes for self-powered portable electronics. *Energy Reports* 8: 1687–1695. https://doi.org/10.1016/J.EGYR.2021.12.072.

CETESB (2013). *Companhia Ambiental Do Estado de São Paulo*. Solo: poluição. São Paulo. Available at http://cetesb.sp.gov.br/solo/poluicao (accessed 30 May 2022).

Chakraborty, S.C., Qamruzzaman, M., Zaman, M.W.U. et al. (2022). Metals in e-waste: occurrence, fate, impacts and remediation technologies. *Process Safety and Environmental Protection* 162: 230–252. ISSN 0957-5820. https://doi.org/10.1016/j.psep.2022.04.011.

Feng, M., Hu, X., Zhou, L. et al. (2021). Real-world vehicle volatile organic compound emissions and their source profile in Chengdu based on a roadside and tunnel study. *Atmosphere* 12 (7): https://doi.org/10.3390/atmos12070861.

Fichtner, M., Edström, K., Ayerbe, E. et al. (2021). Baterias recarregáveis do futuro–-o estado da arte na perspectiva da BATERIA 2030+. *Materiais Avançados de Energia* https://doi.org/10.1002/aenm.202102904.

Grant, K.M., Goldizen, F.C., Sly, P. et al. (2013). Health consequences of exposure to e-waste: a systematic review. *The Lancet Global Health* 1 (6): e350–e361. https://doi.org/10.1016/S2214-109X(13)70101-3.

Han, Z., Zeng, D., Li, Q. et al. (2019). Public willingness to pay and participate in domestic waste management in rural areas of China. *Resources, Conservation and Recycling* 140: 166–174. ISSN 0921-3449, https://doi.org/10.1016/j.resconrec.2018.09.018.

Heacock, M., Kelly, C.B., Asante, K.A. et al. (2016). E-waste and harm to vulnerable populations: a growing global problem. *Environmental Health Perspectives* 124: 550–555. 2016. doi: 10.1289/ehp.1509699 Available at: https://pubmed.ncbi.nlm.nih.gov/26418733 (accessed 10 February 2022).

Jiang, Y.L., Fei, J., Cao, P. et al. (2022). Serum cadmium positively correlates with inflammatory cytokines in patients with chronic obstructive pulmonary disease. *Environmental Toxicology* 151–160. https://doi-org.ez2.periodicos.capes.gov.br/10.1002/tox.23386.

Kapoor, N., Sulke, P., and E Badiye, A. (2021). E-waste forensics: an overview. *Forensic Science International*

Animals and Environments 1: 100034. https://doi.org/10.1016/J.FSIAE.2021.100034.

Khanna, R., Park, M., Cayumil, R. et al. (2018). Global perspectives on regulatory measures and collection procedures for electronic waste: a critical assessment. *Open Access Journal of Waste Management & Xenobiotics* 1 (1): 000101.

Li, K., Zhang, S., Tan, Q. et al.(2021). Insulator in photocatalysis: essential roles and activation strategies. *Chemical Engineering Journal* 426: 130772. https://doi.org/10.1016/J.CEJ.2021.130772.

Li, W. and E Achal, V. (2020). Environmental and health impacts due to e-waste disposal in China --a review. *Science of the Total Environment* 737: 139745. https://doi.org/10.1016/J.SCITOTENV.2020.139745.

Liang, B., Jiehua, L., Bibai, D. et al. (2022). E-waste recycling emits large quantities of emerging aromatic amines and organophosphites: a poorly recognized source for another two classes of synthetic antioxidants. *Environmental Science & Technology Letters 2022* 9 (7): 625–631. https://doi.org/10.1021/acs.estlett.2c00366.

Lou, J., Yu, S., Feng, L. et al. (2021). Environmentally induced ribosomal DNA (rDNA) instability in human cells and populations exposed to hexavalent chromium [Cr (VI)]. *Environment International* 153: 106525. https://doi.org/10.1016/J.ENVINT.2021.106525.

Lv, Z., Liu, X., Wang, G. et al. (2021). Sector-based volatile organic compounds emission characteristics from the electronics manufacturing industry in China. *Atmospheric Pollution Research* 12 (7): 101097. https://doi.org/10.1016/J.APR.2021.101097.

Murthy, V. and Ramakrishna, S. (2022). A review on global e-waste management: urban mining towards a sustainable future and circular economy. *Sustainability* 14 (2): 647. https://doi.org/10.3390/su14020647.

Nascimento, T.A.D.J., Sousa, J.K.F., Miranda, S.B.A. et al. (2021). Os impactos e soluções para os resíduos eletrônicos: estudo de caso em uma instituição federal de ensino. *Research, Society and Development* 10 (9): 13. https://doi.org/10.33448/rsd-v10i9.16321.

Parajuly, K., Kuehr, R., E Awasthi, A.K. et al. (2019). *Future E-waste Scenarios*. StEP (Bonn), UNU Life-SCYCLE (Bonn), and UNEP IETC (Osaka).

Purushothaman, M., Gousuddin Inamdar, M., and Muthunarayanan, V. (2020). Socio-economic impact of the e-waste pollution in India. *Materials Today: Proceedings* 37 (Part 2): 280–283. https://doi.org/10.1016/j.matpr.2020.05.242.

Qianchi, M., Liu, Y., Yang, X. et al. (2022). Effect-directed analysis for revealing aryl hydrocarbon receptor agonists in sediment samples from an electronic waste recycling town in China. *Environmental Pollution* 308: 119659. ISSN 0269-7491. https://doi.org/10.1016/j.envpol.2022.119659.

Qiao, L.N., Ma, W.L., Liu, L.Y. et al. (2022). Particle/gas partitioning behavior of polychlorinated biphenyls (PCBs) in global atmosphere: equilibrium or steady state? *Atmospheric Environment* 270: 118926. https://doi.org/10.1016/J.ATMOSENV.2021.118926.

Sitaramaiah, Y. and Kusuma-Kumari, M. (2016). National seminar on impact of toxic metals, minerals and solvents leading to environmental pollution. *Journal of Chemical and Pharmaceutical Sciences* 3 (3): 39–42.

Sommerville, R., Zhu, P., Rajaeifar, M.A. et al. (2021). A qualitative assessment of lithium ion battery recycling processes. *Resources, Conservation and Recycling* 165: 105219. https://doi.org/10.1016/J.RESCONREC.2020.105219.

Viegas, V., Burzlaff, A., Brock, T.O. III et al. (2022). A tiered approach to investigate the inhalation toxicity of cobalt substances. Tier 3: inflammatory response following acute inhalation exposure correlates with lower tier data. *Regulatory Toxicology and Pharmacology* 130: 105127. https://doi.org/10.1016/J.YRTPH.2022.105127.

Wang, T., Tu, Y., Zhang, G. et al. (2020). Development of a benchmark dose for lead-exposure based on its induction of micronuclei, telomere length changes and hematological toxicity. *Environment International* 145: 106129. https://doi.org/10.1016/J.ENVINT.2020.106129.

Xia, D., Wang, B., Yu,Y. et al. (2014). Combination of magnetic parameters and heavy metals to discriminate soil-contamination sources in Yinchuan—a typical oasis city of Northwestern China. *Science of the Total Environment* 485–486 (1): 83–92. https://doi.org/10.1016/J.SCITOTENV.2014.03.070.

Xiao, H., Leng, X., Qian, X. et al. (2022). Prediction of heavy metals in airborne fine particulate matter using magnetic parameters by machine learning from a metropolitan city in China. *Atmospheric Pollution Research* 13 (3): 101347. https://doi.org/10.1016/J.APR.2022.101347.

Yüksel, B., Arıca, E., and E Söylemezoğlu, T. (2021). Assessing reference levels of nickel and chromium in cord blood, maternal blood and placenta specimens from Ankara, Turkey. *Journal of the Turkish-German Gynecological Association* 22 (3): 187–195. https://doi.org/10.4274/jtgga.galenos.2021.2020.0202.

Zhao, X., Yang, X., Du,Y. et al. (2021). Polybrominated diphenyl ethers in serum from residents living in a brominated flame retardant production area: occurrence, influencing factors, and relationships with thyroid and liver function. *Environmental Pollution* 270: 116046. https://doi.org/10.1016/J.ENVPOL.2020.116046.

8

The Environmental Fate of E-waste

Its Impact on Environmental Samples (Soil, Water, and Air)

Olusola Olaitan Ayeleru[1,], Helen Uchenna Modekwe[2], Oyetayo Olaoluwa Adefiranye[3], Bimbo Lolade Fafowora[4], Matthew Adah Onu[1], Tarhemba Tobias Nyam[1], Ismaila Adejare Adesiyan[2], and Peter Apata Olubambi[1]*

[1] *Centre for Nanoengineering and Advanced Material, University of Johannesburg, Doornfontein Campus, Johannesburg, South Africa, 2028, South Africa*
[2] *Renewable Energy and Biomass Research Group, Department of Chemical Engineering, University of Johannesburg, Johannesburg, 2028, South Africa*
[3] *Department of Microbiology, Faculty of Science, University of Lagos, Lagos, Nigeria*
[4] *School of Journalism and Media Studies, Rhodes University, South Africa*
[*] *Corresponding author*

8.1 Introduction

Electronic waste (e-waste) contributes to a large extent to the volume of municipal solid waste (MSW) generated every year globally. This is due to factors like rapid development, industrialization, high living standards, etc. (Saha et al. 2021). Other factors contributing to the incessant generation of MSW include the rapid, short lifespan of appliances (which contributes to the rise in e-waste), improvement in designs, advanced features or technology of new electronic and electrical gadgets, etc.(Forti et al. 2020). E-waste generation has been rising over time, with 54 million metric tons produced in 2019 and 75 million metric tons projected to be produced by 2030 (Ghimire and Ariya 2020). E-waste is usually disposed of at landfill sites (LSs) or is incinerated at the end of its lifespan since the rate of recycling is usually very low (Saha et al. 2021). Therefore, for the purpose of protecting public health and the environment, effective handling and management of e-waste is necessary. However, proper management of e-waste is lacking in low-income countries (LCs) owing to constraints in needed infrastructure, budgeting, or funding for waste management services (Kiddee et al. 2014). Failure to handle e-waste properly leads to soil pollution and degradation, the discharge of pollutants into water bodies, and hazardous material release into the environment that may have a direct or indirect negative influence on the environment or public health (Akram et al. 2019; Amoabeng Nti et al. 2020; Khan et al. 2021; Rautela et al. 2021). Despite accounting for only ~5% of MSW generated, e-waste contributes about 70% of soil toxicity on an annual basis (Rautela et al. 2021). Heavy metals, which include chromium (Cr), lead (Pb), beryllium (Be), gallium (Ga), palladium (Pd), cadmium (Cd), etc., are found in e-waste. These contaminants can possibly disrupt the food chain (Akram et al. 2019; Liu et al. 2020). Other pollutants in e-waste comprise esters, polyaromatic hydrocarbons (PAHs), etc. These pollutants affect both living and non-living things (Zhang et al. 2015). When e-waste is properly recycled, precious metals like silver (Ag), gold (Au), copper (Cu), etc. can be recovered (Khanna et al. 2020). Due to the variations in the material content of each electronic and electrical product, they are usually discarded and recycled in different ways. The prevalence of hazardous halogenated compounds and toxic metals in e-waste makes waste handling difficult and complicated, and failure to follow the recommended disposal, recycling, or recouping protocol poses serious environmental and health challenges. Distinctive components of e-waste consist of about 40% inorganic components such as base metals, heavy metals, noble metals, radioactive metals, and rare earth metals; 30% organic components (such as plastics, glass fibers, flame retardants, etc.); and 30% ceramics (e.g. mica, silica, etc.) (Kaya 2016; Rautela et al. 2021). Collected and recycled e-waste accounts for about 17.4% while a larger percentage (about 82.6%) of e-waste is either dumped in landfills or disposed of through illicit trade (Forti et al. 2020; Rautela et al. 2021). Informal recycling approaches toward e-waste involve open burning, acid stripping of metals, acid bath, and incineration, resulting in the generation of secondary pollutants that have

Electronic Waste Management: Policies, Processes, Technologies, and Impact, First Edition. Edited by Sunil Kumar and Vineet Kumar.

harmful impacts on health (Forti et al. 2020; Rautela et al. 2021). Therefore, what then are the consequences of e-waste mismanagement and poor treatment methods on the environment (land, air, and water) and on public health?

This chapter provides information on the fate of e-waste on air, water resources, and soil and its generalized impact on human health (including that of children) in an effort to contribute to a better understanding of the effect on human health and harm caused by e-waste to the lithosphere, hydrosphere, and atmosphere. The main sources and classification of e-waste are highlighted, with various disposal and management approches available in some developed and developing economies also detailed.

8.2 Sources and Types of E-waste

Electrical and electronic waste are kinds of waste that run on power (electricity or batteries), and they have the possibility of impacting the environment negatively when the method of managing it is poor. Various sources of e-waste comprise big and little home appliances, lighting equipment, information technology and telecommunications (ITTC) equipment, etc. (Saha et al. 2021). The largest sources of e-waste globally come from mobile phones and personal computers (Chakraborty et al. 2022). E-waste is usually made of metals, printed circuit boards, cables, a metal-plastic mixture, plastics, and other contents. According to Kaya (2016), e-waste consists of 30% washing machines, air conditioners, dryers, coffee machines, vacuum cleaners, and irons; 10% televisions; 30% computers, CDs, telephones, DVDs, printers; 20% refrigerators; and 10% monitors (Kaya 2016; Saha et al. 2021). Figure 8.1 shows the percentages of the electronic waste contents generated globally. According to the European Union's (EU) classification, EEE is divided into 54 different product-centric categories, which are categorized based on the component's density, similar material constituents, similar functionalities, and/or similar end-of-life features (Baldé et al. 2017; Forti et al. 2020). EEE is further subdivided into six generic groups based on their related waste management features (Baldé et al. 2017; Forti et al. 2020). These categories are: temperature exchange equipment (freezers, refrigerators, air conditioners, etc.); screens and monitors (e.g., laptops, televisions, monitors, tablets, etc.); lamps (such as LED lamps and fluorescent lamps, etc.); small equipment (such as electric kettles, calculators, microwaves, etc.); small IT and telecommunication equipment (personal computers, printers, global positioning system (GPS), routers, telephones, mobile phones); and large equipment (for example, dishwashing machines, photovoltaic panels, washing machines, printing machines, photocopying machines, electric stoves) etc. (Baldé et al. 2017; Forti et al. 2020; Saha et al. 2021).

8.3 E-waste Disposal and Management Methods in Both Developing and Developed Economies

The degree of consumption has an impact on the generation and management of e-waste in any economy, consumers' disposal behavior, public awareness, the need for a

Figure 8.1 Characteristic contents of e-waste Adapted from Jain, 2021. (*Source:* https://electricalfundablog.com/e-waste-electronic-waste-sources-composition).

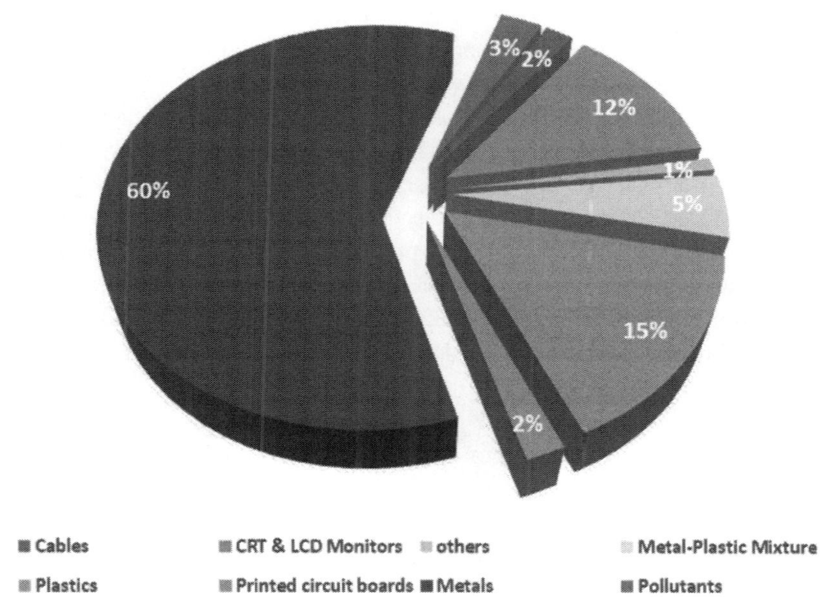

greener economy, and sustainable development. Most developed countries have implemented various feasible approaches to the disposal and management of e-waste to safeguard public health and the environment for a sustainable future. For instance, in China, collection and recovery businesses play a major role in the recycling and recovery of e-waste. Ideally, the major preferred mode of e-waste disposal, especially for electronic appliances, is through their resale (Chi et al. 2014). Consumers believe that damaged electronic appliances have secondhand value; therefore, they sell their electronic waste to hawkers within their vicinity who then direct the collected electronics waste to the secondhand market, where they are eventually refurbished and returned back into the market for resale (Chi et al. 2014).

Japan, for example, has recorded 70% recovery of e-waste generated from households (Borthakur and Govind 2017; Zoeteman et al. 2010). This milestone was achieved due to the 2001 Home Appliances Recycling Laws, which mandated manufacturers to take the responsibility of recycling e-waste. The law allows consumers to return their malfunctioning or damaged products to the retailers. The retailers are then in charge of transporting this waste to a designated stockyard and disposing of it (Borthakur and Govind 2017). Similarly, consumers can discard their e-waste themselves or can take it directly to the nearest municipality that would then dispose of it at the stockyard. This approach discourages consumers from paying and coming up with the transportation and recycling fees; rather, they resort to exportation of their outdated and old-fashioned electronics (Borthakur and Govind 2017). In Korea, e-waste collection and disposal fees are the responsibility of both the households (consumers) and the suppliers or retailers. Consumers can return their outdated or waste products to retailers, suppliers, or manufacturers and, in exchange, buy a new product from the supplier in replacement of the returned product; the supplier then bears the burden of the collection and recycling fees (Jang 2010). Retailers or suppliers send collected e-waste to designated disposal facilities set up across the country by manufacturers of electrical and electronics equipment, such as off-site treatment facilities, private e-waste recycling facilities, manufacturer recycling centers, and community reuse centers (Borthakur and Govind 2017; Jang and Kim 2010). Consumers also pay collection fees to the local government or municipality collection system, while other municipalities or local authorities collect local's household e-waste from curbside collection containers and dispose of it at treatment facilities.

In certain European countries, such as the United Kingdom (UK) and Switzerland, waste management and disposal routes for electrical and electronic equipment (EEE) are formally established through the principle of extended producer responsibility (Owusu-Sekyere et al. 2018) and take-back schemes (Borthakur and Govind 2017). In Switzerland, the manufacturers and producers of EEE bear the financial and handling responsibilities of e-waste collection, disposal, recycling, and eco-friendly treatment. However, consumers still bear some financial responsibility through the advanced recycling fee (ARF), which they pay when they buy their EEEs. It is also the responsibility of the consumers to return their obsolete appliances to the collection centers or retailer shops (Borthakur and Govind 2017). The UK EEEs waste management and disposal policies are also based on the Extended Producer Responsibility (EPR) initiatives. Producers and retailers of a product are under an obligation to provide their customers with the means of disposing of their former household appliances when new ones are sold to them (Borthakur and Govind 2017).

Prior to 2003, e-waste disposal in Spain was primarily collected by retailers for consumers to take new models of electronic appliances and disposed of to the metal managers, while some are also disposed of in domestic waste bins and end up in collection outlets inside waste storage and eventually at the landfill sites (LSs) (Queiruga et al. 2012). In 2005, the Royal Decree (RD) 208/2005 on EEE-waste management was established and reformed into a national legislation from the European Directive. The directives mandated producers to be responsible for their EEE waste management (take-back) and financing in terms of the collection, disposal, and recycling of new and old e-waste without charging additional fees to consumers (Queiruga et al. 2012).

In the African context, developing countries (DCs) such as Nigeria do not have recognized legislation specifically for EEE waste or a well-established integrated system for e-waste control and management for the collection, separation, transportation, disposal, storage, and treatment (Nnorom and Osibanjo 2008). E-waste handling and management systems in Nigeria are largely informal, involving local recyclers and open dumping on land and water bodies and other methods (Nnorom and Osibanjo 2008), except for the first registered e-waste recycling facility established by HP in 1989 (Aremu et al. 2020). It was reported that ~5 million personal computers (PCs) alone and other unaccounted data of electronic waste made up of secondhand and junk EEE such as computer accessories and monitors were imported into Nigeria annually (Nnorom and Osibanjo 2008). On average, about 500 shipping containers of secondhand electronics (comprising both functional and non-functional electronics products such as PCs, CPUs, monitors, etc.) enter Lagos port monthly (Nnorom and Osibanjo 2008). Data on the local e-waste generation in Nigeria is rarely available. Recently, electronics manufacturers such as Dell, Hewlett Packard, and others formed a recycling alliance. The purpose of the alliance was to purchase electronic scraps from consumers for recycling purposes. The initiative was able to reclaim some of the metal ores, but other important elements such as mercury and precious metals (gold, copper, etc.) were not recovered due to lack of the necessary infrastructure (Aremu et al. 2020).

Earlier in the United States of America (USA), all costs (collection, transportation, recycling, and disposal) associated with the management of e-waste were borne only by the consumers of EEEs through the control of municipality and states. Here, extended consumer responsibility (ECR) is the most practiced regulatory tool for e-waste management (recycling and disposal), while storage is the most preferred method of managing end of life e-waste (Lepawsky 2012; Rautela et al. 2021). Consumers of electronics waste pay a fee to discard their damaged or obsolete products. The fees are used for the transportation and recycling of the e-waste. Later on, a modified form of the EPR tool was initiated, whereby the management costs relating to EEEs waste collection, transportation, and recycling were shared among the producers, consumers, and municipalities (Borthakur and Govind 2017; Wagner 2009).

E-waste disposal and management in Australia is mostly carried out in drop-off facilities free of charge; however, recycling fees are charged to the consumers at the time of purchase of their new products. Therefore, EPR and advanced refund fees (ARFs) are the major tools used in financing the management of e-waste in Australia. Consumers' electronic waste such as televisions, computers, and other e-waste, is collected and recycled by municipalities, charitable organizations, and producer business organizations (Borthakur and Govind 2017).

From the e-waste disposal and management methods practiced by different countries as discussed earlier, it is obvious that four major ways of dealing with e-waste by these countries are through the following, as reported by (Kumar et al. 2017):

1) Official take-back policies (through EPR and ARF policies)
2) Disposal of leftover mixed waste
3) Informal e-waste collection outside the official take-back/EPR systems
4) Unofficial EEE trash recycling and collection in poor nations

8.4 Environmental Fates due to E-waste

Improper management of e-waste could lead to contamination of the human food chain via the entry of toxic content from e-waste into the water body, atmosphere, and soil (Zhao et al. 2008). Several studies have shown that the long-term presence of e-waste in soil leads to the presence of contaminants such as polyaromatic hydrocarbons (PAH), heavy metals, polybrominated diphenyl ethers, and polyvinyl chloride phthalates in the soil. These have a substantial negative impact on the variety, content, and number of bacterial cells in the soil as well as on the health of higher organisms. They also disrupt the soil's phylogeographic configuration and functionality (Saha et al. 2021; Salam and Varma 2019; Tang et al. 2013). Another study found that disposal facilities with e-waste as part of the MSW deposited in them have a higher occurrence of poisonous substances than disposal facilities without e-waste (Singh et al. 2020).

8.4.1 Effect of E-waste on the Environment

Waste electrical and electronic equipment (WEEE) comprises both useful and dangerous materials and, therefore, requires special treatment and recycling procedures in order to avert harmful effects on people and adverse environmental impacts. In some developing countries (DCs) where informal sector recycling is practiced, they employ crude methods such as open burning, manual disassembling, acid stripping, etc. to separate reusable components, and retrieve metals and precious metals (Borthakur and Govind 2017; Nnorom and Osibanjo 2008). These crude activities, especially when not properly monitored, expose the public (workers and residents) and the environment (soil, water, and air) to numerous contaminations (Gangwar et al. 2019). Burning of e-waste at LSs prior to disposal results in increased mobility and leaching of heavy metals into both the underground and surface waters, and causes toxicological effects on aquatic animals (Gaidajis et al. 2010). The burning of plastic cover insulators in cables pollutes the environment with the emission of dioxins, as well as the burning of refrigerators and air conditioners that contains refrigerants such as chlorofluorocarbons (CFCs) and produce polycyclic aromatic hydrocarbons (PAHs), which pollute the environment. CFCs are considered lethal, and over time exposure to the atmosphere may result in the depletion of the ozone layer (Kumar et al. 2017). Again, polybrominated diphenylethers (PBDEs), a fire retardant, influence soil microorganisms, metabolic functioning, etc. (Kumar et al. 2017). A typical example of the long-term hazardous effects of e-waste's improper disposal and recycling on the environment and associated health issues is seen in the case of Guiyu town in China (Nnorom and Osibanjo 2008). A variety of negative environmental effects result from the disposal of e-waste in landfills or through incineration, but proper recycling of e-waste poses little or no risk to the environment when the right waste management technology is used.

Some of the dangerous substances and heavy metals found in e-waste and some of their associated risk effects on human and animal health are presented in Tables 8.1 and 8.2, respectively.

Table 8.1 Occurrences and dosage of some environmental pollutants produced from e-waste during management processes.

Substance	Occurrence in e-waste	Average concentration in e-waste (mg/kg)	Global emissions (tons)
***Halogenated compounds*:**			
Polychlorinated biphenyls (PCB)	Condensers, transformers	~14.1	~281.1
Polybrominated biphenyl (PBB), polybrominated diphenyl ethers (PBDE), and tetrabromobisphenolA (TBBA)	Fire retardants found in plastics for example thermoplastic types (polyvinyl chloride, polycarbonate, etc.)	–	–
Chlorofluorocarbon (CFC)	Cooling (refrigerant) unit in refrigerators and air conditioners, insulation foam	–	–
Polyvinyl chloride (PVC)	Insulation for cables	–	–
***Heavy metals and other metals*:**			
Antimony	Fire retardant, plastics	~1,801	~35,001
Arsenic	Gallium arsenide, in trace amounts, is a component in light-emitting diodes	–	–
Barium	Getters found in cathode ray tubes (CRT)	–	–
Beryllium	Power supply units with X-ray lenses and silicon-controlled rectifiers	–	–
Cadmium	Printer drums, rechargeable nickel cadmium batteries, fluorescent layer in cathode ray tubes (CRT) displays, printer inks and toners, photocopying and printing equipment	~181	~3,601
Chromium	Data tapes, floppy disks	~9901	~200,001
Copper	Cables	~41 001	~820,001
Lead	Found in CRT screens, printed wiring (circuit) boards, and batteries	~2901	~59,001
Lithium	Lithium batteries (Li-batteries)	–	–
Mercury	Fluorescent lighting used in some alkaline batteries, mercury wetted switches, and liquid crystal displays (LCDs) (thermostats)	~0.69	~13.7
Nickel	Rechargeable nickel cadmium batteries or NiMH-batteries, electron gun in cathode ray tubes (CRT)	~10 301	~207,001
Rare earth elements	Fluorescent layer in cathode ray tubes (CRT-screen)	–	–
Selenium	Found in older photocopying and printing machines (photo drums)		
Tin	Soldering metal glues, liquid crystal display (LCD)	~2401	~50,001
Zinc sulphide	Found in interiors of cathode ray tube (CRT) screens, mixed with rare earth metals	~5100	~102,001
***Others*:**			
Toner dust	Photocopier and laser printer toner and ink cartridges	–	–
Radioactive substances	Medical devices, smoke detectors with active sensing components, and fire detectors	–	–

(*Source:* Adapted from Gaidajis et al. 2010, with some modifications.)

8.4.2 Effect of E-waste on Soil

Environmental pollution resulting from crude e-waste recycling alters the soil microorganisms' activity and population, which may adversely influence the overall soil properties, soil mineral contents, and plant growth, and largely affect their ecosystem functions. A study by Sanusi (2015) examined the outcome of burning waste electrical and electronic (WEE) materials on the chemical, physical, and microbiological features of subsurface soil obtained from an e-waste dumpsite. The study reported increased concentrations of heavy metals (Pb, Zn, Mn, Ni, Co, Cr,

Table 8.2 Detrimental effects of some substances found in e-waste on exposure to humans, animals, and plants.

Substances	Effect on exposure (on humans, plants, and animals)
Antimony	Causes severe skin related problems and other health concerns.
Cadmium	Causes deficit in learning abilities and neuro-motor skills in children. Damage to kidneys, lungs (when inhaled) and bone structure; could accumulate in human body over time.
Polychlorinated biphenyls	Leads to suppression of human immune system, liver damage, promotes cancerous cell growth, damage to nervous system, behavioral changes, and damage to male and female reproductive system.
Polychlorinated naphthalene	Toxic to animals and human, harmful effects on liver, nervous systems, and reproductive systems, impacts on skin.
Triphenyl phosphate	Inhibits some important enzymes functioning in the human blood; can result in contact dermatitis, very toxic to aquatic life, and a potential endocrine disruptor

(*Source:* Adapted from Kumar et al. 2017, with some modifications.)

and Cd), organic matter, moisture, and organic contents in the subsurface soil samples collected and analyzed from dumpsites where e-waste was incinerated compared to the control soil (that is, soil collected from a non-e-waste dumpsite). In addition, low soil pH and organic phosphorus were also reported. Important soil microorganisms such as *Articulospora, Bacillus, Candida, Enterobacter, Proteus, Staphylococcus, Varicosporium*, and *Zoospora* were isolated from the subsurface soil samples from e-waste dumpsites. Soil fungal and bacterial populations in the examined soil samples were also altered by the e-waste burning activities at the dumpsite (Sanusi 2015). Also, laboratory studies by Liu et al. (2015) investigated how the composition and diversity of microbes and microbial assemblages affect the soil properties in a set of e-waste-polluted soil samples collected from Guiyu, China. Their results revealed high variation (about 70.3%) in microbial assemblages and composition for the e-waste-contaminated soils compared to uncontaminated soil. The high variation was attributed to the interaction effects of soil physicochemical variables (such as pH, total organic carbon, and so on), soil moisture content, copper, and toxic organic compounds; decabromodiphenyl ether (BDE-290) present in e-waste, which pollutes the soil (Liu et al. 2015).

8.4.3 Effect of E-waste on Water (Both Surface and Underground)

The technological advancement in electronic equipment is meant to make lives better and easier, but the rate of e-waste production due to innovations, obsolescence, and damaged devices being dumped away in solid waste (SW) streams pose environmental and health challenges (Needhidasan et al. 2014). The production of e-waste will continue to grow as technological evolution continues to influence the behavior of people in electronic device utilization and disposal due to the short lifespan of electronic equipment and changes in consumer tastes (Needhidasan et al. 2014). Most of these materials are transported to DCs as secondhand goods where they are dismantled and recycled. Workers and the environment are exposed to dangerous components of electronic devices such as lead, cadmium, and other halogenated flame-retardant compounds, etc., when e-waste recycling operations take place in unorganized settings with limited precautionary measures by those carrying out the recycling activities (Heacock et al. 2018; Parvez et al. 2021). As a result, situating recycling operations near residential, educational, and recreational facilities often results in dramatically elevated levels of pollutants in the soil, road dust, air, and water (Zheng et al. 2016). Inadequate handling of e-waste results in a variety of damaging effects on the environment, such as soil degradation and contamination, water source pollution and toxicity, and the production of poisonous gases (from the incineration of e-waste), which has a direct detrimental impact on the health of living things (Saha et al. 2021). In a study by Otache et al. (2014) in Lagos, Nigeria that was intended to determine the impact of e-waste emanating from Alaba International Market, where e-waste processing is prominent, it was revealed that the heavy metal content of the soil within the vicinity of the e-waste dumpsite was above the permitted levels for lead, cadmium, and zinc (Otache et al. 2014). This study also showed that the toxic contents of e-waste can percolate through the soil and contaminate underground water if not properly addressed. Disposal of hazardous e-waste in landfills and waterways contaminates groundwater (Twagirayezu et al. 2022). A similar study carried out on the eastern coast of Guangdong province in China, where e-waste recycling is well practiced, showed that the concentration of lead (Pb) in the water samples analyzed was found to be 2400 times higher than the set permissible value as given by the World Health Organization (Guo et al. 2009; Xu et al. 2015). Many nations that rely on groundwater as a natural water resource are concerned about water quality degradation and aquifer storage reductions. Hazardous heavy metals are produced by e-waste, and when they are disposed of in landfills, they leach into the groundwater. Percolation of such pollutants into the groundwater would definitely make the water unsuitable for irrigation, drinking, or other uses due to the high contaminants and toxicity level (Gupta and Nath 2020).

8.4.4 Effect of E-waste on Air

Hazardous materials are frequently discharged during recycling processes, particularly the rudimentary activities of informal recyclers. When the e-waste is incinerated in the open air for material extraction, toxic and cancer-causing chemicals such as furans, dioxins, and polycyclic aromatic hydrocarbons are released, thereby causing environmental pollution (Heacock et al. 2018). E-waste can be a channel for air pollution when it is treated by informal recyclers with inadequate recycling processes/technologies and poor regulatory systems, as is seen in the case of some developing economies. The respiratory health of employees without adequate respiratory protection is substantially and chronically impacted in these informal sectors. E-waste is frequently disassembled and shredded in a rudimentary manner, releasing dust and particulates into the environment (Saha et al. 2021). The informal sector's crude and unsophisticated e-waste processing techniques have considerably contributed to the elevated air particulate matter levels, specifically the levels of heavy metals, in an illegitimate e-waste incinerating zone. Burning of plastics that contain flame retardants results in the emission of harmful and toxic dioxins, which are classified as persistent organic pollutants (POPs). Polychlorinated biphenyl compounds, usually found in plasticizers and coolants of condensers in refrigerators, air conditioners, etc., could result in the depletion of the ozone layer (Rautela et al. 2021). Continuing exposure to air pollution resulting from the processing of e-waste has led to frightening levels of toxicity due to heavy metal exposure to the residents within the vicinity, which in turn has been linked to a high prevalence of cardiovascular illnesses such as hypertension, etc. (Gangwar et al. 2019). The amount of fine and coarse particle pollutants produced by burning of e-waste would significantly increase. In their studies, Bungadaeng et al. (2019) established the intensity of particulate matter (PM) inhaled by e-waste workers in Thailand. It was discovered that the average $PM_{2.5}$ mass intensities were greater than 50% of the entire PM_{10} massintensities, indicating that $PM_{2.5}$ was both the primary kind of particulate matter produced from open burning activities and the main component of PM_{10}. The amount of incinerated e-waste and the form of incineration activities are two important determining factors that can have an adverse effect on the employees subjected to particles emanating from e-waste burning activities through inhalation (Bungadaeng et al. 2019).

8.5 Impact of E-waste on Public Health

Effective waste management, especially for hazardous materials, is crucial for worldwide sustainable development (Maphosa and Maphosa 2020). Waste can be treated or disposed of using a variety of techniques, including landfilling, burning, composting, and recycling. Due to its low cost, landfilling continues to be the most frequently used method in DCs, but improper management of this waste can cause serious health and environmental issues (Yang et al. 2018). Most developing nations lack the technical capabilities and infrastructure required for safe disposal of hazardous waste, and in these nations, e-waste has been connected to several health issues, including cancer, impaired neurological and respiratory conditions, birth defects, etc. (Bhutta et al. 2011). The industrialized nations adopted legislation for the management (collection, treatment, and disposal) of electrical and electronic waste because of the detrimental effects of its inadequate processing on both humans and the environment (Maphosa and Maphosa 2020). In underdeveloped countries, most e-waste handlers process them using crude methods such as open burning in landfills, which impacts the most vulnerable populations of the general public (for instance, children and pregnant women) who live close to the informal recycling dumpsites (Bakhiyi et al. 2018). Breathing issues (lung pain, choking, coughing, pneumonitis), neuropsychiatric issues, tremors, convulsions, comas and even death are among the human health concerns associated with e-waste (Kumar et al. 2020). Table 8.3 shows the lists of various e-waste sources, compositions, and health effects.

The informal sectors typically carry out the majority of the tasks involved in collecting, handling, disassembling, and reprocessing of e-waste because they lack the necessary technical and physical infrastructure as well as little or no knowledge of the negative impacts that inappropriate management of e-waste may pose on public health and the environment (Gangwar et al. 2019). When e-waste is indiscriminately discarded into the environment, its hazardous components usually seep through porous soil and eventually combine with groundwater, causing pollution of the water and soil and posing serious health issues to the public via the food chain distortion (Twagirayezu et al.

Table 8.3 E-waste compositions, sources, and adverse health effects.

E-waste compositions	Sources	Health effects
Lead	Soldering in printed circuit board, glass panels, and computer display gaskets.	Damage to the kidneys, circulatory system, central and peripheral nerve systems, adverse effects on children's brain development.
Cadmium	Chip resistors, semi-conductors, infrared detectors, printer inks and toners.	Permanent toxic consequences on human health, particularly kidney damage.
Mercury	Relays and switches, batteries, backlight bulbs or lamps, and printed circuit boards.	Acute brain injury, skin and respiratory conditions brought on by bioaccumulation in fish.
Nickel	Circuit (printed) boards, cathode ray tubes, Ni batteries, and computer housing.	Can result in lung cancer, pneumonia, allergic reactions, and impaired lung function.
Arsenic	Light-emitting diodes employ gallium arsenide.	Has long-term impacts that lead to nerve signaling impairment, skin disease, and lung cancer.
Brominated flame retardants	Plastic housing for electrical devices and circuit boards.	Interferes with endocrine system operations.
Beryllium	Motherboards, power supply boxes.	Carcinogenic (lung cancer) dust and fume inhalation causes beryllicosis or chronic beryllium illness and skin diseases such as warts.
Antimony	A melting agent in cathode-ray tube (CRT) glass, plastic computer housings, and soldering alloy in cabling.	Categorized as carcinogen. Inhalation of this substance might result in stomach pain, diarrhea, vomiting, and stomach ulcers.
Plastics including PVC	Cabling and computer housing.	Dioxin is produced through burning; it causes issues with reproduction and development. Damages immune system.

Adapted from Kumar et al., 2020.

2022). Although humans may be exposed to e-waste through a variety of complex channels, the type of exposure source, the interval or duration of exposure, and any potential additive, synergistic, or inhibitory effects of multiple exposures can all have an impact on health consequences (Parvez et al. 2021).

8.5.1 Impact of E-waste on DNA and Fetus

E-waste recycling is a major source of income for many people, especially the underprivileged residing in urban parts of developing countries (DCs). However, this activity usually results in severe health risks as people are exposed to toxic substances associated with e-waste reprocessing (Amankwaa et al. 2017). When exposed to heat, the inherent heavy metals and toxic chemicals in e-waste are let out into the air, soil, and water, thereby seeping into underground water and contaminating the soil, plants, and eventually poisoning fish and human (Bakhiyi et al. 2018). Apart from exposure to heat, e-waste stored outside poses environmental pollution when it is soaked by a heavy downpour or flooded into the nearest river. Run-off from the dumpsites would have access to local waterways and finally contaminate the groundwater (Lovo and Rawlings 2021). Some of the associated toxic substances include cadmium, barium, polybrominated flame retardants, arsenic, lithium, mercury, arsenic, chromium, copper, and lead. Their concentration has been confirmed to increase in human blood through some environmental vectors such as contaminated air, water, soil, plants, seafood, etc. (Chen et al. 2011; Saha et al. 2021). This trend continues to happen when tools such as extended producer responsibility (EPR), material flow analysis (MFA), multi-criteria analysis (MCA), and life cycle assessment (LCA), which have been established to manage e-waste, are not duly implemented (Kiddee et al. 2013). The above-mentioned toxic substances are usually recalcitrant, mutagenic, and carcinogenic (Wang et al. 2012). Their voyage into the human body is also through ingestion, inhalation, and dermal contact to blood and organs, and is traceable to contaminated air humans breathe, leaching into community water regularly consumed, as well as soil used for farming (Grant et al. 2013; Saha et al. 2021). Studies have established that there is an increase in the concentrations of heavy metals and organic contaminants in marine life (Bandowe et al. 2014; Hosoda et al. 2014). These toxic substances have been confirmed to be capable of reaching the fetus by crossing the placental barrier from the mother, and this is indirectly responsible for many detrimental effects in neonates (Sanders et al. 2014). Maternal plasma has been reported to contain cell-free DNA (deoxyribonucleic acid) fragments that is derived from both the mother and placenta; hence, babies in the womb are not safe from the outcome of their mothers' exposure to heavy metals and other toxic components in e-waste (Canick et al. 2013). Although gene sets between mother and fetus DNA were reported to be distinct, functional similarities were observed at the biological path as well as an enhancement of genes that code for proteins controlling transcriptional adjustment and apoptosis (Sanders et al. 2014). Similarly, conserved DNA prototypes, which have structural resemblance to a given transcription component binding site, were observed inside the CpG islands of the gene sets. This was confirmed by a study carried out as a proof for distinctive arrays of DNA "footprints" in fetal and maternal DNA traceable to cadmium contact (Sanders et al. 2014). In the mammalian genome, DNA methylation is described as an epigenetic tool that entails the transfer of a methyl group to the C5 position of the cytosine, forming 5-methylcytosine (Moore et al. 2013). Abnormalities in DNA methylation lead to the loss of normal imprinting, which is responsible for several inherited genetic diseases. Hazardous components of e-waste are also present in the form of halogenated compounds like tetrabromobisphenolA (TBBPA), carcinogenic polychlorinated biphenyls (PCBs), and polybrominated biphenyls (PBBs), as well as mercury and its compounds, and they have the capacity to intercalate the DNA strands leading to DNA damage resulting from unrepairable single-strand breakage of the DNA and eventually mutation in affected sites (Alabi et al. 2021; Orisakwe et al. 2020). They are also responsible for the formation of superoxide radicals and the depletion of reduced glutathione in healthy cells, as well as cytotoxicity in nerve cells (Alabi et al. 2021). High mercuric concentration has also been implicated in low DNA integrity, motility, and acrosomal status of human spermatozoa (Alabi et al. 2021; Balali-Mood et al. 2021; Kumar et al. 2020). The genotoxic effect of lead, manganese, copper, and cadmium on DNA integrity in plant cells, which serve as environmental and biotic indicators to quantify potential effects of pollutants, was analyzed by the application of random amplified polymorphic DNA (RAPD) in regularly consumed kidney bean (*Phaseolus vulgaris*), leading to confirmation of DNA damage and mutation in the kidney-bean seedlings treated with different concentrations of these heavy metals (Enan 2006). Contact with the toxic chemicals in e-waste is also responsible for chromosomal aberrations (Sousa et al. 2022). Workers at informal e-waste establishments are exposed to toxic components of e-waste due to a lack of proper protective equipment, putting their health and that of their families at risk. Prenatal exposure to the noxious components of informal recycling of electronic waste is attributed to the high rate of adverse birth outcomes such as low birth weight, stillbirth, and an increase in neonatal and infant death rates by 4.6% and 6.6%, respectively (Baldé et al. 2017; Grant et al. 2013; Lovo and Rawlings 2021; Xu et al. 2012). More than 70% of children and neonates were confirmed to possess higher levels of these toxic chemicals in their blood, and the placenta of pregnant women has been confirmed to possess blood lead levels of above 10 μg/dL; at the lowest doses, these toxic substances have been described as being highly detrimental to human and fetal health (Chakraborty et al. 2022; Xu et al. 2012). Pregnant women have a tendency to face spontaneous abortions, premature death, fetal loss, abnormal thyroid function, neurobehavioral disturbances, gene toxicity, and congenital malformations as a result of their exposure to several recalcitrant organic pollutants and polycyclic aromatic hydrocarbons in e-waste (Xu et al. 2012). Samples of breast milk taken from women residing close to an e-waste dumpsite in Ghana were confirmed to contain abnormally high concentrations of pollutants (Daum et al. 2017). Heavy metals have been confirmed to act as endocrine disruptors, displaying either an antagonistic or agonistic effect on the endocrine system by interfering with biosynthesis and endogenous hormone activities (Georgescu et al. 2011). It has been demonstrated that the zinc atom from the Zn fingers of the estrogen receptor can be substituted with numerous heavy metal molecules, and that this substitution can interfere with the binding of the DNA hormone-responsive elements in the cell (Georgescu et al. 2011). Although metal ions play vital roles in living systems, such as their catalytic involvement in traces or ultra-traces of quantity, they also act as cofactors during several important biochemical reactions. These metals become poisonous to human cells when they are present in high concentrations or at levels that exceed the permissible optimal levels. Copper ions are recognized to initiate the development of reactive oxygen species, which could harm biomolecules including DNA and chromatin (Linder 2012). This is due to its mobilization and redox activity, which enables the formed reactive oxygen species (ROS) to bind rapidly to the DNA, causing harm by splitting the DNA strands or modifying the base and/or deoxyribose, resulting in carcinogenesis and other diseases (Chakraborty et al. 2022). Some neurologic diseases related to copper metabolism disorders involve the occipital horn syndrome, Menkes disease, and distal motor neuropathy (Woimant and Trocello 2014). Maternal contact with polybrominated diphenyl ethers (PBDE) and other flame retardants has been connected to some ill-health outcomes such as thyroid hormone dysregulation, neurotoxicity, endocrine disruption, and poor semen quality (Buttke et al. 2013). Maternal exposures can be responsible for fetal exposures because it has been confirmed that PBDEs pass directly into the placental circulation (Buttke et al. 2013).

8.5.2 Impact of E-waste on Reproductive Organs

Reproductive hazards resulting from exposure to heavy metals and other toxic components of e-waste are responsible for various reproductive dysfunctions in humans. Heavy metal exposure causes structural and functional cellular damages in reproductive systems (Alabi et al. 2021; Machado-Neves 2021). These toxic constituents have been found to be responsible for possible health conditions caused by chronic or acute heavy metal toxicity, which include immunodeficiencies, osteoporosis, neurodegeneration, and organ failure (Rzymski et al. 2015). They have been directly implicated in estrogen-dependent diseases such as endometrial cancer, breast cancer, endometriosis, spontaneous abortions, and hypotrophy (Rzymski et al. 2015). In both females and males, low-level exposure to heavy metals like cadmium inhibits the biological activities of steroid hormones in the reproductive organs (Georgescu et al. 2011). An example is found in arsenic, which possesses the capability to bind the glucocorticoid receptor, leading to disruption in glucocorticoid biological activity. However, mercury may trigger modifications in male and female fertility by inhibiting the functions of the hypothalamo-pituitary-thyroid axis or hypothalamo-pituitary-adrenal axis, thereby disrupting the biogenesis of steroid hormones (Georgescu et al. 2011; Machado-Neves 2021). Heavy metal accumulation in male reproductive organs is responsible for reproductive disorders, particularly in the epididymis, which plays an important role in sperm development (Machado-Neves 2021). Research has established that exposure to heavy metals and toxic compounds is responsible for reduced epididymal weight, reduced sperm motility, and low sperm count (Machado-Neves 2021). Similarly, in women, this has been confirmed to cause hormonal imbalance, follicular atresia, and a delay in oocyte maturation (Bhardwaj et al. 2021). The mechanism described for the activity of heavy metal-induced toxicity is the production of reactive oxygen species (ROS), which leads to oxidative damage and other health-associated adverse effects (Rehman et al. 2018). On the other hand, sodium arsenate and lead acetate are implicated in the generation of oxidative stress caused by unevenness between ROS production and scavenging (Machado-Neves 2021). Oxidative stress is the main causative factor of infertility, as it is the outcome of an imbalance between the ROS and antioxidants leading to sperm damage, deformity, and finally, male infertility (Bansal and Bilaspuri 2011; Kumar et al. 2020). Low-concentration ROS secreted during energetic metabolism via the respiratory chain play important physiological roles, but their excessive amount produced under oxidative stress is extremely hazardous to the central nervous system (CNS) and is specifically vulnerable to oxidative stress (Li et al. 2013). ROS, on the other hand, have been reported to cause sperm pathology (ATP depletion), which leads to inadequate axonemal phosphorylation, lipid peroxidation, and loss of motility and viability (Bansal and Bilaspuri 2011). Cadmium chloride is implicated in bringing about a rise in pH level of luminal fluid (from 6.4 to 7.5), leading to a reduction in sperm viability (Machado-Neves 2021). Subacute contact with heavy metals at strengths of around 0.1 mg/L starts a dyshomeostasis of cadmium, etc., which interrupts sperm considerations, leading to a reduction in the weight of the epididymis, which becomes worse with time and exposure to higher doses (Machado-Neves 2021). Metals have been observed to interfere with the gametogenic cells, Leydig cells, or spermatozoa directly in the semen resulting in reduced fertility and other genetic diseases caused by congenital malfunction (Alabi et al. 2021; Machado-Neves 2021). Exposure to heavy metal has also been suspected to be responsible for the characteristics of heat stress protein (HSP), androgen-binding protein (ABP), cadherin, and several other stressor proteins, as well as ROS and neuro-endocrine mechanisms (Kumar et al. 2020). Persistent polybrominated diphenyl ethers (PBDEs) used as flame resistants in e-waste have shown bioaccumulative properties and have been linked to various hormonal disorders such as those in thyroid hormones, neonatal health, and reproductive stress (Singh et al. 2022). One of the well-recognized groups of enzymes known as metallothioneins is described as being responsible for the detoxification of heavy metals, but their activity in individuals may vary depending on life habits, level of exposure to heavy metals, age, dietary habits, genetic variations, and body population-related factors (Joneidi et al. 2019).

8.5.3 Impact of E-waste on Children

Children are usually exposed to the hazardous constituents of e-waste via the ingestion of contaminated food, water, breastmilk, soil, and dust. It could also be via transplacental exposure, inhalation of contaminated air, or skin exposure (Chakraborty et al. 2022). Various pathways through which children are exposed to various toxic substances from e-waste are shown in Figure 8.2. The World Health Organization confirmed that more than 18 million children and adolescents between the ages of 5 and 17 are globally exposed to e-waste, and statistics have revealed that some of the most affected children are under the age of 5 (World

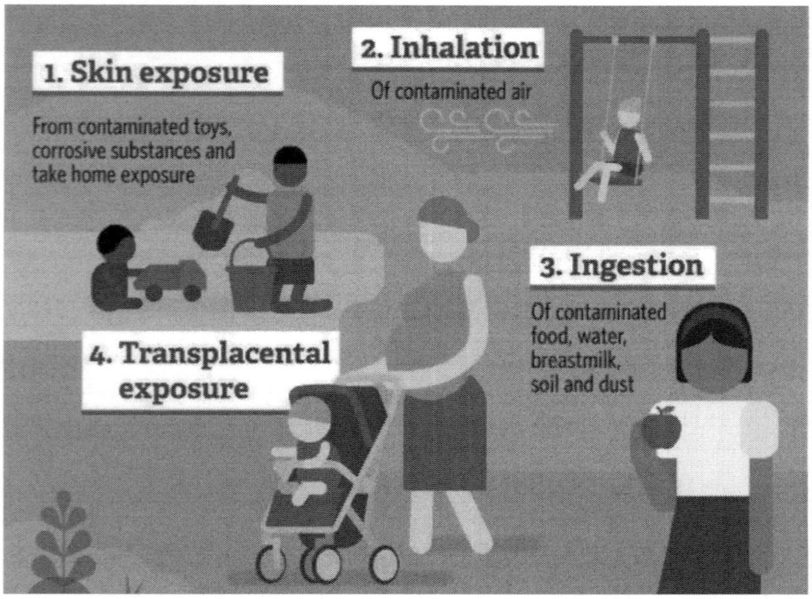

Figure 8.2 Ways in which children are exposed to e-waste toxic substances (World Health Organization 2022). (*Source:* https://www.who.int/multi-media/details/Children).

Health Organization 2021). Children are more susceptible to the toxic effects of e-waste exposure because of their hand to mouth activities and physiological discrepancies from adults. This is due to the fact that they require higher air and food intake per body weight, thereby absorbing more pollutants in comparison to the tenderness of their immature organs, rapid growth rate, and size, which have much less capability to digest or exterminate poisonous substances from their bodies (Perlroth and Branco 2017).

Exposure of children to lead (Pb) whose source has been traced to activities of e-waste recycling has been reported. A change in neurological functioning has been observed in newborns, as well as an increase in the rates of attention deficit/hyperactivity disorder (ADHD), abnormal changes in child temperament, behavioral problems, difficulty in sensory integration, changes in lung function, a respiratory disorder, and an increased risk of some chronic diseases such as diabetes, risk of cancer, neuronal damage, renal injuries, impaired thyroid function, and cardiovascular disorders later in life (Rehman et al. 2018; Scott and Pocock 2021). In children and adults, e-waste exposure is one of the major causes of IQ deficits, attention dysfunction, slowed growth, asthma, whooping cough, tuberculosis, birth defects, weight loss, cardiovascular diseases, childhood cancer, low birth weight, preterm delivery, chronic obstructive pulmonary disease (COPD), and many other infectious diseases (Alabi et al. 2021; Balali-Mood et al. 2021; Orisakwe et al. 2020; Rautela et al. 2021). E-waste has also been reported as the cause of headaches, skin damage, nausea, vertigo, chronic gastritis, and gastroduodenal ulcers, and children who experience higher exposures to flame retardants display poorer skills in academics and social life (Xu et al. 2012).

8.6 Conclusion

In this study, an overview of the fate of e-waste in the environment, and its adverse impact on public health and the environment (soil, air, and water) have been discussed. Inadequate management of electronic waste causes pollution of soil and water bodies, which consequently impacts the lives of marine organisms negatively. Exposure to the hazardous constituents of e-waste leads to contamination of food, water, and breastmilk. Poor e-waste handling could also disrupt the food chain in humans. It was also established that exposure to heavy metals and toxic compounds from e-waste could result in reduced epididymal weight, sperm motility, low sperm count and hormonal imbalance, follicular atresia, and delay in oocyte maturation in women.

Acknowledgment

This work is based on the research supported in part by the National Research Foundation of South Africa (Ref Numbers: SASPD22091357540).

References

Akram, R., Fahad, S., Hashmi, M.Z. et al. (2019). Trends of electronic waste pollution and its impact on the global environment and ecosystem. *Environmental Science and Pollution Research* 26 (17): 16923–16938.

Alabi, O.A., Adeoluwa, Y.M., Huo, X. et al. (2021). Environmental contamination and public health effects of electronic waste: an overview. *Journal of Environmental Health Science and Engineering* 19: 1209–1227.

Amankwaa, E.F., Tsikudo, K.A.A., and Bowman, J.A. (2017). 'Away' is a place: the impact of electronic waste recycling on blood lead levels in Ghana. *Science of the Total Environment* 601: 1566–1574.

Amoabeng Nti, A.A., Arko-Mensah, J., Botwe, P.K. et al. (2020). Effect of particulate matter exposure on respiratory health of e-waste workers at Agbogbloshie, Accra, Ghana. *International Journal of Environmental Research and Public Health* 17 (9): 3042.

Aremu, S.A., Olukanni, D.O., Mokuolu, O.A. et al. (2020). Circular economy: Nigeria perspective. In: *Circular Economy: Global Perspective* (ed. S. Ghosh), 279–297. Singapore: Springer.

Bakhiyi, B., Gravel, S., Ceballos, D. et al. (2018). Has the question of e-waste opened a Pandora's box? An overview of unpredictable issues and challenges. *Environment International* 110: 173–192.

Balali-Mood, M., Naseri, K., Tahergorabi, Z. et al. (2021). Toxic mechanisms of five heavy metals: mercury, lead, chromium, cadmium, and arsenic. *Frontiers in Pharmacology* 12: 643972.

Baldé, C.P., Forti, V., Gray, V. et al. (2017). *The Global E-waste Monitor 2017: Quantities, Flows and Resources*. United Nations University, International Telecommunication Union, and International Solid Waste Association.

Bandowe, B.A.M., Bigalke, M., Boamah, L. et al. (2014). Polycyclic aromatic compounds (PAHs and oxygenated PAHs) and trace metals in fish species from Ghana (West Africa): bioaccumulation and health risk assessment. *Environment International* 65: 135–146.

Bansal, A.K. and Bilaspuri, G.S. (2011). Impacts of oxidative stress and antioxidants on semen functions. *Veterinary Medicine International* 2011.

Bhardwaj, J.K., Paliwal, A., and Saraf, P. (2021). Effects of heavy metals on reproduction owing to infertility. *Journal of Biochemical and Molecular Toxicology* 35 (8): e22823.

Bhutta, M.K.S., Omar, A., and Yang, X. (2011). Electronic waste: a growing concern in today's environment. *Economics Research International* 2011: 474230.

Borthakur, A. and Govind, M. (2017). Emerging trends in consumers' E-waste disposal behaviour and awareness: a worldwide overview with special focus on India. *Resources, Conservation and Recycling* 117: 102–113.

Bungadaeng, S., Prueksasit, T., and Siriwong, W. (2019). Inhalation exposure to respirable particulate matter among workers in relation to their e-waste open burning activities in Buriram Province, Thailand. *Sustainable Environment Research* 29 (1): 1–12.

Buttke, D.E., Wolkin, A., Stapleton, H.M., and Miranda, M.L. (2013). Associations between serum levels of polybrominated diphenyl ether (PBDE) flame retardants and environmental and behavioral factors in pregnant women. *Journal of Exposure Science & Environmental Epidemiology* 23 (2): 176–182.

Canick, J.A., Palomaki, G.E., Kloza, E.M. et al. (2013). The impact of maternal plasma DNA fetal fraction on next generation sequencing tests for common fetal aneuploidies. *Prenatal Diagnosis* 33 (7): 667–674.

Chakraborty, S.C., Qamruzzaman, M., Zaman, M.W.U. et al. (2022). Metals in e-waste: occurrence, fate, impacts and remediation technologies. *Process Safety and Environmental Protection* 162: 230–252.

Chen, A., Dietrich, K.N., Huo, X., and Ho, S. (2011). Developmental neurotoxicants in e-waste: an emerging health concern. *Environmental Health Perspectives* 119 (4): 431–438.

Chi, X., Wang, M.Y.L., and Reuter, M.A. (2014). E-waste collection channels and household recycling behaviors in Taizhou of China. *Journal of Cleaner Production* 80: 87–95.

Daum, K., Stoler, J., and Grant, R.J. (2017). Toward a more sustainable trajectory for e-waste policy: a review of a decade of e-waste research in Accra, Ghana. *International Journal of Environmental Research and Public Health* 14 (2): 135.

Enan, M.R. (2006). Application of random amplified polymorphic DNA (RAPD) to detect the genotoxic effect of heavy metals. *Biotechnology and Applied Biochemistry* 43 (3): 147–154.

Forti, V., Balde, C.P., Kuehr, R., and Bel, G. (2020). *The Global E-waste Monitor 2020: Quantities, Flows and the Circular Economy Potential*. United Nations University/United Nations Institute for Training and Research–Co-hosted SCYCLE Programme. International Telecommunication Union & International Solid Waste Association.

Gaidajis, G., Angelakoglou, K., and Aktsoglou, D. (2010). E-waste: environmental problems and current management. *Journal of Engineering Science and Technology Review* 3 (1): 193–199.

Gangwar, C., Choudhari, R., Chauhan, A. et al. (2019). Assessment of air pollution caused by illegal e-waste burning to evaluate the human health risk. *Environment International* 125: 191–199.

Georgescu, B., Georgescu, C., Dărăban, S. et al. (2011). Heavy metals acting as endocrine disrupters. *Scientific Papers Animal Science and Biotechnologies* 44 (2): 89–93.

Ghimire, H. and Ariya, P.A. (2020). E-wastes: bridging the knowledge gaps in global production budgets, composition, recycling and sustainability implications. *Sustainable Chemistry* 1 (2): 154–182.

Grant, K., Goldizen, F.C., Sly, P.D. et al. (2013). Health consequences of exposure to e-waste: a systematic review. *The Lancet Global Health* 1 (6): e350–e361.

Guo, Y., Huang, C., Zhang, H., and Dong, Q. (2009). Heavy metal contamination from electronic waste recycling at Guiyu, Southeastern China. *Journal of Environmental Quality* 38 (4): 1617–1626.

Gupta, N. and Nath, M. (2020). Groundwater contamination by E-waste and its remedial measure–a literature review. *Paper presented at the Journal of Physics: Conference Series* 1531 (1): 012023. IOP Publishing.

Heacock, M., Trottier, B., Adhikary, S. et al. (2018). Prevention-intervention strategies to reduce exposure to e-waste. *Reviews on Environmental Health* 33 (2): 219–228.

Hosoda, J., Ofosu-Anim, J., Sabi, E.B. et al. (2014). Monitoring of organic micropollutants in Ghana by combination of pellet watch with sediment analysis: e-waste as a source of PCBs. *Marine Pollution Bulletin* 86 (1–2): 575–581.

Jain, N. (2021). *E-waste––sources, composition, effects, treatment, disposal system*. Source: https://electricalfundablog.com/e-waste-electronic-waste-sources-composition (accessed 6 February 2023).

Jang, Y. (2010). Waste electrical and electronic equipment (WEEE) management in Korea: generation, collection, and recycling systems. *Journal of Material Cycles and Waste Management* 12 (4): 283–294.

Jang, Y. and Kim, M.S. (2010). Management of used & end-of-life mobile phones in Korea: a review. *Resources, Conservation and Recycling* 55 (1): 11–19.

Joneidi, Z., Mortazavi, Y., Memari, F. et al. (2019). The impact of genetic variation on metabolism of heavy metals: genetic predisposition? *Biomedicine & Pharmacotherapy* 113: 108642.

Kaya, M. (2016). Recovery of metals and nonmetals from electronic waste by physical and chemical recycling processes. *Waste Management* 57: 64–90.

Khan, J., Kumar, A., Giri, A. et al. (2021). Impact of electronic waste pollutants on underground water. In: *Groundwater Geochemistry: Pollution and Remediation Methods* (eds. S. Madhav and P. Singh), 265–281. John Wiley & Sons Ltd.

Khanna, R., Mukherjee, P.S., and Park, M. (2020). A critical assessment on resource recovery from electronic waste: impact of mechanical pre-treatment. *Journal of Cleaner Production* 268: 122319.

Kiddee, P., Naidu, R., and Wong, M.H. (2013). Metals and polybrominated diphenyl ethers leaching from electronic waste in simulated landfills. *Journal of Hazardous Materials* 252: 243–249.

Kiddee, P., Naidu, R., Wong, M.H. et al. (2014). Field investigation of the quality of fresh and aged leachates from selected landfills receiving e-waste in an arid climate. *Waste Management* 34 (11): 2292–2304.

Kumar, A., Holuszko, M., and Espinosa, D.C.R. (2017). E-waste: an overview on generation, collection, legislation and recycling practices. *Resources, Conservation and Recycling* 122: 32–42.

Kumar, V., Kamboj, N., Payum, T. et al. (eds.) (2020). *Advances in Environmental Pollution Management: Wastewater Impacts and Treatment Technologies*, 1. India: Agro Environ Media, Publication Cell of Agriculture and Environmental Science Academy (AESA).

Lepawsky, J. (2012). Legal geographies of e-waste legislation in Canada and the US: jurisdiction, responsibility and the taboo of production. *Geoforum* 43 (6): 1194–1206.

Li, J., O, W., Li, W. et al. (2013). Oxidative stress and neurodegenerative disorders. *International Journal of Molecular Sciences* 14 (12): 24438–24475.

Linder, M.C. (2012). The relationship of copper to DNA damage and damage prevention in humans. *Mutation Research/Fundamental and Molecular Mechanisms of Mutagenesis* 733 (1–2): 83–91.

Liu, J., He, X.X., Lin, X.R. et al. (2015). Ecological effects of combined pollution associated with e-waste recycling on the composition and diversity of soil microbial communities. *Environmental Science & Technology* 49 (11): 6438–6447.

Liu, Y., Zhang, J., Yang, X. et al. (2020). Efficient recovery of valuable metals from waste printed circuit boards by microwave pyrolysis. *Chinese Journal of Chemical Engineering* 40: 262–268.

Lovo, S. and Rawlings, S. (2021). *Garbage in, garbage out: the impact of e-waste dumping sites on early child health*. Discussion paper no. 2021-07, Department of Economics, University of Reading, UK.

Machado-Neves, M. (2021). Effect of heavy metals on epididymal morphology and function: an integrative review. *Chemosphere* 291: 133020.

Maphosa, V. and Maphosa, M. (2020). E-waste management in Sub-Saharan Africa: a systematic literature review. *Cogent Business & Management* 7 (1): 1814503.

Moore, L.D., Le, T., and Fan, G. (2013). DNA methylation and its basic function. *Neuropsychopharmacology* 38 (1): 23–38.

Needhidasan, S., Samuel, M., and Chidambaram, R. (2014). Electronic waste––an emerging threat to the environment of urban India. *Journal of Environmental Health Science and Engineering* 12 (1): 1–9.

Nnorom, I.C. and Osibanjo, O. (2008). Electronic waste (e-waste): material flows and management practices in Nigeria. *Waste Management* 28 (8): 1472–1479.

Orisakwe, O.E., Frazzoli, C., Ilo, C.E., and Oritsemuelebi, B. (2020). Public health burden of E-waste in Africa. *Journal of Health & Pollution* 9 (22): 1–12.

Otache, M.Y., Musa, J.J., Animashaun, I.M., and Oji, D.M. (2014). Evaluation of the effects of electronic waste on topsoil and groundwater. *International Journal of Science, Engineering and Technology Research (IJSETR)* 3 (12): 6.

Owusu-Sekyere, E., Peprah, K., and Demuyakor, R.M. (2018). Exploring the dynamics of E-waste disposal strategies in

Tamale, Ghana. *Ghana Journal of Development Studies* 15 (2): 168–192.

Parvez, S.M., Jahan, F., Brune, M. et al. (2021). Health consequences of exposure to e-waste: an updated systematic review. *The Lancet Planetary Health* 5 (12): e905–e920.

Perlroth, N.H. and Branco, C.W.C. (2017). Current knowledge of environmental exposure in children during the sensitive developmental periods. *Jornal de Pediatria* 93: 17–27.

Queiruga, D., Benito, J.G., and Lannelongue, G. (2012). Evolution of the electronic waste management system in Spain. *Journal of Cleaner Production* 24: 56–65.

Rautela, R., Arya, S., Vishwakarma, S. et al. (2021). E-waste management and its effects on the environment and human health. *Science of the Total Environment* 773: 145623.

Rehman, K., Fatima, F., Waheed, I., and Akash, M.S.H. (2018). Prevalence of exposure of heavy metals and their impact on health consequences. *Journal of Cellular Biochemistry* 119 (1): 157–184.

Rzymski, P., Tomczyk, K., Poniedzialek, B. et al. (2015). Impact of heavy metals on the female reproductive system. *Annals of Agricultural and Environmental Medicine* 22 (2).

Saha, L., Kumar, V., Tiwari, J. et al. (2021). Electronic waste and their leachates impact on human health and environment: global ecological threat and management. *Environmental Technology & Innovation* 24: 102049.

Salam, M.D. and Varma, A. (2019). A review on impact of e-waste on soil microbial community and ecosystem function. *Pollution* 5 (4): 761–774.

Sanders, A., Smeester, L., Rojas, D. et al. (2014). Cadmium exposure and the epigenome: exposure-associated patterns of DNA methylation in leukocytes from mother-baby pairs. *Epigenetics* 9 (2): 212–221.

Sanusi, A.I. (2015). Impact of burning e-waste on soil physicochemical properties and soil microorganisms. *British Microbiology Research Journal* 8 (2): 434–442.

Scott, N.B. and Pocock, N.S. (2021). The health impacts of hazardous chemical exposures among child labourers in low-and middle-income countries. *International Journal of Environmental Research and Public Health* 18 (10): 5496.

Singh, N., Duan, H., and Tang, Y. (2020). Toxicity evaluation of E-waste plastics and potential repercussions for human health. *Environment International* 137: 105559.

Singh, V., Cortes-Ramirez, J., Toms, L. et al. (2022). Effects of polybrominated diphenyl ethers on hormonal and reproductive health in e-waste-exposed population: a systematic review. *International Journal of Environmental Research and Public Health* 19 (13): 7820.

Sousa, A.H., Pereira, J.P.G., Malaquias, A.C. et al. (2022). Intracellular accumulation and DNA damage caused by methylmercury in glial cells. *Journal of Biochemical and Molecular Toxicology* 36: e23170.

Tang, X., Qiao, J., Chen, C. et al. (2013). Bacterial communities of polychlorinated biphenyls polluted soil around an e-waste recycling workshop. *Soil and Sediment Contamination: An International Journal* 22 (5): 562–573.

Twagirayezu, G., Irumva, O., Huang, K. et al. (2022). Environmental effects of electrical and electronic waste on water and soil: a review. *Polish Journal of Environmental Studies* 31: 2507–2525.

Wagner, T.P. (2009). Shared responsibility for managing electronic waste: a case study of Maine, USA. *Waste Management* 29 (12): 3014–3021.

Wang, Y., Tian, Z., Zhu, H. et al. (2012). Polycyclic aromatic hydrocarbons (PAHs) in soils and vegetation near an e-waste recycling site in South China: concentration, distribution, source, and risk assessment. *Science of the Total Environment* 439: 187–193.

Woimant, F. and Trocello, J. (2014). Disorders of heavy metals. *Handbook of Clinical Neurology* 120: 851–864.

World Health Organization (2021). *Children and digital dumpsites: E-waste exposure and child health: summary for policy-makers. World Health Organization* (License: CC BY-NC-SA 3.0 IGO.). https://apps.who.int/iris/handle/10665/341730 (accessed 6 February 2023).

World Health Organization (2022). Preventing impacts of electronic waste on child health. https://www.who.int/multi-media/details/children-are-exposed-to-e-waste-toxicants-through-4-major-pathways (accessed 6 February 2023).

Xu, X., Yang, H., Chen, A. et al. (2012). Birth outcomes related to informal e-waste recycling in Guiyu, China. *Reproductive Toxicology* 33 (1): 94–98.

Xu, X., Yekeen, T.A., Liu, J. et al. (2015). Chromium exposure among children from an electronic waste recycling town of China. *Environmental Science and Pollution Research* 22 (3): 1778–1785.

Yang, H., Ma, M., Thompson, J.R., and Flower, R.J. (2018). Waste management, informal recycling, environmental pollution and public health. *Journal of Epidemiology and Community Health* 72 (3): 237–243.

Zhang, Y., Luo, X., Mo, L. et al. (2015). Bioaccumulation and translocation of polyhalogenated compounds in rice (Oryza sativa L.) planted in paddy soil collected from an electronic waste recycling site, South China. *Chemosphere* 137: 25–32.

Zhao, G., Wang, Z., Dong, M.H. et al. (2008). PBBs, PBDEs, and PCBs levels in hair of residents around e-waste disassembly sites in Zhejiang Province, China, and their potential sources. *Science of the Total Environment* 397 (1–3): 46–57.

Zheng, X., Xu, X., Yekeen, T.A. et al. (2016). Ambient air heavy metals in PM2. 5 and potential human health risk assessment in an informal electronic-waste recycling site of China. *Aerosol and Air Quality Research* 16 (2): 388–397.

Zoeteman, B.C.J., Krikke, H.R., and Venselaar, J. (2010). Handling WEEE waste flows: on the effectiveness of producer responsibility in a globalizing world. *The International Journal of Advanced Manufacturing Technology* 47 (5): 415–436.

9

Reuse and Recycling of Electronic Waste from a Global Solution Perspective

Karpagaraj A[1]*, Gopikrishnan T*[2,*]*, and Shri Krishna Singh*[2]

[1] *Department of Mechanical Engineering, National Institute of Technology Puducherry, Karaikal - 609609, Puducherry, India*
[2] *Department of Civil Engineering, National Institute of Technology Patna, Patna - 800005, Bihar, India*
* *Corresponding author*

9.1 Introduction

Electrical and electronic products have changed the way we live in the world. Products evolve with technological advancements, so the previous products become outdated, which may not be affordable in context with power consumption, or are sometimes unusable. The technological advancements in electrical and electronic products result in waste electrical and electronic equipment that ranges from industrial to household equipment (Robinson 2009).

One of the most significant sectors in the world is the electronics sector. It has been expanding significantly over the past few decades, creating a large number of jobs, advancing technology, and fueling a strong demand for raw material resources that are regarded as rare or precious. Developed or developing nations are thought to be the primary producers and also generate electronic trash. Europe and the United States of America (USA) produce the majority of the world's electronic garbage (Ongondo et al. 2011). China, Eastern Europe, and South America are now major producers of electronic garbage. According to estimates, 40 million metric tons of electronic and electrical waste are produced annually, or nearly 5% of all the solid waste produced worldwide. Based on a report from Greenpeace, it is observed that almost 20 to 50 million metric tons of electrical and electronic equipment are annually disposed of as waste. Asian countries' contributions amount to 12 million per year. Additionally, Greenpeace predicted that in 2010, the combined global production of computers, mobile phones, and televisions would total 5.5 million metric tons of garbage. Additionally, it was anticipated that this number would be 9.8 million metric tons in 2015. The author estimates that the volume of electronic garbage may be equivalent to 8% of all urban waste produced. The amount of dumped computer monitors each year is not quantified. However, it was projected that 3.2 million from the USA, 5 million from China, and 1 million from Taiwan may be the number of discarded televisions and monitors until 2015 (Sivaramanan 2013). The Computer Industry Almanac claims that for computer systems, there is no specific figure for abandoned equipment, although it can be presumed that it is related to the quantity of computers used globally. Globally, 4.8 million computers were used by the 1980s, and 553 million computers were estimated to be used by the year 2000. It is estimated that 2020 to 2070 million computers will be used in the year 2015. Until the end of 2011, the United States had 310.6 million, China had 195.1 million, Japan had 98.1 million, Germany had 71.5 million, and India had 57 million computers operational, as the countries with the most computers used. According to statistics, globally, 372 million personal computers were sold in 2011; by 2015, sales reached 517 million units. In 2011, China, the USA, and Brazil accounted for high sales that year (Juliussen and Petska-Juliussen 1996).

Because of improvements in electronic circuits, appealing consumer-oriented designs, marketing strategies, and compatibility issues with old models, the service period of many electronic devices has been significantly shorter. For instance, the typical lifespan of a computer system has fallen and is continuing to fall. It was 4.5 years in 1992 and reduced to 2 years in 2005. The amount of electronic waste produced per person in developing nations was measured in 2005 using data provided by the United Nations. As per this survey, Brazil and Mexico produce the most electronic trash from computers among developing nations, with roughly 0.5 kg/person/year and 0.45 kg/person/year, correspondingly, surpassing South Africa's 0.4 kg/person/year and China's 0.2 kg/person/year (Widmer et al. 2005).

Mobile phones are another category of electrical and electronic equipment with significant volumes of manufacturing and sales. An Environmental Protection Agency assessment in the USA revealed that 65 000 tons of electronic waste from around 125 million cell phones is

Electronic Waste Management: Policies, Processes, Technologies, and Impact, First Edition. Edited by Sunil Kumar and Vineet Kumar.

discarded each year. According to the same poll, a cell phone seems to have a typical lifespan of 9 to 18 months. According to a different survey, there are currently over 500 million outdated mobile phones in the world, and that figure is increasing quickly. In November 2013, there were 271 million cell phones in use in Brazil, far more than the 200 million people who call that country home (Nelson et al. 2018). There are no specific statistics on recycling rates for mobile phone devices worldwide, but a study done by the company Nokia in 13 nations revealed that just 3% of cell phones are sent for recycling.

The 53.6 million metric tons of e-waste that were generated in 2020 were 44.75 million metric tons of e-waste in 2019 (Rautela et al. 2021). At present, computers, cell phones, cathode ray tubes (CRT), displays, printers, and televisions account for the lion's share of the market of electronic trash. However, as newer telecommunications devices like liquid crystal displays (LCD) have replaced old gadgets, at the same time, new e-waste is being produced. Most of the e-waste is processed in informal trash management operations without any safety or guidance by utilizing basic techniques like combustion and acid washes in nations such as India and China due to the lowest cost and ease of processing, for retrieving metals of economic importance. In order to retrieve valuable metals, many businesses have released articles on the method of refining metals (Kaur et al. 2020). The colossal quantities of worldwide Waste Electrical and Electronic Equipment (WEEE) are critical issues that need immense attention through research. More research is required for carrying out efficient recycling of e-waste. The research must proceed from the consumer point of view and improvising recycling of e-waste. Almost 20 rare earth elements and other non-ferrous metals are recovered during the processing of various industrial wastes and non-ferrous metal byproducts, particularly Printed Circuit Boards (PCBs). A complicated flow chart outlines the various steps in the processes, which also involve pyrometallurgical, hydrometallurgical, and electrochemical methods and procedures.

The objective of this chapter is to provide comprehendable research question that researchers may pursue to find answers. The research questions are a combination of existing reclamation methods with an environmentally friendly process to provide efficient reclamation methods. Research also needs to proceed toward psychological aspects of consumer practices of purchasing new electrical and electronic gadgets without any proper purpose. For example, buying a new smartphone to have a newly launched smartphone model, buying new electrical appliances to replace old, good, working condition electrical appliances. On the basis of the previous e-waste records, the status of the current e-waste recycling industry of China is analyzed. The current recycling capability of China is lacking behind the required and expected level. The feasibility of a sustainable level of recycling and e-waste optimization using legal and possible public practices are discussed in this literature (Yu et al. 2023). The authors discuss the essential need to use paper-based PCBs through sustainable End of Life cycle (EoL) use. The discussion is supported by Life Cycle Assessment (LCA) carried out between paper-based circuit boards and PCBs. A recycling design for WEEE favors sustainable recycling through the use of PCBs, but the biodegradation and recyclable ability of paper-based PCBs has the advantage over PCBs for sustainable recycling. The regulations of manufacturing WEEE need to be changed in favor of paper-based printed PCBs (Sudheshwar et al. 2023). The effects of WEEE on human and environment are discussed by the authors, which sheds knowledge about the secondary, tertiary effects (Ghulam et al. 2023).

9.2 Problems due to Waste Electrical and Electronic Equipment (WEEE)

WEEE contains chemical components that may be poisonous and damaging to the environment and human health. Many agencies responsible for environmental protection in various nations classify hazardous waste as being present in WEEE. According to each product, WEEE has a different chemical content. For instance, stoves and microwaves contain more metals than polymer, and Light Emitting Diode (LED) screens have a higher proportion of polymer (Li et al. 2013). The WEEE variety, the manufacturing year, the producer's brand, and the place of origin are only a few of the variables that affect the chemical composition. WEEE are primarily classified by category in accordance with environmental agency requirements, including Regulation 2002/96/EC of the European Union (Directive 2012), as part of a variety of polymeric materials, which include polypropylene, polyethylene, polyurethane, and other types. The WEEE typically consists of a mixture of metals, including copper, iron, aluminum, brass, and even precious metals like gold, silver, and palladium. Glass, ceramics, organic, inorganic, and sometimes radioactive elements may potentially be included in WEEE. WEEE is regarded as hazardous to the health of the environment and humans because, most of the time, it contains both organic and inorganic toxicants. The organic toxicants are chlorofluorocarbons, polyhalogenated substances, polycyclic aromatic hydrocarbons, and polychlorinated biphenyls. The inorganic toxicants include chromium, mercury, cadmium, antimony, nickel, and arsenic. Toxic materials have been used less frequently or omitted altogether in electronic equipment. The European Union's Regulation 2002/95/EC, commonly recognized as the Restriction of Certain Hazardous Substances (RoHS), was perhaps the most significant (Directive and Others 2013). The European Union's Regulation 2015/863/EU, called Restriction of Certain Hazardous Substances (RoHS 3), identified four extensively used phthalates such as bis(2-ethylhexyl) phthalate (DEHP), butyl benzyl phthalate (BBP), dibutyl phthalate (DBP), diisobutyl phthalate (DIBP) with presence of a minimum 0.1% (Wilson et al. 2015). Consumer gadgets that contain hexavalent chromium, lead, mercury, cadmium, polybrominated diphenyl ethers, and polybrominated biphenyls are no longer allowed to be manufactured or sold in the European Union. As a result, several nations with trading ties to the EU were compelled to produce goods devoid of harmful compounds.

A gas processing system is the main method for limiting the discharge of hazardous gases and pollutants during the burning of WEEE. For the businesses that provide these services, though, these are incredibly pricey pieces of machinery. As a consequence, many businesses burn WEEE without providing proper gas treatment. According to EPA research on WEEE disposal in the United States of America (USA), in 2001 there were roughly 2 million metric tons of WEEE produced, of which about 90% were transported to landfills. In 2009, 2.37 metric tons were produced, but in 2001, recycling rates reached 25%, which is significantly better. The recycling methods of WEEE are a global practice that is expanding (Vazquez and Barbosa 2016). The tactics and technology used to collect and process e-waste are essential for an effective recycling process. And even though many nations, especially those in the European Union, Japan, and the USA, have laws in place addressing e-waste issues, the majority of nations still have inadequate systems in place for managing this garbage.

9.3 Managing Waste Electrical and Electronic Equipment (WEEE) Problems

Extensive research is now being done on electronic waste management and to address the issues on a local and international level. Extended Producer Responsibility, Multi-Criteria Analysis, Material Flow Analysis, and Life Cycle Assessment (LCA) are some of the techniques that have been evolved and used in e-waste management. These management methods, along with local legislation, can assist in improving the way electronic waste is disposed of globally, boosting material reuse and lowering environmental impacts (Kiddee et al. 2013). Numerous legal papers have been created and/or put into effect in Europe, mandating producers and other participants to use an ecological approach to develop and evaluate the ecological attributes of products over the course of their life cycle. The major law guidelines and control regulations are the WEEE and Restriction of the Use of Certain Hazardous Substances (RoHS) Directives, developed by the European Union. Manufacturers are now liable for the acquiring, reclaiming, and disposition of electronic and electrical device waste under the WEEE initiative (Goosey 2004). Based on the RoHS Directive from July 2006, six hazardous substances cannot be used for the production of electronic products. These include hexavalent chromium, lead, mercury, cadmium, polybrominated biphenyls (PBB), and polybrominated diphenyl ethers (PBDE).

The amount of WEEE produced in Asia is rising as a result of the region's fast economic development. Approximately 4000 metric tons of WEEE are disposed of globally every hour; 80% of this waste iss shipped to Asian countries, with China taking 90% of the shipped waste. China in March 2007 adopted legislation to govern the pollution brought on by electronic goods, focusing on prohibitions on the use of hazardous materials in electronic equipment (Lu et al. 2015). In April 2001, Japan developed and implemented the Home Appliance Recycling Law (HARL) to reduce WEEE's negative environmental effects with the goal of reducing WEEE's negative environmental effects. The law was passed for the four primary WEEE consumer sources in Japan: air conditioners, TVs, washing machines, and refrigerators. The law was updated in 2008 to include Plasma and Light Emitting Diodes (LEDs), Liquid Crystal Display (LCD) screens, as well as laundry dryers. The replacement of harmful materials, improvement of recyclability, promotion of recycling, and prohibition of unsuitable deposits are the fundamental tenets of Japan's HARL (Kumar et al. 2017).

On May 5, 1992, the Basel Convention became effective; the law was framed for the regulation of waste hazardous materials movement and disposal across international borders. Hazardous waste must be managed by the origin nation based on the convention decision. E-waste is classified as perilous by the Base Convention, and its export may be permitted only under specific circumstances (Khan 2020).

In terms of Africa, it is well recognized that several African countries receive used equipment. The majority of electrical equipment exported to Africa is not functionally checked. As a result, determining whether these shipments are defined by law as toxic materials under the Basel Convention is impossible. Data on WEEE recycling in Africa is limited, although in South Africa there is documented proof for electronic waste recovery. Despite operating in often appalling conditions, Africa's local collection agents, equipment dismantlers, and waste recyclers are becoming more significant in the WEEE industry (Dias et al. 2022).

Electronic devices sales in South American nations have risen abruptly, reaching those in industrialized countries, resulting in increased WEEE output. Despite the fact that a growing number of traditional metal recycling factories have discovered the WEEE recycling markets in nations like Brazil, Colombia, Peru, Argentina, and Chile, processed volumes are still at a low level. The National Policy on Solid Waste was adopted in Brazil in August 2010, which states that the details of the electronic products design and parts must be recorded by the manufacturers, importers, distributors, and resellers of electronic products (Mathias 2021). After the use of the electronic products and parts the collection, logistics, and their recycling will be carried out.

With 7.5 kg per person per year, the United States (US) produces the most electronic waste. Several US states have WEEE programs in place, but the particular law is uncommon. Federal law adoption initiatives have not yet been effective.

It is really concerning how all of this global garbage will be disposed of. Recycling the materials included in this garbage is the best approach to manage it. The chemical makeup of WEEE is quite varied, fluctuating, among other things, based on the variety of equipment, the year it was made, the manufacturer, and the place of origin. In actuality, it is possible to salvage metals, ceramics, and polymers from electronic waste. Due to their high economic importance, the various metals here attract the most interest. These elements may be found in various kinds of compounds, in varying proportions, either in their purest form or as alloys. Electronic devices contain predominantly printed circuit boards (PCBs), which are a valuable source of raw materials for reclamation of copper, palladium, silver, and gold (Barnwal and Dhawan 2020). The copper and precious metals reclamation rate is high when compared to the extraction of these metals from their respective ores using traditional methods. This encourages the reclamation of these metals, preserving their availability for use in the future and mitigating any environmental effects associated with their initial extraction.

Due to high metals recovery capability, WEEE recycling-oriented management has increased in all nations. In industrialized and underdeveloped nations, e-waste management is handled differently. According to the types of processes used, two major kinds of facilities are available for recycling of WEEE. Mechanical methods of e-waste processing are classified as the first type of group. The second group recovers metals using metallurgical procedures. In comparison, the industry for recycling e-waste operating in underdeveloped nations is largely uncontrolled, and precious components are frequently recovered from e-waste in tiny workshops using crude recycling techniques (Sarvar et al. 2015). The informal recycling of e-waste that is common in poor nations is linked to serious environmental degradation and occupational exposure to chemicals originating from e-waste. Using scaled-up methods facilitates the reclamation of significantly higher levels of essential minerals, like rare metals, than home garden recycling operations, despite the fact that recycling costs in developing nations' typical backyards can be lower (Oguri et al. 2018). This improved yield typically outweighs the cost disadvantages for pertinent elements like PCBs.

9.4 Methods of Waste Electrical and Electronic Equipment (WEEE) Processing

Due to the high degree of variation in its content and its poor environmental compatibility, the processing of electronic trash is particularly complicated. Typically, manual disassembly is the initial phase, during which specific parts are isolated. Thermal, chemical, and mechanical methods are implemented for the reclamation and recovery e-waste processing after disassembly. Mechanical or physical methods, hydrometallurgy, electrometallurgy, and pyrometallurgy are the four basic methods for recovering metals (Krishnan et al. 2021). The general method of WEEE processing is shown in Figure 9.1.

9.4.1 Waste Electrical and Electronic Equipment (WEEE) Processing by Mechanical Methods

E-waste materials are chosen and separated via mechanical processing, and the segregation takes place depending on phased processing procedures of minerals. Methods of mechanical processing typically involve a preprocess of the segregation of e-waste in relation to scraps and are associated with various stages of WEEE component separation. The methods of mechanical processing of e-waste that consists of a variety of types of materials is the subject of numerous studies at this time (Roy et al. 2022). This process includes classification, separation and comminution, and separation by difference in density, size of particles, magnetic, and electrical characteristics. Following mechanical processing, the metal fraction is transported to electro-, hydro-, and pyrometallurgical processes. Specific methods are used for the ceramic and polymer fractions.

9.4.2 Waste Electrical and Electronic Equipment (WEEE) Processing by Hydrometallurgical Methods

A series of acidic or caustic treatments are used in the first stages of hydrometallurgical processing to disintegrate the hard solid materials. The solutions are then put through

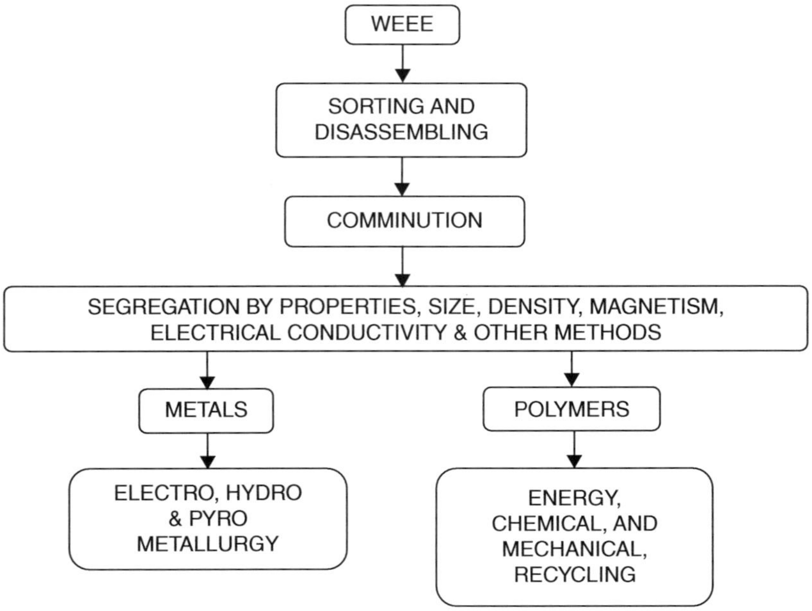

Figure 9.1 General waste electrical and electronic equipment processing procedures.

separation techniques such as cementation, distillation, solvent extraction, ion exchange, precipitation, and filtration in subsequent phases in order to separate and concentrate the desired metals. The primary benefits of hydrometallurgical treatment of electronic waste include lower air pollution risk, more specificity to metals, reduced processing costs such as less electricity usage, and recurrent use of chemicals used for e-waste processing. The drawbacks include trouble in complex processing of e-waste, the requirement of preliminary mechanical treatment of the e-waste toward volume reduction (Kononova et al. 2013), the metal surface exposure for chemical disintegration, requirements of vast quantities of chemical solutions, harms caused by utilized waste chemicals corrosive nature, and the new solid waste emergence as result of entire e-waste processing.

9.4.3 Waste Electrical and Electronic Equipment (WEEE) Processing by Biotechnological Methods

In particular, the literature has discussed the employment of microorganisms in the dissolving of metallic components and extracting gold from e-waste. Although the ability to implement this process on a broad scale is currently fairly limited, various research is being carried out that ought to be able to generate methods having fewer operational costs and inexpensive equipment investments, in addition to producing little waste, wastewater, or harmful gases. According to the analyses, biohydrometallurgical processes involve diverse bacterial cultures leaching metals. In the tests discussed here, a solution comprising 10 grams per liter of ferric ion and a colony of YTL-2 bacteria culture was used to treat electronic trash at temperatures from 20°C to 35°C and a pH less than 2.5. In addition to reproducing the bacteria and recycling the solution, it was feasible to recover 97% of the gold after 2 days (Kour et al. 2021). In a study, microbiological techniques were used to treat electronic waste. It was discovered that this kind of residue promotes the growth of fungi like *Aspergillus niger* and *Penicillium simplicissimum* as well as bacteria like *Thiobacillus thiooxidans* and *ferroxidans*. High levels of metals, including nickel, aluminum, copper, and zinc, were found leaching in this study. The primary drawbacks of methods of biohydrometallurgical process involve prolonged leaching times.

9.4.4 Waste Electrical and Electronic Equipment (WEEE) Processing by Electrometallurgical Methods

The majority of electrometallurgical procedures involved in recycling electronic trash are phases of the electrowinning procedure, which aims to recover a pure metal at the end. Metals can be recovered from many forms of trash via electrochemical procedures, which are typically carried out in liquid electrolytes or molten salts. On the cathode, metal concentrates obtained through hydrometallurgical procedures such as dissolution of selected metals, exchange of ions, or solvent extrication may be electroplated. Fewer stages, greater selectivity for target metals, reusing electrolytes, and obtaining pure metals are benefits of electrometallurgical processes (Gunarathne et al. 2022). The primary restriction is the requirement for a pretreatment, which is often dependent on mechanical and hydrometallurgical procedures.

9.4.5 Waste Electrical and Electronic Equipment (WEEE) Processing by Pyrometallurgical Methods

Metal recovery from e-waste has traditionally been accomplished by pyrometallurgical procedures, most notably smelting. Most other materials are effectively rejected in a slag and gas phase while metals are concentrated in a metallic phase in the traditional pyrometallurgical processing mechanism. Polymers and other insulation materials contribute to pollution of air through the production of furans and dioxins. Through the volatilization of their chlorides, several metals can be lost. Ceramic and glass scrap components enhance the quantity of slag produced in the furnace, which increases rare and base metal losses (AliAkbari et al. 2020). Some metals, such as tin and lead, are difficult to recover, and some are nearly impossible to do so, such as aluminum and zinc.

9.5 Processing of Printed Circuit Boards (PCBs)

Among other industries, the telecom, information and technology, robotics, entertainment devices, and automobile sectors frequently use PCBs, a fundamental part of digital technological products. A PCB is made up of printed conductive circuits, electrical components, and a non-conductive substrate. A non-mounted board, or a board that simply contains the substrates and the printed conducting circuit, is sometimes referred to as a PCB (Lim et al. 2021). Different kinds of metals and non-metal materials are used to make PCBs. The typical mass of PCB components is roughly 30% polymer, 30% refractories oxide, mostly alumina, sand, and precious metal oxide, and 40% other metallic substances, including aluminum, tin, ferrous, copper, lead, nickel, and silver and gold. Because there is no standard design for PCBs, every PCB is made specifically to meet the requirements of the intended product's functioning methods. A typical PCB is shown in Figure 9.2.

PCBs can be divided into three categories according to the technology they use: single-sided PCB board, which has circuits that work only on one side of the board; double-sided PCB board, which has dual-sided circuits; and several layered, which has circuit connection on both sides of the board (Ta et al. 2015). PCBs can also be categorized according to the substance that was used to create their substrate. Commercially, the boards are classified as Flame Retardant type 1 (FR-1), Flame Retardant type 2 (FR-2), Flame Retardant type 3 (FR-3), and Flame Retardant type 4 (FR-4), and the boards made of CEM-x and flame-retardant materials. The FR-4 type PCB is manufactured using fiberglass and synthetic resins, whereas the FR-1 and FR-2 boards are manufactured using boards of cellulose and treated with the resins of phenol. The CEM-x board types are constructed from composite materials composed of epoxy resin treated with cellulose paper and fiberglass (Czél et al. 2010). Polytetrafluoroethylene and polyester polyethylene terephthalate can also be used to make boards. FR-4 boards are typically used in computers, mobile phones, and various telecommunication devices, whereas Flame Retardant type 2 (FR-2) boards are used in TVs and home appliances.

Copper is the primary material used in conductive circuits, but nickel, aluminum, chrome, and other metals may also be employed in certain applications. The electronic components that may be harvested mechanically in a waste PCB are capacitors, transistors, transformers,

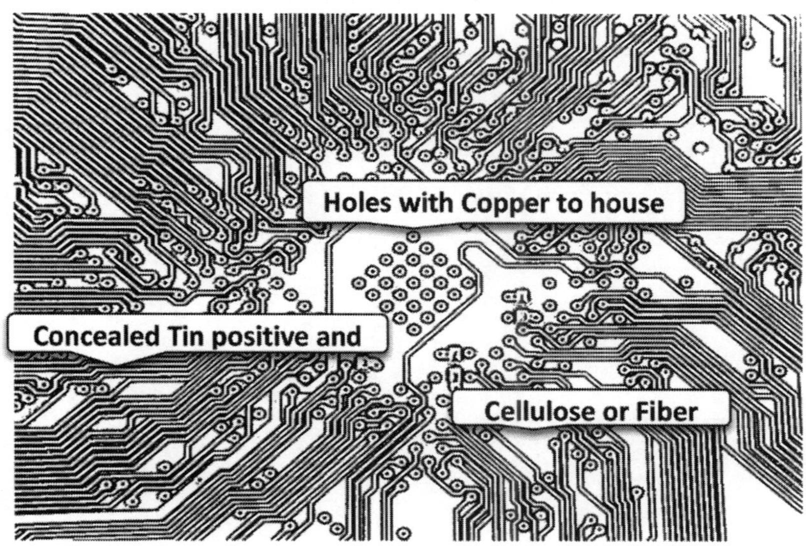

Figure 9.2 A typical printed circuit board.

chips, etc. Integrated circuit chips use metals including lead, iron, nickel, tin, aluminum, and rare metal substances like gold, silver, and palladium, which may be present in these components. Other substances include: tantalum, which is used in capacitors; tellurium, germanium, arsenic, gallium, indium, silicon, selenium, and titanium, which are used in integrated circuits and the rest of the components; germanium, cadmium, and tin, which are employed in solders. Silicon and gallium are also used for semiconductor manufacturing (Huang et al. 2022). Without a pollution control system, improper disposal or incineration of PCBs can result in major environmental issues, especially when cadmium, mercury, arsenic, lead, furans, and dioxins are released into the environment. A typical capacitor is shown in Figure 9.3, and a typical Negative-Positive-Negative (NPN) transistor is shown in Figure 9.4.

These materials eventually contaminate the air, the soil, or the water. Many of these toxins biologically accumulate in the food chain and endanger both human health and the environment, and they can linger in the environment for long periods of time. The guidelines on collecting, processing, procedure of recycling, waste disposition, and the Restriction of Certain Hazardous Substances (RoHS) of waste electronic and electrical devices must be followed by the waste electrical and electronic equipment processors throughout Europe (WEEE) (Goodship et al. 2019). Manufacturers are responsible for the collecting, waste recycling, and disposition of electronic and electrical equipment waste under the WEEE guideline. According to the RoHS directive, six hazardous compounds may not be used in the production of electronics and electrical devices. On the basis of the RoHS, lead, mercury, hexavalent chromium, polybrominated biphenyls (PBB), and polybrominated diphenyl ethers (PBDE) are prohibited, and the percentage of cadmium must not be beyond 0.01% in identical material of electric and electronics devices waste.

Concerning issues with waste from electronics and electrical devices, various nations, including Japan, China, Brazil, and the rest of the nations, follow unique policies. In recent years, the electronics industry has improved its environmental performance by using methods like analysis of life cycle (LCA), ecological design, and end-of-life (EOL) scrutiny procedures to minimize hazardous emissions and regulate environmental pollution. These methods enable the evaluation of the financial losses and environmental effects connected to their manufacturing, consumption, and disposal. They also make it possible to create actions

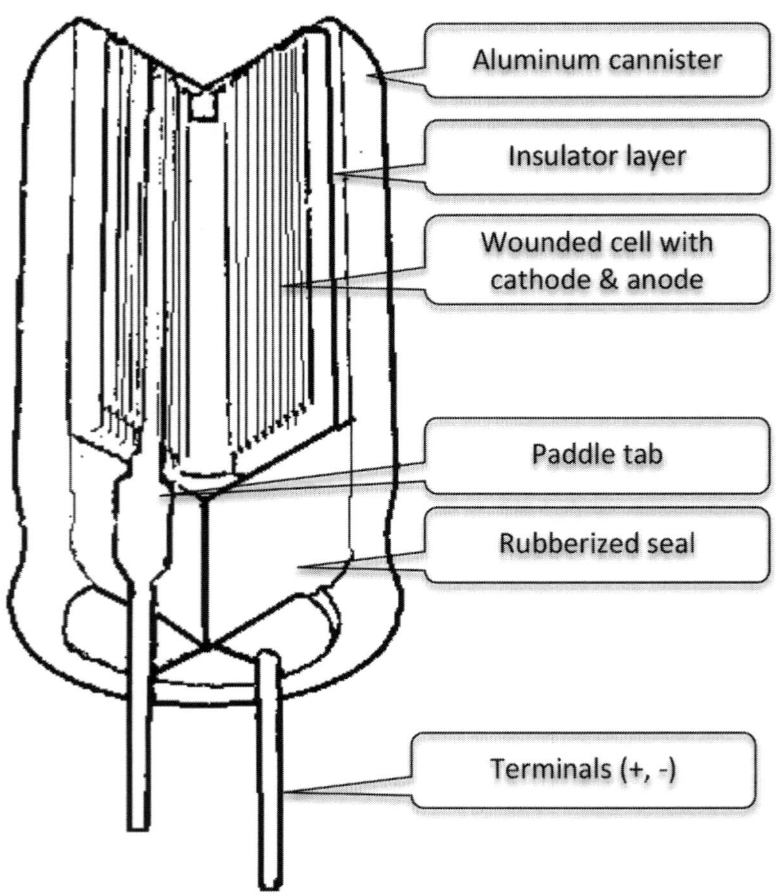

Figure 9.3 General capacitor.

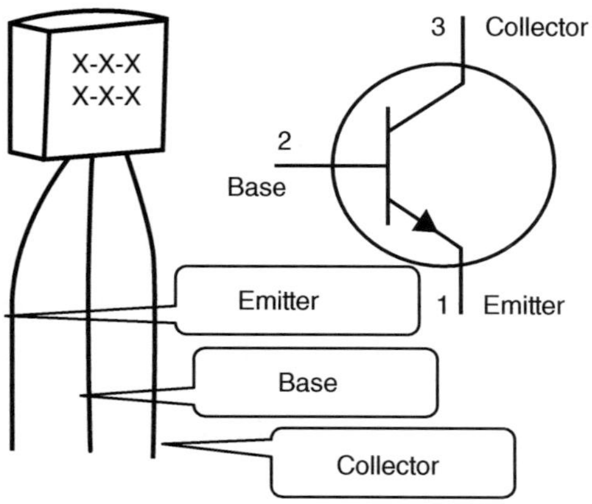

Figure 9.4 A typical negative-positive-negative transistor.

that lengthen the life of equipment and techniques that reuse and recycle its parts (Sharma et al. 2020). Around 40 million tons of electronics and electrical devices waste is reportedly produced worldwide each year. Between 3% and 6% of the overall mass of e-waste is made up of circuit boards. For instance, 10 300 tons of garbage from PCBs were gathered in Germany in 2009.

PCBs typically have precious metal concentrations that are significantly higher than those found in ores. PCBs of desktop personal computers (PCs) might consist of about 250 grams per ton of gold and 110 grams per ton of palladium, although the mineral ores from which palladium and gold are currently extracted have less than 10 grams per ton (Işildar 2018). In addition to this clear economic benefit, it is important to remember that the manufacturing of precious metals has significantly fewer negative environmental effects than the mining process used to obtain these metals. Evidently, the amounts of precious metallic substances in PCBs differ significantly, which is dependent upon the device and the purpose of design. In general, PCB waste in a PC has 7% iron substances, 5% aluminum, 20% copper, 1.5% lead, 1% nickel, 3% tin, and 25% compounds of organic forms, including silver, gold, and palladium in parts per million. There are also minute amounts of barium, antimony, arsenic, bromine, and bismuth. A mobile phone's PCB may typically contain 20% to 30% copper, 6% to 8% ferrous, roughly about 5% nickel, about 2.5% tin, about 1.1% zinc, about 250 mg silver, about 24 mg gold, and 9 mg of palladium, among other metals.

The product and year of manufacturing affect the percentage of materials (Yaashikaa et al. 2022). The inclusion of rare metals, silver, gold, and copper renders PCBs an appealing raw material for recycling, even though recycling them is highly difficult, especially due to the range of components included in them. The fact that more than 60% by weight of PCBs is made of non-metallic materials poses the biggest obstacle to their recycling. The major components of PCB non-metallic materials are thermosetting adhesives and glass fibers. Due to their network structure, thermosetting adhesives could not be melted down or reformed. Because inorganic fillers like glass fiber greatly lower the fuel efficiency, incinerating non-metallic materials is not the greatest option.

Currently, the primary approach for handling the non-metallic components of PCBs is disposal in landfills (Huang et al. 2021). However, this approach has the potential to waste resources and generate secondary contamination. Some researchers have recommended using a portion of this material, roughly 20%, along with resins to create various kinds of plates or, in smaller quantities, as filler to create other plastic materials. The unauthorized repurposing of PCBs, the incineration of wastes to get rare earth metals without emission control, or the use of acids for leaching without managing effluents are all thought to be severe environmental and public health risks. The higher amounts of lead, polybrominated dioxins or furans, and polychlorinated dioxins or furans, in places where non-industrious recyclable methods are adapted results in a significant economical support, but it causes serious environmental and health problems (Peecher et al. 2022). Disruption of the endocrine system and neural toxic effects, which could last for many families, are the main issues associated with this kind of pollution. Therefore, it is crucial that PCB recycling be done properly to prevent environmental pollution. When recycling PCBs, mechanically processing methods, hydrometallurgically processing methods, pyrometallurgically processing methods, and bioleaching processing methods may all be employed separately or in combination.

9.5.1 Mechanical Processing of Printed Circuit Boards (PCBs)

Mechanical treatment has been utilized in primary metallurgy as one of the phases in the processing and treatment of rare metals and minerals. Mechanically processing the waste is a pretreatment in the repurposing of PCBs' metal percentage. Disassembly, grinding, classification, and material separation are actions that are a part of mechanical processing (Hsu et al. 2019). Materials can be separated using variations in their particle size, weight, density, and magnetic characteristics. The purpose of crushing is to liberate metal compounds from the PCB waste. Because it gives materials the right qualities for future separation procedures, grinding is particularly significant. Ineffective grinding can lead to issues like inadequate dissociation of metals and non-metals, extensive spraying of particles, insufficient grinding, material clogging, and equipment fever. The grinding is typically done in two processes. The boards are first processed to break them down to a size of about 10 mm, and then the boards are further broken-processed to 2 mm to release more metals in the second processing stage. Because very small particles result in less segregation between metals and non-metals, the size of particles should not be lower than 0.3 mm. Following the grinding process, a crucial step is to create a size-based particle segregation before using various segregation techniques, because particle size directly affects various segregation procedures (Yuan et al. 2018). The metallic component of PCBs can be separated from the non-metallic fraction using techniques including dense medium separation and gravimetric (or concentration) separation. Gravity segregators employ zig-zag separators resulting in a collection of metals up to 95% in the heavier fragments while used on materials with particle sizes greater than 0.6 mm. However, the portion of metals collected gets restricted to 60% when the size of the particles is less than 0.6 mm.

The zig-zag model extraction machine was utilized in an experiment that integrates mechanical processing with vacuum pyrolysis methods such as breaking down, particle-based segregation, and segregation by gravity in the reclamation of metal from dismantled PCBs (Prasad and Vithanage 2019). An organic PCB fraction was recycled using pyrolysis, and copper was released during mechanical processing in order to be recovered afterward. The findings reveal a 99% recovery rate for the copper in PCBs. Studies demonstrating the segregation of fine particles (1.0 mm) using Falcon-type concentrators show that this form of separator can segregate about 92% of metal in the heavier sizes, allowing a reclamation quantity of 97%. The dense media segregation method was tested for metal concentration with some degree of success. In this study, the ability to segregate the non-metal lightweight floating particles in the heavier metal particles using a mixture of tetrabromoethane and 27% acetone is investigated. Adopting thick layer segregation has drawbacks to using costly and organic hazardous solutions.

The method that adapts electrostatic and magnetic segregation is the one that is most frequently used in the concentration and separation of metallic substances in the recovery of PCBs. Ferromagnetic metals like nickel, iron, and various metallic substances are segregated in the first step of magnetic separation (Benzaazoua and Taha 2020). Metals that can be segregated through methods of electrostatic induction, such as lead, tin, and copper, are segregated in the other elements through a second-level electrostatic separation. The distribution of the electromagnetic product and the particle sizes are the two most crucial factors to take into account when selecting the electromagnetic segregator. To improve the segregation of ferrometals in the magnetic fractioning, it may be possible to utilize two levels of electromagnetic segregation; the first level is to use a low-intensity magnet, and the second level is to use a high-intensity magnet. One stage utilizes a low-intensity magnetic field and the next process utilizes a high-intensity magnetic field. PCBs could be separated from concentrates of metals and non-metals by electrostatic separation, and the Corona isolators are generally used (Franke et al. 2020). Aluminum scrap was recovered from PCBs using the electrostatic separation process using eddy current or Foucault current. More than 90% of the aluminum, with a purity of 85%, could be recovered with this technique.

Some authors have successfully tested a mixture of mechanical processing, including crushing and particle diameter separation, as well as electrostatic isolation and magnetic separation, in the metal concentrations of PCBs. The tests produced magnetic particles with a mean iron content of 45%, an electrostatic fraction with between 50% and 60% copper, 24% tin, and 8% lead, and non-metals like ceramics and polymers. These techniques provide a high rate of recovery for these metals (Kan et al. 2021).

9.5.2 Hydrometallurgical Processing of Printed Circuit Boards (PCBs)

In the recycling of PCBs, hydrometallurgical methods have some benefits over pyrometallurgical processes, which are low investment, no smoke, selective operation, and suitability for microscale industries. Because of the benefits, hydrometallurgical procedures are a substitute for the processing of PCB waste. The leaching of metals of commercial relevance, including tin, silver, gold, copper, and nickel, among others, is the main focus of study employing hydrometallurgical processes for the recovery of metals from waste, especially electrical and electronic waste. There are typically two steps of leaching involved in the hydrometallurgical treatment of PCBs. Base metals are recovered in the first step, while rare metallic substances are recovered in the second level. In the initial step, mineral acids like hydrochloric, sulfuric, nitric, and perchloric acids might be employed. In order to improve metal extraction, hydrogen peroxide, a potent oxidizing agent, is frequently added to the acids. Solutions including thiosulfate, aqua regia, cyanide, halides, or thiourea can be used to leach valuable metals. In most instances, a pretreatment is carried out prior to the beginning of the hydrometallurgical operations used to recover metals from PCBs (Zhang et al. 2016). This preparation may involve mechanical processing or selective deconstruction. The elements and equipment can be broken down into different fractions during disassembly, such as microchips, semiconductor devices, connections, capacitor varieties, and circuit boards and then handled differently based on the individual properties. With the help of this pretreatment, the efficiency of the process will be increased, and less solvent will be used during the leaching.

Numerous researchers have explored the usage of acids of nitrate, hydrochloride, or sulfurs in the dissolving of the basic metal found in PCBs. In an experiment, copper and nickel were recovered from PCBs that are unmounted types, which represent PCBs devoid of electrical items, using two stages of leaching at various nitric acid concentrations. They produced a nickel-rich liquid with modest amounts of copper in the first stage by using dilute nitric acid (0.1 M). In the second phase, a stronger solution of nitric acid (1.0 M) was utilized to produce a copper-rich, nearly impurity-free solution. Other researchers looked at the impact of ionic copper on dissolving copper using hydrochloric acid and chloride. Studies show that the reclamation of 97% tin in PCBs was made possible by the dissolving of lead and tin using hydrochloric acids and nitric acids and the consequent is tin precipitates in metastannic acid. Over 90% of iron, copper, silver, and nickel, as well as 58% of palladium, may be leached by using a solution of sulfuric acid, copper sulfate, and sodium chloride. Studies have shown that utilizing a pyrolysis stage and processing by a leaching stage can improve the extraction of tin and copper, but it also results in a decrease in tin recovery. The findings show that the quantity of diffused copper found in pyrolyzed samples was higher than that found in other processing methods, whereas the amount of tin in diffused form found in pyrolyzed samples was lower than that found in other processing methods. The authors utilized leaching agent aqua regia for valuable metal extraction such as silver, palladium, and gold, achieving a rate of leaching of 93% of palladium, 98.7% of silver, and 97.7% of gold. Nevertheless, despite the fact that the gold separation in PCBs is exceedingly efficient, the adapted method seems to be having few real-world applications because, at the time of leaching, the aqua regia emits nitrogen oxides and chlorine fumes that could poison people and the environment. There are various benefits, including minimal expenditure, low consumption of reagent, and working in an environment with alkalinity.

For more than a century, cyanide remained as a leaching solution of silver and gold by the sectors of the mining industry. The silver and gold included in PCBs are also dissolved using cyanide-containing solutions. In this instance, the cyanide complex is reduced in order to recover the gold. However, because of how harmful these substances are to both humans and the environment, researchers are now looking at leaching procedures that adapt other eco-friendly leaching options. Thiosulfate, thiourea, and halides leaching were used experimentally to recover silver and gold in the PCBs of cell phones, solutions comprising 24 grams of thiourea per liter and ferric iron about 0.6% resulted in leaching rates of 50% for silver and 90% for gold. Gold's ability to dissolve is substantially impacted by the high copper concentrations found in PCBs. As a result, higher gold dissolving rates are possible, for instance, in solutions containing thiosulfate or thiourea, when the copper is first recovered using a preprocessing method using sulfuric acid and hydrogen peroxide. More than 90% of the copper and roughly 70% of the gold can be recovered by using a preprocessing phase for the extraction of copper and then dissolving the gold with thiourea. Thiourea costs more when compared to cyanide, but the easy oxidizing nature of thiourea has increased its use. Extraction of gold using thiourea is underdeveloped, which hinders the commercial deployment of thiourea for leaching silver and gold.

How to extract gold from PCBs in electronic waste using ammonium thiosulfate as a leaching agent was also investigated. In a study, ammoniacal thiosulfate was used as a leaching agent on electronic PCB scrap, resulting in 95% gold leaching within 48 hours and 100% silver leaching within 24 hours. Similarly, using solvents having ammonium thiosulfate of 0.12 M, copper of 20 M, and ammonia of 0.2 M, it was possible to achieve 98% gold leaching within two hours processing of PCBs in waste keyboards and cell phones. This method has achieved 90% leaching within 10 hours for other kinds of PCBs of cell phones. High selectivity, non-toxicity, and non-corrosivity are advantages of thiosulfate leaching. The high reagent usage during extraction is the key issue with thiosulfate leaching.

Hydrometallurgical processing requires the recovery of valuable metals from leach liquors. Numerous methods, including cementation, electrometallurgy, activated carbon adsorption, solvent extraction, and exchange of ions, can be utilized in this recovery. The chemicals employed in leaching, the metal quantity, and the presence of contaminants are the main considerations in the selection and development of metal recovery methods. Electrochemical techniques are one method for recovering copper from leach liquors. The electrochemical procedure to extract copper from solutions derived from metal samples that have been mechanically concentrated by steps such as size of particle segregation, grinding, electrostatic and magnetic segregation, results in copper reclamation surpassing 92%. Two recovery techniques were examined for the metal recovery from the pickled fluids used in the production of PCBs. The second technique combines copper electrowinning with precipitation of metal for lead and tin, the previously explained first procedure exclusively used in the electrochemical extraction of metals. A pyrometallurgical procedure was used to extract the metals from the precipitate.

One of the most efficient and cost-effective methods for recovering metals from hydrometallurgical methods, in addition to the extraction of precious metals in leached liquors, is cementation. The potential to recover priceless metals, relative simplicity, and ease of process control are all benefits of the cementation process. Zinc cementation is commonly called the Merrill-Crowe technique, which was employed in gold extraction from solutions of gold-cyanide. Two key processes that take place during this process are the anodic corrosion of zinc and the accretion of gold in the cathode. Copper can be employed in the cementation of silver and gold liquors comprising cyanide or thiourea because it is relatively less expensive and has low reducing capability when compared to silver and gold. However, it has a larger reducing capability when compared to various basic metals. Silver was fully recovered from the cement experiments for extracting silver and gold with copper powder, while gold was recovered to a level of roughly 80%.

On a large scale, gold and silver from the leach fluid have been reduced using the reduction-precipitation method with metal hydrides. Sodium borohydride is the most widely used metal hydride reductant. Using liquids containing 12% sodium tetrahydridoborate and 40% sodium hydroxide, reduction-precipitation was effectively used to extract gold from liquids of thiourea, thiosulfate, or thiocyanate. The outcomes showed that even at room temperature, gold ions from acidic solutions may be successfully converted to metallic gold. Both highly diluted and more concentrated solutions can be used to carry out this reaction. By the principle of adsorption, silver and gold adsorbs on polymers, carbon in activated form, and also on various adsorbing materials available naturally, such as rice husk carbon and gel made of persimmon tannin.

Some authors have tested the successful metal recovery through the leaching solution using solvent extraction. To selectively reclaim gold using PCBs through aqua regia liquid, toluene extraction is succeeded by cementation using dodecanethiol and sodium borohydride. The commercial solvent LIX984 was used to extract copper through a solution consisting of nitric acid, nickel, and copper. While Cyanex 921 is suitable for all power of hydrogen (pH) ranges, the industrial solvent LIX79 could be used to extract gold from cyanide-containing alkaline solutions. Gold may be recovered from leaching solutions using ion exchange resins. The extraction of gold from thiosulfate liquids has been successfully tested using industrial resins such as Amberlite IRA-410, Dowex G51, and Dowex 21 K (Nicol and Kganyago 2020). The resins contain quaternary ammonium functional groups and a gel-type matrix made of polystyrene divinylbenzene.

9.5.3 Biohydrometallurgical Processing of Printed Circuit Boards (PCBs)

Many mineralogy and metallurgy firms are particularly interested in the extraction of metals by bioleaching, which is a key technique tried in recent decades to obtain metals. It can be used on many different metals, including gold, silver, copper, nickel, cobalt, and zinc. A number of experiments have been carried out recently to examine the effectiveness of using such methods to extract metals from ores. However, there have not been many investigations on the viability of employing this method to recover metals in PCBs.

In essence, biohydrometallurgical processes are hydrometallurgical processes that enhance the dissolving of metals from ores, concentrates, or trash by using microorganisms, such as bacteria, archaebacteria, and fungus archaea. Microorganisms are employed depending on the capability to oxidize inorganic and organic substances producing the leaching reagent to carry out metal dissolving. The microorganisms *Acidithiobacillus thiooxidans* and *Acidithiobacillus ferrooxidans*, archaea, are used in experiments in the recovery of metals. In some circumstances, it may be feasible to use microbes like the cyanogenic bacteria *Chromobacterium violaceum*, which can produce cyanide ions and aid in the dissolving of rare metals in the forms of complex cyanomolecules. Leaching columns are employed on various occasions. These instances include grinding off pieces of PCBs that are placed in the columns. The sulfuric acid under regulated pH through a type of irrigation flow is then used to irrigate the column bed. The solution percolates through the layer of electronic waste that includes bacteria, archaea, and nutrient sources like potassium, phosphorus, and nitrogen. For ensuring the presence of carbon dioxide and oxygen throughout the experimental setup, air is pumped at the bottom of the column. There is the potential for bioleaching of metallic substances that are present in the PCBs employing various acidophilic bacteria, including *Sulfobacillus thermosulfidooxidans*. Nickel leaching levels are identified to be 81%, copper leaching rates are 89%, aluminum leaching rates are 79%, and zinc leaching rates are 83%. Metal substances in PCBs are removed through a mixed culture of acidophilus bacteria. In an ideal experiment setup, the achieved leaching efficiencies are 96.8% of copper after about 45 hours, 88.2% of aluminum and 91.68% of zinc after about 98 hours. The ideal circumstances in this example were starting pH 2.00, 12 grams per liter Fe, 12 grams per liter metal concentrate, 10% dose immune quantity, and particle sizes between 60 mm and 80 mm sieve size (Jorjani and Ghahreman 2017). The utilization of microorganisms for extracting metallic substances in the e-waste could be a more cost-effective option than pyrometallurgical processing because it requires less initial investment and uses less energy.

9.5.4 Pyrometallurgical Processing of Printed Circuit Boards (PCBs)

The most established method for recovering non-ferrous and rare earth metals from used electronics is pyrometallurgical processing. Sintering, smelting in a plasma arc furnace or blast furnace, incineration, melting, drossing, and processing in a gaseous state at extreme temperatures are a few of the pyrometallurgical processes. For removal of polymer-based materials, the trash is incinerated in a furnace or in a melting tub during these processes. The metals of significance can be extracted in this phase as the metal and refractory oxides enter a slag phase. The acceptance of any type of electronic trash, the lack of a pretreatment phase, and the simplicity of the process are the key benefits of pyrometallurgical operations. The main drawbacks include the potential for harmful gas emission levels; the evaporation of metal by heat; the increased quantity of slag due to glass and ceramic components, which results in increased loss of rare metals and basic metallic components; and the reduced amounts of extraction of other metal substances such as tin, lead, and the improbability of extracting various metal substances, such as aluminum and zinc.

The most popular method for disposing of polymeric and other organic compounds found in PCBs is incineration. Impure metal alloys are created when the material is melted, and these alloys can then be refined using electrolytic or pyrometallurgical techniques. The thermal energy emitted by such materials, in this instance, is not captured. It can be extremely harmful to burn PCBs in a conventional combustor made for municipal solid trash. When brominated flame retardants, which are used in the majority of PCBs, are burned, copper, for instance, acts as a catalyst to cause the production of dioxins. The low burning efficiency of PCBs due to the presence of inorganic components like glass fiber is another issue with their incineration.

One alternate strategy for recycling PCBs is pyrolysis. The organic material is broken down during the pyrolysis process, which means heating without oxygen, creating lower molecular mass liquids or gases as products that can be utilized as fuels or sources of chemicals. Some scientists have researched the utilization of pyrolysis as the reclamation source of metallic substances from the (WEEE) waste electrical and electronic equipment. According to certain investigations, pyrolysis can be accomplished in an autoclave oven at 500°C for 30 minutes. Another source says that the complete pyrolysis occurs at temperatures of 700 to 800°C. Oils, gases, and solid trash were the end results of this pyrolysis. Fiberglass and metal fragments were present in the pyrolysis's solid waste and were simple to separate. With modest amounts of precious metals like bismuth, gallium, silver, and gold, the metallic fraction was primarily composed of zinc, copper, iron, nickel, calcium, and aluminum.

The treatment of discarded waste PCBs by the authors involved centrifugal vacuum separation and pyrolysis. In this investigation, the first stage was pyrolyzing the PCBs at 600°C for 30 minutes while under vacuum. The acquired residue was heated in the subsequent phase under vacuum until the solder melted, and it was then spun apart from the pyrolysis byproducts. In this manner, it would be possible to separate the oil, gases, and solid debris generated in the process of pyrolysis and soldering, such as metals and fiberglass. The solder in this situation can be recovered, reused, or used as a valuable supply of tin and lead for several other procedures. The recovery of a copper-rich fraction has proven to be successful when using salts in molten form, particularly liquid with lowest melting point like potassium hydroxide and sodium hydroxide for dissolving glass oxides, with polymers without oxidizing the more precious metals. According to the gaseous emissions, the major portion of the emitted gas throughout the operation was found to be trapped in the molten salt. This innovative process had a positive energy balance since it produced a vast amount of hydrogen that could be utilized as synthesis gas or fuel gas.

Some metal recyclers use the Noranda method, which immerses the materials entering the reactor in a 1250°C bath of molten metal that is stirred by a blend of supercharged air having 39% oxygen. Iron, lead, and zinc impurities are transformed into oxides during the oxidation zone's agitation and then cemented in a slag made of silica. Precious metal-containing copper matte is taken out and put into the converter equipment. Liquid copper blister is purified through the furnaces of anode and electrocoated over anode having 99.1% purity after being upgraded in converters. The remaining 0.9% is made up of recoverable metals like nickel, selenium, and tellurium as well as the precious metals gold, silver, platinum, and palladium. These marketable metals are then recovered by electrorefining the anodes. The burning of plastic materials and other combustible items lowers the cost of electricity.

The extraction of precious metals is the primary goal of the processing done in Belgium-based company Umicore at Hoboken, New Jersey, USA. The Precious Metals Operations unit completes the first phase of the metal recovery process. In the processing units, the process of smelting is done in a furnace named IsaSmelt. The process starts when coke is partially replaced by the plastic or other organic materials in the feed as a reduction reagent and energy source. Precious metals in copper bullion are separated by the smelter from practically all other metals concentrated in lead slag, which is then further processed in the Base Metals Operations unit (Okajima et al. 2020). The precious metals are then extracted from the copper bullion during copper leaching and electrowinning. The main disadvantage of this kind of procedure is the significant financial outlay required by these units, especially in light of the requirement to build pollution control systems.

9.6 Processing of Monitors

Various displays like liquid crystal displays (LCDs) and light emitting diode (LED) panels are progressively taking the place of cathode ray tubes (CRT). Global repurposing and disposal systems are impacted by the change of monitors. The traits and potential for recycling of these monitors are discussed in this chapter.

9.6.1 Processing of Cathode Ray Tubes (CRT)

Oscilloscopes of electrostatic deflection are cathode ray tube monitors. Numerous scientists and innovators are credited for helping develop cathode ray tube technology, which started in the year 1855. In the 1930s, television gained popularity across a number of nations. This item became necessary to enable user contact with the central processing unit with the advent of personal computers. The original CRT monitor screens utilized in PCs had a limited resolution and were monochromatic, measuring roughly 140 × 192 pixels, but numerous enhancements to the CRT display were made possible by the emergence of new technologies.

In 1987, monitors were able to display 640 × 480 pixels in 256 colors thanks to the introduction of IBM's Personal System computer and the Video Graphics Array technology. The impact of electrons on a fluorescence panel powers CRT monitors. Typical CRT components are shown in Figure 9.5. The picture we see on the computer monitor is made up of a mixture of numerous brilliant points in three different hues. The electron cannon, located at the back of the monitor, is where the picture generation process begins and where it concludes. The generation of the electron beam, focalization, and energy conversion processes may all be broken down into three separate steps. The cathode is preheated from 700 to 1200°C to speed up the electron release process. The cathode is typically constructed of nickel with only a surface coating of strontium oxide and a thin coating layer of barium. For creating a picture inside the tube, the electron gun should concentrate the electron ray. The portion of area that the electron particles approach over the screen of phosphor and the color intensity, which affect the resolution of CRTs, depend on the focusing point and electron ray's energy. Coils of wires forming electromagnetic waves are used to focus the electron ray or beam. The majority of CRT monitors use electrostatic grids due to their cost-effectiveness. When the electron beam leaves the cathode, three anodes guide the electron beam which are named the early accelerating, the targeting, and the accelerating anode. The process of deviation of the beam of electrons starts after the beam of electrons has been targeted toward the middle beam. This procedure directs the beam of electrons to a specific location on the fluorescent display. The system of deflection might be coil-based electromagnetic or electrostatic, just like the focusing system. Electromagnetic deflection is the technology most frequently utilized in CRT monitor screens and television sets in this scenario. The electron beam is directed toward an appropriate angle to penetrate into the mask of shadow and strike the illumination screen on a specific portion by the coil's magnetic field, which is generated in two opposite

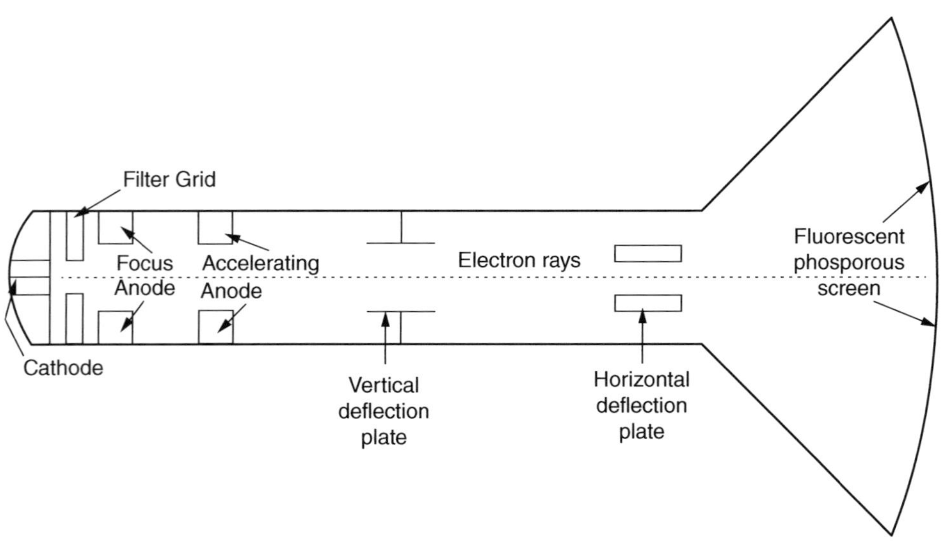

Figure 9.5 Typical cathode ray tube components.

directions at right angles, the x axis in the horizontal direction and the y axis in the vertical direction. The conversion of energy takes place at the end of the process. This entails transforming the electron beam's energy into light that is visible to the human eye. Fluorescence and phosphorescence procedures are used by the phosphorescent components on the front glass to emit light. The phenomenon of light emission during an electron bombardment is known as fluorescence, and the process of emitted light following an electron bombardment is known as phosphorescence.

CRT monitors have a number of parts made of a variety of materials, including metals, ceramics, and plastics. The electron gun and three different types of glass combine to make the conical cathode ray tube. Because of the image generation process happening in the CRT, it has become the main part of the monitor. The cathode ray tube is its heaviest part and accounts for 60% of the monitor's weight.

The printed circuit board (PCB), which supports the electrical components, is in charge of managing the image formation process. Due to the significant variability of materials used in their construction, the PCBs in the monitors are highly complicated parts. Recycling procedure for the polymeric components, which include rubber and casings, is quite well established despite their tiny mass percentage. Polycarbonates, styrene-butadiene, high-impact polystyrenes, and acrylonitrile-butadiene-styrene are among the main polymers used in CRTs. The front display releases radiation during the creation and acceleration of electrons for display of pictures. The phosphor layer is struck by the electron beam of a CRT, which emits radiation with an X-ray wavelength. Lead is incorporated into the glass's formula to stop radiation leakage from the tube. Lead can be absorbed by radiation that is ionizing. Lead is added to the glass of the funnel and neck. Barium has been added to the panel glass. Due to its density, barium, like lead, can absorb some amount of radiation. Lead must be added to the glass tube to limit the emitted radiation levels. The radiation measured at 30 cm distance from the display must not exceed 10 mR/h. Manufacturers of CRT monitors worldwide often adhere to these parameters, which are restrictions set by the United States Food and Drug Administration. The prospect of the lead included in CRT tubes seeping poses the greatest environmental risk. According to some authors, lead contained in glass can seep in a variety of conditions and at levels higher than those permitted by law. Other dangerous metals added to glass are strontium, barium cadmium, that is present in the phosphor layer; these also pose a risk of leaching.

A crucial stage in the recovery of electronic trash is the recovery of CRT waste. Because the CRT display is the biggest and heaviest part of a computer, and because it is primarily made of glass and leaves behind hazardous waste, it is not of great interest to WEEE recycling firms. Systems for closed-loop and open-loop recovery are used for CRT recycling. The recovery of CRTs within the supply chain itself is referred to as a closed-loop system. In this instance, the older CRT tubes are utilized for the production of newer CRT displays. In the method of open-loop, recycled CRT glass is utilized in the manufacturing of various newer goods. Both systems have challenges and barriers. Recycling CRTs provides significant logistical challenges for the closed-loop system. This recycling procedure in a particular country is logistically and financially feasible due in large part to the existence of producers in the nation. Even though there are CRT manufacturers in numerous nations, the market for recycled CRT tubes would outweigh the requirement for new CRT tubes, primarily because the requirement for new CRTs is dropping. The inability of glass producers to accept CRT glass is one obstacle to the system of open-loop recycling. The inclusion of this sort of glass in the production of new glassy products is hampered by the difference in the content of the material from monitor to monitor.

The collection, short-term storage, and cleaning of displays is where the WEEE recycling firms' CRT recycling circuits begin. CRT monitors present a collecting and storage challenge because they are regarded as hazardous residue, necessitating strict compliance with all hazardous waste management standards during transit and storage. The next step is to disassemble the CRT's individual parts. The CRT monitor screen is carefully opened, and each individual component is taken out one at a time. The CRT assembly, polymer materials, the metallic substances, and the PCBs are four categories into which the components are typically sorted. The next stage involves depressurization and glass type separation. The three kinds of glass must be separated since they each have a different chemical makeup and require a different recycling technique. Typically, the anode is drilled into the glass funnel during the depressurization process, or the apex of the collar glass is broken. Because there is a chance of fracturing the glass, the tube separation must be done mechanically. The hot wire technique and the diamond saw technique are the two approaches that are most frequently employed for this purpose. With the use of small tubes, both methods are mechanized (Coyne 2006). Other technologies include the thermal system shock, the water jet segregation technique, and laser cutting procedures. The wire made of nickel-chromium alloy that is wound tightly on the CRT between the funnel junction and window panel is subjected to an electric current by the hot wire system. Glass is melted by a heat wire, and the wire is cooled by air flow. The interface glass is broken by the thermal shock caused by rapid heating and cooling. The operator must then manually finish the full isolation between the funnel and the faceplate glass. This approach has the benefit of being inexpensive and requiring little upfront money. The link issues include incorrect wire positioning that results in an uneven cut surface, the interfaces with sharp edges, and any potential fumes from heating.

The frit interface is sliced through using two or more diamond saw razors in the diamond saw process. Wet or dry environments are used in this procedure. Water is pushed in the wet state to lessen heat and friction. Aeration is employed when the weather is dry. This technique may use revolving diamond saw blades on a CRT. To achieve a clean cut, the blades and frit interface must be properly aligned. The key benefits of this technology are its low operating costs and broad application. However, an automated machine must do the right alignment, and a modest initial expenditure of capital is required. Different recycling pathways are made possible by the glass separation. The standard glass recycling technique is typically used for panel glass. It is delivered to glass smelters (businesses) where it is combined with certain other kinds of glassware, like soda-lime glass, to create items made of glass without the use of heavy metals. Since the panel glass does not contain any lead, no cleaning procedure is required before recycling it. To avoid any contamination between the different types of glass, the glass segregation process must be closely watched.

Panel glass cannot be reclaimed directly if it has a particular quantity of lead. Leaded glass can be recycled to make items like optically dense barium crown glass, jewelry, radioactive glass, crystal glass, and electron gun and funnel glass. Lead content changes in glass types from 1% in jewelry to about 65% in electromagnetic wave insulation applications. Albeit, due to the difference in glass composition across CRTs and the abundance of components that could potentially obstruct the process, glass producers generally do not need to accept CRT glass. This fact acts as a roadblock to leaded glass recycling. Leaded glass can also be recycled for items like foam glass, bricks, clay, and heaps, which consist of a tiny composition of lead. Many studies advocate for the ceramics sector to recycle CRT glass, particularly leaded glass. Because of the potential for contamination, production of leaded glass items for human use is banned. The encapsulation method of lead must be guaranteed regardless of the final product made from the leaded glass. Leaded glass recycling can also be done by ceramics businesses. Lead decontamination is a technique used by the Sweeep Kuusakoski Company and the Nulife Glass Company. To eliminate the lead from the glass, they use a furnace and chemicals. The method is centered on reducing lead and removing it from the matrix of glass. A lead bar and deleaded glass are the end results. Although only a small number of businesses worldwide employ this technology, the procedure offers a fresh approach to handling this residue.

9.6.2 Processing of Liquid Crystal Displays (LCDs) and Light Emitting Diodes (LEDs)

There are numerous varied applications for liquid crystal displays, including cell phones, televisions, digital cameras, monitors, calculators, and equipment display. Cathode ray tube (CRT) screen replacement is one of the goals of this technology. The LCD is made up of dual plates of glass having electrodes that are clear and a thick crystalline layer of 4 microns to 10 microns. The screen's construction also contains PCBs, a light source, sheet diffuser, sheet for reflecting, perspex plate for guiding light, and polarizers. In 1888, Friedrich Reinitzer of Austria made the discovery of liquid crystals. He experimented with the organic fluid cholesteryl benzoate and discovered that the milky white liquid turns transparent when heated. The process is explained by a liquid change to an isotropic state from being an electrical and optical anisotropic state. The crystal structures while in the isotropic liquids, the liquid crystals remain in a transitional state. A deformable crystalline structure is formed in the isotropic liquids that has specific behavior and qualities. The liquid change behavior and anisotropy are related to the crystalline nature employed in the functioning of the monitors. Typically, 0.6 mg/cm^2 of liquid crystals are present per unit area of the glass layer in LCD displays. One 15-inch monitor screen contains about 400 mg of liquid crystal overall. In the 1980s, passive matrix LCD monitors gained popularity. The monitors are preferred for laptops screens, but because of the technology's slow screen-refreshing speed, the images were distorted. The weight and price were the key benefits. A grid is utilized in a passive matrix to deliver a load to particular monitor pixels. Two glass plates, referred to as substrates, are used to create this grid. This substrate has transparent conductive columns and lines, commonly constructed of indium tin oxide (ITO). Electrodes made from ITO that are transparent are designed into smaller segments to display limited data with alphanumeric outputs that are commonly used for electronic calculators and timepieces. The integrated circuit produces a charged signal for the designated column's layer and a ground charged signal to the other layer to activate a pixel. The voltage needed to untangle the liquid crystals at that specific pixel is provided by the intersection of the column and row at the location of that pixel. Transparent conductive electrodes are structured as stripes in two opposing glass layers of the passive LCD matrix, perpendicular to one another. This enables external connections M rows + N columns to handle pixels as rows (M) and columns (N) in the form of arrays resulting in M rows and N columns of pixels. The voltages that pass through each pixel in this system are not separately controlled. The pixels in the passive matrix form are better described from a forward angle of viewing because seeing the display from an oblique position causes color distortion, darkening, or other issues, which is one drawback of the technology. As a result, this technology is only useful for displays with limited information. Large numbers of elements or pixels in displays with a lot of relevant information make it impossible to individually control each pixel. With active-matrix technology, each pixel's substrate contains transistors that are recorded there. The active-matrix individual pixels may be controlled independently thanks to Thin Film Transistors (TFTs) on the glass layer. For example, 1120 TFT transistors would be needed for construction of pixels in a passive matrix with 640 pixels × 480 pixels resolution, compared to 307 200 TFT transistors for an active matrix. Attaching the electrode to the substrate is thus vitally necessary. The glass substrate utilized for the pixels of the active matrix will have 307 200 TFT electrodes at this resolution, compared to just 1120 electrodes for the passive matrix. Active-matrix displays are more expensive when compared to the passive matrix displays having the similar resolution, because the manufacturing of active-matrix monitor displays is more complicated.

In the active-matrix LCD display the pixels are maintained by circuits with short-lived memory provided by the capacitor and the TFT screen. When compared to CRTs, this circuit increases the price and complexity of LCDs, but when the total cost is taken into account, it must be remembered that the CRT's size and weight present far greater shipping, handling, and disposal challenges than LCDs. The light source for an LCD screen is found in the back. This light source is paired with the reflective light path sheet of Perspex. This acts as a diffusing agent enabling light to be dispersed uniformly throughout the LCD panel. Small grooves separate the glass plates from one another, and each groove has electrodes attached to the transistors. The purpose of the groove is to coordinate the orientation of liquid crystal molecules of the helix type. Polarizing coatings on the glass screen are perpendicular to one another. The emitted light inside the display is polarized by the primary film for polarizing, which makes connection to the sheets for diffusing. Only when the liquid crystal is present will this light be able to flow through a different polarizing film. A transparent conductor of electricity is required in LCDs for directing electron beams and for connecting liquid crystals to transfer electricity, which in turn creates an image. This property is present in oxides of indium when doped using oxides of tin, which is a combination of tin oxide with the indium oxide in suitable ratios (Feng et al. 2023). According to the tin oxide concentration, the ITO layer has transparency of about 90% and behaves with the characteristics of a solid material while having strong conductivity toward electricity, electrical UV resistance, reflection of infrared radiation, and a translucent pale yellow appearance. The orientation of the light that is polarized when passing through the film used for polarizing is altered when the crystal particles present in the liquids are arranged as a helix shape. The white mode is the name given to this setup. The crystals in the liquids are oriented toward the transmitted light via the primary polarizer in the screen of glass while electricity is passed. Since the light does not change direction after passing through the first polarizer, it does not transit through the second polarizer. The black mode is the name given to this setup. A transistor controls each subpixel individually. The transistor functions as a switch that enables line-by-line updating of the screen. Each subpixel of these lines can acquire a video stream from the top of the screen when a voltage is applied to a certain row. While just one line could receive the data at once, each line receives data so quickly that the display appears to display a flowing image. Red, green, and blue filters are used in each of the screen's unique subpixels. According to the additive color theory, which underpins color representation, the three subpixels that make up each pixel will effectively mix their colors at normal viewing distances. The color of the pixel will be white when all three subpixels are transmitting. The transmissivity and the unique color of the subpixel's color range determine the precise color coordinates of the spot with the monitor's color temperature.

The illumination source distinguishes an LCD monitor from an LED display. Cold cathode fluorescent lamps (CCFLs) are typically used as the backlight on LCD active matrix displays, in the back of television panels, monitors, and even at the edge of the laptop's display in the majority of the displays. One or more lamps with inside coatings of red, blue, and green phosphor make up the backlight CCFLs. One of its most widely used types of illumination is the LED array. Behind the LCD panel, a thin diffuser sheet can be illuminated by an LED array to create even background lighting. The LED lighting can be simmered or switched off through adjusting the voltage supplied. The LEDs are available in primary, secondary, and tertiary colors. With a usable life of more than 50 000 hours, the backlighting in LEDs has significantly the longest lifespan when compared to previous electroluminescent equipment. The major LCD manufacturers have adapted source lighting by using the LEDs in the place of CCFLs, and they are adopting them more and more. LEDs are superior to CCFLs in practically every regard due to their higher manufacturing efficiency, which results in less heat production and energy consumption.

The liquid crystals in In-Plane Switching (IPS) monitors are positioned flatter than vertically similar to conventional twisted nematic monitors. In contrast to the twisted nematic cell, the electrodes in the IPS cell are only placed over one end of the layers of glass, creating a varying electric charge that is roughly parallel to the surface of the substrate and has lines of equipotentiality that are at right angles to the surface of the glass layer. In comparison to twisted nematic monitors, the particles are twisted in the substrate's plane, which significantly reduces the amplitude of transmitted light's reliance on the viewing angle. The fundamental advantage of IPS technology and what makes it so fascinating for displays is the enhancement in viewing angle dependency that can be attained without the use of additional optical compensating material. The single electrode configuration and layer orientation result in another unique feature of the IPS cell. While the alignment layer and liquid crystal's comparable electronic circuits in twisted nematic cells are connected in series, the IPS's structure offers a parallel coupling of the two components.

The front cover, plastic frame, LCD screen, PCB, and back cover make up the basic structure of every LCD monitor. The display is made up of a number of different parts, including the parts in the front portion, the monitor's parts devoid of plastic framing, the front and back portion of the display, fluorescent lamp, screen made of glass, sheets for diffusing, sheets of reflection, and the back-end monitor cover. Typical LCD panels are made up of 51% polymers, 16% metals, 22% LCD glass, 9% PCB, 0.34% backlight, and 1.7% wire, natural rubber, plastics that are laminated, and other substances.

An estimated 2.5 billions of LCDs are nearing the end of their useful life. The author claims that incineration is inefficient, costly, and bad for the environment. The greatest priority is given to reclamation and repurposing of the LCDs, especially in light of cautious estimates for the sales growth of LCDs of 16–28% per 5 years. While monitors are not made for simple disassembly, dangerous materials like CCFLs must be physically removed. However, this is not a widespread practice at the moment. Larger metal components and plastic casings need to be manually removed. To get material fractions like ferrous materials, polymeric materials, and non-ferrous metallic substances, the retained quantity is often powdered and subjected to varying levels of segregation processes. An LCD monitor's latches, which include clips, screws, and adhesives, are tiny and sometimes difficult to reach. Additionally, the designs of products created by various manufacturers vary, including the kinds of screws and varying parts. Due to these obstacles, designing a procedure of disassembly that can fit the many models in the marketplace is challenging. Due to these challenges, many businesses choose to ship complete displays to nations that recycle using technically incorrect methods that are environmentally harmful.

An LCD unit may have some components that can be taken out and managed separately. Mercury lamps, PCBs, and LCD screens are some of these. As a result, glass LCD panels may become unpopular and may end up in the waste stream. This garbage can either be landfilled or used by secondary processors, such as businesses that recycle polymers or operate smelters, depending on the recycling company. Currently, two different kinds of glass substrate are used to make LCDs. They are made of borosilicate glass and soda-lime glass. It might be possible to merge processes that selectively separate contaminants. The recovered glass from this procedure may subsequently be applied to the ceramics or cement industries.

The fluorescent lights of an LCD unit contain roughly 4 mg of molecular mercury, according to the EPA, and should be handled carefully. An unbroken lamp of mercury can be taken out by hand. However, in order to retrieve the mercury, specialized machinery is needed that smashes the lamps and safely segregates the glasses, mercury, and various parts. To carry out this procedure, an enclosed negative pressure method that stops mercury release is required. As a result, mercury recovery from lamps is often only possible by specialized businesses. For all the 80 millions of LCDs in global use in the year 2010, the estimation is that the mercury disposal content of about 290–480 kg will take place in the environment. The mercury is obtained via hydrometallurgical techniques, after which it can be collected as a solid substance. The following requirements must be met by humid chemical mercury treatment methods: The applied method must create a product that is marketable, which may be reinstated in the manufacturing methods, and its waste products must either not contain any environmentally hazardous compounds or be treated until they do not pose a threat to the environment.

When smaller internal parts can potentially be retrieved through fragmenting and segregating, the bigger metal components can be manually removed. LCD monitor framing parts are made of steel and aluminum. On printed circuit boards (PCBs), copper, precious metals, and lead solder are all used in the circuit (PCBs). While tiny parts could be reclaimed using crushing and segregation methods, large metallic parts could be manually reclaimed. Typically, PCBs are broken down and then incinerated in a furnace, leaving behind a concentrated metal-rich solid residue. Burning plastics produce either volatile chemicals or slag, which contain refractory oxides. Some metals may act like collections of other metals during this process. As a result, only alloys with unusual compositions, not pure metals, are obtained. In some nations, rare earth metals are also extracted using crude refining methods with fewer worker safety and environmental restrictions. It is technically possible to recover crystals in liquid from outdated or unusable LCDs. The liquid crystal quantity in every LCD is meagre, 0.6 mg/cm^2 of monitor with a thickness of 5 micrometers. Large quantities of solvent are necessary to recycle the crystal liquid, but using the solvent is constrained by regulations and considerations based on the economy. The above factors and some others significantly limit the viability of recycling. According to Merck research, recovery procedures cost more than creating the same quantity of fresh material. Therefore, it does not seem that recycling liquid crystal mixes would be profitable right now.

Because it is a rare metal that is crucial to the solar energy sector, flat displays, and other modern technological uses, indium is categorized as a critical metal by many studies. There are no alternatives that provide the same performance for most of these applications. The transparent conductive sheet used in LCD panels contains indium tin oxide (ITO), which Sharp stated it had devised a method to recycle. Glass shards made from the LCD screen are crushed and then treated in an acid solution. The consortium of Sharp claims that this process is relatively straightforward because it recovers indium that is incredibly pure while using readily available chemicals and little energy. Indium is required to be utilized more frequently in the electrodes of transparent monitors despite being a scarce metal. From main and secondary raw resources, increased indium has been generated since 2007. Indium tin oxide accounts for more than half of the world's indium use. It was estimated that 425 tons of indium were utilized to make ITO in 2011. By decomposing ITO in hydrochloric acid and precipitation using sodium hydroxide, the sputtered waste indium could be reclaimed. Only 5.5 mg of indium can be reclaimed from a cell phone screen. Compared to larger LCD, which uses 260 mg, personal computer monitors require 79 mg and 82 mg of indium. There are a few options for recycling indium. Although the percentage of EOL recycling is low, secondary metal makes up between 25 and 50% of the overall metal input used in metal production. Where ITO is produced and sputtering occurs, suitable processes of recycling are practiced in nations such as Japan, China, and South Korea (Azimi et al. 2020). The authors have patented a process that recovers indium from a nitric acid liquid via solvent extraction. It is expensive to use nitric acid, and it contains contaminants such as heavy metals. Clearly, a more effective way to extract indium from used LCDs needs to be developed. Indium from LEDs can be partially recovered during typical pyrometallurgical smelting processes, such as copper scrap recycling and e-waste disposal.

9.7 Conclusion

E-waste is a fast-growing problem since it is being produced at an exponentially rising rate and is a hazardous waste stream. Legislation has been created in numerous nations to promote reusing, recycling, and many other types of waste recovery. When e-waste was characterized, it became clear that it is extremely diverse and complicated with respect to the many types of parts and materials. But for the majority of the e-waste samples, copper and rare earth metals account for more than 80% of the value. This suggests that the resurgence of rare metals and copper may continue to be a significant economic driver for a considerable amount of time. The hierarchy of e-waste treatment promotes the utilization of the entire piece of equipment first, followed by remanufacturing, material recovery through recycling, and, as a final option, disposition by burning and open dumps. Generally speaking, there are three main steps in the recycling of e-waste: (a) sequential dismantling, identifying, and segregating out hazardous or valuable modules for specific treatment; (b) mechanical and metallurgical procedures to enhance preferable components; and (c) refinement of retrieved components, which are retreated or improved using chemical or metallurgical processing so that they are appropriate for further utilization in their original application.

References

AliAkbari, R., Marfavi, Y., Kowsari, E., and Ramakrishna, S. (2020). Recent studies on ionic liquids in metal recovery from E-waste and secondary sources by liquid-liquid extraction and electrodeposition: a review. *Materials Circular Economy* 2: 1–27.

Azimi, G., Forsberg, K., Ouchi, T. et al. (eds.) (2020). *Rare Metal Technology*. Springer Nature.

Barnwal, A. and Dhawan, N. (2020). Recycling of discarded mobile printed circuit boards for extraction of gold and copper. *Sustainable Materials and Technologies* 25: e00164.

Benzaazoua, M. and Taha, Y. (2020). *Recent Trends in Phosphate Mining and Beneficiation and Related Waste Management*. MDPI.

Coyne, G.S. (2006). *The Laboratory Companion: A Practical Guide to Materials, Equipment, and Technique*. New York: Wiley-Interscience.

Czél, G., Kelemen, A., and Sólyom, D. (2010). Flame retardant polyethylene terephthalate fibers added with melamine based FR additive. In: *Materials Science, Testing and Informatics V: 7th Hungarian Conference on Materials Science* (eds. P.J. Szabó and T. Berecz). Materials Science Forum 659, Trans Tech Publications Ltd. https://doi.org/10.4028/www.scientific.net/MSF.659.135.

Dias, P.R., Cenci, M.P., Bernardes, A.M., and Huda, N. (2022). What drives WEEE recycling? A comparative study concerning legislation, collection and recycling. *Waste Management & Research* 40 (10): 1527–1538.

Directive, E.C. (2012). Directive 2012/19/EU of the European Parliament and of the Council of 4 July 2012 on waste electrical and electronic equipment, WEEE. *Official Journal of the European Union L* 197: 38–71.

Directive, E.U. and others (2013). Restriction of the use of certain hazardous substances in electrical and electronic equipment (RoHS). *Official Journal of the European Communities* 46: 19–23. Available at. https://www.telecomconnections.ltd.uk/wp-content/uploads/2019/06/ROHS-3-Statement.pdf.

Feng, J.J., Sun, X.F., and Zeng, E.Y. (2023). Predicted health and environmental hazards of liquid crystal materials via quantitative structure-property relationship modeling. *Journal of Hazardous Materials* 446: 130592.

Franke, D., Suponik, T., Nuckowski, P.M. et al. (2020). Recovery of metals from printed circuit boards by means of electrostatic separation. *Management Systems in Production Engineering* 28 (4): 213–219. https://doi.org/10.2478/mspe-2020-0031.

Ghulam, S.T. and Abushammala, H. (2023). Challenges and opportunities in the management of electronic waste and its impact on human health and environment. *Sustainability* 15 (3): 1837. https://doi.org/10.3390/su15031837.

Goodship, V., Stevels, A., and Huisman, J. (eds.) (2019). *Waste Electrical and Electronic Equipment (WEEE) Handbook*. Woodhead Publishing.

Goosey, M. (2004). End-of-life electronics legislation––an industry perspective. *Circuit World* 30 (2): 41–45. https://doi.org/10.1108/03056120410512235.

Gunarathne, V., Rajapaksha, A.U., Vithanage, M. et al. (2022). Hydrometallurgical processes for heavy metals recovery from industrial sludges. *Critical Reviews in Environmental Science and Technology* 52 (6): 1022–1062. https://doi.org/10.1080/10643389.2020.1847949.

Hsu, E., Barmak, K., West, A.C., and Park, A.H.A. (2019). Advancements in the treatment and processing of electronic waste with sustainability: a review of metal extraction and recovery technologies. *Green Chemistry* 21 (5): 919–936. https://doi.org/10.1039/C8GC03688H.

Huang, J., Deng, Y., Han, Y. et al. (2022). Toxic footprint and materials profile of electronic components in printed circuit boards. *Waste Management* 141: 154–162. https://doi.org/10.1016/j.wasman.2022.01.019.

Huang, K., Zheng, J., Yuan, W. et al. (2021). Microwave-assisted chemical recovery of glass fiber and epoxy resin from non-metallic components in waste printed circuit boards. *Waste Management* 124: 8–16. https://doi.org/10.1016/j.wasman.2021.01.010.

Işildar, A. (2018). *Metal Recovery from Electronic Waste: Biological versus Chemical Leaching for Recovery of Copper and Gold*. CRC Press.

Jorjani, E. and Ghahreman, A. (2017). Challenges with elemental sulfur removal during the leaching of copper and zinc sulfides, and from the residues; a review. *Hydrometallurgy* 171: 333–343. https://doi.org/10.1016/j.hydromet.2017.06.011.

Juliussen, E. and Petska-Juliussen, K. (1996). *Computer Industry Almanac*, 8e. Computer Industry Almanac, Inc. Available at: http://www.c-i-a.com (accessed 30 June 2022).

Kan, Y., Zheng, F., and Zhang, R. (2021). Comparative study of pyrolytic carbons prepared from printed circuit boards by magnetic and electrostatic separation. *RSC Advances* 11 (53): 33490–33499. https://doi.org/10.1039/d1ra05287j.

Kaur, P.J., Pant, K.K., Chauhan, G., and Nigam, K.D.P. (2020). *Sustainable Metal Extraction from Waste Streams*. John Wiley & Sons.

Khan, S.A. (2020). Clearly hazardous, obscurely regulated: lessons from the Basel Convention on waste trade. *American Journal of International Law* 114: 200–205. https://doi.org/10.1017/aju.2020.38.

Kiddee, P., Naidu, R., and Wong, M.H. (2013). Electronic waste management approaches: an overview. *Waste Management* 33 (5): 1237–1250. https://doi.org/10.1016/j.wasman.2013.01.006.

Kononova, O.N., Melnikov, A.M., and Demitrichenko, D.S. (2013). Simultaneous ion exchange recovery and subsequent separation of platinum (II, IV), rhodium (III), and nickel (II) from chloride and sulfate-chloride solutions. *Solvent Extraction and Ion Exchange* 31 (3): 306–319. https://doi.org/10.1080/07366299.2012.757133.

Kour, G., Kothari, R., Mohan Singh, H. et al. (2021). Microbial leaching for valuable metals harvesting: versatility for the bioeconomy. *Environmental Sustainability* 4: 215–229. https://doi.org/10.1007/s42398-020-00143-9.

Krishnan, S., Zulkapli, N.S., Kamyab, H. et al. (2021). Current technologies for recovery of metals from industrial wastes: an overview. *Environmental Technology & Innovation* 22: 101525. https://doi.org/10.1016/j.eti.2021.101525.

Kumar, A., Holuszko, M., and Espinosa, D.C.R. (2017). E-waste: an overview on generation, collection, legislation and recycling practices. *Resources, Conservation and Recycling* 122: 32–42. https://doi.org/10.1016/j.resconrec.2017.01.018.

Li, J., Liu, L., Zhao, N. et al. (2013). Regional or global WEEE recycling. Where to go? *Waste Management* 33 (4): 923–934. https://doi.org/10.1016/j.wasman.2012.11.011.

Lim, C.H.J., Suchand Sandeep, C.S., Murukeshan, V.M., and Kim, Y.J. (2021). Direct laser writing of graphene-based electrical interconnects for printed circuit board repair. *Advanced Materials Technologies* 6 (12): 2100514. https://doi.org/10.1002/admt.202100514.

Lu, C., Zhang, L., Zhong, Y. et al. (2015). An overview of e-waste management in China. *Journal of Material Cycles and Waste Management* 17: 1–12. https://doi.org/10.1007/s10163-014-0256-8.

Mathias, R.V. (2021). Management of solid waste from health services according to the national solid waste policy: a study conducted in the South of Brazil. *Gestão & Produção* 28 (4): 5727. https://doi.org/10.1590/1806-9649-2021v28e5727.

Nelson, V., Rueda, X., and Vermeulen, W.J. (2018). Challenges and opportunities for the sustainability transition in global trade (Introduction). *Business Strategy and the Environment* 27 (2): 173–178. https://doi.org/10.1002/bse.2008.

Nicol, M.J. and Kganyago, M.P. (2020). Extraction of gold from chloride solutions using dibasic esters: a structure-reactivity study. *Solvent Extraction and Ion Exchange* 38 (3): 340–349. https://doi.org/10.1080/07366299.2020.1729326.

Oguri, T., Suzuki, G., Matsukami, H. et al. (2018). Exposure assessment of heavy metals in an e-waste processing area in northern Vietnam. *Science of the Total Environment* 621: 1115–1123. https://doi.org/10.1016/j.scitotenv.2017.10.115.

Okajima, N., Takei, T., and Usui, S. (2020). Pb and other impurities recovery from cu smelting residues in JX Nippon Mining & Metals. In: *PbZn 2020: 9th International Symposium on Lead and Zinc Processing. The Minerals, Metals & Materials Series* (ed. A. Siegmund, S. Alam, J. Grogan, U. Kerney, and E. Shibata). Cham, Switzerland: Springer. https://doi.org/10.1007/978-3-030-37070-1_45.

Ongondo, F.O., Williams, I.D., and Cherrett, T.J. (2011). How are WEEE doing? A global review of the management of electrical and electronic wastes. *Waste Management* 31 (4): 714–730. https://doi.org/10.1016/j.wasman.2010.10.023.

Peecher, J.S., Schecter, A.J., Lu, H. et al. (2022). Biomonitoring of polybrominated dioxins & furans, polychlorinated dioxins & furans, and dioxin-like polychlorinated biphenyls in Vietnamese female electronic waste recyclers. *Journal of Occupational and Environmental Medicine* 64 (9): 742–747. https://doi.org/10.1097/JOM.0000000000002506.

Prasad, M.N.V. and Vithanage, M. (eds.) (2019). *Electronic Waste Management and Treatment Technology*. Butterworth-Heinemann.

Rautela, R., Arya, S., Vishwakarma, S. et al. (2021). E-waste management and its effects on the environment and human health. *Science of the Total Environment* 773: 145623. https://doi.org/10.1016/j.scitotenv.2021.145623.

Robinson, B.H. (2009). E-waste: an assessment of global production and environmental impacts. *Science of the Total Environment* 408 (2): 183–191. https://doi.org/10.1016/j.scitotenv.2009.09.044.

Roy, J.J., Rarotra, S., Krikstolaityte, V. et al. (2022). Green recycling methods to treat lithium-ion batteries E-waste: a circular approach to sustainability. *Advanced Materials* 34 (25): 2103346. https://doi.org/10.1002/adma.202103346.

Sarvar, M., Salarirad, M.M., and Shabani, M.A. (2015). Characterization and mechanical separation of metals from computer printed circuit boards (PCBs) based on mineral processing methods. *Waste Management* 45: 246–257. https://doi.org/10.1016/j.wasman.2015.06.020.

Sharma, M., Joshi, S., and Kumar, A. (2020). Assessing enablers of e-waste management in circular economy using DEMATEL method: an Indian perspective. *Environmental Science and Pollution Research* 27 (12): 13325–13338. https://doi.org/10.1007/s11356-020-07765-w.

Sivaramanan, S. (2013). E-waste management, disposal and its impacts on the environment. *Universal Journal of Environmental Research & Technology* 3 (5): 531–537.

Sudheshwar, A., Malinverno, N., Hischier, R. et al. (2023). The need for design-for-recycling of paper-based printed electronics––a prospective comparison with printed circuit boards. *Resources, Conservation and Recycling* 189: 106757. https://doi.org/10.1016/j.resconrec.2022.106757.

Ta, T., Fukumoto, M., Narumi, K. et al. (2015 September). Interconnection and double layer for flexible electronic circuit with instant inkjet circuits. In: *Proceedings of the 2015 ACM International Joint Conference on Pervasive and Ubiquitous Computing* 181–190. https://doi.org/10.1145/2750858.2804276.

Vazquez, Y.V. and Barbosa, S.E. (2016). Recycling of mixed plastic waste from electrical and electronic equipment. Added value by compatibilization. *Waste Management* 53: 196–203. https://doi.org/10.1016/j.wasman.2016.04.022.

Widmer, R., Oswald-Krapf, H., Sinha-Khetriwal, D. et al. (2005). Global perspectives on e-waste. *Environmental Impact Assessment Review* 25 (5): 436–458. https://doi.org/10.1016/j.eiar.2005.04.001.

Wilson, D.C., Rodic, L., Modak, P. et al. (2015). Global waste management outlook. UNEP. www.rohsguide.com. RoHS 2 vs RoHS 3 (EU 2015/863) [online]. Available at: https://www.rohsguide.com/rohs3.htm (accessed 14 February 2023).

Yaashikaa, P.R., Priyanka, B., Kumar, P.S. et al. (2022). A review on recent advancements in recovery of valuable and toxic metals from e-waste using bioleaching approach. *Chemosphere* 287: 132230. https://doi.org/10.1016/j.chemosphere.2021.132230.

Yu, Z., Gao, C., Yang, C., and Zhang, L. (2023). Insight into quantities, flows, and recycling technology of E-waste in China for resource sustainable society. *Journal of Cleaner Production* 393: 136222. https://doi.org/10.1016/j.jclepro.2023.136222.

Yuan, Z., Huang, Z., Ruan, J. et al. (2018). Contact behavior between cells and particles in bioleaching of precious metals from waste printed circuit boards. *ACS Sustainable Chemistry & Engineering* 6 (9): 11570–11577. https://doi.org/10.1021/acssuschemeng.8b01742.

Zhang, J., Zhao, B., and Schreiner, B. (2016). *Separation Hydrometallurgy of Rare Earth Elements*. Springer.

10

Circular Economy and the Development of Sustainable Products Through the Application of E-waste

William Pinheiro[1,2], Edson Pablo da Silva[1,3,4], and Flávio Augusto de Freitas[1,2,5,]*

[1] *Centro de Biotecnologia da Amazônia – CBA, Manaus, Amazonas, Brazil*
[2] *Chemistry Graduate Program – PPGQ, Federal University of Amazonas – UFAM, Manaus, Amazonas, Brazil*
[3] *Biotechnology Graduate Program – PPGBIOTEC, Federal University of Amazonas – UFAM, Manaus, Amazonas, Brazil*
[4] *Instituto de Tecnologia e Educacao Galileo da Amazonia, Manaus, Amazonas, Brasil*
[5] *Universidade Federal do Amazonas, Programa de Pós-Graduação em Química–PPGQ, Manaus, Amazonas, Brasil*
[*] *Corresponding author*

10.1 Circular Economy

The circular economy model involves designing a plan for the systematic management and reuse of materials in a particular system. This approach aims to minimize waste in industrial processes through optimized material, product, and system design (Akcil et al. 2019; Sagnak et al. 2021). This requires deliberate planning of resource extraction, fabrication processes, and waste management. Minimized material use and maximized material recovery are critical components of a healthy circular economy in the industrial world. Using recovered waste materials to design new products allows the materials to retain their value for a more extended period before they reach the end of their useful life (Gautam et al. 2022; Xiao et al. 2022). To begin designing a sustainable circular economy for a given product, the entire life cycle—from raw material extraction and production to the end of useful life—must be thoroughly studied and modeled (EPA 2021).

Countries such as China, the United States of America (USA), India, Japan, and Brazil are considered to be the largest producers of electronic waste in the world (in ascending order) (Van Yken et al. 2021). Mainly, these countries aim at different technologies to use the materials present in the electronic waste (e-waste) generated to reapply the extracted compounds and elements and, thus, apply the circular economy concept.

10.2 Electrical and Electronic Waste (E-waste)

The term e-waste encompasses any products with electrical circuits or components with a conventional power supply—for example, via battery or solar energy—that are discarded as waste and are no longer useful in their current state (Baldé et al. 2022). Nearly everything used in daily life—from ovens, cars, and cell phones to electrical grid components, satellites, and data server centers—has these characteristics. Other examples include toys, musical devices, tools used in construction, large industrial machines, and sophisticated research and medical equipment. Increasingly, traditionally tech-free products like picture frames, furniture, and clothing are becoming integrated into the so-called Internet of Things (IoT) via enhancement with electronic devices such as sensors, Light-Emiting Diodes (LEDs), circuits, and batteries (Baldé et al. 2022). All of these products will, one day, be e-waste. Therefore, in the coming decades, humanity must devise a plan to manage this massive and increasing quantity of generated e-waste.

Six general categories of e-waste were created after separating consumer products based on function, material composition, average weight, and end-of-life characteristics through statistical evaluation and qualitative comparison (Forti et al. 2020). Figure 10.1 shows some examples of e-waste in each category. These six categories of electrical and electronic equipment (EEE) are divided into 54 different categories. This new categorization is known as the United Nations University Keys (UNUKEYs), which consist of more specific keys to distinguish e-waste to obtain a more excellent resolution of the statistics that encompass the generation of these types of waste (Forti et al. 2018).

A significant problem with e-waste is that these waste products are made up of many elements that must be separated and recycled in different ways (Pan et al. 2022; Parajuly et al. 2020).

Electronic Waste Management: Policies, Processes, Technologies, and Impact, First Edition. Edited by Sunil Kumar and Vineet Kumar.

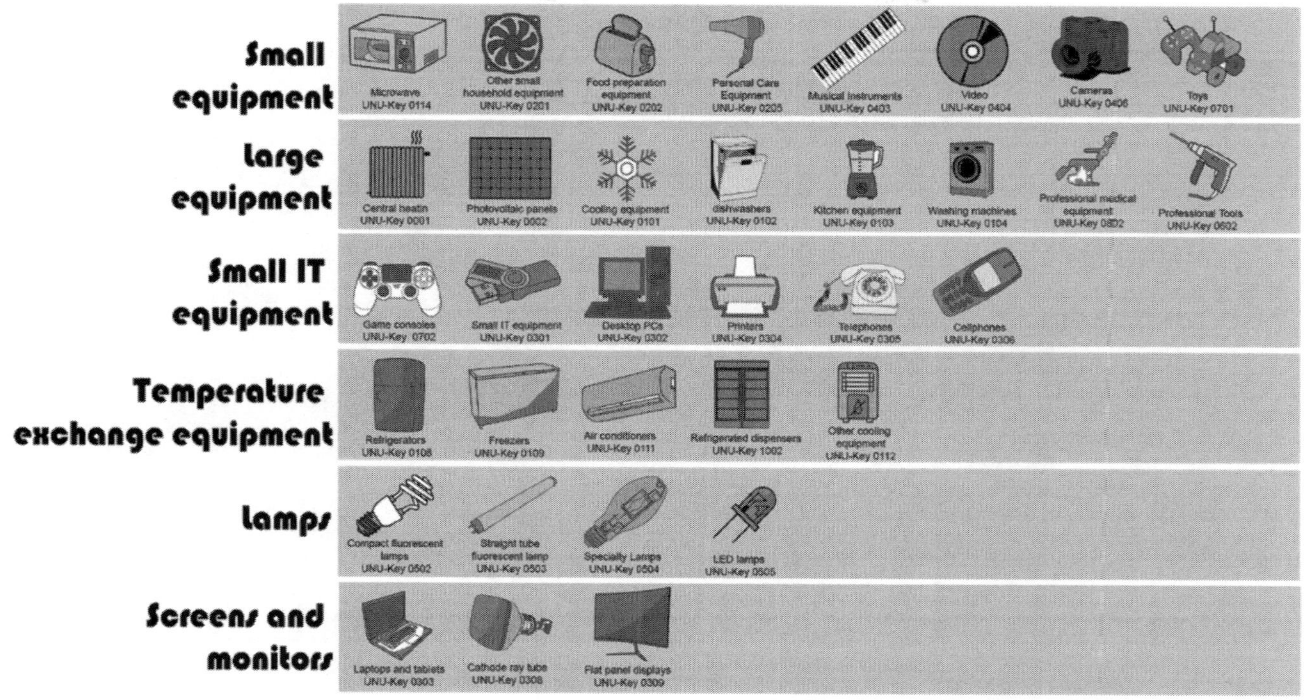

Figure 10.1 Some examples for each of the top six categories of e-waste with their respective categorization in the UNUKEY code.

E-waste contains about 69 species from the periodic table in the metal fraction (MF) present in the waste, including precious metals, critical raw materials, and non-critical metals (Figure 10.2), as well as a variety of organochlorine and organophosphate compounds that cause harmful effects to fauna, flora, and humans (Forti et al. 2020).

In addition to the metallic fraction, e-waste has a majoritarian fraction, called the non-metallic fraction (NMF), which is mainly present in printed circuit boards (PCBs) (Gundupalli et al. 2018). This fraction comprises plastics, thermosetting resins (epoxy), fiberglass, brominated flame retardants, and additives, representing about 70% of the waste weight (Yang et al. 2019). Some authors cite that the recycling of NMF is one of the most important challenges (Jadhao et al. 2022a; Yang et al. 2019).

Usually, these residues are thrown in landfills or incinerated, which will cause waste of resources and environmental damage caused by the toxic gases generated in the combustion of these materials (Sengupta et al. 2022). Therefore, much remains to be done to recycle these materials, which is also a consequence of the greater attention given to the metallic fraction of e-waste.

Because they have a complex structure with metallic and non-metallic parts, recycling waste from electrical and electronic equipment (also called WEEE) requires four steps: collection, sorting, disassembly by mechanical ways, and chemical processing to recover the elements of interest. Of this entire process, the second step is the most critical to detect potential resources in end-of-line products and safeguard their technical and economic value (Awasthi et al. 2022). However, in many countries, the first stage (collection) is still an arduous and primitive process, which depends on human labor to search for e-waste in dumps.

10.3 Recovery of Elements Present in E-waste

In the last monitoring carried out by the United Nations Institute for Training and Research (UNITAR), it was estimated that in 2019 about 39 million tons (Mt) of raw materials were needed to produce new electronic equipment. Of this volume, approximately 14 Mt were made up of new raw material from mining activities, with the other 25 Mt being in charge of recycling electronic waste (Forti et al. 2020).

Among the 69 elements potentially present in e-waste, some have better recovery capacity, either because they are in higher concentration or have less complex extraction processes. The main metallic elements recovered are gold (Au), silver (Ag), copper (Cu), palladium (Pd), ruthenium (Ru), rhodium (Rh), germanium (Ge), osmium (Os), iridium (Ir), platinum (Pt), cobalt (Co), indium (In), bismuth (Bi), antimony (Sb), aluminum (Al), and iron (Fe). The potential value of these raw materials alone has reached US$57

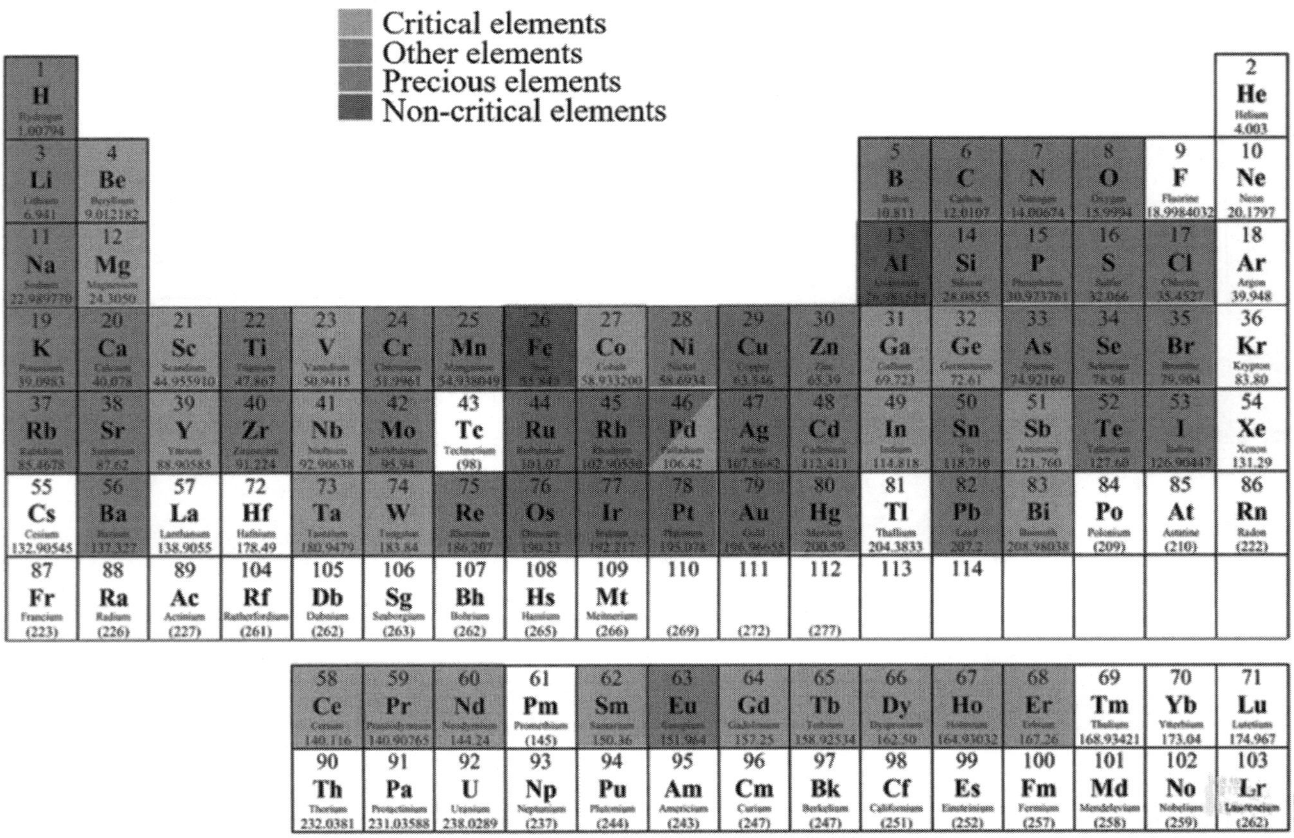

Figure 10.2 Elements found in e-waste presented in the periodic table.

billion, a value associated with the enormous environmental benefits following the recycling effort (Table 10.1).

Three processes widely used to recover these metallic elements from e-waste are pyrometallurgy, hydrometallurgy, and bioleaching. Through these techniques, metals are recovered and can be applied, entering the circular economy concept. Let us understand more about each one.

a) Pyrometallurgy

Pyrometallurgy consists of the extraction and purification of metals using heat. In a generic process of recovering non-ferrous and precious metals from e-waste, crushing, incineration, and smelting steps are performed using a plasma arc furnace, slag, sintering, and melting at high temperatures (Ma 2019). Precious metals such as Ag, Au, Pt, Pd, Rh, Ir, and Ru can be separated through subsequent electrorefining or leaching (Faraji et al. 2022).

These processes are generally cost-effective and provide for the recovery of a sufficient amount of precious metals. The separation is performed through the different melting points and interactions between chemical species (Faraji et al. 2022; Ma 2019).

Pyrometallurgy encompasses different techniques used around the world. Some of these methods are shown in Figure 10.3.

Table 10.1 Main elements recovered from e-waste. Adapted from Forti et al., 2020.

Element	Quantity produced in 2019 (kt)	Thousand (US$)
Ag	1.2	579
Sb	76	644
Al	3046	6062
Ru	0.0003	3
Au	0.2	9481
Rh	0.01	320
Bi	0.1	1.3
Pt	0.002	71
Co	13	1036
Pd	0.1	3532
Cu	1808	10 960
Os	0.01	108
Fe	20 466	24 645
Ir	0.001	5
Ge	0.01	0.4
In	0.2	17

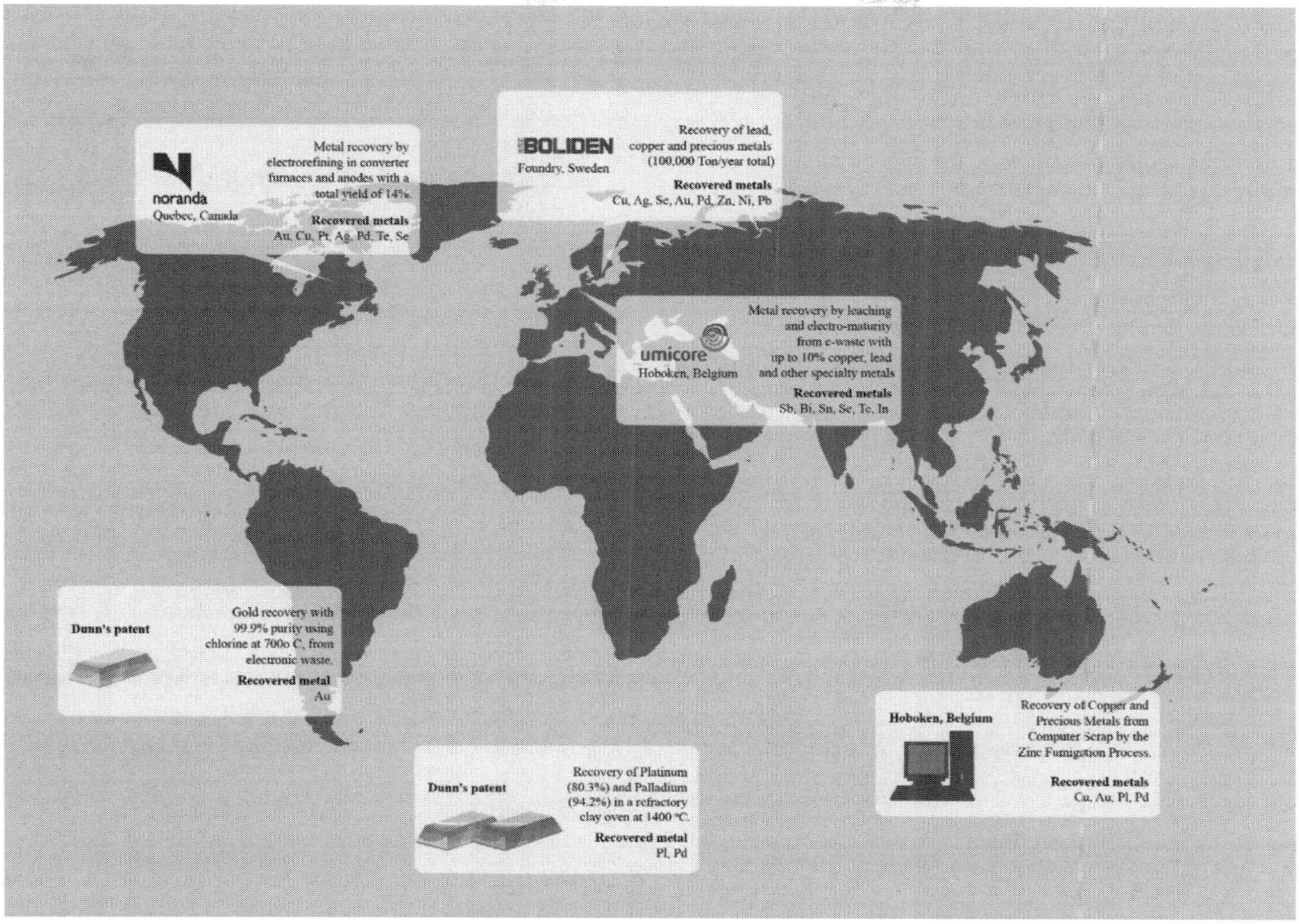

Figure 10.3 Summary of typical pyrometallurgical methods for the recovery of electronic waste metals by industry in the sector (according to Cui and Roven 2011).

Although pyrometallurgy is an efficient process, there are several disadvantages, including significant monetary investment, large energy demand, and release of toxic gases into the environment. In addition, the difficulty in separating mixed metals from pure metals and alloys due to their high stability demands more energy (Cui and Roven 2011). However, the research seeks to optimize pyrometallurgy processes using a vacuum. Vacuum metallurgy is a method by which metallic mixtures are separated using different vapor pressures for different metals. There are several benefits to this method. Under vacuum conditions, i.e., in a system that does not contain air, environmental pollution is significantly reduced. This approach eliminates the need for secondary exhaust treatment and sewage treatment equipment. Notably, the vacuum environment means that easily oxidized heavy metals do not become oxidized. High vapor pressure metals and toxic metals, such as lead and cadmium, are also safely recoverable using this method (Ma 2019).

b) Hydrometallurgy

Hydrometallurgy consists of a group of metal extraction processes where the foremost step involves dissolving the metal in an aqueous medium. Among the traditional applications of hydrometallurgy is the production of alumina, gold, uranium, zinc, nickel, copper, titanium, and rare earths, among others (Sronsri et al. 2022).

These processes have been employed to recover metals from e-waste using acidic or caustic leaching of selective dissolution of the target metals in solid e-waste. Then the resulting solution, which contains metals and impurities, is further refined. Next, the metals are concentrated via solvent extraction, adsorption, or ion exchange processes. Finally, the metals are fully recovered through electrorefining or electrochemical reduction reactions to recover solid metals from e-waste (Ashiq et al. 2019).

Hydrometallurgy has been considered an up-and-coming alternative method due to pyrometallurgy's high-energy demand. Hydrometallurgy is often employed as an auxiliary method for pyrometallurgy, improving the extraction of dissolved elements in coke (Ashiq et al. 2019; Sronsri et al. 2022).

c) Chemical leaching
Chemical leaching is a method that involves the reaction of a solid when it comes into contact with a liquid. Leaching is the method that uses chemicals and metal complex formation techniques to target a valuable metal using acids, metal-bound ligands, various leaching agents such as alkalis, cyanide, and thiosulfate, and many acids such as sulfuric, hydrochloric, and nitric acid (Ashiq et al. 2019).

In the leaching process, the combination of different leaching solutions (Table 10.2) is carried out in a large group of ores. Then the concentrated solution containing the target element is enriched and sedimented by gravity, accumulating for later processing and recovery of metals (Ashiq et al. 2019).

Although this process is very efficient in recovering metals from WEEE, it uses residual toxic substances such as cyanide, thiosulfate, thiourea, and halides in addition to acid leaching (Jadhao et al. 2022a). Thus, this process has been highly discussed since less-polluting processes are sought.

d) Bioleaching
Certain extreme acidophilus has been widely used to promote the recovery of precious metals from e-waste in an environmentally friendly and cost-effective manner. However, oxidizing microbes of Fe^{2+}, S_0, pyrite, or other reduced sulfur substances generate Fe^{3+} and H_2SO_4 as leaching agents. The process of converting solid metals into ions in an aqueous solution employing a leaching agent produced by the catalysis of microorganisms is called bioleaching (Zhao and Wang 2019).

Bioleaching is a promising and sustainable technology that is low-cost, low-energy, and simple. Moreover, bioleaching is not limited to high-grade sulfide minerals; instead, it is widely used in recovering metals from solid waste such as e-waste, sludge, sediment, batteries, catalysts, and low-grade ore (Zhao and Wang 2019).

Several investigations into new bioleaching are underway for many metals, such as indium, vanadium, and cobalt. Some applications of extreme acidophiles used in e-waste bioleaching are shown in Table 10.3.

Although bioleaching is ecologically correct and uses microorganisms to recover metals, it has slow kinetics, easy contamination, and microorganisms sensitive to changes in temperature and pH (Jadhao et al. 2022b). These factors make its application on a commercial scale difficult, and therefore, many studies have yet to be developed to overcome these obstacles.

10.4 Plastic

Plastics constitute a significant fraction of electronic waste, representing about 20% of all produced volume. However, recycling plastics from e-waste presents more challenges

Table 10.2 Leaching solutions used in the industry for metal extraction in electronic waste.

Leaching solution (leaching agent)	Optimal conditions for extraction	Recovery
Halogenated leaching	Solid:liquid (S:L) ratio of 1:10, exposure time 4 h, pH 7; 1.1% iodine;	97.5% Au
	Sulfuric acid-hydrogen peroxide and iodine (SAHPI) leaching with sulfuric acid (H_2SO_4), H_2O_2, and I_2;	95.27% Cu
	Iron (III)/(II) in acidic solution, increase in temperature and ferric chloride ($FeCl_3$) ratio causes an increase in leaching.	98% Au
Thiourea leach	Thiourea (CH_4N_2S) at a concentration of 20 g/L, ferric oxide (Fe_2O_3) 6 g/L in 10 g/L of H_2SO_4 at 600 rpm for 3.5 h.	82% Au
Thiosulfate leaching	0.12 M thiosulfate ($H_2S_2O_3$), 0.2M ammonia (NH_3), and 20 mM Cu solutions;	98% Au
	0.2M $(NH_4)_2SO$, 0.02M $CuSO_4$, NH_4OH solutions (0.4M) at 40°C/48 h for Au and 24 h for Ag;	98% Au
	S:L of 1:50 (v:v), 80°C, 3M HCl for 90 min.	100% Ag, Au > 95%
Acid leaching	S:L of 1:50 (v:v), 80°C, 3M HCl for 90 min.	99.4% Li
	1.25M ascorbic acid ($HC_6H_7O_6$), 70°C, S:L of 25 g/L, 20 min;	94.8% Co, 98.5% Li
	HCl/NaCl with HNO_3/H_2O_2 leaching.	93–95% Pd

Table 10.3 Extreme acidophile application techniques used in metal mining through e-waste bioleaching. Adapted from ZHAO and WANG, 2019.

E-waste	Microorganisms	Metal
Printed circuit boards	*At. ferrooxidans*	Cu
	At. thiooxidans,	Cu
	S. thermosulfidooxidans + acidophilic heterotroph	Cu, Al, Zn, Ni
	Acidiphilium acidophilum	Cu
Liquid crystal displays (LCDs)	*At. thiooxidans*	In
E-waste	*At. thiooxidans*	Cu
Dust from e-waste shredding	*At. thiooxidans*	Cu, Ni, Al
Li-ion batteries	*At. ferrooxidans*	Li, Co

than plastic recycling from more traditional sources due to brominated flame retardants.

Currently, plastic recycling technologies from electronic waste are divided into four categories: primary, secondary, tertiary, and quaternary recycling.

10.4.1 Primary and Secondary Recycling

Primary recycling consists of the mechanical processing of polymers from electronic waste. The waste is first crushed, followed by sorting, which can occur in various ways, such as manual sorting, magnetic separators, and optical separators for plastics. The sorting method will be determined based on the final product desired. Once granulated and adequately separated, the plastics are then thermally processed via methods such as melting and extrusion, especially for pellet manufacturing or injection molding, depending on the desired product. Although the primary and secondary processing steps are the same, these two characteristics are distinguished by the final quality of the product obtained. While in primary processing, the final product has physical properties equivalent to the original product, in secondary processing, the final products have properties inferior to the original product (Sahajwalla and Gaikwad 2018).

10.4.2 Tertiary Recycling

In a sense, tertiary-grade recycling employs primary and secondary processing. However, this time there is a chemical treatment step with solvents to extract the retarding substances from e-waste plastics (Sahajwalla and Gaikwad 2018).

The Fraunhofer IVV institute in Germany developed CreaSolv, a process for the tertiary recycling of polymers. This process is considered revolutionary for separating and reapplying polymers using selective dissolution. Because of that, several companies looking to apply the circular economy in producing plastic packaging have adopted it. In this process, the polymers are dissolved in a combination of solvents and precipitated in a solution. The precipitated polymer, free of brominated flame retardants, is thermally processed by means of extrusion (primary and secondary processes). Several modifications of the CreaSolv process have been investigated worldwide to optimize the recycling of various species of plastic waste from e-waste, with the application of supercritical CO2, ionic liquids, and other solvents (Sahajwalla and Gaikwad 2018).

10.4.3 Quaternary Recycling

Quaternary recycling consists of the production of energy from polymers. Although there are several technologies capable of converting polymers into fuels, recycling e-waste polymers for this purpose comes up against the presence of brominated flame inhibitors. The main concern when using e-waste plastics is the formation of dioxins, a highly toxic and environmentally harmful by-product. Several research fronts are promising as they present emerging technologies capable of solving the problem of dioxin production, such as the KDV technology (Alphakat, Germany), which promises catalytic depolymerization below 250°C and can be applied to various types of plastics e-waste (Sahajwalla and Gaikwad 2018).

10.5 Environmental Impact of Electronic Waste

The continuous development of new technologies and their immediate updates contribute to an increase in electronic product production to meet the demands of an increasingly demanding public regarding design, performance, and quality. This modern phenomenon creates the need to commercialize a considerable variety of e-products on a large scale (Mohammadi et al. 2021). As a side effect, there is an increase in the volume of electronic waste generated, including substances with very high harmful and polluting potential.

Improper disposal of e-waste, as well as a poorly designed recycling process, can pollute the environment with hazardous materials that are detrimental to human, plant, and animal life (Figure 10.4). Exposure of e-waste to weathering causes the emergence of leachates, which in turn can potentially transport toxic substances to groundwater. On the other hand, in a poorly designed recycling process, combustion in an incinerator can emit toxic gases into the atmosphere causing respiratory, water, and food contamination (Song and Li 2015).

While there is an increasing amount of information on the health effects of exposure to e-waste, there is limited data available on the number of people who suffer from its effects (Song and Li 2015). The challenges in obtaining health statistics related to e-waste, addressing potential health outcomes, and studying chemical mixtures are significant. The lack of confirmed exposure–outcome relationships and the long latency periods of some diseases increase the difficulty of establishing disease prevention measures resulting from exposure to e-waste substances. Internationally standardized indicators obtained through the monitoring of electronic waste can help to quantify the number of people at risk of health effects related to this waste, and to monitor their trends over time (Forti et al. 2020).

10.6 Monitoring of Electronic Waste

In 2019, humanity produced a record amount of 53.6 Mt of electronic waste. The main constituents of these materials were small equipment, with around 17.4 Mt, followed by large equipment, with 13.1 Mt; refrigerators and heaters,

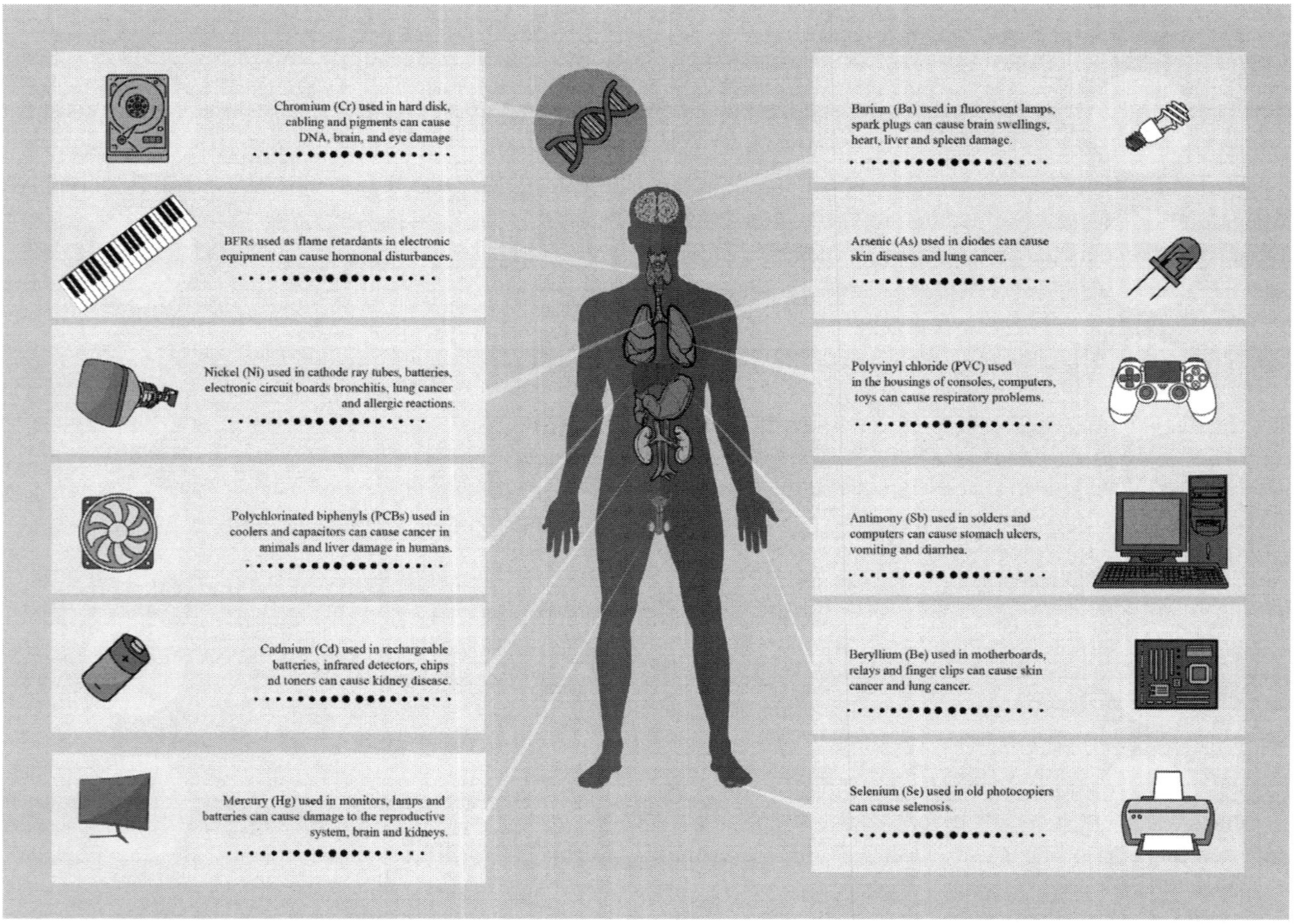

Figure 10.4 Some chemical substances that make up electronic waste and its respective harmful effects on human health. Adapted from Kiddee et al., 2013.

with 10.8 Mt; monitors, with 6.7 Mt; IT and telecommunications equipment, with 4.7 Mt; and light bulbs with around 0.9 Mt. The *Global E-waste Monitor 2020* published by UNITAR (United Nations Institute for Trading and Research) reports an increase of 9.2 Mt (metric tons) in 5 years, and with a projection of up to 21.1 Mt more by 2030, reaching the mark of 74.7 Mt of e-waste produced annually. Only 17.4% of all this e-waste has been collected and recycled, leaving behind a large amount of valuable metals and environmentally harmful substances (Forti et al. 2020).

Understanding the domestic and cross-border amounts and flows of e-waste provides a basis for monitoring, controlling, and, ultimately, preventing the illegal transport, dumping, and inappropriate handling of e-waste. Monitoring the quantities and flow of e-waste is a fundamental tool for assessing technological evolution over the years, establishing sustainable strategies, besides developing the circular economy. The development of public policy and regulations for management can only be achieved with better data on electronic waste (Baldé et al. 2022; Forti et al. 2020; Jadhao et al. 2022a).

10.7 Application of Compounds or Elements Extracted from E-waste

In WEEE, the most targeted elements are a metal fraction (MF), which has high-value metals such as palladium, ruthenium, platinum, gold, and copper. However, the other essential elements are also extracted and reapplied in new products. These elements can be applied in similar electronic products, such as batteries for electric cars or cell phones and electrical wires, or others, such as the applying of high purity silica for the production of zeolite-based catalysts, among other applications. It is essential to mention that if these metals were extracted correctly, they would not lose their properties.

The non-metallic fraction (NMF) has also been applied in some studies. Due to being little explored and containing polluting compounds, studies seek to apply this residue to reduce these environmental impacts and use the circular economy. For example, alumina's application to remove bromine is used as a flame retardant in NMF on printed circuit boards (PCBs) (Ali et al. 2022). NMFs have also been applied in building materials, mainly because they have a lower density when compared to sand and cement and have a finer granulometry. These characteristics make the microstructures more resistant, in addition to containing thicker glass fibers, increasing the mechanical strength of the materials (Colangelo et al. 2017). NMFs are also being applied to replace modeling materials such as plastic and plaster, among others, in addition to being applied in the development of new printed circuit boards (Yang et al. 2019).

From 15 to 20% of all e-waste is plastic (Jadhao et al. 2022a; Sahajwalla and Gaikwad 2018). Thus, several studies have sought alternatives to work with this material. Physical processes can treat this plastic (recovering the plastic), and chemical processes (turning it into fuel) (Kumar et al. 2021; Yang et al. 2019). As a result, several chemical recycling technologies have been developed, such as gasification, depolymerization in the supercritical fluid, pyrolysis, and catalytic pyrolysis (Jadhao et al. 2022b; Yang et al. 2019).

10.8 Conclusions

Electronic waste (e-waste) is often seen as a problem, as it contains several metals that can cause environmental damage and human health problems. However, correct techniques such as halogenated leaching, thiosulfate leaching, acid leaching, and thiourea leaching can recover the metallic fraction (MF) present in these residues almost wholly, establishing a circular character for the improvement of these strategic substances for the high-tech industry. Additionally, we have the non-metallic fraction (NMF) that can be applied in various composites, such as building or modeling materials, and even in developing new printed circuit boards. Plastics can also be extracted from this fraction, representing much of this waste. Processes for recovering metals and plastics can be cheaper than obtaining metals from finite resources such as minerals and oil fractions. However, some extraction processes cited may be more polluting than others, although they are very efficient. One of the great advantages of extracting metals is that they do not lose their properties, unlike thermoplastic materials and other polymers. Given this, public policies must be rethought since our resources are finite. Therefore, we must reduce the runaway production and consumption of electronic products, in addition to the execution of necessary actions, before the final disposal of electronic waste in the environment, such as selective collection. In this context, the circular economy emerges as a possible solution to this problem. Finally, using the circular economy concept, e-waste can be considered a new way of mining precious metals, enabling the reapplication of compounds extracted from waste in new products.

References

Akcil, A., Agcasulu, I., and Swain, B. (2019). Valorization of waste LCD and recovery of critical raw material for circular economy: a review. *Resources, Conservation and Recycling* 149: 622–637.

Ali, L., Mousa, H.A., Al-Harahsheh, M. et al. (2022). Removal of bromine from the non-metallic fraction in printed circuit board via its co-pyrolysis with alumina. *Waste Management* 137: 283–293. https://doi.org/10.1016/j.wasman.2021.11.025.

Ashiq, A., Kulkarni, J., and Vithanage, M. (2019). Hydrometallurgical recovery of metals from E-waste. In: *Electronic Waste Management and Treatment Technology* (eds. M.N.V. Prasad and M. Vithanage), 225–246. Elsevier.

Awasthi, A.K., Awasthi, M.K., Mishra, S. et al. (2022). Evaluation of E-waste materials linked potential consequences to environment in India. *Environmental Technology and Innovation* 28: 102477.

Baldé, C.P., D'Angelo, E., Luda, V. et al. (2022). *Global Transboundary E-waste Flows Monitor*. Bonn, Germany: United Nations Institute for Training and Research.

Colangelo, F., Messina, F., Di Palma, L., and Cioffi, R. (2017). Recycling of non-metallic automotive shredder residues and coal fly-ash in cold-bonded aggregates for sustainable concrete. *Composites Part B: Engineering* 116: 46–52. https://doi.org/10.1016/j.compositesb.2017.02.004.

Cui, J. and Roven, H.J. (2011). Chapter 20—Electronic Waste. In: *Waste* (eds. T.M. Letcher and D.A. Vallero), 281–296. Academic Press.

EPA (2021). United States Environmental Protection Agency [WWW document]. https://www.epa.gov/recyclingstrategy/what-circular-economy (accessed 06/12/22).

Faraji, F., Golmohammadzadeh, R., and Pickles, C.A. (2022). Potential and current practices of recycling waste printed circuit boards: a review of the recent progress in pyrometallurgy. *Journal of Environmental Management* 316: 115242. https://doi.org/10.1016/j.jenvman.2022.115242.

Forti, V., Baldé, C.P., Kuehr, R., and Bel, G. (2020). *The Global E-waste Monitor 2020: Quantities, Flows and the Circular Economy Potential*. Bonn/Geneva/Rotterdam: United Nations University.

Forti, V., Baldé, K., and Kuehr, R. (2018). *E-waste Statistics: Guidelines on Classifications, Reporting and Indicators*, 2e. Bonn: United Nations University.

Gautam, A., Shankar, R., and Vrat, P. (2022). Managing end-of-life solar photovoltaic e-waste in India: a circular economy approach. *Journal of Business Research* 142: 287–300.

Gundupalli, S.P., Hait, S., and Thakur, A. (2018). Classification of metallic and non-metallic fractions of e-waste using thermal imaging-based technique. *Process Safety and Environmental Protection* 118: 32–39.

Jadhao, P.R., Ahmad, E., Pant, K.K., and Nigam, K.D.P. (2022a). Advancements in the field of electronic waste recycling: critical assessment of chemical route for generation of energy and valuable products coupled with metal recovery. *Separation and Purification Technology* 289: 120773.

Jadhao, P.R., Vuppaladadiyam, A.K., Prakash, A., and Pant, K.K. (2022b). Co-pyrolysis characteristics and kinetics of electronic waste and macroalgae: a synergy study based on thermogravimetric analysis. *Algal Research* 61: 102601.

Kiddee, P., Naidu, R., and Wong, M.H. (2013). Electronic waste management approaches: an overview. *Waste Management* 33: 1237–1250.

Kumar, A., Holuszko, M.E., and Janke, T. (2021). Analysis of rejects from waste printed circuit board processing as an alternative fuel for the cement industry. *Waste Management & Research: The Journal for a Sustainable Circular Economy* 39 (6): 841–848. https://doi.org/10.1177/0734242x20952847.

Ma, E. (2019). Recovery of waste printed circuit boards through pyrometallurgy. In: *Electronic Waste Management and Treatment Technology* (eds. M.N.V. Prasad and M. Vithanage), 247–267. Elsevier.

Mohammadi, E., Singh, S.J., and Habib, K. (2021). Electronic waste in the Caribbean: an impending environmental disaster or an opportunity for a circular economy? *Resources, Conservation and Recycling* 164: 105106.

Pan, X., Wong, C.W.Y., and Li, C. (2022). Circular economy practices in the waste electrical and electronic equipment (WEEE) industry: a systematic review and future research agendas. *Journal of Cleaner Production* 365: 132671.

Parajuly, K., Fitzpatrick, C., Muldoon, O., and Kuehr, R. (2020). Behavioral change for the circular economy: a review with focus on electronic waste management in the EU. *Resources, Conservation and Recycling: X* 6: 100035.

Sagnak, M., Berberoglu, Y., Memis, İ., and Yazgan, O. (2021). Sustainable collection center location selection in emerging economy for electronic waste with fuzzy Best-Worst and fuzzy TOPSIS. *Waste Management* 127: 37–47.

Sahajwalla, V. and Gaikwad, V. (2018). The present and future of e-waste plastics recycling. *Current Research in Green and Sustainable Chemistry* 13: 102–107.

Sengupta, D., Ilankoon, I.M.S.K., Dean Kang, K., and Nan Chong, M. (2022). Circular economy and household e-waste management in India: integration of formal and informal sectors. *Minerals Engineering* 184: 107661. https://doi.org/10.1016/j.mineng.2022.107661.

Song, Q. and Li, J. (2015). A review on human health consequences of metals exposure to e-waste in China. *Environmental Pollution* 196: 450–461. https://doi.org/10.1016/j.envpol.2014.11.004.

Sronsri, C., Panitantum, N., Sittipol, W. et al. (2022). Optimization of selective gold recovery from electronic wastes through hydrometallurgy and adsorption. *Process Safety and Environmental Protection* 163: 659–668.

Van Yken, J., Boxall, N.J., Cheng, K.Y. et al. (2021). E-waste recycling and resource recovery: a review on technologies, barriers and enablers with a focus on Oceania. *Metals (Basel)* 11: 1313.

Xiao, Z., Pramanik, A., Basak, A.K. et al. (2022). Material recovery and recycling of waste tyres—a review. *Cleaner Materials* 5: 100115.

Yang, J., Wang, H., Zhang, G. et al. (2019). Recycling organics from non-metallic fraction of waste printed circuit boards by a novel conical surface triboelectric separator. *Resources, Conservation and Recycling* 146: 264–269. https://doi.org/10.1016/j.resconrec.2019.03.008.

Zhao, F. and Wang, S. (2019). Bioleaching of electronic waste using extreme acidophiles. In: *Electronic Waste Management and Treatment Technology* (eds. M.N.V. Prasad and M. Vithanage), 153–174. Elsevier. https://doi.org/10.1016/B978-0-12-816190-6.00007-8.

11

Recovery of Non-precious Metals from WEEE Using Emerging Leachants

Carlos Perea[1,2,*], *Christian Ihle*[1,2], *Humberto Estay*[2], *and Laurence Dyer*[3]

[1] *Department of Mining Engineering, Universidad de Chile. Beauchef 850, 8370448, Santiago, Chile*
[2] *Advanced Mining Technology Center, Universidad de Chile, Tupper 2007, Santiago, Chile*
[3] *Department of Mining Engineering and Metallurgical Engineering, Western Australian School of Mines, Curtin University, Kalgoorlie, Western Australia*
[*] *Corresponding author*

11.1 Introduction

One of the fastest-expanding solid waste streams worldwide is WEEE, due to fast technological advances, continuous market expansion, and lifespan. It has become necessary to regularly replace and upgrade cell phones, computers, and appliances to carry out daily activities, forcing the population to discard equipment that has been declared obsolete. Global WEEE production in 2019 was predicted to be around 54 million tons (MT) (Pekarkova et al. 2021). It has caused great concern worldwide and has led several countries to generate regulatory standards and integral management plans for WEEE control and disposal. However, there is still a large accumulation of WEEE in landfills, which represents a potentially serious threat to the environment and human health due to the leaching of metals and the generation of acid drainage that contaminates soils and water sources. For instance, about 25–40% of WEEE in the European Union is handled via the official system; the remainder is exported to underdeveloped nations or turned into municipal waste. In the case of developing countries, most WEEE is treated in the informal recycling system by artisanal processes, such as open burning or acid leaching, which generate low metal recovery and serious environmental problems (Zhang et al. 2017).

In addition to the risk that these wastes represent, they also attract attention due to their economic potential because they contain critical metals that, due to their unique physical and chemical properties, play a significant role in current electrical and electronic devices. Computer chips, for example, are made with more than 60 different elements today compared to 12 in the 1980s. As a consequence, electric and electronic equipment (EEE) consists of highly heterogeneous mixed materials, comprising metals in printed circuit boards (PCBs), germanium in fiber optic cable, lithium and cobalt in Li-ion batteries, rare earth elements in permanent magnets, indium in liquid crystal display (LCD) displays, etc. (Zhang et al. 2017). Due to increased consumer and industrial market penetration, adoption of industry 4.0 and the "internet of things," and increased use of low-carbon technologies to achieve the Paris Agreement's goals, the demand for these metals has been steadily rising (Charles et al. 2020). Due to their significant relative economic importance and supply risk, the European Union (EU) has designated several metals as essential raw resources, including germanium (Ge), indium (In), tungsten (W), niobium (Nb), gallium (Ga), cobalt (Co), rare earth elements (REEs), tantalum (Ta), graphite (C), beryllium (Be), antimony (Sb), magnesium (Mg), and the platinum group metals (PGMs). As a result of production being concentrated in a few nations, access to supplies is typically a problem greater than geological abundance when it comes to critical metal availability. Implementing the circular economy is one of the suggestions for addressing problems with material criticality (Charles et al. 2020). In addition, WEEE has a greater metal concentration than primary sources like minerals (Kumar et al. 2017), representing potential economic benefits, such as lower costs and lower energy consumption compared to conventional mining operations. The value of the secondary raw materials included in the total WEEE produced in 2019 is estimated to be over US$57 billion (Pekarkova et al. 2021).

To address the WEEE problem, multiple studies have been carried out over the decades, resulting in the development of numerous methods of metal extraction. These can be divided into mechanical, pyrometallurgical, biometallurgical, electrometallurgical, and hydrometallurgical processes. However, these technologies still pose technical limitations due to the complexity of WEEE. For example, using comminution, the recovery is inefficient and has high energy consumption. The most common technique for recycling metals from WEEE has always been pyrometallurgy. It requires specialized equipment to

Electronic Waste Management: Policies, Processes, Technologies, and Impact, First Edition. Edited by Sunil Kumar and Vineet Kumar.

operate at high temperatures, generating toxic gases and high slag volume. Also, the policy of reducing global warming and strict environmental regulations make the implementation of this technique increasingly complex and costly (Hao et al. 2020). Biometallurgical processes have some advantages, such as their low cost and eco-friendliness. Nevertheless, these methods have some disadvantages, such as low selectivity in the biosorption process, application only at laboratory scale, and low recovery rate (Yu et al. 2020). Contrarily, hydrometallurgy offers many benefits, including high leaching efficiency, a favorable operating environment, and the opportunity to eliminate secondary wastes (Hao et al. 2020). However, most of the non-precious metals in WEEE are extracted using strong acids, such as HCl, H_2SO_4, and HNO_3, which have low selectivity and whose improper handling can cause serious environmental problems. For this reason, multiple studies have been carried out in recent years with alternative leaching systems such as chelating agents, ionic liquids, organic reagents, and organic acids to develop methods to extract the metals from WEEE. The purification and recovery steps can be accomplished using chemical precipitation/cementation, solvent extraction, ion-exchange resins, and electro-refining.

Several processes have been developed for recycling non-precious metals from WEEE (base metals especially) through hydrometallurgical or hybrid routes (generally combining pyro- and hydrometallurgical processes). During the metal recovery and purification stage of hybrid processes, the hydrometallurgical technique is still used since it can treat several metals present in the solution, even at low concentrations (a few parts per million scale). This technique is used for recycling WEEE, such as PCBs, photovoltaic modules (PV), batteries, and light-emitting diodes (LEDs). Hydrometallurgical processing, however, has the difficulty of constantly being in continual adaptation and research due to the ongoing change in the composition and amount of non-precious metals in waste created each year (Espinosa et al. 2020).

11.2 Categorization and Characterization of WEEE

According to the EU Directive 2012/19/EU, EEE is a term used to refer to all objects that depend on electromagnetic fields or electric current to function. This definition also includes the apparatus required to produce, transmit, and measure such currents and fields. Once these devices are considered useless and discarded, they become WEEE (Pérez-Belis et al. 2015). Approximately 50% of WEEE is composed of electrical waste, including household appliances such as refrigerators, air conditioners, toasters, microwave ovens, etc. The remaining 50% consists of electronic waste, which includes televisions, monitors, computers, printers, cell phones, and other electronic devices such as medical instruments, etc. (Gopikrishnan et al. 2020). This EU Directive classified EEE into six categories based on the size of the equipment and typical WEEE collection streams, as shown in Table 11.1.

Table 11.1 Categorization of e-waste with targets of EU WEEE Directive, 2012/19/EU. (Adapted from Pérez-Belis et al. 2015).

		Target from August 2018	
Categories	**EEE considered in the category**	**Recovered (%)**	**Prepared for reuse or recycled (%)**
Temperature exchange equipment	Air conditioners, dehumidifiers, heat pumps, oil radiators, and other temperature-exchange systems that use fluids other than water. Refrigerators, freezers, automatic cold-delivery appliances, refrigerators, freezers, and other appliances that do the same.	85	80
Screens, monitors, and equipment containing screens that have a surface greater than 100 cm^2	Displays, LCD digital photo frames, monitors, portable computers, including laptops and tablets, and televisions.	80	70
Lamps	LED lamps, low-pressure sodium lamps, fluorescent lamps, high-intensity discharge lamps, including pressure sodium and metal halide lamps, straight fluorescent lamps, and compact fluorescent lamps.	–	80

(Continued)

Table 11.1 (Continued)

Categories	EEE considered in the category	Target from August 2018	
		Recovered (%)	Prepared for reuse or recycled (%)
Large equipment (any external dimension greater than 50 cm)	Washing machines, dryers, dishwashers, stoves, electric stoves and ovens, electric hotplates, luminaires, sound or image reproducing apparatus, music equipment (aside from pipe organs installed in churches), knitting and weaving machines, large computers, large printers, copiers, large slot machines, large sanitary products, large monitoring and control instruments, large apparatuses that supply products and money automatically, and photovoltaic panels.	85	80
Small equipment (without any external dimension greater than 50 cm).	Vacuum cleaners, sewing machines, luminaires, microwave ovens, ventilation equipment, irons, toasters, electric knives, electric kettles, watches, electric razors, scales, hair and body care equipment, calculators, video cameras, radio equipment, video recording equipment, hi-fi systems, musical instruments, sound or image reproduction equipment, electric and electronic toys, sporting goods, computers for cycling, diving, running, rowing, etc., smoke detectors, heating regulators, thermostats, small electric and electronic tools, small sanitary products, small monitoring and control instruments, small appliances that supply products automatically, small appliances with integrated photovoltaic panels.	75	55
Small computers and telecommunication devices (no external dimensions greater than 50 cm)	Smartphones, GPS, calculators in pockets, routers, PCs, printers, and telephones.	75	55

The major contributors to the WEEE are small and large equipment producing 12.8 and 11.8 Mt per year, respectively, followed by temperature exchange equipment with 7 Mt and screens and monitors with 6.3 Mt. A newly added kind of waste to the WEEE list is photovoltaic panels, which cause great concern due to the increase in the implementation of new technologies to produce energy with zero carbon emission. It is estimated that in 2016 between 43 500 and 250 000 metric tons (Mt) of waste were generated from photovoltaic panels, and it is expected that these will reach 5.5–6 Mt by 2050 (Kumar et al. 2017).

11.2.1 Physical and Chemical Composition of WEEE

The composition of WEEE is non-homogeneous and can contain more than 1000 different substances, in which ferrous and non-ferrous metals, plastics, glass, wood and chipboard, printed circuit boards, ceramics, rubber, and other materials can be found. As can be seen in Figure 11.1, approximately 60% of WEEE is mainly composed of metals, followed by plastics with 15%, cathode ray tubes (CRT) and LCD screens with 12%, and the remaining 13% other constituents (Wang and Xu 2014).

WEEE contains at least 60 elements of the periodic table. Of this list, 40 are elemental metals: Fe; basic metals, such as Cu, Al, Ni, and Pb; precious metals such as Au, Ag, Pd, and Pt; rare metals, such as Co, Ga, Ge, and In; and also rare earth elements (REEs) such as Ce, La, Eu, and Sm (Pérez-Martínez et al. 2019). Some elements that are considered hazardous, such as Cd, Hg, Se, As, among others, can also be found. In addition to metals, WEE also has other chemical components, such as plastics and brominated flame retardants, whose treatment is considerably expensive (Zhang and Xu 2016).

The metals come from non-renewable resources, and global ore grades are declining. Mines have found it necessary to excavate more complex ore deposits to meet the global metal demand (Kumar et al. 2017). Due to the high amount of base metals and REEs in WEEE, it may be a source of secondary resources for metal recovery. Tables 11.2 and 11.3 show the concentration of base metals and REEs in some electronic devices and an average grade in the mined ores.

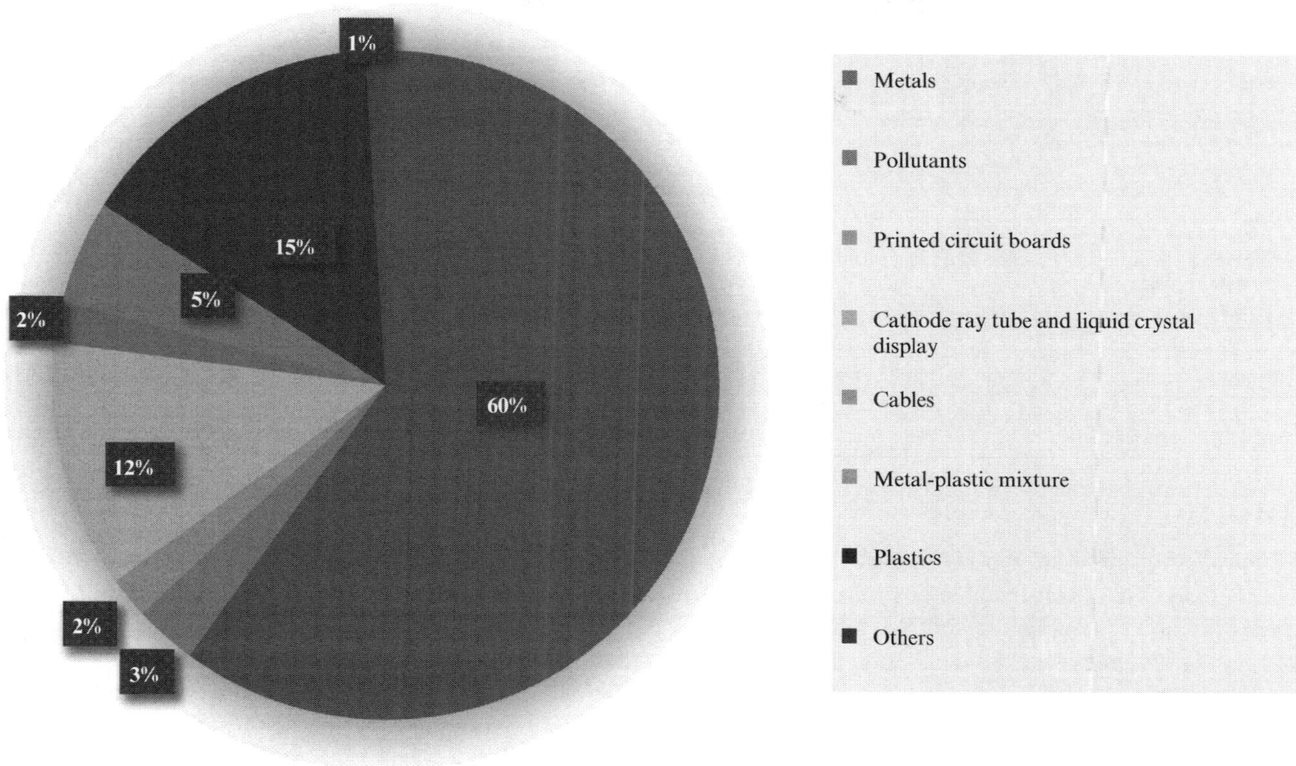

Figure 11.1 Average composition of WEEE.

Table 11.2 Average content of base metals for different WEEE. Values reported in % (wt/wt).

Electronic waste	Fe	Cu	Al	Pb	Ni	Sn	Co	Li	References
TV board	28	10	10	1	0.3	1.4	–	–	(Jadhav and Hocheng 2015)
PC board	1.2–8	6–27	2–7.2	1–4.2	0.3–5.4	1–5.6	<0.4	–	(Kaya 2016)
Mobile phones	5	13	1	0.3	0.1	0.5	19		(Jadhav and Hocheng 2015: Charles et al. 2020)
Portable audio	23	21	1	0.14	0.03	–	–	–	(Zhang and Xu 2016)
DVD players	62	5	2	0.3	0.05				(Zhang and Xu 2016)
Calculators	4	3	5	0.1	0.5	–	–	–	(Zhang and Xu 2016)
PC mainboard	4.5	14.3	2.8	2.2	1.1	–	–	–	(Zhang and Xu 2016)
Printed circuit boards	12	10	7	1.2	0.85	–	–	–	(Zhang and Xu 2016)
Electronic scrap	8	20	2	2	2	–	–	–	(Zhang and Xu 2016)
PC scrap	20.5	6.93	14.2	6.3	0.85	–	–	–	(Tabelin et al. 2021b)
Laptops loudspeakers	63–66.4	0.05–0.2	0.54–1	–	–	–	0.03–1.3	–	(Lixandru et al. 2017)
Lithium-ion batteries	–	26–76	20	–	–	–	2–30	1–15	(Zeng et al. 2015)
Nickel metal hydride batteries	0.59	0.25	1.2	–	34.5	–	4.2	–	(Lie and Liu 2021)
Typical Cu ore	16	0.4	4.5	0.1	0.001	0.002	–	–	(Li et al. 2018)
Salt lake brines	–	–	–	–	–	–	–	0.017–0.15	(Tabelin et al. 2021a)
Pegmatite hard-rock minerals	–	–	–	–	–	–	–	1–4	(Tabelin et al. 2021a)

Table 11.3 Average content of REE for different WEEE (values reported in wt%).

Electronic waste	Y	Eu	La	Ce	Tb	Gd	Pr	In	Nd	References
LCD monitors	<0.01	<0.01	<0.01	<0.01	<0.01	<0.01	<0.01	2.61	-	(Birloaga and Vegliò 2018)
LCD TVs	<0.01	<0.01	<0.01	<0.01	<0.01	<0.01	<0.01	1.82	-	
Notebooks	<0.01	<0.01	<0.01	<0.01	<0.01	<0.01	0.015	2.42	0.12	
Laptops loudspeakers	-	-	-	-	-	0.1–0.2	4.4–8	-	21.7–24.7	(Lixandru et al. 2017)
Nickel metal hydride batteries	-	-	8.96	2.69	-	-	-	-	2.64	(Lie and Liu 2021)
REE ores in Bayan Obo deposit	<0.18	<0.13	0.01–15.4	0.01–33.4	<0.02	<0.27	<3.5	-	0.002–11.1	(Gao et al. 2021)

Improper disposal of the hazardous components of WEEE represents a considerable environmental concern when these are disposed of in a landfill or burned. Recycling WEEE might thereby lessen the pollution caused by these materials, the energy needed to produce one unit of metal, and the overall global demand for new metal production, all of which can contribute to a reduction in greenhouse gas emissions. For instance, recycling 10 kg of aluminum can save 90% of the energy needed to produce it while also preventing the production of 13 kg of bauxite residue, 20 kg of carbon dioxide, and 0.11 kg of sulfur dioxide gas (Kumar et al. 2017).

11.3 Metallurgical Processing of WEEE

The recovery of metal content is a crucial component of WEEE management from an economic and environmental standpoint. Given the complexity of WEEE, it is necessary to implement innovative scientific approaches that allow the selective and sustainable recovery of metals contained in these materials. For this reason, processes have been studied and developed that combine mechanical-physical and metallurgical techniques to extract metals from WEEE as seen in Figure 11.2. Depending on the metal of interest, the available technologies, and the type of WEEE, the steps can be modified. Generally, metal recovery from WEEE starts with physical-mechanical processes such as dismantling, comminution, froth flotation, and magnetic, electrostatic, and gravity separation. These processes are performed for size reduction and particle separation, allowing the separation of the non-metallic fraction and ferrous materials from the base metals and rare earth elements. Often these processes are an additional prerequisite for subsequent extraction stages, such as pyro-, hydro-, and electrometallurgical processes, in addition to hybrid or combined routes (Espinosa et al. 2020).

During the recovery stage, elements of interest are extracted using pyrometallurgical techniques such as calcination, roasting, smelting, and refining, or hydrometallurgical techniques such as acid leaching (H_2SO_4, HCl, HNO_3, and aqua regia), alkaline leaching (NH_4OH), bioleaching, and emerging leaching (chelation, organic acids, ionic liquids, and amino acids). Subsequently, a stage of recovery/purification of the metal of interest can be carried out through selective chemical precipitation or cementation, solvent extraction, ion exchange, biosorption, bioelectrochemical, bioprecipitation, and electrowinning/electrorefining methods.

11.3.1 Pyrometallurgical Processes

Pyrometallurgical processes have been the traditional methods most used to recover non-precious metals from WEEE. Incineration is one of the most used methods and is often considered simple and effective. Waste is heated in a reactor under aerobic conditions at high temperatures, the organic components are decomposed and combusted, generating water (H_2O) and carbon dioxide (CO_2). Fiberglass and metal oxides are obtained from the incineration residues, and these can be recovered through physical and chemical treatments. Incineration is a basic recycling technique that has a simple process and low operational cost to recover some base metals like copper, iron, and aluminum. However, it represents an environmental risk due to the pollutants and emissions generated in the process (Wang et al. 2017).

Pyrometallurgical technology has substantially advanced recently in terms of the treatment of contaminants and the effectiveness of metal recovery. For this

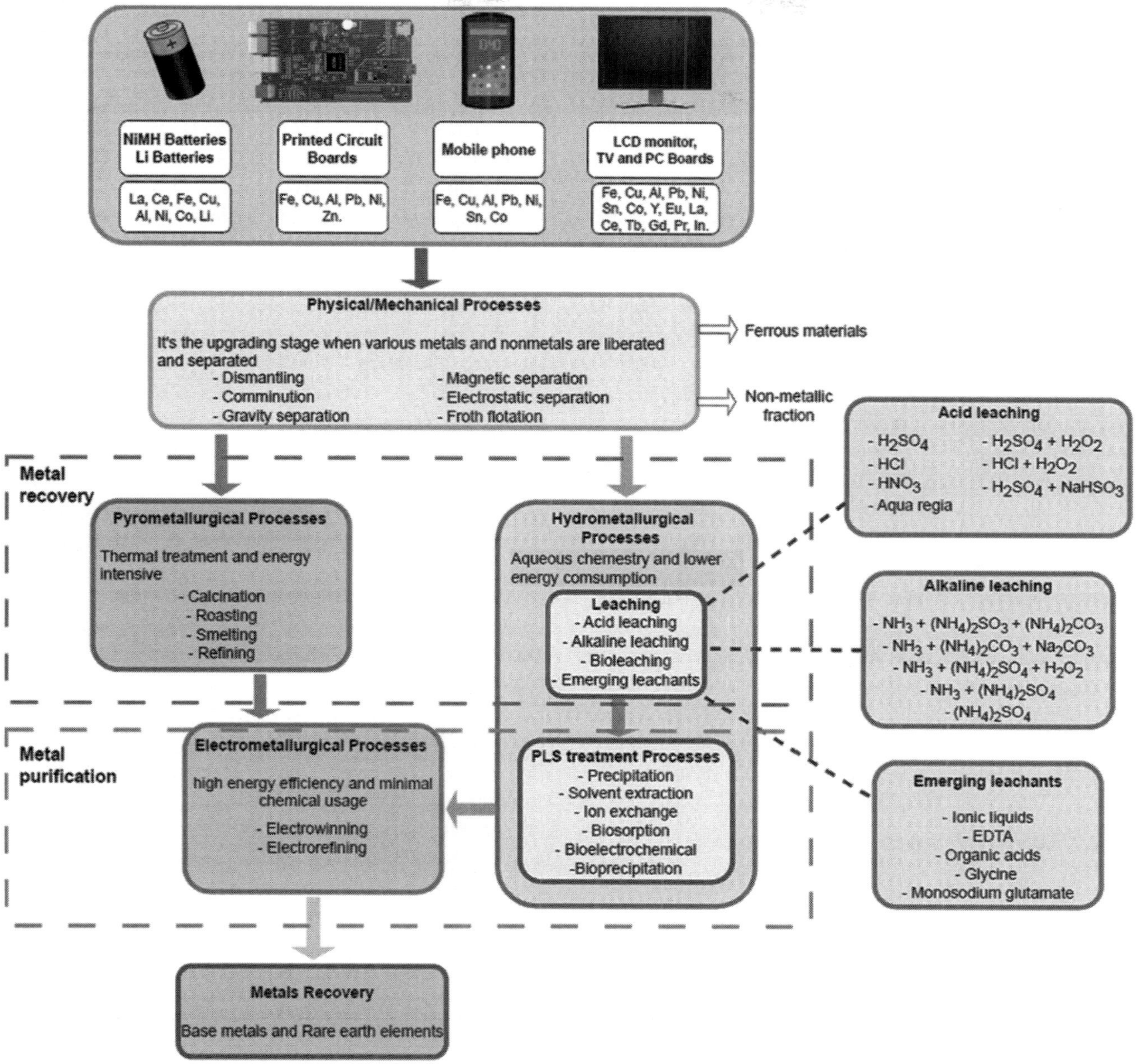

Figure 11.2 Flow chart of metallurgical processes employed in the metal recovery from Waste from Electrical and Electronic Equipment (WEEE).

reason, pyrometallurgical processes can be successfully implemented on an industrial scale through a sequence of procedures that may include disassembling, smelting in a blast furnace or plasma arc furnace, drossing, sintering, melting, and high-temperature gas phase processes. During this procedure, WEEE must first undergo physical-mechanical pretreatment such as disassembling major components, size-reduction shredding/grinding operations, gravity separation, magnetic and electrostatic separation, etc. After metal-plastic separation and subsequent furnace feed, metal ingots are obtained. In smelting, flash smelting and bath smelting are widely applied processes. Finally, the last procedure is refining to separate the various metals and obtain high-purity metals in the bullion or oxide phases (Zhang and Xu 2016).

Currently, most of the WEEE are treated using a pyrometallurgical treatment. The main advantage of this process route is its capacity to receive wastes of any form. However, pyrometallurgical processes have certain drawbacks, including gas generation, the production of significant amounts of slag, difficulties in recovering important metals, and even the loss of some valuable metals. Multiple methods have been developed for the extraction of non-precious metals. For example, in 1997, Berlin Technical

University created a recycling method for WEEE such as printed circuit boards (PCB) and transformed them into a Cu-Ni-Si alloy and a mixed oxide consisting primarily of Pb and Zn using a top-blown reactor. In Sweden, at Boliden Ltd. Ronnskar Smelter, WEEE is directly fed into a Cu converter to recover Cu, Ni, Se, Zn, and precious metals. Meanwhile, the dust is processed separately for the recovery of Pb, Sb, In, and Cd (Kaya 2016). Umicore's integrated metal smelter and refinery uses an integrated process that combines a pyro-, hydro-, and electrometallurgical stage for treating WEEE, obtaining precious metals and specialty metals like Sb, Bi, Sn, Se, Te, and In, as well as base metals like Al, Li, Ni, Co, and Cu (Kaya 2016; Zhang and Xu 2016).

The Ames Laboratory in the United States has developed a liquid metal extraction process using Mg as an extractant to selectively recover REEs from WEEE, such as NdFeB magnet scrap. This process can selectively dissolve some REEs, such as Nd, Pr, and Dy, at about 1000°C. The Anhui University of Technology in China developed an integrated process to selectively dissolve Nd and other REEs from the magnet scrap using molten chloride salt (MgCl-KCl) at temperatures of 700–1000°C. According to an alternative slag treatment procedure, the REEs can be extracted from the magnet scrap at temperatures above 1500°C using $CaO\text{-}SiO_2\text{-}Al_2O_3$ or $CaO\text{-}CaF_2$ fluxes. This technique is efficient since it can recover more than 99% of REEs from End-of-Life (EOL) magnets scrap, like hard disk drive (HDD) leftovers (Yang et al. 2017).

11.3.2 Biometallurgical Processes

In the last 30 years, biotechnology has been one of the most studied technologies for the recovery of metals. Biometallurgy is one of the technologies with the greatest advances and potential for application as an alternative to complex minerals and low-grade minerals processing for metal recovery. For this reason, the mining industry has been studying and developing the application of biometallurgy for the recovery of base metals like Cu, Ni, Co, and Zn as well as precious metals (Zhang and Xu 2016). According to estimates, using biometallurgical processes to remove metals from WEEE could be a more convenient, cost-effective, and environmentally responsible recycling alternative than using conventional recycling techniques (Kumar and Li 2017). However, most of these processes for recovering metals from WEEE are still being tested in laboratories. Two of the most studied and important processes within biometallurgy for metal recovery are bioleaching and biosorption (Zhang and Xu 2016).

As with pyrometallurgical processes, the physical-mechanical pretreatment must be the first stage in treating WEEE with bioleaching because it is more economical and efficient to treat a smaller mass of material. Therefore, manual dismantling is necessary to separate the different components such as metals, polymers, capacitors, batteries, solder, and liquid crystal displays (LCDs). These components can be crushed and milled to separate metals and non-metals by gravity separation, optical sorting, magnetic separation, or electrostatic separation (Kumar and Li 2017).

Generally, for the bioleaching and bioaccumulation of metals, three microbial groups are mainly used including autotrophic bacteria (such as *Thiobacillus* species), heterotrophic bacteria (such as *Pseudomonas* species and *Bacillus* species), and heterotrophic fungi (such as *Aspergillus* specie, *Penicillium* spp.) (Kumar and Li 2017). All these different kinds of microorganisms can transport or bind metal ions from the surrounding environment into the cell for use in a variety of intracellular processes. The recovery of metals could be either more selective or less selective by this interaction. However, this leaching process depends on bacterial metabolism, which depends on the bacteria's health and productivity, pH, temperature, energy source, and solid and solution composition, which generates an unstable recovery efficiency of metals of interest (Kaya 2016; Sun et al. 2021; Zhang and Xu 2016). Table 11.4 summarizes the information on the use of microorganisms for metal recovery from WEEE.

11.3.3 Electrometallurgical Processes

Generally, electrometallurgical processes are carried out in subsequent stages to hydrometallurgical and pyrometallurgical processes in order to selectively recover the metal of interest and obtain a final product with a high percentage of purity through the electrochemical deposition of solid metal using technologies such as electrowinning or electrorefining. Electrowinning has been used extensively in the primary extraction of base metals from metals-containing solutions or leachates. Electrorefining is used to treat metal anodes obtained from pyrometallurgical stages such as smelting and fire refining, in which the metal of interest is dissolved from the anode and then selectively electrodeposited from the electrolyte to obtain a product with a purity percentage close to 99.99% (Verbruggen et al. 2022).

Due to its industrial scale, low energy requirements, minimal environmental impact, and high recovery efficiency, electrometallurgical methods have been extensively utilized in the mining sector to extract base metals and some precious metals (Ciro et al. 2020). For these reasons, these processes are being used for the recovery of base metals from WEEE, for example, to treat black copper produced by smelting high content feeds of WEEE, pyrometallurgical conversion, and electrorefining or leaching with sulfuric acid under pressure and electrowinning (Verbruggen et al. 2022). In addition to Cu, the application of electrowinning has been studied to extract Ni, Zn, In, and REE from WEEE.

Table 11.4 Results compilation of extraction and recovery of base metals from WEEE by bioleaching processes.

Metals of interest	Microorganisms	Reaction conditions	Recovery (wt%)	References
Cu, Mg, Ti, Mn, and Zn	*Aspergillus niger* culture supernatant	Two-step leaching with citric acid conc. 20 g/L; H_2O_2 conc. 3.18%; pH = 1.97; leaching temp. 80°C; pulp density: 10 g/L.	99% Cu, 99% Mg, 98% Ti, 84% Mn, and 82% Zn	(Chauhan et al. 2018)
Cu, Zn, Pb, and Ni	*Acidithiobacillus ferrooxidans*	Sample size 0.075–1 mm and leaching with citric acid conc. 7.5 g/L for 18 days.	94% Cu, 92% Zn, 64% Pb, and 81% Ni	(Hao et al. 2020)
Cu	*Acidithiobacillus* genus	Sample size <80 mesh; 0.25 g/L $FeSO_4 \cdot 7H_2O$; 50 g/L PCBs.	95% Cu	(Hao et al. 2020)
Cu, Ni, and Zn	*Acidithiobacillus ferrooxidans*	First WEEE are milling at 340 r/min using ball material ratio = 10/1 (w/w) for 2 h. Then, microorganisms are added at 28°C and 125 r/min.	94.33% Cu, 90.69% Ni, and 90.78% Zn	(Hao et al. 2020)
Cu	*Leptospirillum ferriphilum*	Sample size = 19 cm x 15 cm, room temperature for 200 h.	99% Cu	(Hao et al. 2020)
Cu, Li, Mn, Al, Co, and Ni	*Aspergillus niger*	One-step, two-step, and spent medium bioleaching; 1% w/v pulp density.	100% Cu, 95% Li, 70% Mn, 65% Al, 45% Co, and 38% Ni in spent medium	(Chauhan et al. 2018)
Ga	*Cellulosimicrobium funkei* isolated from contaminated soil	Indirect (by growth supernatant) and direct bioleaching at pH 7.64–8.37 for 15–30 days.	92% Ga with indirect leaching; 78% Ga with direct leaching	(Chauhan et al. 2018)
Ni, Co, Li, and Mn	*Acidithiobacillus thiooxidans* and *Leptospirillum ferriphilum*	Stirring rate: 120 r/min; temperature: 30°C.	96% Co, 97% Ni, 95% Li, and 96% Mn	(Kumar and Li 2017)
Ni, Cu, Al, and Zn	*Sulfobacillus thermosulfidooxidans*	Scrap conc: 10 g/L; pH = 2.0; stirring rate: 180 r/min; temperature: 45°C; time: 18 days and sample size 50–150 μm	81% Ni, 89% Cu, 79% Al, and 83% Zn	(Kaya 2016)

However, there are additional processing difficulties related to electrolytes, choosing the right electrode combination, and cell design. Therefore, it is necessary to consider in electrochemical treatments the influence of operating parameters such as the quality of the processed material, the current density, the cathode material, the electrolyte on the specific energy consumption (SEC) and current efficiency (CE) (Ciro et al. 2020). In the case of WEEE, they contain some impurity elements (e.g. arsenic, antimony, and bismuth) that dissolve into the electrolyte used for the production of metals, which can contaminate the final product.

11.3.4 Hydrometallurgical Methods

Hydrometallurgical methods are metal recovery processes that involve a series of chemical reactions carried out in aqueous or organic solutions. Compared to pyrometallurgical processes, these have certain advantages, such as the ability to control the level of impurities, relatively low capital cost, lower environmental impact, and the potential for high metal recovery. Due to these factors, hydrometallurgical techniques have emerged as a leading WEEE treatment solution. There are multiple steps in a hydrometallurgical process, including physical-mechanical pretreatment, leaching of metals, purification of the pregnant leach solution, and metal recovery.

Since the metallic elements are covered or encapsulated by various plastics or ceramic materials, it is necessary to first perform a physical-mechanical pretreatment stage in order to expose the mineral of interest to the action of the leaching reagent and improve extraction efficiency. In addition, for the selection or development of a suitable leaching process for WEEE, it is necessary to consider that the main characteristic of these wastes is the presence of metals in native form and/or as alloys. The extraction of base metals and REE must therefore be accomplished via an oxidative leaching procedure. After the leaching stage, metals are purified and recovered using conventional techniques such as precipitation, solvent extraction, ion exchange, adsorption, and electrowinning. Figure 11.3 shows the most common leaching agents used for the recovery of base metals and WEEE.

Leaching is the stage where the solid wastes are converted into free metallic and non-metallic ions. Therefore, it is

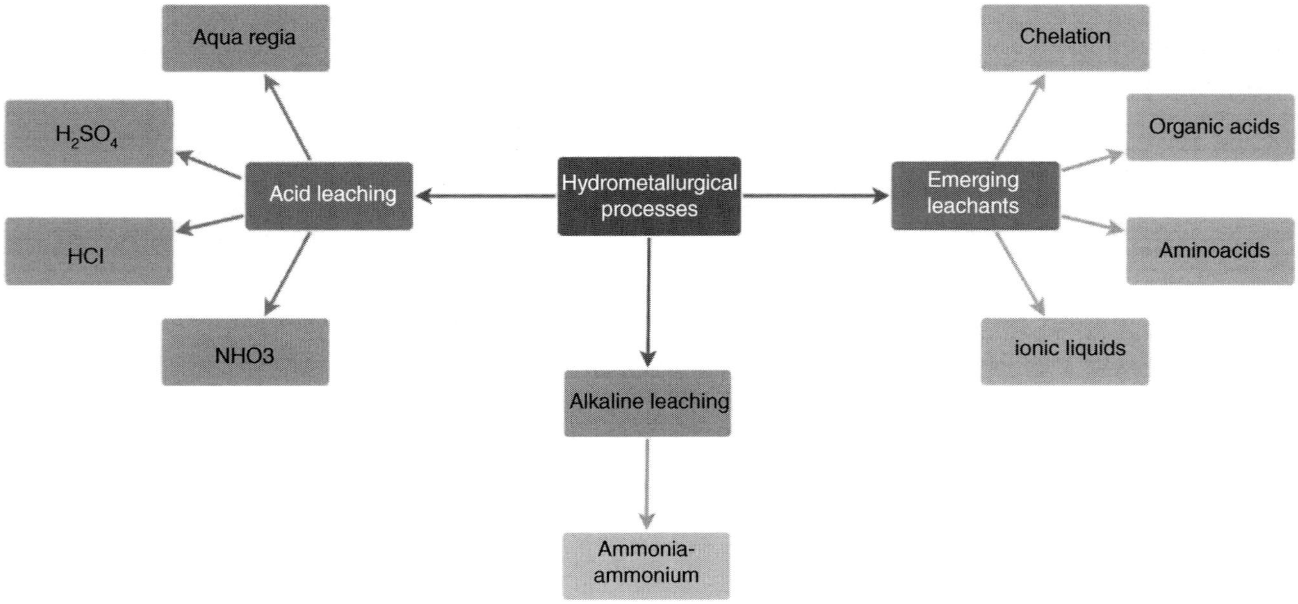

Figure 11.3 Most used leaching agents for the recovery of base metals and REE from WEEE.

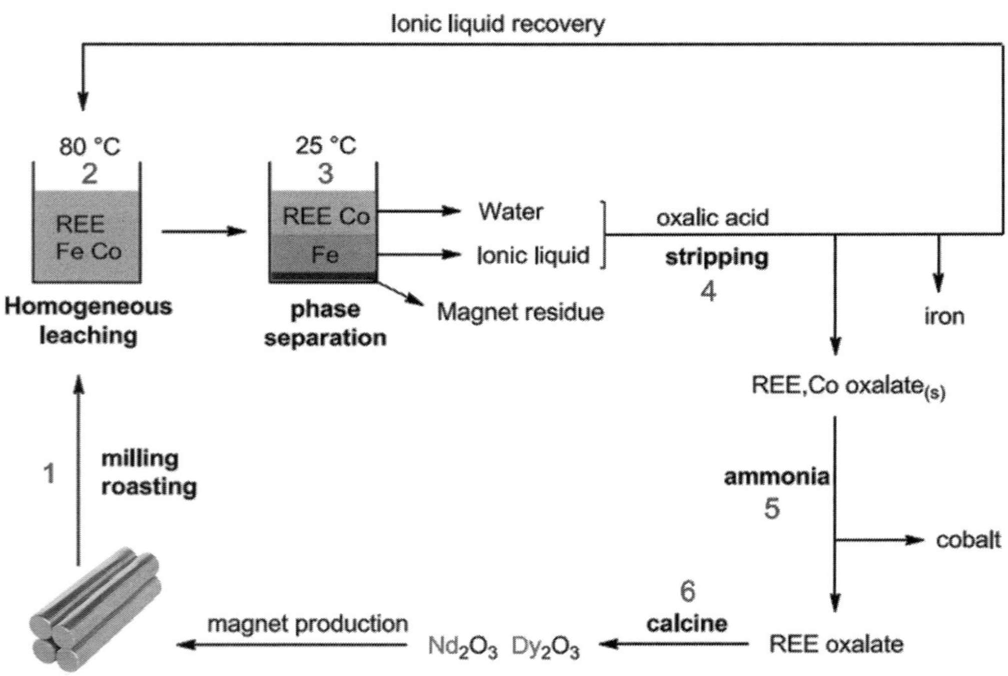

Figure 11.4 Overview of the [Hbet][Tf_2N]-H_2O recycling process for roasted NdFeB magnets. (1) The magnets are crushed and roasted. (2) Magnets that have been roasted are leached at 80°C using a 1:1 wt/wt [Hbet][Tf_2N]-H_2O solution. (3) Automatic metal distribution between the two phases and biphasic settling at ambient temperature. (4) Oxalic acid is used to strip two distinct phases, removing the iron while generating a mixed REE/Co oxalate and regenerating the ionic liquid. (5) Cobalt removal by aqueous ammonia treatment. Finally, (6) REE oxalate is calcined to produce oxides with a purity of better than 99.9%. (David Dupont & Koen Binnemans. 2015 / Royal Society of Chemistry / CC BY 3.0.)

necessary to employ a suitable leachant or leaching agent at this stage for the effective extraction of the metals of interest. The most commonly used leaching agents considered traditional and the most used emerging leaching agents for the recovery of base metals and REE are shown in Figure 11.4. Solubilization also consists of the reaction with acids or alkalis. Previous studies have shown efficient leaching of metals from WEEE in acidic and alkaline solutions.

Table 11.5 gives a series of examples of the process of leaching base metals and REE from WEEE.

Table 11.5 Results compilation for extraction and recovery from WEEE by leaching processes.

Electronic waste	Acid/alkaline reagent	Reaction conditions	Metal recovery (wt%)	References
PCB	$H_2SO_4 + H_2O_2$	First stage: sulfuric acid solution 1M, pH: 0–0.5, temperature 75°C, ORP: 220–280 mV, for 4 h. Second stage: sulfuric acid solution 1M, 240 mL of H_2O_2, pH: 0–4.0, temperature 75°C, ORP: 550–700 mV, for 4 h.	First stage: 90% Al, 40% Zn, and 8.6% Sn Second stage: 100% Cu, 50% Zn, and 10% Al	(Kumar and Li 2017)
PCB	$HCl + H_2O_2$	Pre-treatment with supercritical water. Then dilute HCl solution (1 mol/L) was added. Weight of sample: 1 g, the volume of leaching reagent: 100 mL, stirring: 300 r/min and temperature: 60°C.	99.8% Cu, 80% Pb, (up to 90%) Sn, Zn, Cr, Cd, and Mn	(Kumar and Li 2017)
Spent lithium-ion batteries	H_2SO_4, HNO_3	Acid baking at 300°C using H_2SO_4, 25% pulp density, 60 min leaching at 75°C; reductive leaching using mixture of H_2SO_4 and HNO_3 in 45 min.	93.2% Li, 90.52% Co, 82.8% Ni, and 77.7% Mn	(Chauhan et al. 2018)
Spent lithium-ion batteries	$H_2SO_4 + H_2O_2$	Pre-treatment by vacuum pyrolysis followed by acid leaching of the $LiCoO_2$, with H_2SO_4 2M, 5% H_2O_2, solid-liquid ratio: 50 g/L, for 60 min at 80°C.	99% Co	(Bertuol et al. 2016)
Spent lithium-ion batteries	$H_2SO_4 + NaHSO_3$	H_2SO_4 1M, with $NaHSO_3$ as an oxidant, solid-liquid ratio: 20 g/L, for 240 min at 95°C.	96.7% Li, 91.6% Co, 96.4% Ni, and 87.9%	(Sun et al. 2021)
Fluorescent lamp production waste	H_2SO_4	2 mol/L H_2SO_4, 8 h leaching at 100°C.	80.4% Y, 82.2% Eu, 81.4% Ce, and 80% Tb	(Wu et al. 2014)
Waste fluorescent powders	H_2SO_4	Solid-liquid ratio: 20%, 4 N H_2SO_4 concentrations and under the temperature 90°C.	85% Y	(Wu et al. 2014)
Ni-MH batteries	H_2SO_4	2M H_2SO_4, 75°C temperature and 120 min of time at a pulp density of 100 g/L.	98.1% Nd, 98.4% Sm, 95.5% Pr, 89.4% Ce, and more than 90% of base metals (Ni, Co, Mn, and Zn)	(Omodara et al. 2019)
Lamp phosphor waste	HNO_3	4 N HNO_3, with a solid-liquid ratio of 50 g/L at 25°C for 120 min.	99% La, 87.3% Ce, and 86.3% Tb	(Omodara et al. 2019)
Spent lithium-ion batteries	$NH_3 + (NH_4)_2SO_3 + (NH_4)_2CO_3$	1M NH_3, 0.5M $(NH_4)_2SO_3$, 1M $(NH_4)_2CO_3$, temperature: 80°C, time: 1 h, solid-liquid ratio: 10 g/L and stirring speed: 300 r/min.	100% Cu and 80% Co	(Jadhao et al. 2021)
Spent lithium-ion batteries	$NH_3 + (NH_4)_2CO_3 + Na_2CO_3$	4M NH_3, 1M $(NH_4)_2CO3$, 0.3M Na_2CO_3, temperature: 80°C, time: 5 h, liquid-solid ratio: 100 mL/g and stirring speed: 300 r/min.	78.1% Li, 86.4 Co, and 85.3% Ni	(Jadhao et al. 2021)
PCB	$NH_3 + (NH_4)_2SO_4 + H_2O_2$	90 g/L NH_3, 180 g/L $(NH_4)_2SO_4$, 0.4M H_2O_2, time: 4 h, liquid-solid ratio: 20 mL/g, temperature: 80°C and stirring speed: 700 r/min.	100% Cu and 90% Ni	(Jadhao et al. 2021)
Spent lithium-ion batteries	$NH_3 + (NH_4)_2SO_4$	4M NH_3, 1.5M $(NH_4)_2SO_4$, time: 5 h, solid-liquid ratio: 10 g/L, temperature: 80°C and stirring speed: 500 r/min.	95.3% Li, 80.7% Co, 89.8% Ni, and 4.3% Mn	(Sun et al. 2021)
Spent lithium-ion batteries	$(NH_4)_2SO_4$	3M $(NH_4)_2SO_4$, time: 2 h, solid-liquid ratio: 83 g/L, temperature: 180°C.	98% Li, 81% Co, 98% Ni, and 92% Mn	(Sun et al. 2021)

11.4 Emerging Leachants

Performing the extraction of metals from WEEE not only has a positive impact from the environmental and economic point of view but also implies a lower cost, energy, and natural resource savings compared to direct extraction, which is due to the fact that its recycling is carried out in urban areas, where such waste is generated and collected. For this reason, the procedures used for metal recycling must adhere to strict safety requirements to safeguard the nearby communities (Li et al. 2020). Thus, hydrometallurgical processes are the most suitable methods for the recovery of metals from this type of waste. However, although the technical feasibility of traditional hydrometallurgical processes in acid and alkaline media has been established, the effluent produced by these operations could be hazardous. Additionally, it offers some challenges to the recovery process and is difficult to treat (Huang et al. 2014). Therefore, in recent years some emerging methods or lixiviants have been studied for the recycling of metals from WEEE, such as ionic liquids and organic leachants. The theory and technology behind these are already known. Some of these even have commercial applications in other areas. However, the use of these methods in WEEE recycling is still in a laboratory-scale study stage (Li et al. 2018). Thus, it is necessary to study them in more detail before they can have any industrial application.

11.4.1 Ionic Liquids

Ionic liquids are substances that exist in a liquid state consisting only of ions that are not in a crystalline state at temperatures below 100°C or even at room temperature (25°C). These liquids have at least one delocalized charged ion and one organic component, usually consisting of an organic cation and an inorganic or organic anion, which does not allow the formation of a stable crystalline lattice. Ionic liquids have the advantage that these can be used in a structure of freely synthesized anions and cations according to their purpose. In addition, ionic liquids also have the advantage of having a low melting point (<100°C), negligible volatility, high thermal stability, non-measurable vapor pressure, high conductivity, and wide electrochemical window. For these reasons, ionic liquids have multiple applications as solvents, refrigerants, absorbents, lubricants, and electrochemical electrolytes (Kim et al. 2018). In recent years, several studies have been carried out using ionic liquids for the extraction of base metals and REE from WEEE.

11.4.1.1 Base Metals

In 2012, Zhu et al. first used ionic liquids (1-ethyl-3-methylimizadolium tetrafluoroborate [$EMIM^+$][BF_4^-]) to treat WEEE, such as printed circuit boards (PCBs), which was achieved when the temperature reached 260°C, separation, and recovery of copper, glass fiber, and solder. After that, Huang et al. in 2014 reported that it was possible to dissolve almost all copper (99%) from PCBs using 10% v/v ionic liquid acid (1-butyl-3-methyl-3-methyl-imidazolium hydrogen sulfate [Bmim]HSO_4) in water, 5 ml H_2O_2 at a temperature of 50°C for 2 hours (Chen et al. 2015b).

In 2015, Chen et al. examined five other typical liquid ionic acids such as 1-(4-sulfobutyl)pyridinium hydrosulfate ([BSO_3HPy]HSO_4), 1-(4-sulfobutyl)-3-methylimidazolium hydrosulfate ([BSO_3HMIm]HSO_4), 1-(4-sulfobutyl)-3-methylimidazolium trifuoromethanesulfonate ([BSO_3HMIm] OTf), 1-methylimidazolium hydrogen sulfate ([MIm] HSO_4), and 1-(4-sulfobutyl)pyridinium trifuoromethanesulfonate ([BSO_3HPy]OTf), achieving full Cu recovery in the first three cases, 75% using [MIm]HSO_4 and 90% of Cu using [BSO_3HPy]OTf. In all cases, the ionic liquids were used with 5 mL of H_2O_2 at 50°C for 2 hours (Chen et al. 2015a). In 2018, Zhang et al. also studied the copper leaching behavior from PCBs. According to the experimental findings, the copper leaching rate increases as particle size increases from 0.071 to 0.50 mm. In addition, as the temperature rises, the leaching rate also increases. The leaching time, hydrogen peroxide dosage, and the ratio of solid to liquid all have an impact on the leaching rate. Particle size >0.50 mm, 8.5 mL 90% (v/v) ionic liquid ([CM-MIM] HSO_4), 1.5 mL hydrogen peroxide, a solid-to-liquid ratio of 1/20, a temperature of 80°C, and a leaching time of 2 hours were the ideal operating parameters that Zhang et al. used to achieve 98.31% copper recovery (Barrueto et al. 2022).

While the dissolved oxygen concentration in the solution drops due to a decrease in solubility at high temperatures, the leaching reaction and hydrogen peroxide breakdown generally accelerate with rising temperature. These can lead to the decomposition of the ionic liquid or increase its viscosity, conditions that are considered responsible for affecting copper leaching. This may occur because the mass transfer mechanisms in ionic liquids depend on the diffusion coefficients of the cations and anions, which are justified by the values of their viscosities, like the higher the viscosity, the lower the diffusion coefficient. Analyzing that the effect of anion variation in ionic liquids is closely related to viscosity, it is possible to classify the ionic liquids in ascending order according to this at 25°C as follows: [Bmim]HSO_4 < [Bmim]PF_6 < [Bmim]Cl < [Bmim]BF_4 (Barrueto et al. 2022).

It is simpler to leach copper from WEEE using acidic ionic solutions that include the HSO_4 anion, because they release their HSO_4^- ions and form stable complexes with the Cu ions present in the solution. Although it was mentioned above that the physical properties of the ionic liquids depend on the anion, in the case where they have a common anion, there are differences in the viscosity values between them. It follows that the cation and chain length also affect viscosity. At room temperature (25°C), [BSO_4 HPy]HSO_4 is the least dense ionic liquid, followed by [Bmim]HSO_4, [MIm]HSO_4, and [BSO_3 HMIm]HSO_4. Additionally, utilizing this kind of acidic ionic liquid is similar to switching the hydrogen ion in the sulfuric acid molecule for an imidazolium cation, which lowers the concentration of hydrogen ions involved in the reaction (Barrueto et al. 2022).

In 2021, Barrueto et al. evaluated the leaching of copper and cobalt from PCBs with different ionic liquids, obtaining the highest recovery working with conditions such as 60% w/w ionic liquids and 31.2% hydrogen peroxide at 60°C, achieving extractions with [Bmim]HSO_4 of 86.19% Cu and 96.65% Co; with [Hmim]HSO_4 76.62% Cu and 79.90% Co; with [Bmim]Br 4.4% Cu and 6.9% Co; and with [Bmim]Cl 13.9% Cu and 13.9% Co. These continue to demonstrate that acidic ionic liquids with HSO_4 anions are more efficient for leaching copper from WEEE, as well as having good efficiency for the extraction of cobalt from these wastes (Barrueto et al. 2021).

In addition to the above studies, Table 11.6 shows other research that has been conducted to evaluate the use of ionic liquids in the leaching of zinc, lead, and indium from WEEE.

Comparing the variation of anions in the ionic liquids between HSO_4^- and CF_3SO_3, for the case of [BSO_3HMIm] HSO_4, the extraction percentages of copper and zinc are much higher compared to the extractions obtained with [BSO_3HMIm]OTf. This is because the acidic nature of [BSO_3HMIm]HSO_4 and the standard oxidation potential difference favors the extraction of copper and zinc over the lead, while [BSO_3HMIm]OTf has better extraction results for lead. In both cases, the leaching kinetics are diffusion-controlled (Barrueto et al. 2022). One of the metals that can be dissolved by the protonated carboxyl-functionalized ionic liquid betainium bis(trifluoromethylsulfonyl)imide [Hbet][Tf_2N] is indium, which has a high leaching efficiency from some WEEE like LCDs, in which the leaching process is mainly controlled by chemical reactions. Some researchers recommend the [Hbet][Tf_2N]-ascorbic acid system because it not only allows leaching and separation of indium but also improves selectivity by effectively reducing the proportion of iron in the ionic liquid (Luo et al. 2019).

11.4.1.2 Rare Earth Elements (REE)

In addition to base metals, the use of ionic liquids for the leaching of REE from WEEE has also been studied. One of the most widely used ionic liquids has been the protonated carboxyl-functionalized ionic liquid betainium bis(trifluoromethylsulfonyl)imide ([Hbet][Tf_2N]), which in addition to dissolving indium oxides as mentioned above, also can dissolve other metal oxides such as rare earth oxides, as can be seen in Equation (11.1) (Dupont and Binnemans 2015).

$$Cu^{2+} + 2(C_5H_8NO_4Na) \leftrightarrow Cu(C_5H_8NO_4)_2 + 2Na^+, \text{LogK} = 15.78 \qquad (11.1)$$

In addition, some oxidized metals such as iron and cobalt oxides are very poorly soluble in this ionic liquid. This may favor the selective leaching of REE oxides from roasted WEEE. Dupont and Binnemans (2015) proposed a process

Table 11.6 Summary of research with ionic liquids for the leaching of zinc, lead, and indium from WEEE.

Electronic waste	Ionic liquid	Reaction conditions	% Metal recovery	References
PCB	[BSO_3HMIm] HSO_4	60% IL and 5 mL H_2O_2 at 50°C for 2 h.	100% Zn	(Barrueto et al. 2022)
		10% IL and 25 mL H_2O_2 at 40°C for 2 h.	10% Pb	
	[BSO_3HMIm] OTf	10% IL and 5 mL H_2O_2 at 50°C for 2 h.	90% Zn and 40% Pb	(Barrueto et al. 2022)
	[BSO_4HPy] HSO_4	40% IL and 5 mL H_2O_2 at 50°C for 2 h.	95% Zn	(Barrueto et al. 2022)
		10% IL and 5 mL H_2O_2 at 50°C for 2 h.	10% Pb	
	[MIm]HSO_4	10% IL and 5 mL H_2O_2 at 50°C for 2 h.	100% Zn	(Barrueto et al. 2022)
	[BSO_3HPy]OTf	80% IL and 5 mL H_2O_2 at 50°C for 2 h.	75% Zn	(Barrueto et al. 2022)
		60% IL and 5 mL H_2O_2 at 50°C for 2 h.	10% Pb	
		80% IL and 0.25 mL H_2O_2 at 50°C for 2 h.	75% Zn and 8% Pb	
LCDs	[Hbet][Tf2N]	50% (v/v) ionic liquid/ascorbic acid, 20 g/L of solid/liquid ratio, and 90°C operation temperature for 24 h.	99.75% In	(Luo et al. 2019)

to recycle REE from roasted NdFeB magnets. In this process, valuable elements such as Nd, Dy, and Co are separated from Fe with high separation factors, managing to recover about 99% of the rare earth metals in the NdFeB magnet with a purity greater than 99.9%. These results are because this process combines the leaching and extraction stages as can be seen in Figure 11.4 (Dupont and Binnemans 2015).

Additionally, a new method to recycle phosphorus from lamps was put out by Dupont and Binnemans in 2015, showing that the important red phosphorus Y^2O^3:Eu^{3+} (YOX), which makes up 80% of the weight of REEs, can be selectively leached. This process also combines leaching and extraction stages, using only [Hbet][Tf_2N] and oxalic acid, allowing the ionic liquid to be completely regenerated and reused. This process resulted in a 100% recovery of YOX using [Hbet][Tf_2N] with 5% wt H_2O at 90°C for 24 hours (Barrueto et al. 2022).

In addition to the NdFeB magnet leaching method suggested by Dupont and Binnemans, Orefice et al. (2020) also did research. Nd and Dy can be entirely separated from Fe by optimizing the roasting process utilizing an argon environment at PO_2 10–2 atm with 5 wt% carbon for 2 hours at 1400°C. Because Nd does not completely transform into Nd_2O_3 when roasting in air and at temperatures above 500°C it instead forms the more difficult-to-dissolve ternary phase $NdFeO_3$. Subsequently, the roasted material was leached using 10% wt [Hbet][Tf_2N]-H_2O at 80°C, achieving full recovery of Nd and Dy. This optimization process allowed reducing the process duration time (roasting + leaching) to only 2.3 hours, which is 17 times less than the time taken by the process proposed by Dupont and Binnemans (Barrueto et al. 2022).

11.4.2 Organic Leachants

11.4.2.1 Chelation

Chelation is one of the most environmentally friendly technologies that has been put to use for a long time in a variety of home and commercial applications and has gained interest recently in the area of heavy metal extraction from soils and industrial waste treatment to minimize the negative impacts that pollutants can have on the environment. Chelation is the process by which stable, soluble-in-water metal complexes are formed. In other words, it is the process of replacing monodentate ligands in metal complexes with ligands of a multidentate character. As a result, the foundation of this idea is the coordinated introduction of a metal ion into a heterocyclic ring structure. There are several organic chelators, like ethylenediaminetetraacetic acid (EDTA), ethylenediamine-N, N'-disuccinic acid (EDDS), and nitrilotriacetic acid (NTA), which can form stable complexes with some metal ions such as Cu, Co, Mo, Cr, Ni, Zn, Sn, and Pb (Chauhan et al. 2018).

In general terms, the process of metal extraction using chelating agents can be explained using the unreacted nucleus theory, where the formation of metal-ligand complexes on the particle surface begins with the mass transfer of the leaching agent from the solid-liquid interface to the interior of the particle, and where the mass transfer of the complex from the particle surface to the solution maximizes the complex's thermodynamic stability. These mass transfer processes are completely controlled by the diffusion mechanism of metals present both on the surface and inside the particle, as can be seen in Figure 11.5 (Chauhan et al. 2015).

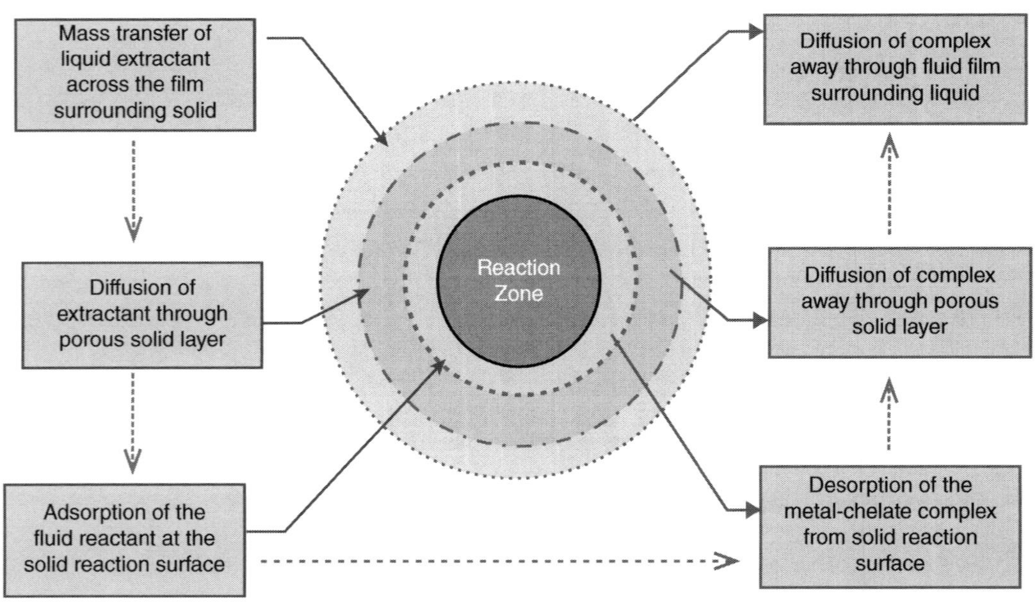

Figure 11.5 Mechanism of metal leaching process with chelating agents.

During the extraction process, several reactions occur on the particle surface between the chelating agents and the target metals. These reactions can be divided into primary and secondary reactions depending on how the chelating chemicals interact with the target metals and other elements on the surface of the solid. Primary reactions are those in which the ligand substitution mechanism at the solid-liquid interface interacts with the metal (Equations 11.2, 11.3, and 11.4). This is the case for M_i^{m+} target metals that are present as oxides and hydroxides or that are bonded to surface hydroxyls, while by secondary reactions, some of the metal-ligand complexes can adhere to the particle surface, either because they can be re-adsorbed by complex formation on the surface (Equation 11.5) or replaced with different M_{ii}^{m+} metal ions (Equations 11.6 and 11.7). These will depend on the stability constants of the metal-ligand complexes. Also, it is possible that the dissociated metal M_i^{m+} may be present as free metal ions in solution or may be partially re-adsorbed on the particle surface, as can be seen in Equations (11.8) and (11.9) (Chauhan et al. 2015).

Primary reactions

$$[\mathrm{HDS}]-\mathrm{O}-M_i^{m+}+\mathrm{L}^{n-}+\mathrm{H_2O}\leftrightarrow[\mathrm{HDS}]-\mathrm{OH}+M_i-\mathrm{L}^{m+n-}+\mathrm{OH}^- \tag{11.2}$$

$$M_i(\mathrm{OH})_m+\mathrm{L}^{n-}\leftrightarrow M_i-\mathrm{L}^{m+n-}+\mathrm{mOH}^- \tag{11.3}$$

$$(M_i\mathrm{O})^{m+2-}+\mathrm{L}^{n-}+\mathrm{H_2O}\leftrightarrow M_i-\mathrm{L}^{m+n-}+2\mathrm{OH}^- \tag{11.4}$$

Secundary reactions

$$[\mathrm{HDS}]-\mathrm{OH}+M_i^{m+}-\mathrm{L}^{n-}\leftrightarrow[\mathrm{HDS}]-\mathrm{L}^{n-}-M_i^{m+}+\mathrm{OH}^- \tag{11.5}$$

$$[\mathrm{HDS}]-\mathrm{L}-M_i+[\mathrm{HDS}]-\mathrm{O}-M_{ii}+\mathrm{H}^+\leftrightarrow[\mathrm{HDS}]-\mathrm{L}-M_{ii}+[\mathrm{HDS}]-\mathrm{OH}+M_{ii}^{m+} \tag{11.6}$$

$$[\mathrm{HDS}]-\mathrm{L}^{n-}-M_{ii}^{m+}+\mathrm{OH}^-\leftrightarrow[\mathrm{HDS}]-\mathrm{OH}+M_{ii}-\mathrm{L}^{m+n-} \tag{11.7}$$

$$[\mathrm{HDS}]-\mathrm{OH}+M_i^{m}\leftrightarrow\{[\mathrm{HDS}]-\mathrm{O}\}-M_i^{m+2-}+\mathrm{H}^+ \tag{11.8}$$

$$M_i^{m+}+\mathrm{pOH}^-\leftrightarrow M_i(\mathrm{OH})_p{}^{m+p-} \tag{11.9}$$

There have been very few studies on the use of chelating agents for the recovery of metals from WEEE, despite the high affinity of chelating agents to form stable complexes with various metals and information in the literature on multiple applications of chelating agents in medicine, microbial growth, soil remediation, and metal recovery from industrial waste. Jadhao et al. (2016) evaluated the leaching of Cu from WPCBs using 0.5M EDTA at 100°C and a pH 7 for 3 hours. They obtained, as a result, a recovery of 84% Cu, which compared to acid leaching of this same waste is equivalent to almost 50% more Cu extraction. Sharma et al. used a similar idea in 2017 to recover Cu and Co from portable batteries. Additionally, they used the Box–Behnken design and central composite design matrices with response surface methods to conduct statistical optimization. With this model, they found that it is possible to achieve Cu extractions higher than 90% operating within the optimal range of process parameters such as EDTA concentration between 0.5–0.6M, solid/liquid ratio 1/19–1/23 g/mL, pH 8–9, temperature 83–89°C, and a reaction time of about 130–148 minutes. Similarly, the distribution curve suggested an optimal EDTA concentration range between 0.33–0.41M, solid/liquid ratio 1/25–1/28 g/mL, temperature 87–92°C, pH 8.3–8.8, and reaction time of 163 to 178 minutes to extract more than 94% Co from the mobile batteries (Chauhan et al. 2018; Li et al. 2018).

11.4.2.2 Organic Acids

The majority of organic acids are generated by microorganisms found in the environment. As a result, they are not regarded as a pollutant that poses a risk to the environment. These acids' functional groups determine how acidic they are (e.g. carboxylic or sulfonic). It is feasible to divide them into several categories based on their characteristic carbon chains, saturation levels, substitution patterns, and functional groups since they do not all have the same quantity of carboxylic groups, hydroxyl groups, and carbon-carbon double bonds in their structure. Organic acids are considered green leaching agents with the advantage of not generating secondary contamination, causing less corrosion in equipment due to their low acidity, having less potential risk for operators, and can selectively leach metals. Also, they are biodegradable organic agents, do not produce gases, and the residues resulting from the leaching process are easily manageable compared to organic acids (Golmohammadzadeh et al. 2020).

Some organic acids such as oxalic acid ($C_2H_2O_4$), DL-malic acid ($C_4H_6O_5$), tartaric acid ($C_4H_6O_6$), trichloroacetic acid ($C_2HCl_3O_2$), aspartic acid ($C_4H_7NO_4$), succinic acid ($C_4H_6O_4$), acetic acid ($C_2H_4O_2$), and citric acid ($C_6H_8O_7$) have similar leaching performance of base metals and REE as inorganic acids, which means that these eco-friendly leachants can recover these metals from WEEE efficiently (Meshram et al. 2020). However, although they are less potent than inorganic acids, organic acids are more thermally stable, can form stable complexes, and have less environmental impact. Generally, the relative stability of the acid's conjugate base and other groups, including the –OH, -OS, alcohol, thiol, enol, and phenol groups, affects the strength of organic acids. This stability is determined by the pKa, which is the pH at which the acid and anion concentrations are equal; the lower the pKa, the stronger the acid (Golmohammadzadeh et al. 2020).

The dissolution of metals using organic acids occurs through acidolysis and complexolysis mechanisms. Hydrogen ions attack the metals and deploy them into the solution during the acidolysis mechanism. While in complexolysis, the formation of complexes by the organic molecules stabilizes the metal ions. The strength of the hydrogen hydroxyl and oxygen link in the carboxylic acids influences the rate of complex formation. The carboxylic acids' hydrogen hydroxyl and oxygen bond strength affect how quickly complexes form. In addition, acids with more hydroxyl groups tend to react faster. In the following chemical reactions, the formation of metal ion complexes with some organic acids such as acetic acid (Equation 11.10), ascorbic acid (Equations 11.11 and 11.12), malic acid (Equations 11.13 and 11.14), oxalic acid (Equations 11.15 and 11.16), and citric acid (Equations 11.17, 11.18, and 11.19) can be observed (Golmohammadzadeh et al. 2020):

$$n\left[C_2H_3O_2^-\right] + M^{n+} \rightarrow M\left[C_2H_3O_2\right]_n \tag{11.10}$$

$$n\left[HC_6H_6O_6^-\right] + M^{n+} \rightarrow M\left[HC_6H_6O_6\right]_n \tag{11.11}$$

$$n\left[C_6H_6O_6^{2-}\right] + 2M^{n+} \rightarrow M_2\left[C_6H_6O_6\right]_n \tag{11.12}$$

$$n\left[HC_4H_4O_5^-\right] + M^{n+} \rightarrow M\left[HC_4H_4O_5\right]_n \tag{11.13}$$

$$n\left[C_4H_4O_5^{2-}\right] + 2M^{n+} \rightarrow M_2\left[C_4H_4O_5\right]_n \tag{11.14}$$

$$n\left[HC_2O_4^-\right] + M^{n+} \rightarrow M\left[HC_2O_4\right]_n \tag{11.15}$$

$$n\left[C_3O_4^{2-}\right] + 2M^{n+} \rightarrow M_2\left[C_2O_4\right]_n \tag{11.16}$$

$$n\left[H_2C_6H_5O_7^-\right] + M^{n+} \rightarrow M\left[H_2C_6H_5O_7\right]_n \tag{11.17}$$

$$n\left[HC_6H_5O_7^{2-}\right] + 2M^{n+} \rightarrow M_2\left[HC_6H_5O_7\right]_n \tag{11.18}$$

$$n\left[C_6H_5O_7^{3-}\right] + 3M^{n+} \rightarrow M_3\left[C_6H_5O_7\right]_n \tag{11.19}$$

Acetic acid is a weak acid and has the characteristic that it dissociates in a single step, so some studies have been carried out on the recycling of lithium-ion batteries (LIBs) using this acid. In 2017, Golmohammadzadeh et al. reported recovery of approximately 75% of lithium and 30% of cobalt in the presence of acetic acid at a solid to liquid (S/L) ratio of 30 g/L (Golmohammadzadeh et al. 2020). Gao et al., in 2018, also studied the leaching of LIB using 3.5M acetic acid, 4% H_2O_2 as a reducing agent, and a pulp density of 40 g/L at 60°C for 60 minutes, obtaining a recovery of 99.97% Li, 93.97% Co, 96.32% Mn, and 92.67% Ni (Meshram et al. 2020). Similarly, Natarajan et al. (2018) employed H_2O_2 as a reducing agent and acetic acid as a leaching agent to dissolve mobile phone batteries. It was discovered that 99.9% Li, 98.7% Co, and 99.5% Mn were leached with 3M acetic acid, 7.5 vol.% H_2O_2, and a pulp density of 20 g/L at 70°C for 40 minutes. The reaction of this type of cathode material in the presence of acetic acid and H_2O_2 can be represented by Equation (11.20) (Meshram et al. 2020):

$$\begin{aligned} &Li_2CoMn_3O_{8(S)} + 10CH_3COOH_{(aq)} + 10H_2O_{2(aq)} \\ &\rightarrow 2CH_3COOLi_{(aq)} + Co(CH_3COO)_{2(aq)} \\ &+ 3Mn(CH_3COO)_{2(aq)} + 8H_2O + 3O_2 \end{aligned} \tag{11.20}$$

Ascorbic acid is a 2,3-enediol-L-gulonic acid, which can release H^+ from the enediol group in solution, which is the reason for its strong acidity. In addition, enodiols are considered to be good reducing agents. Therefore, ascorbic acid can act as both a leaching and a reducing agent (Golmohammadzadeh et al. 2020). In 2012, Li et al. developed a hydrometallurgical process for the recovery of lithium and cobalt from LIBs using ascorbic acid, recovering nearly 98.5% Li and 94.8% Co using 1.25M ascorbic acid and a pulp density of 25 g/L at 75°C in 20 minutes (Meshram et al. 2020). The results showed that ascorbic acid functioned as a leaching agent and as a reducing agent, which facilitates the extraction of cobalt by converting the insoluble Co^{3+} compound into soluble Co^{2+} (Golmohammadzadeh et al. 2020). The leaching reaction is represented in Equation (11.21) as follows (Meshram et al. 2020):

$$\begin{aligned} &4C_6H_8O_6 + 2LiCoO_2 \rightarrow C_6H_6O_6 + C_6H_6O_6Li_2 \\ &\quad + 2C_6H_6O_6Co + 4H_2O \end{aligned} \tag{11.21}$$

Malic acid has two carboxylic functional groups and high water solubility, which allows this acid to be able to dissolve metals from spent LIBs. In addition, waste solutions of this acid can be easily treated because both aerobic and anaerobic conditions allow malic acid to degrade. Thus, it is an environmentally friendly recycling technique for recovering metals from WEEE (Golmohammadzadeh et al. 2020). Results obtained from the leaching of cathodes with malic acid showed only a 54% recovery of lithium and 37% of cobalt using 1.5M malic acid with a pulp density of 20 g/L at 90°C for 30 minutes. However, by adding 2% H_2O_2, the recovery rate increased considerably to 99% Li and 90% Co. Equation (11.22) shows the reaction of LIBs in the presence of H_2O_2. Recently, it was reported that the concentration of malic acid did not affect the leaching and that at a 1M acid concentration, a pulp density of 10 g/L, and a temperature of 95°C, it is possible to recover 96% Li, 98% Co, and 99% Ni in 30 minutes (Meshram et al. 2020).

$$2LiCoO_{2\,(S)} + 6C_4H_6O_{5\,(aq)} + H_2O_{2\,(aq)} \rightarrow 4LiC_4H_5O_{5\,(aq)} + 2CoC_4H_5O_{5\,(aq)} + 4H_2O_{(l)} + O_{2\,(g)} \quad (11.22)$$

Oxalic acid is highly inclined to release hydrogen ions ($pKa_1 = 1.25$), which makes it the most acidic of all organic acids (Golmohammadzadeh et al. 2020). In addition, this acid can act as a leaching and precipitating agent in some hydrometallurgical processes. For this reason, oxalic acid has been a widely used leaching agent for the dissolution and extraction of metals from different types of secondary waste, including WEEE. In 2012, Sun and Qiu reported the combination of oxalate leaching and vacuum pyrolysis for the recovery of lithium and cobalt from LIBs, using 1M oxalic acid with 15% H_2O_2 and a pulp density of 25 g/L at 80°C for 120 minutes, achieving a recovery efficiency of 98%. Also, they reported that due to the differences in the solubility properties of lithium oxalates and cobalt oxalates, the separation was favored since cobalt oxalates were insoluble in the acid and therefore precipitated. While lithium oxalates were soluble, it was possible to precipitate separately at high pH. Similarly, Zeng et al. (2015) studied the extraction of Li and Co from LIBs by synchronous leaching and precipitation using 2M oxalic acid and a pulp density of 15g/L at 95°C for 150 minutes, achieving a recovery of 98% Li and 97% Co (Meshram et al. 2020).

Citric acid can be produced naturally in fermentation processes and is one of the best complexing agents because it can easily form metal complexes in acidic media and is a strong chelating agent. Moreover, it is the only organic acid that dissociates in three steps, i.e., one mole of citric acid provides three moles of H^+ in solution (Golmohammadzadeh et al. 2020). These characteristics have led many studies to investigate how different leaching variables, including acid content, temperature, solid-liquid ratio, reducing agent, etc., affect the recovery of metals from some WEEE, such as used batteries. Li et al. (2010) reported that citric acid can leach 100% Li and 90% Co from LIBs using 1.25M citric acid and 1% H_2O_2 with a pulp density of 20 g/L at 90°C for 30 minutes (Meshram et al. 2020).

In 2014, Chen and Zhou explored the extraction of valuable metals from wasted LIBs using 2M citric acid, 2% H_2O_2, 50 g/L pulp density, 70°C for 80 minutes, and a combined approach of selective precipitation and solvent extraction to separate and recover each metal from the pregnant leach solution (PLS). Dimethylglyoxime ($C_4H_8N_2O_2$) and ammonium oxalate ($(NH_4)_2C_2O_4$) were initially used to selective precipitate nickel and cobalt. Then, manganese was extracted using Na-D2EHPA, and separation was completed using sulfuric acid. Finally, lithium was precipitated with Na_3PO_4 at 0.5M concentration. This process resulted in the recovery of approximately 98% Ni and Mn, 97% Co, and 89% Li (Meshram et al. 2020).

Ma et al. (2017) compared the leaching of metals from battery waste using citric acid and sulfuric acid as leaching agents. With this, they were able to determine that citric acid with hydrogen peroxide as oxidizing agent favored the dissolution of valuable metals such as Co (recovery of 85%) and Ni (100%) but did not promote the leaching of metals such as Fe, Cu, Zn (over 50%), and Mn (28.3%). According to these results, citric acid favors the selective dissolution of metals from residues compared to sulfuric acid. However, thermodynamic calculations showed that metals precipitate more easily in a sulfuric acid system than in a citric acid system (Meshram et al. 2020).

Li et al. (2013) compared citric acid with two other organic acids such as aspartic and malic acid, using hydrogen peroxide as a reducing agent for the dissolution of lithium and cobalt from waste LIBs, obtaining a recovery of 100% Li and >90% Co using citric and malic acid, while with aspartic acid the recovery was very low, which is attributed to its low acidity. Some studies allow comparing organic acids with each other, establishing that in terms of efficiency to recover cobalt and lithium from LIBs, the best sequence is citric acid>DL-malic acid>acetic acid>oxalic acid (Meshram et al. 2020).

Other organic acids have also been studied for the leaching of LIBs, such as succinic acid, aspartic acid, lactic acid, formic acid, trichloroacetic acid, and tartaric acid. Table 11.7 shows some studies performed with these acids and the optimized conditions for the metals leaching from LIBs.

In a more recent study, (Zhou et al. 2019) investigated the recovery of gallium from light-emitting diode waste using organic acids. The results showed that it could extract 83.42% gallium with oxalic acid, 70.62% with citric acid, and 71.69% with DL-malic acid. In addition, they were able to establish that 90°C leaching temperature, 10 g/L pulp density, 0.7M oxalic acid concentration, and a particle size between 48 and 75 μm are the optimum conditions to obtain a gallium recovery efficiency of 90.36% in one hour.

The implementation of organic acids for the recovery of REE from WEEE has also been studied. In 2016, Behera and Parhi used acetic acid to study the kinetics of Nd leaching from NdFeB waste. The experiments used acetic acid

Table 11.7 Summary of research on metal recovery from LIBs using other organic acids.

Organic acid	Leaching condition	Leaching efficiency	References
Succinic acid	1.5M succinic acid, 4% H_2O_2 and 15 g/L pulp density at 70°C for 40 min.	96% Li and 96% Co	(Meshram et al. 2020)
Aspartic acid	1.5M aspartic acid, 2% H_2O_2,with a pulp density of 20 g/L at 90°C for 30 min.	60% Li and 60% Co	(Meshram et al. 2020)
Lactic acid	1.5M lactic acid, 0.5% H_2O_2 and 20 g/L pulp density at 70°C for 20 min.	97.7% Li, 98.9% Co, 98.4% Mn, and 98.2 Ni	(Meshram et al. 2020)
Trichloroacetic acid	3M arichloroacetic acid, 4% H_2O_2, a solid to liquid ratio (S/L) of 50 g/L at 60°C for 30 min.	99.7% Li, 93% Ni, 91.8% Co, and 89.8% Mn	(Sun et al. 2021)
Formic acid	2M formic acid, 2% H_2O_2 and pulp density of 50 g/L at 70°C for 2 h.	99.97% Li, 96.32 Mn, 93.97% Co, and 92.67% Ni	(Meshram et al. 2020)
Tartaric acid	4 vol% H_2O_2, 2M L-tartaric acid, 17 g/L pulp density, at 70°C, for 30 min.	99.07% Li, 99.31% Ni, 99.3% Mn, and 98.64 Co	(Meshram et al. 2020)

concentrations greater than 0.4 mol/L, a 1% solid/liquid ratio, and an 80°C temperature. The results demonstrated recoveries higher than 99% Nd and Fe co-leaching from NdFeB magnets. Menad et al. also reported recoveries greater than 90% of Nd, Dy, and Fe from NdFeB magnets by leaching with acetic acid for 7 hours. A recent study evaluated the leaching of REE from NdFeB magnets using maleic, glycolic, and ascorbic acids. At a maleic acid and glycolic acid concentration of 1 mol/L and a solid/liquid ratio of 1%, 95% recoveries were obtained for Nd, Dy, and Fe after 7 hours. In the case of ascorbic acid, it was less efficient at room temperature; however, when the temperature was increased to 70°C, similar recovery values to those observed with the other two acids were obtained (Belfqueh et al. 2022).

Reisdorfer et al. investigated the use of citric acid and malic acid to leach Nd from roasted and unroasted NdFeB magnets. The effects of temperature (30–90°C), acid concentration (0.2–1.2M), solid/liquid ratio (1:10–1:50) and time (10–900 min) on the leaching kinetics were evaluated. They obtained for both acids that the optimum parameters for the Nd leaching process were at a temperature of 90°C, acid concentration 1M, and solid/liquid ratio 1:20. From unroasted NdFeB powder, it was possible to recover 99% Nd using malic acid for 6 hours and a recovery of 72.8% Nd using citric acid for 1 hour. The Nd recovery efficiency for roasted NdFeB powder was 22% with malic acid and 32.1% with citric acid. The outcomes also demonstrated that acids were no selectivity between Nd and Fe in the leaching of unroasted NdFeB powder, while leaching of roasted NdFeB powder provided better selectivity but lower Nd leaching yield (Belfqueh et al. 2022).

Therefore, the use of organic acids can be an alternative to the use of inorganic acids for the recovery of base metals and REE from WEEE. In addition, since they are considered green leachants because they are of natural origin and biodegradable, they would allow the recycling of metals in an efficient and environmentally friendly manner.

11.4.2.3 Amino Acids

In recent years amino acids have become a very interesting alternative for the leaching of metals from WEEE because they are reagents of organic origin, non-toxic, and operate adequately in an alkaline medium. These allow for efficient, safe, and environmentally friendly processes, conditions that are necessary to carry out recycling processes since this is considered urban mining.

Glycine is one of the amino acids that has been most studied for its application in the recovery of metals from WEEE. It is the simplest amino acid and is naturally derived from microorganisms. It is relatively inexpensive because it is commonly mass-produced as a food additive. In addition, it is environmentally friendly, non-toxic, non-volatile, stable over a wide pH-Eh range, and has a high affinity for some metals (Li et al. 2018). Glycine forms complexes through bidentate chelation or monodentate coordination. Therefore, depending on the pH, glycine can exist in solution as glycinate cation ($H_2(gly)^+$), zwitterion ($^+H_3NCH_2COO^-$), and glycinate anion ($H_2NCH_2COO^-$). Both the zwitterion and the anion can form stable complexes with Cu^+ and Cu^{2+} ions, with the complexes formed with Cu^{2+} showing a larger stability domain, as can be seen in the following reactions and their stability constants (Li et al. 2020) in Equations (11.23) through (11.26):

$$Cu^{2+} + 2(NH_2CH_2COO)^- \leftrightarrow Cu(NH_2CH_2COO)_2,\ \mathrm{LogK} = 15.64 \tag{11.23}$$

$$Cu^{2+} + (NH_2CH_2COO)^- \leftrightarrow Cu(NH_2CH_2COO)^+, \text{LogK} = 8.57 \quad (11.24)$$

$$Cu^+ + 2(NH_2CH_2COO)^- \leftrightarrow [Cu(NH_2CH_2COO)_2]^-, \text{LogK} = 10.1 \quad (11.25)$$

$$Cu(NH_2CH_2COO)^+ + H^+ \leftrightarrow Cu(NH_3CH_2COO)^{2+}, \text{LogK} = 2.92 \quad (11.26)$$

Oraby et al. (2019) developed a two-stage leaching process of PCBs using in the first stage a glycine concentration of 30 g/L, pH 11, 0.4% solids at room temperature for 24 hours, and a second stage under the same conditions as the previous one but adding 300 ppm NaCN, achieving a total recovery of 99.1% Cu (Perea et al. 2021). Li et al. (2020) studied the leaching of PCBs in alkaline glycine solutions obtained under optimum conditions of 0.5M glycine, initial pH of 10.2, 2% solids, particle size <2 mm, and using ambient O_2 as an oxidizing agent at room temperature for 72 hours, for a recovery of 96.5% Cu, 92.5%Zn, 46.8% Pb, 16% Al, 12.6% Ni, 9.3% Sn, and 3.1% Co.

Recently, studies of metal recycling from waste LIBs using glycine have been reported. In 2016, Nayaka et al. developed a hydrometallurgical process to recover cobalt from spent LIBs using glycine and ascorbic acid as leaching reagents, achieving 95% Co recovery by leaching 0.2 g of $LiCoO_2$ with 0.5M glycine and 0.02M ascorbic acid at 80°C for 360 minutes (Nayaka et al. 2016). Also, Chen et al. 2021 developed a process for the recycling of cobalt and lithium from LIBs using 300 g/L glycine, 10% H_2O_2, and 1:100 solute/liquid ratio at 80°C for 7 hours. The leaching rates were 97.07% Co and 90.95% Li. Oghabi et al. 2020 reported a selective Cd leaching process from spent Ni-Cd batteries, achieving more than 86% Cd recovery using optimum leaching parameters such as 1.2M glycine and pH = 9.6 at room temperature for 150 minutes.

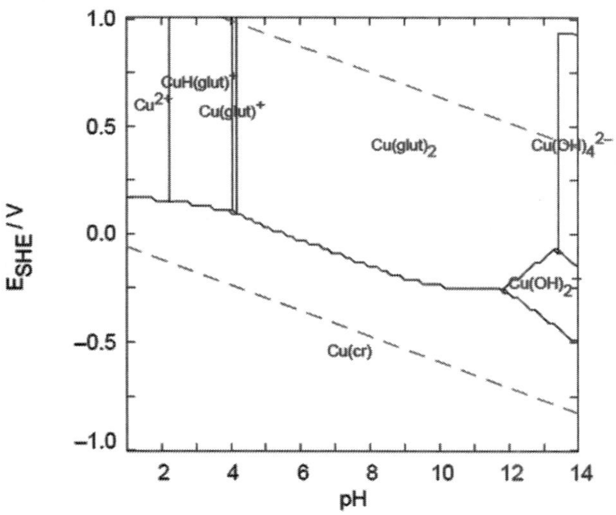

Figure 11.6 Eh-pH stability diagram for the Cu-glut-H_2O system at 25°C, [Cu] = 10^{-5}M and [glut] = 0.5M.

Another amino acid that has recently been used for metal recycling from WEEE is monosodium glutamate. It is one of the most abundant non-essential amino acids in nature and is a derivative of glutamic acid. Glutamate is a non-volatile reagent widely used in the food industry as a seasoning because it has no adverse health effects and is economical. The structure of this amino acid comprises an alcohol group, an amino group, and a carboxylate group, which gives it the ability to bind tridentate form with metal ions. Therefore, like glycine, glutamate can exist in solutions such as glutamate cation ($H_3(glut)^+$, $H_2(glut)^+$), zwitterion (H(glut)), and glutamate anion ($Glut^-$) depending on the pH, with glutamate anion predominating at pH > 9.4. Therefore, Cu dissolution is strongly related to glutamate anion concentration. For this reason, it can also form stable complexes with cuprous (Cu^+) and cupric (Cu^{2+}) ions, as can be seen in the following reactions (Perea et al. 2021) in Equations (11.27) and (11.28):

$$Cu^{2+} + C_5H_8NO_4Na \leftrightarrow Cu(C_5H_3NO_4)^+ + Na^+, \text{LogK} = 9.19 \quad (11.27)$$

$$Cu^{2+} + 2(C_5H_8NO_4Na) \leftrightarrow Cu(C_5H_8NO_4)_2 + 2Na^+, \text{LogK} = 15.78 \quad (11.28)$$

Elemental copper is stable as a species throughout the pH range and at low potentials. However, the copper-glutamate complex is stable in the zone between pH values 4 to 13 and potential values from –0.25 to 1 volt approximately, as can be seen in the Eh-pH diagram for the copper-glutamate system (Perea et al. 2021) shown in Figure 11.6.

Recently, Perea et al. (2021) evaluated the leaching of copper from WEEE using alkaline monosodium glutamate solutions. The effect of oxidizing agents such as hydrogen peroxide (H_2O_2) and potassium permanganate ($KMnO_4$) on the leaching kinetics and comparing the leaching efficiency of copper with glycine have been studied, achieving a maximum recovery of 92% Cu with glutamate and 86% Cu with glycine using both amino acids at a concentration of 0.5M, H_2O_2 0.03M, and pH 9.44 at room temperature. Glutamate has a higher rate of copper dissolution, which may be due to the ability of glutamate to coordinate metal ions in the three groups since it can bind in a tridentate form, as mentioned above, while glycine can only bind in a bidentate form because it has only one amino group and one carboxyl group.

11.5 Future Perspectives

Technology advancements in electrical and electronic products keep the WEEE processing industry changing. The recycling process will need to be renewed to recycle new equipment such as batteries for electric vehicles, compact PCBs, organic light-emitting diodes (OLEDs), and flexible smartphones. Examples are the cobalt, nickel, and rare earth elements that make up a hybrid car battery, which has an average weight of 50 kg and for which a metal recovery process (Espinosa et al. 2020) needs to be developed.

Using ionic liquids (ILs) as a new reaction medium for WEEE treatment is a promising option. Due to their electrochemical properties, such as low volatility, low combustibility, high extractability, and thermal stability, they present great potential for treating this waste in an environmentally friendly way. These liquids have been used in studies to extract base metals and REE from WEEE, showing very good results (Barrueto et al. 2021; Espinosa et al. 2020). However, these solvents' commercial application is hampered by the high initial investment. Even though these systems offer high purity and recyclability, their high cost limits their use as metal extraction systems (Barrueto et al. 2022). In addition, the utility of these liquids in metal extraction may be limited by solubilization losses and the difficulty in recovering the extracted metal ions (Chauhan et al. 2018).

Organic reagents, such as chelation, organic acids, and amino acids, are another novel and potentially green technology that has been studied to extract and recover metals from WEEE. In the case of chelation, some biodegradable ligands like EDTA, EDDS, NTA, and DTPA (diethylene triamine pentaacetate) have been used as chelating agents for the extraction of base metals like Cu, Co, Ni, Mo, Pb, and Zn. It is necessary, however, to further explore this concept of chelation-dechelation chemistry for effective WEEE management and the efficient recycling of chelating agents (Chauhan et al. 2018). In the case of organic acids, these can be an eco-friendly alternative to strong acids, which are traditionally used as leaching agents to base metals and REEs from WEEE (Belfqueh et al. 2022). It has been demonstrated that organic acids play a critical role in extracting REEs (Nd, Pr, and Dy) from NdFeB magnets (Belfqueh et al. 2022) and base metals (Ni, Li, Mn, and Co) from spent LIBs (Meshram et al. 2020). In this last type of waste, it has been possible to show that CO_2 emissions for Co extraction can be reduced from 4000 g using inorganic acids to 500 g using organic acids, which is proof of the environmental benefits of this type of acid. However, it is necessary to redouble efforts to investigate the separation of leached metals from these organic acid media (Meshram et al. 2020). Finally, organic acids are the most recent and interesting alternative for extracting metals from WEEE, mainly base metals such as Cu, Cd, and Zn. Since they are organic and biodegradable reagents that operate at alkaline pH and have shown high efficiency in the extraction of these metals in laboratory studies, however, these studies have only focused on the leaching stage, so it is still necessary to investigate later stages for recovering ions in solution and obtaining the final product.

Therefore, non-precious metal leaching processes from WEEE with emerging leachants are still under development, and more research is required. Future research should focus on process optimization and scaling up current leaching methods with emerging reagents for industrial-scale applications. In general terms, in emerging leachants, the kinetics and efficiency of metal dissolution are affected by operational parameters such as reagent concentration, type of oxidizing agent, pH, and temperature. Therefore, carrying out fundamental studies that allow a clear understanding of the dissolution mechanisms and the controlling stage of the process, as well as the optimization of these parameters, is necessary to achieve a method of leaching and recovery of metals with emerging leachants that is efficient and friendly to the environment. In addition, it is necessary to study the stages after leaching that allow recovering the metals in solution and that adjust to the current processes and infrastructure of the industry.

11.6 Conclusions

WEEE is the waste with the highest annual growth (approximately 4–5% per year), generating great concern globally. It is a situation that has led to a policy of Reduce, Reuse, and Recycle (circular economy) to reduce this waste and to be able to meet the growing demand for metals with technological applications such as base metals and REEs. For this reason, different mechanical, pyro-, and hydrometallurgical processes are currently being used for the recovery of metals from WEEE, which are efficient. But none of these technologies can be considered environmentally friendly given the limitations associated with each one. For example, acid leaching processes can dissolve the highest concentration of metals quickly and efficiently. However, they are not very selective, which generates problems in later recovery stages, and their improper use represents a risk to the environment and human health. On the other hand, biohydrometallurgical processes, despite being considered green and more friendly to the environment, have extraction kinetics that is too slow and are far from being able to be implemented on an industrial scale. Finally, pyrometallurgical processes, despite being the most widely implemented processes on an industrial scale for the recovery of base metals and REE, are undesirable for continued implementation due to the high energy requirement, high cost, generation of toxic gases, and the increasingly strict policies that seek to mitigate global warming.

Therefore, emerging technologies such as ionic liquids, chelation, organic acids, and amino acids can be an excellent alternative for the recovery of non-precious metals from WEEE. Ionic liquids have shown excellent results for both leaching and recovery of metals in solution efficiently and selectively. For its part, chelation has shown a great potential for the extraction of metals in soil recovery and industrial waste treatment. Therefore, progress should be made in its use for the treatment of WEEE. Finally, organic acids and amino acids have also shown satisfactory results for the recovery of base metals such as Cu, Li, Co, Cd, and Zn. In addition, given their nature, they are friendly to the environment and do not represent any risk for their use in urban mining. However, all these technologies still need more research and development efforts to be able to optimize them and achieve their application on an industrial scale.

References

Barrueto, Y., Hernández, P., Jiménez, Y., and Morales, J. (2021). Leaching of metals from printed circuit boards using ionic liquids. *Journal of Material Cycles and Waste Management* 23: 2028–2036. https://doi.org/10.1007/s10163-021-01275-8.

Barrueto, Y., Hernández, P., Jiménez, Y.P., and Morales, J. (2022). Properties and application of ionic liquids in leaching base / precious metals from e-waste. A review. *Hydrometallurgy* 212: https://doi.org/10.1016/j.hydromet.2022.105895.

Belfqueh, S., Seron, A., Chapron, S. et al. (2022). Evaluating organic acids as alternative leaching reagents for rare earth elements recovery from NdFeB magnets. *Journal of Rare Earths* 41: 1–11. https://doi.org/10.1016/j.jre.2022.04.027.

Bertuol, D.A., Machado, C.M., Silva, M.L. et al. (2016). Recovery of cobalt from spent lithium-ion batteries using supercritical carbon dioxide extraction. *Waste Management* 51: 245–251. https://doi.org/10.1016/j.wasman.2016.03.009.

Birloaga, I. and Vegliò, F. (2018). WEEE characterization, division, and regulation. In: *Waste Electrical and Electronic Equipment Recycling: Aqueous Recovery Methods*, (ed. I. Birloaga and F. Vegliò), 1–12. Woodhead Publishing.

Charles, R.G., Douglas, P., Dowling, M. et al. (2020). Towards increased recovery of critical raw materials from WEEE—evaluation of CRMs at a component level and pre-processing methods for interface optimisation with recovery processes. *Resources, Conservation and Recycling* 161: 1–21. https://doi.org/10.1016/j.resconrec.2020.104923.

Chauhan, G., Pant, K.K., and Nigam, K.D.P. (2015). Conceptual mechanism and kinetic studies of chelating agent assisted metal extraction process from spent catalyst. *Journal of Industrial and Engineering Chemistry* 27: 373–383. https://doi.org/10.1016/j.jiec.2015.01.017.

Chauhan, G., Ram, P., Pant, K.K., and Nigam, K.D.P. (2018). Novel technologies and conventional processes for recovery of metals from waste electrical and electronic equipment : challenges & opportunities—a review. *Journal of Environmental Chemical Engineering* 6: 1288–1304. https://doi.org/10.1016/j.jece.2018.01.032.

Chen, M., Huang, J., Ogunseitan, O.A. et al. (2015a). Comparative study on copper leaching from waste printed circuit boards by typical ionic liquid acids. *Waste Management* 41: 142–147. https://doi.org/10.1016/j.wasman.2015.03.037.

Chen, M., Wang, R., Qi, Y. et al. (2021). Cobalt and lithium leaching from waste lithium ion batteries by glycine. *Journal of Power Sources* 482: 1–8. https://doi.org/10.1016/j.jpowsour.2020.228942.

Chen, M., Zhang, S., Huang, J., and Chen, H. (2015b). Lead during the leaching process of copper from waste printed circuit boards by five typical ionic liquid acids. *Journal of Cleaner Production* 95: 142–147. https://doi.org/10.1016/j.jclepro.2015.02.045.

Ciro, E., Era, A.D., Pasquali, M., and Lupi, C. (2020). Indium electrowinning study from sulfate aqueous solution using diff erent metal cathodes. *Journal of Environmental Chemical Engineering* 8: 1–6. https://doi.org/10.1016/j.jece.2020.103688.

Dupont, D. and Binnemans, K. (2015). Recycling of rare earths from NdFeB magnets using a combined leaching/extraction system based on the acidity and thermomorphism of the ionic liquid [Hbet][Tf2N]. *Green Chemistry* 17: 2150–2163. https://doi.org/10.1039/C5GC00155B.

Espinosa, D.C.R., Oliveira, R.P.D., and Martins, T.A.G. (2020). Recycling technologies—hydrometallurgy. In: *Electronic Waste Recycling and Reprocessing for a Sustainable Future*, (eds. M. Holuszko, A. Kumar, and D.C.R. Espinosa), 165–188. Straive, Chennai: WILEY-VCH.

Gao, Z., Liu, Y., Jing, Y. et al. (2021). Mineralogical characteristics and Sr–Nd–Pb isotopic compositions of banded REE ores in the Bayan Obo deposit, Inner Mongolia, China: implications for their formation and origin. *Ore Geology Reviews* 139: 1–16. https://doi.org/10.1016/j.oregeorev.2021.104492.

Golmohammadzadeh, R., Faraji, F., and Rashchi, F. (2020). Recovery of lithium and cobalt from spent lithium ion batteries (LIBs) using organic acids as leaching reagents: a review. *Resources, Conservation & Recycling* 136: 418–435. https://doi.org/10.1016/j.resconrec.2018.04.024.

Gopikrishnan, V., Vignesh, A., Radhakrishnan, M. et al. (2020). Microbial leaching of heavy metals from e-waste. In: *Biovalorisation of Wastes to Renewable Chemicals and Biofuels*, (eds. N.K. Rathinam and R. Sani), 189–216. K. Elsevier.

Hao, J., Wang, Y., Wu, Y., and Guo, F. (2020). Metal recovery from waste printed circuit boards: a review for current status and perspectives. *Resources, Conservation and Recycling* 157: 1–15. https://doi.org/10.1016/j.resconrec.2020.104787.

Huang, J., Chen, M., Chen, H. et al. (2014). Leaching behavior of copper from waste printed circuit boards with Brønsted acidic ionic liquid. *Waste Management* 34: 483–488. https://doi.org/10.1016/j.wasman.2013.10.027.

Jadhao, P., Chauhan, G., Pant, K.K., and Nigam, K.D.P. (2016). Greener approach for the extraction of copper metal from electronic waste. *Waste Management* 57: 102–112. https://doi.org/10.1016/j.wasman.2015.11.023.

Jadhao, P.R., Pandey, A., Pant, K.K., and Nigam, K.D.P. (2021). Efficient recovery of Cu and Ni from WPCB via alkali leaching approach. *Journal of Environmental Management* 296: 1–11. https://doi.org/10.1016/j.jenvman.2021.113154.

Jadhav, U. and Hocheng, H. (2015). Hydrometallurgical recovery of metals from large printed circuit board pieces. *Scientific Reports* 5: 1–10. https://doi.org/10.1038/srep14574.

Kaya, M. (2016). Recovery of metals and nonmetals from electronic waste by physical and chemical recycling processes. *Waste Management* 57: 64–90. https://doi.org/10.1016/j.wasman.2016.08.004.

Kim, B., Joo, E., Kang, Y., and Lee, J. (2018). Application of ionic liquids for metal dissolution and extraction. *Journal of Industrial and Engineering Chemistry* 61: 388–397. https://doi.org/10.1016/j.jiec.2017.12.038.

Kumar, A., Holuszko, M., and Espinosa, D.C.R. (2017). E-waste: an overview on generation, collection, legislation and recycling practices. *Resources, Conservation and Recycling* 122: 32–42. https://doi.org/10.1016/j.resconrec.2017.01.018.

Kumar, A. and Li, J. (2017). An overview of the potential of eco-friendly hybrid strategy for metal recycling from WEEE. *Resources, Conservation & Recycling* 126: 228–239. https://doi.org/10.1016/j.resconrec.2017.07.014.

Li, L., Ge, J., Wu, F. et al. (2010). Recovery of cobalt and lithium from spent lithium ion batteries using organic citric acid as leachant. *Journal of Hazardous Materials* 176: 288–293. https://doi.org/10.1016/j.jhazmat.2009.11.026.

Li, L., Dunn, J.B., Zhang, X.X. et al. (2013). Recovery of metals from spent lithium-ion batteries with organic acids as leaching reagents and environmental assessment. *Journal of Power Sources* 233: 180–189. https://doi.org/10.1016/j.jpowsour.2012.12.089.

Li, H., Eksteen, J., and Oraby, E. (2018). Hydrometallurgical recovery of metals from waste printed circuit boards (WPCBs): current status and perspectives—a review. *Resources, Conservation and Recycling* 139: 122–139. https://doi.org/10.1016/j.resconrec.2018.08.007.

Li, H., Oraby, E., and Eksteen, J. (2020). Extraction of copper and the co-leaching behaviour of other metals from waste printed circuit boards using alkaline glycine solutions. *Resources, Conservation and Recycling* 154: 1–12. https://doi.org/10.1016/j.resconrec.2019.104624.

Lie, J. and Liu, J.C. (2021). Selective recovery of rare earth elements (REEs) from spent NiMH batteries by two-stage acid leaching. *Journal of Environmental Chemical Engineering* 9: 1–8. https://doi.org/10.1016/j.jece.2021.106084.

Lixandru, A., Venkatesan, P., Jönsson, C. et al. (2017). Identification and recovery of rare-earth permanent magnets from waste electrical and electronic equipment. *Waste Management* 68: 482–489. https://doi.org/10.1016/j.wasman.2017.07.028.

Luo, D., Zhu, N., Li, Y. et al. (2019). Hydrometallurgy simultaneous leaching and extraction of indium from waste LCDs with acidic ionic liquids. *Hydrometallurgy* 189: 1–8. https://doi.org/10.1016/j.hydromet.2019.105146.

Ma, L.W., Xi, X.L., Zhang, Z.Z. et al. (2017). Hydrometallurgical treatment for mixed waste battery materials. *IOP Conference Series: Materials Science and Engineering* 170. https://doi.org/10.1088/1757-899X/170/1/012024.

Meshram, P., Mishra, A., Abhilash, and Sahu, R. (2020). Environmental impact of spent lithium ion batteries and green recycling perspectives by organic acids—a review. *Chemosphere* 242: 1–16. https://doi.org/10.1016/j.chemosphere.2019.125291.

Natarajan, S., Boricha, A.B., and Bajaj, H.C. (2018). Recovery of value-added products from cathode and anode material of spent lithium-ion batteries. *Waste Management* 77: 455–465. https://doi.org/10.1016/j.wasman.2018.04.032.

Nayaka, G.P., Pai, K.V., Santhosh, G., and Manjanna, J. (2016). Recovery of cobalt as cobalt oxalate from spent lithium ion batteries by using glycine as leaching agent. *Journal of Environmental Chemical Engineering* 4: 2378–2383. https://doi.org/10.1016/j.jece.2016.04.016.

Oghabi, H., Haghshenas, D.F., and Firoozi, S. (2020). Selective separation of Cd from spent Ni-Cd battery using glycine as an eco-friendly leachant and its recovery as CdS nanoparticles. *Separation and Purification Technology* 242: 1–8. https://doi.org/10.1016/j.seppur.2020.116832.

Omodara, L., Pitkäaho, S., Turpeinen, E.-M. et al. (2019). Recycling and substitution of light rare earth elements, cerium, lanthanum, neodymium, and praseodymium from end-of-life applications—a review. *Journal of Cleaner Production* 236: 1–13. https://doi.org/10.1016/j.jclepro.2019.07.048.

Oraby, E.A., Li, H., and Eksteen, J.J. (2019). An alkaline glycine-based leach process of base and precious metals from powdered waste printed circuit boards. *Waste and Biomass Valorization* 11: 3897–3909. https://doi.org/10.1007/s12649-019-00780-0.

Orefice, M., Bulck, A. Van den, Blanpain, B., and Binnemans, K. (2020). Selective roasting of nd–fe–b permanent magnets as a pretreatment step for intensified leaching with an ionic liquid. *Journal of Sustainable Metallurgy* 6: 91–102.

Pekarkova, Z., Williams, I.D., Emery, L., and Bone, R. (2021). Economic and climate impacts from the incorrect disposal of WEEE. *Resources, Conservation and Recycling* 168: 1–10. https://doi.org/10.1016/j.resconrec.2021.105470.

Perea, C.G., Restrepo Baena, O.J., Ihle, C.F., and Estay, H. (2021). Copper leaching from wastes electrical and electronic equipment (WEEE) using alkaline monosodium glutamate: thermodynamics and dissolution tests. *Cleaner Engineering and Technology* 5: 1–7. https://doi.org/https://doi.org/10.1016/j.clet.2021.100312.

Pérez-Belis, V., Bovea, M.D., and Ibáñez-Forés, V. (2015). An in-depth literature review of the waste electrical and electronic equipment context: trends and evolution. *Waste Management and Research* 33: 3–29. https://doi.org/10.1177/0734242×14557382.

Pérez-Martínez, S., Giro-Paloma, J., Maldonado-Alameda, A. et al. (2019). Characterisation and partition of valuable metals from WEEE in weathered municipal solid waste incineration bottom ash, with a view to recovering. *Journal of Cleaner Production* 218: 61–68. https://doi.org/10.1016/j.jclepro.2019.01.313.

Sun, S., Jin, C., He, W. et al. (2021). Management status of waste lithium-ion batteries in China and a complete closed-circuit recycling process. *Science of the Total Environment* 776: 1–11. https://doi.org/10.1016/j.scitotenv.2021.145913.

Tabelin, C.B., Dallas, J., Casanova, S. et al. (2021a). Towards a low-carbon society: a review of lithium resource availability, challenges and innovations in mining, extraction and recycling, and future perspectives. *Minerals Engineering* 163: 1–23. https://doi.org/10.1016/j.mineng.2020.106743.

Tabelin, C.B., Park, I., Phengsaart, T. et al. (2021b). Copper and critical metals production from porphyry ores and E-wastes: a review of resource availability, processing/recycling challenges, socio-environmental aspects, and sustainability issues. *Resources, Conservation and Recycling* 170: 105610. https://doi.org/10.1016/j.resconrec.2021.105610.

Verbruggen, F., Ostermeyer, P., Bonin, L. et al. (2022). Electrochemical codeposition of arsenic from acidic copper sulfate baths: the implications for sustainable copper electrometallurgy. *Minerals Engineering* 176: 10. https://doi.org/10.1016/j.mineng.2021.107312.

Wang, H., Zhang, S., Li, B. et al. (2017). Recovery of waste printed circuit boards through pyrometallurgical processing: a review. *Resources, Conservation & Recycling* 126: 209–218. https://doi.org/10.1016/j.resconrec.2017.08.001.

Wang, R. and Xu, Z. (2014). Recycling of non-metallic fractions from waste electrical and electronic equipment (WEEE): a review. *Waste Management* 34: 1455–1469. https://doi.org/10.1016/j.wasman.2014.03.004.

Wu, Y., Yin, X., Zhang, Q. et al. (2014). The recycling of rare earths from waste tricolor phosphors in fluorescent lamps: a review of processes and technologies. *Resources, Conservation and Recycling* 88: 21–31. https://doi.org/10.1016/j.resconrec.2014.04.007.

Yang, Y., Walton, A., Sheridan, R. et al. (2017). REE recovery from end-of-life NdFeB permanent magnet scrap: a critical review. *Journal of Sustainable Metallurgy* 3: 122–149. https://doi.org/10.1007/s40831-016-0090-4.

Yu, Z., Han, H., Feng, P. et al. (2020). Recent advances in the recovery of metals from waste through biological processes. *Bioresource Technology* 297: 122416. https://doi.org/10.1016/j.biortech.2019.122416.

Zeng, X., Li, J., and Liu, L. (2015). Solving spent lithium-ion battery problems in China: opportunities and challenges. *Renewable and Sustainable Energy Reviews* 52: 1759–1767. https://doi.org/10.1016/j.rser.2015.08.014.

Zhang, L. and Xu, Z. (2016). A review of current progress of recycling technologies for metals from waste electrical and electronic equipment. *Journal of Cleaner Production* 127: 19–36. https://doi.org/10.1016/j.jclepro.2016.04.004.

Zhang, S., Ding, Y., Liu, B., and Chang, C.C. (2017). Supply and demand of some critical metals and present status of their recycling in WEEE. *Waste Management* 65: 113–127. https://doi.org/10.1016/j.wasman.2017.04.003.

Zhou, J., Zhu, N., Liu, H. et al. (2019). Recovery of gallium from waste light emitting diodes by oxalic acidic leaching. *Resources, Conservation & Recycling* 146: 366–372. https://doi.org/10.1016/j.resconrec.2019.04.002.

12

Waste Reduction Strategy

E-waste Recycling and Reuse Protocol

Olusola Olaitan Ayeleru[1,], Helen Uchenna Modekwe[2], Matthew Adah Onu[1], Bimbo Lolade Fafowora[3], Ismaila Adejare Adesiyan[2], Tarhemba Tobias Nyam[1], and Peter Apata Olubambi[1]*

[1] *Centre for Nanoengineering and Advanced Material, University of Johannesburg, Doornfontein Campus, Johannesburg, South Africa, 2028, South Africa*
[2] *Renewable Energy and Biomass Research Group, Department of Chemical Engineering, University of Johannesburg, Johannesburg, 2028, South Africa*
[3] *School of Journalism and Media Studies, Rhodes University, South Africa*
* *Corresponding author*

12.1 Introduction

High order levels for gadgets owing to urbanization, economic, technological, and industrial development have resulted in the production of innovative products to meet consumer demand and keep pace with this trend in the world economy (Petridis et al. 2017). The term electronic waste (e-waste) denotes old-fashioned electrical and electronic (EE) appliances running on electricity or a battery, such as lab electrical devices, laptops, desktop computers, televisions, MP3 players, etc. that have been dumped by their owners (Dzah et al. 2022). E-waste consists of rejects from production or refurbishing activities that would not be useful for the original purposes and will need to be thrown away (Kumari and Samadder 2022). E-waste generation is the corollary of technological innovativeness that renders some products obsolete and inoperable in addition to damaged electronic equipment that is disposed of (Gupta et al. 2014). Studies have shown electronic gadgets to be one of the waste streams with the most rapid rate of growth in the modern world (Bungadaeng et al. 2019; Forti et al. 2020). According to some studies, it was estimated that ~54 million tons of e-waste were generated globally in 2020 (Fu et al. 2018; Ghosh et al. 2022; Turaga et al. 2019). By 2040, it is anticipated that this waste stream, which is expanding quickly, will have contributed ~15% of the world's greenhouse gas (GHG) emissions, and by 2050 it will weigh a total of ~120 million tons annually (United Nations Environment Programme Press Release 2019). Although the raw materials in this waste have the potential to be worth up to ~$57 billion, only ~17.5% of e-waste was properly recycled in 2019 (Stefanakis et al. 2021; Wibowo et al. 2021). While the last two decades have witnessed the development of specific "take-back" and processing systems to handle e-waste collected from owners and process it in professional treatment facilities as prescribed by policymakers, manufacturers, and recyclers (Forti et al. 2020), the amount of e-waste being collected and treated in a cutting-edge manner is still limited, and most nations do not even have e-waste management systems. This situation is prevalent in developing countries (DCs) that employ crude and inefficient techniques of extracting components from e-waste (Asante et al. 2019). The objective of this study is to provide an overview of the waste reduction strategy with a special focus on e-waste, reuse, and recycling protocol as a basis for the attainment of a circular economy and measures for complete diversion of e-waste from landfills.

12.2 Recovery of Value-Added Materials from E-waste and the Public Perception

Valuable raw materials are recovered from e-waste recycling, but this process remains underused owing to the complexity of collection and recycling (Shahabuddin et al. 2022). Even though it is becoming a global phenomenon to combat e-waste, new strategies are being sought for appropriate management of this waste. The strategies include use of state-of-the-art design, extended producer responsibility (EPR), development of new ethics and labels, innovative recycling, remanufacturing, reuse, and redesign, alongside the sensitization of the populace on suitable management

Electronic Waste Management: Policies, Processes, Technologies, and Impact, First Edition. Edited by Sunil Kumar and Vineet Kumar.

methods (Kumar and Singh 2014). Social awareness of e-waste can provide an excellent approach that can change the perception of people on reuse, recycling, and innovation that can bring about a reduction in e-waste (Ghosh et al. 2022; Naik and Eswari 2022). Awareness creation could be tailored in a manner that would encourage the protocol of e-waste management among the various players (formal and informal) in the e-waste management scheme. The major stages of e-waste salvaging include production, collection, storage, processing/recovering, transportation, and disposal (Ugwu et al. 2021). Thus, the development and implementation of an efficient system of e-waste are imperative (Tangaran et al. 2021). Hazardous materials are embedded in different e-waste components that affect the environment and humans when it is inappropriately handled (Madanhire et al. 2020). Considering this, the management of e-waste is not solely a government mandate or an individual obligation, but rather a group effort involving collaboration among those with diverse areas of interest to safeguard the ecosystem for both the current and next generation. This requires the enactment of strict rules on the management and disposal of e-waste to curtail the extent of e-waste damage to the environment (Yang et al. 2018). And most importantly, the knowledge and perception of the public with regard to e-waste reuse, recycling, and component recovery via awareness campaigns is very crucial (Nuwematsiko et al. 2021).

12.3 Overview of Waste Reduction Strategy with a Special Focus on E-waste

Solid waste reduction approaches cover the application of appropriate management schemes that promote the collection, sorting, and recycling for the conservation of materials (Ugwu et al. 2021). The exponential rise in e-waste production has attracted concerns in the global community, placing demands on governments of nations to formulate policies that can effectively address issues associated with e-waste management (Murthy and Ramakrishna 2022). These will include designing systems that promote the concept of circular economy (CE) and policies that restrict trans-boundary transportation of e-waste. CE is a substitute for the current linear economy (LE) model of a take-make-consume-dispose system of material utilization and is otherwise known as recycling (Nordic Council of Ministers 2018). Recycling is an inexpensive method of waste utilization requiring a smaller amount of subventions and lower taxes (Ayeleru et al. 2020). Formal recycling of e-waste accounted for ~17.5% globally in 2019 (Van Yken et al. 2021), and recycling and remanufacturing of the waste items to like-new are approaches to waste reduction (Zeng et al. 2016). Rudimentary recycling of e-waste is a common practice among market economy in countries like China, Ghana, India, and Nigeria (Awasthi et al. 2019). Although this process is accompanied by the generation of toxic residues and emissions to the environment, it is still one of the most viable options toward resource utilization (Ayeleru et al. 2020).

The policies of the United Nations (UN) on controlling the movement of hazardous waste and its disposal such as the Basel Convention have been influential in the process of e-waste management (Awasthi et al. 2019; Baldé et al. 2017). Recycling and remanufacturing enhance the use of recovered materials and components to limit raw materials consumption and increase the worth of waste materials. However, other approaches have been outlined in the literature such as life cycle assessment (LCA), material flow analysis (MFA), multi-criteria analysis (McAllister), and EPR for the effective reduction and management of e-waste (Abalansa et al. 2021; Al-Rahmi et al. 2018; Saha et al. 2021). The EPR approach relies heavily on various fees and taxes, such as advance recycling fees (ARFs), product take-back obligations, raw material taxes, pay-as-you-throw (PAYT), and landfill bans (Cai and Choi 2019). These emerging approaches aim to effectively tackle e-waste management for the overall improvement of the recycling of recovered waste materials. For instance, India has launched an aggressive awareness campaign on e-waste through the creation of the Swachh Bharat Mission, the Digital India Mission that has brought great improvement in e-waste management among the various key players (Debnath et al. 2022). In the Netherlands, a nonprofit organization called Wecycle is responsible for the collection and recycling of e-waste on behalf of the producers and importers, which makes the recycling process very efficient (Golsteijn and Valencia Martinez 2017). Development of such structures across different countries will enhance the recycling process efficiency for optimal resource utilization and overall health and environmental benefits (Lamma 2021).

12.4 Circular Economy and E-waste Management

The principle of a circular economy (CE) is to begin to view what was once regarded as "waste" as a vital resource in the production process (Golsteijn and Valencia Martinez 2017). Recycling, repair, and remanufacturing are examples of "circular economy" practices. The outdated "extract, make, use, and dispose of" linear model is replaced. Embracing the CE concept lowers waste production and material extraction. The adoption of a CE is envisaged to give rise to ~6 million new jobs globally and can serve as a source of job creation and re-creation and lead to net job gains

(ILO 2019). Exploring the resourcefulness of waste is done via urban mining, a term commonly employed to describe resource recovery from e-waste since the metals have great economic worth (Van Yken et al. 2021). The concept of CE is the reuse and recycling of materials to minimize raw materials' continuous extraction (Debnath et al. 2022). A market creation concept and a single, shared global agenda is needed for sustainable innovation, characterized as the introduction of sustainable technologies, goods, and services to the market (Virlanuta et al. 2020). The linear pattern of take-make-use-dispose does not provide producers with sufficient incentives to make their products more circular. The CE model takes back waste generated and reuses it in the production process (Geissdoerfer et al. 2017). CE is an emerging way of creating value and wealth thereby extending the lifespan of products through improved design, servicing, and taking back waste from the end of the supply chain to the beginning of the process (Stahel 2016). In CE, items are made to be durable, reusable, and recyclable, and materials for new products are sourced from old ones. Every item is recycled, reused as much as possible, and finally disposed of at its extreme life end (Patwa et al. 2021). CE puts both the economy and environment into consideration for a sustainable system of production, consumption, and recovery (Xavier et al. 2021). Figure 12.1 depicts CE strategies for the e-waste management approach for sustainable development. Urban mining is the newest development in e-waste processing, and implementing it in a formal system is crucial to the development of a sustainable CE (Murthy and Ramakrishna 2022). If efficient technology is deployed into the recycling stream, it will not only produce the finest form of valuables but will also assist in the maintenance of the ecosystem and natural resource conservation (Ahirwar and Tripathi 2021).

12.4.1 Reuse Protocol as a Basis for the Attainment of a Circular Economy

Reuse of e-waste is one of the strategies in maintaining safe environment and attaining a CE. The reuse of e-waste comprises the direct use of secondhand gadgets that have been discarded due to modification or advancement in the technological functions (Saha et al. 2021). Research shows that many electronic gadgets such as mobile phones, computers, and home appliances could be reused by those with lower or no purchasing power for newly made gadgets. This process will not only give them the privilege of having a taste of those gadgets but would also reduce the amount of e-waste production and create economic value from the e-waste. Reuse of e-waste also stands as a good alternative to recycling of e-waste and does not consume energy, which means it is environmentally and economically friendly (Forti et al. 2020). According to OECD (2021), reuse of e-waste extends the products' lifespan, slows the flow of constituent materials through the economy, and the underused equipment could be shared to reduce demand for new gadgets and raw materials for the processing of brand-new gadgets (OECD 2021; Sinha et al. 2018).

12.4.2 Recycling Protocol as a Basis for the Attainment of a Circular Economy

Recycling of e-waste is another route to a CE (Thakur and Kumar 2021). It is a vibrant step toward stopping the

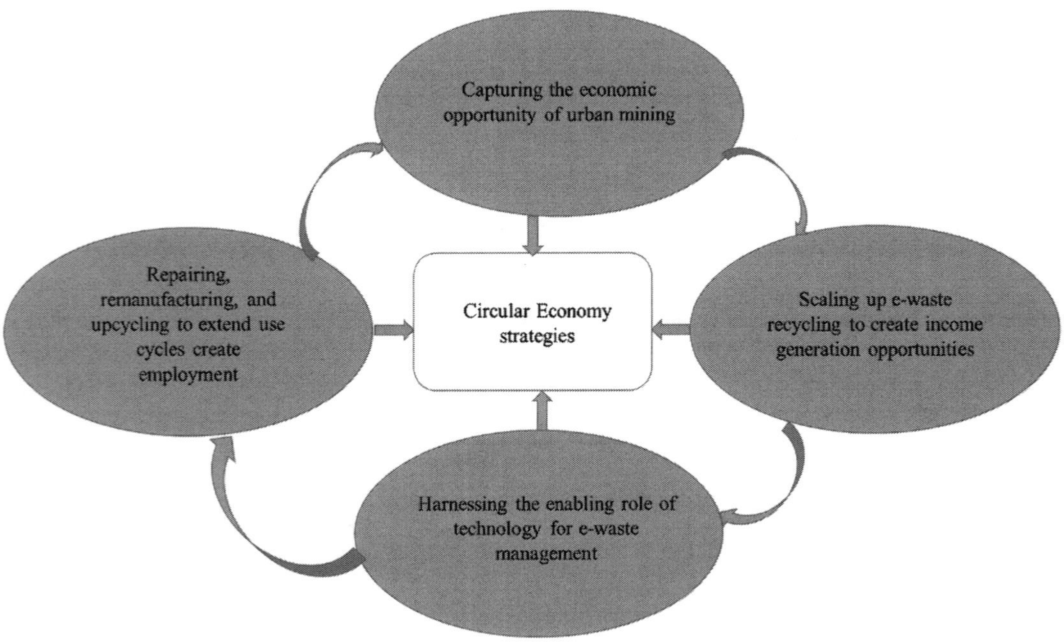

Figure 12.1 Circular economy strategies for electronic waste management.

exploitation of the limited natural resources. The benefits of e-waste recycling include recovery of valuable materials from waste electronic items, safeguarding of limited natural resources, generation of jobs, and reduction of greenhouse gases (GHGs) emission (Ayeleru et al. 2020). With e-waste recycling to recoup valuable components from waste electronics as an alternate to exploitation of the earth's limited natural resources, GHGs emission to the environment is minimized up to ~90% (OECD 2018). Figure 12.2 shows the avoided GHGs emission from various materials. There are three major methods to recycle e-waste based on the kind and makeup of components, and these include chemical recycling, thermal recycling, and the mechanical method of recycling (Kaya 2016; Saha et al. 2021). Figure 12.3 shows the stages in the recycling of e-waste, and Figure 12.4 shows the e-waste management pyramid.

12.4.3 Challenges of the E-waste Recycling Process

Developing nations are continuously faced with toxicity from e-waste due to growing technological advancement, which renders electrical and electronic gadgets obsolete within a very short period, leaving e-waste recycling as the only sustainable option left (Hossain et al. 2015). In developing countries (DCs), the e-waste recycling process is also confronted with diverse challenges including dearth of technological development, etc. The large amount of e-waste generated is mostly managed primitively, which involves manual dismantling, open burning, and acid leaching, which is contributing to release of pollutants into the environment (Han et al. 2019). One other serious issue connected with e-waste recycling is lack of safety and health awareness of recycling

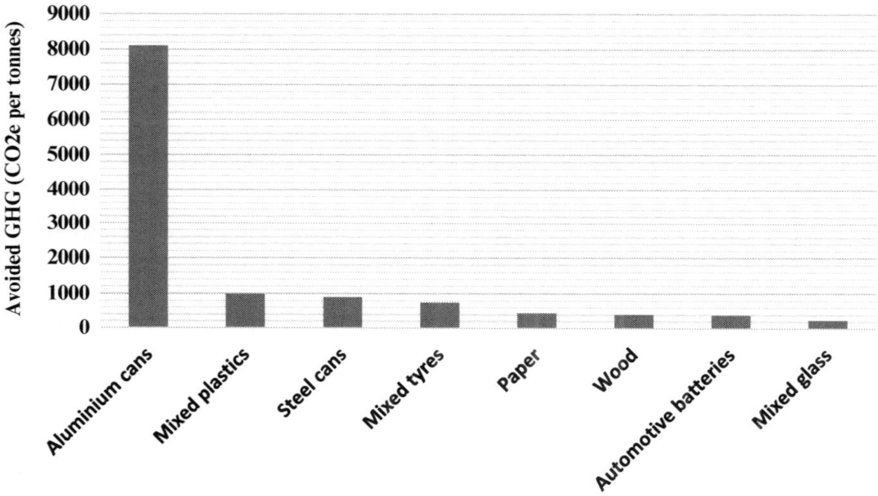

Figure 12.2 Avoided GHG emissions per ton of recycled materials. [This work is shared under a Creative Commons Attribution-Non-Commercial-No Derivative works license, and copyright has been granted by the authors.] (OECD 2018).

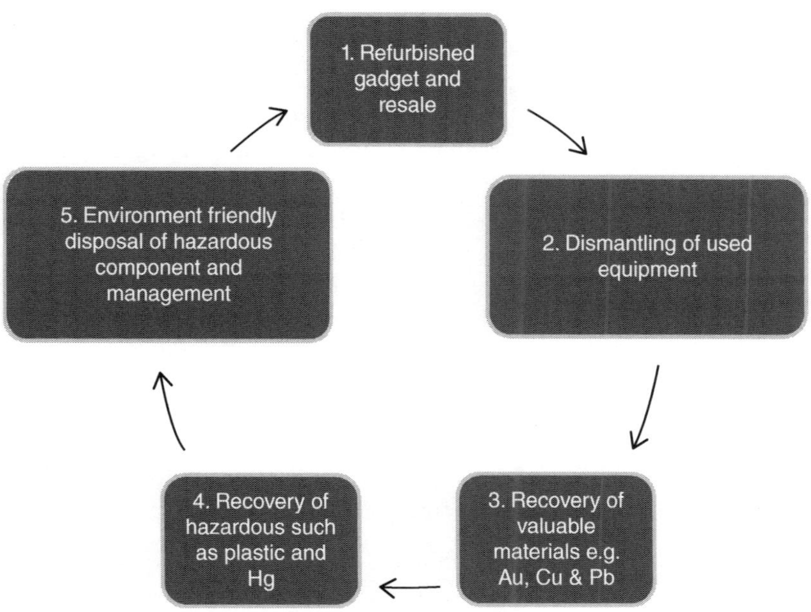

Figure 12.3 Procedures involved in the recycling of e-waste. (Adapted from Saha et al., 2021).

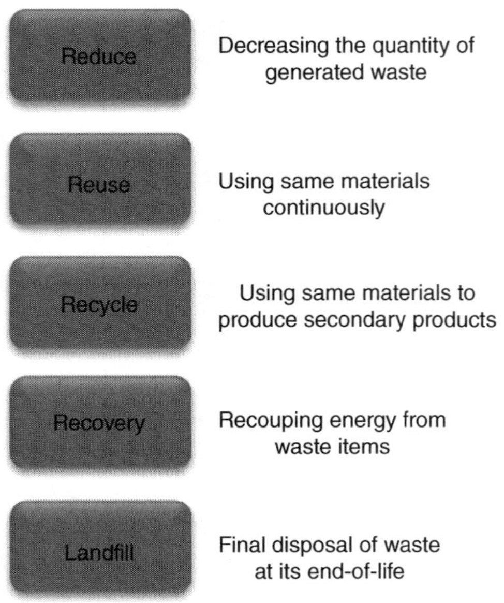

Figure 12.4 E-waste management pyramid. (Reprinted from Naik and Eswari 2022, copyright 2022, with permission from Elsevier Inc).

processes, which exposes recyclers and the public to toxic contaminants from the e-waste recycling process (Bazargan et al. 2012). Also, the fast-growing market of electronic gadgets and upsurge in importation of used gadgets with no proper documentation of the amounts are huge challenges confronting the e-waste management process in DCs. Poor capital investment in the e-waste recycling business and the absence of viable policy and legislation with reference to e-waste handling are other major issues militating against the e-waste recycling process (Nivedha and Sutha 2020).

12.5 Measures for Complete Diversion of E-waste from Disposal Sites

The continuous growth in the information and communication technology (ICT) sector is driven by the advancement in technology and the production of new and advanced electronics products (Roztocki et al. 2019). This has led to the unmanageable global growth rate in e-waste generation. E-waste has been described as the fastest growing waste stream found in municipal solid waste (MSW) (Nnorom and Osibanjo 2008). The lifespan of most household electronic products is relatively short and tends to depreciate in features and capacities due to continuous usage and evolution of better and newer models, translating to a huge amount of e-waste (Gaidajis et al. 2010; Nnorom and Osibanjo 2008). E-waste consists of hazardous materials, which when not properly managed may result in serious environmental and health issues. Therefore, many countries and governments of nations globally have drafted legislation and regulations to reduce and divert e-waste from illegal importation and informal disposal and to encourage sustainable management options such as reuse, recycling, and recovery of components from old-fashioned electronic products (Wirtu and Tucho 2022). Proceeding from the various global experiences, any comprehensive attempt to address e-waste issues or divert e-waste should focus on economy, technical feasibility, eco-friendliness, and sustainability. Evidently, substantial social support from consumers should also be encouraged for such program to be a success (Nnorom and Osibanjo 2008). It has been argued that the convenience of collection/recycling facilities at households is one of the most effective ways of encouraging consumers to recycle their waste electrical and electronic equipment (WEEE) (Shevchenko et al. 2019). The convenient curb-side pickup and door-to-door waste collection by municipalities/recyclers are practiced in some countries, and they have proven to be effective toward the realization of circularity in the waste management (WM) sector (Borthakur and Govind 2017; Shevchenko et al. 2019).

12.5.1 Incentive Models for Increased Recycling

Various approaches have been adopted by the developing and developed economies to motivate consumers against the disposal of WEEE on uncultivated land, water bodies, or landfill sites (LSs), but instead to return the end-of-life (EoL) electronic gadgets to retailers, manufacturers, and recyclers. This idea is aimed at improving the recycling and management of end-of-life electrical and electronics equipment via an incentive mechanism. Incentives have been viewed as an intervention mechanism to increase consumers' involvement in WM and a regulated recycling rate (Dixit and Vaish 2015). Incentives are grouped under economic incentives and non-economic incentives. These instruments are aimed at influencing consumers' behavior and participation toward effective disposal or diversion of e-waste from landfills. Incentive models also promote recycling rates and eventually increase the recovery of valuables from end-of-life electrical and electronics equipment (Gibovic and Bikfalvi 2021). Economic incentives, also known as green economic incentives as defined by Sajid et al. (2022), are "the monetary benefits provided by the government in return for active participation in e-waste recycling programs" (Sajid et al. 2022).

12.5.2 Extended Producer Responsibility (Producer Take-Back) Model

The extended producer responsibility (EPR) model provides incentives to producers/manufacturers to improve their products and processes, and to redesign their products with minimal environmental impacts so that the products can be easily recycled or reused (OECD 2016). EPR policy complies with the polluter-pays principle (PPP), and is characterized by shifting of responsibility away from municipalities by factoring cost of product treatment and disposal into the price of the products, thereby reflecting the environmental impacts of the product (Widmer et al. 2005). It shows that manufacturers' responsibility extends to the various segments of their products' lifespan including the collection, recovery, and final disposal of their products (Kosior and Crescenzi 2020). EPR aims at sustainable development (SD) via environmentally sustainable product development and product recovery; the EPR concept changes the balance of responsibility among the actors-producers, users, authorities, and waste managers in products' life cycle with more attention on the products' end of life (Nnorom and Osibanjo 2008). The elements of the EPR model have been identified and discussed extensively by many studies in the literature as follows (Lindhqvist 2000; Nnorom and Osibanjo 2008; Thompson and Oh 2006):

- Economic responsibility––this means the producer will cover all or part of the cost related to their manufactured product in terms of the collection, recycling, or final disposal/treatment of the products. These expenses could be paid directly by the producer or by a special fee.
- Physical responsibility––this stipulates that manufacturers are involved in the physical management, collection, and disposal of their products. This ranges from simply developing the necessary technology, to managing the total take-back system for collecting and management/disposal of the products.
- Ownership––the manufacturer may retain the ownership of the products throughout their life cycle, and consequently will be linked to the environmental problems of the products. Hence, the producer assumes both physical and economic responsibilities, consequently, the product seems to be leased to the consumer, who only purchases the power to use the products.
- Liability––this describes the manufacturers' responsibility in terms of verifiable environmental damages caused by their products. The producer/manufacturer is, therefore, responsible for any environmental damage/harm caused by their products. The extent of the liability is determined by legislation and may account for the different segments through the products' life cycle, including product usage and final disposal.
- Informative responsibility––the producer is responsible for providing information about the product or its effects at the various stages of its life cycle. That is, the producer provides information on components or materials out in the market, provides third-party audits reports and performance measurements, publishes annual reports, invests in public awareness campaigns, etc.

Different countries employ various strategies/systems in the management of WEEE. A range of financing models operate in the US and Canada such as the visible and invisible fees system (Lepawsky 2012). E-waste management programs in all Canadian provinces have imposed visible Environmental Handling Fees (EHFs) at the point of electronic product purchase (Leclerc and Badami 2020). Revenues obtained from EHFs are used in financing the Producer Responsibility Organisation (PRO) operations such as the cost for research and communications, and cost for collecting, consolidating, and recycling of e-waste (Leclerc and Badami 2020). This shows that consumers are indirectly bearing the financial burden for e-waste management. The producer's real cost of managing their share of their product at end of life is distributed across their products and then shifted completely to the consumers (Leclerc and Badami 2020; Lepawsky 2012). Some state's regulations, like those in Québec, mandate that fees imposed on Original Equipment Manufacturers (OEMs) for their products should be based on the products' recyclability, toxicity, and impact on the environment, life cycle, and recovery/reclamation process. The EHFs fees are unified for a similar EEE category, and there is no cross-financing, so that one category cannot be charged a higher price to reduce the end-of-life management cost of another (Leclerc and Badami 2020).

In China, nationally e-waste recycling and management are funded by producers and other stakeholders. Producers are not responsible for the physical take-back and recovery of their products; rather, retailers, distributors, and formal recyclers are responsible for the e-waste physical collection. Recycling funds sourced from the producers' financial obligation are used for subsidies to formal recyclers based on the amount of e-waste recycled (Kojima et al. 2009). However, part of the recycling cost as mandated in the legislation is borne by the consumer from the time of purchase through the advanced recycling fees (ARFs) and advanced disposal fees (ADFs) (Borthakur and Govind 2017). Generally, in China and other developing nations, consumers are unwilling to shoulder any financial responsibility particularly in terms of disposal services and waste recycling fees because e-waste is traded at a positive value market, where consumers' can benefit directly by selling their outdated and damaged EEE (Kojima et al. 2009).

12.5.2.1 Economic Instruments

Economic instruments that fall under the EPR approach include different types of product fees and taxes, such as advanced recycling fees (ARFs), mandated product take-back, virgin material taxes/subsidies, pay-as-you-throw

(PAYT), waste collection charges, advanced disposal fees (ADFs), landfill bans, etc. (Nnorom and Osibanjo 2008; Widmer et al. 2005). The advanced landfill disposal fees and advanced recycling fees (ARF) are effective WM fees (Hong and Ke 2011). They are taxes/fees often assessed on product sales to encourage separation of waste at sources and to discourage landfill disposal and incineration by making the landfill tax more expensive than the recycling tax (Nnorom and Osibanjo 2008; Widmer et al. 2005). Collected disposal taxes and fees are used to further finance and improve WM in line with the waste hierarchy (Leclerc and Badami 2020; Widmer et al. 2005). ARF and ADF are charged based on potential toxicity of the product or based on weight (Widmer et al. 2005). The PAYT scheme charges waste producers based on the actual amount of waste they generate and provides incentives to consumers for separating their waste at source or for reducing mixed waste. This scheme has recorded success in some countries, especially regarding packaging waste (Sajid et al. 2022; Widmer et al. 2005).

Non-economic incentives are also used to stimulate consumers to return their old-fashioned EEE products to the manufacturer, supplier, or retailers, who then retrieve the products for repurposing, recycling, or reclamation. Non-economic incentives are classified as financial incentives, social incentives, and environmental incentives (Dixit and Vaish 2015).

- Financial incentives: These are in the form of monetary incentives offered to consumers for opting to offer their product for recycling. Producers, suppliers, or recyclers also offer discounts to consumers on the purchase of new products with the return of former appliances. Financial incentives are mostly in the form of cash back, discount in exchange/trade-in for buying another product, discount coupons, gift cards, etc.
- Social incentives offered to consumers could be in the form of donations in support of societal needs, research for treatment, or funds for charity and nonprofit organizations, and in exchange for the consumers' end-of-life product.
- Environmental incentives: In environmental incentives, consumers donate their old electronics to manufacturing or recycling organizations in which money raised is used in environmentally sustainable projects such as in the planting of trees.

Other viable repair incentives have been practiced in some developed countries such as repair bonuses and free repair cafes. The repair bonuses have been practiced in Austria and in Thuringia State in Germany. The initiative is designed to repay consumers the repair cost of their faulty appliances, and also to reward consumers for choosing to repair and maintain their faulty appliances rather than disposing of them. The repair café is usually undertaken by volunteers, and consumers are assisted with free repair of their defective and damaged electrical and electronic devices (Bateman 2022).

It is often argued that non-economic incentive systems can only function in developed countries where the producers' and manufacturers' business model is already operating with circular products and design (Shevchenko et al. 2019, 2021). These businesses have already included circularity strategies (refurbishing and remaking) in their products' initial design, thereby, they are willing to offer monetary incentives to consumers to ensure consistent and proper collection of end-of-life devices (Shevchenko et al. 2021).

12.5.3 Donation of Secondhand Electronic Gadgets

EEE has substantial value, and there are possibilities that EEE devices are being transferred from one user to another more than once, usually beyond their official lifespan until they eventually end up with the informal recyclers, in storage bins, or at landfills (Yedla 2016). Although the various material flows make it difficult to estimate the actual material flow of EEE and e-waste generation in a city or country, a study by Yedla (2016) has described the various stages of electrical and electronics gadget usage before they reaches their final stage of disposal in terms of "their primary use (life time), secondary use (first degree extension of life), third-hand use (second degree extension of life), donation, retention at the respective places (without discarding), fraction shifted to scrap vendor, and the components reaching the final dumpsite from various end points of use" (Yedla 2016). Donation of outdated EEE appliances to individuals or organizations is an important disposal strategy that gives the poor and disadvantaged in society the opportunity to be part of the new digitalized society and globalized interconnected world, which is in line with the e-waste strategy as outlined by the integrated solid waste management (ISWM) framework (Owusu-Sekyere et al. 2018). Developed countries also use this avenue to dispose of their outdated devices to the developing countries (DCs) as "donations" or "aides" with the aim of making the technologies available to the countries that cannot import new devices due to financial limitations (Owusu-Sekyere et al. 2018). Nevertheless, this is a breach of the Basel Convention agreement, which prohibits cross-border trade of e-waste and hazardous waste (Owusu-Sekyere et al. 2018). In line with the control of trans-boundary trade and movement of hazardous and e-waste, the Basel Convention inaugurated the Partnership for Action on Computing Equipment (PACE) to assist with the management of used and EoL of computing equipment (Perkins et al. 2014). The PACE provided guidelines on functional equipment considered suitable for charitable donation, which when followed accordingly could reduce the volume of damaged and WEEE, especially computers and computer components that were branded and exported to Asia and Africa as donated functional equipment (Perkins et al. 2014).

Singh et al. (2018) conducted e-waste management assessment in Chandigarh, India. In their quantitative survey study, they observed that the majority of e-waste disposal pattern for used EEE devices that were still in good working condition by most high and middle income earners' households in Chandigarh area were often through donations to friends, relatives, or nonprofit organizations (NGOs), and through selling to secondhand dealers, stored, or traded in for another device. The disposal channel for damaged devices were via selling to service centers, dumping in dustbins, or selling to scrap dealers, etc. The disposal strategy through donation of consumers' appliances to the poor, relatives, and NGOs could be due to the donators' affluence and ability to purchase the latest version of gadgets without earning an income from the sale of the old ones (Singh et al. 2018).

12.6 Conclusion

In this study, an overview of e-waste reduction strategies with a special attention on reuse and recycling protocols as the basis for the attainment of a circular economy (CE) has been discussed. In the CE concept, waste is eliminated from the process and pollution is reduced completely. E-waste is best managed under the concept of CE since more of the outdated electronics will be recycled and components recovery rates shall increase. To achieve CE, EPR, incentive models, donating secondhand electronic gadgets to nonprofit organizations or friends are very crucial to achieving complete diversion of e-waste from landfills, and in that way, sustainability will be attained in the e-waste management sector.

Acknowledgement

This work is based on the research supported in part by the National Research Foundation of South Africa (Ref Numbers: SASPD22091357540).

References

Abalansa, S., El Mahrad, B., Icely, J., and Newton, A. (2021). Electronic waste, an environmental problem exported to developing countries: the GOOD, the BAD and the UGLY. *Sustainability* 13 (9): 5302.

Ahirwar, R. and Tripathi, A.K. (2021). E-waste management: a review of recycling process, environmental and occupational health hazards, and potential solutions. *Environmental Nanotechnology, Monitoring & Management* 15: 100409.

Al-Rahmi, W., Alias, N., Ali, A. et al. (2018). Strategies to manage electronic waste approaches: an overview in East Asia. *International Journal of Engineering & Technology* 7 (4): 3268–3275.

Asante, K.A., Amoyaw-Osei, Y., and Agusa, T. (2019). E-waste recycling in Africa: risks and opportunities. *Current Opinion in Green and Sustainable Chemistry* 18: 109–117.

Awasthi, A.K., Li, J., Koh, L., and Ogunseitan, O.A. (2019). Circular economy and electronic waste. *Nature Electronics* 2 (3): 86–89.

Ayeleru, O.O., Dlova, S., Akinribide, O.J. et al. (2020). Challenges of plastic waste generation and management in sub-Saharan Africa: a review. *Waste Management* 110: 24–42.

Baldé, C.P., Forti, V., Gray, V. et al. (2017). *The Global E-waste Monitor 2017: Quantities, Flows and Resources*. Bonn/Geneva/Vienna: United Nations University (UNU), International Telecommunication Union (ITU), and International Solid Waste Association (ISWA).

Bateman, K. (2022 January 28). E-waste is booming, but repair incentives can help keep goods out of landfills. Retrieved from https://www.greenbiz.com/article/e-waste-booming-repair-incentives-can-help-keep-goods-out-landfills (accessed 20 August 2023).

Bazargan, A., Lam, K.F., and McKay, G. (2012). Challenges and opportunities of e-waste management. In: *E-Waste: Management, Types and Challenges* (eds. Y.C. Li and B.L. Wang), 39–66.

Borthakur, A. and Govind, M. (2017). Emerging trends in consumers' E-waste disposal behaviour and awareness: a worldwide overview with special focus on India. *Resources, Conservation and Recycling* 117: 102–113.

Bungadaeng, S., Prueksasit, T., and Siriwong, W. (2019). Inhalation exposure to respirable particulate matter among workers in relation to their e-waste open burning activities in Buriram Province, Thailand. *Sustainable Environment Research* 29 (1): 1–12.

Cai, Y. and Choi, T. (2019). Extended producer responsibility: a systematic review and innovative proposals for improving sustainability. *IEEE Transactions on Engineering Management* 68 (1): 272–288.

Debnath, B., Das, A., and Das, A. (2022). Towards circular economy in e-waste management in India: issues, challenges, and solutions. In: *Circular Economy and Sustainability: Vol. 2* (eds. A. Stefanakis and I. Nikolaou), 523–543. Elsevier.

Dixit, S. and Vaish, A. (2015). Perceived barriers, collection models, incentives and consumer preferences: an exploratory study for effective implementation of reverse logistics. *International Journal of Logistics Systems and Management* 21 (3): 304–318.

Dzah, C., Agyapong, J.O., Apprey, M.W. et al. (2022). Assessment of perceptions and practices of electronic waste management among commercial consumers in Ho, Ghana. *Sustainable Environment* 8 (1): 2048465.

Forti, V., Baldé, C.P., Kuehr, R., and Bel, G. (2020). *The Global E-waste Monitor 2020*. Bonn, Geneva, and Rotterdam: United Nations University/United Nations Institute for Training and Research, International Telecommunication Union, and International Solid Waste Association.

Fu, J., Zhang, H., Zhang, A., and Jiang, G. (2018). E-waste recycling in China: a challenging field. *Environmental Science & Technology*. https://doi.org/10.1021/acs.est.8b02329.

Gaidajis, G., Angelakoglou, K., and Aktsoglou, D. (2010). E-waste: environmental problems and current management. *Journal of Engineering Science and Technology Review* 3 (1): 193–199.

Geissdoerfer, M., Savaget, P., Bocken, N.M.P., and Hultink, E.J. (2017). The circular economy––a new sustainability paradigm? *Journal of Cleaner Production* 143: 757–768.

Ghosh, B.K., Mekhilef, S., Ahmad, S., and Ghosh, S.K. (2022). A review on global emissions by e-products based waste: technical management for reduced effects and achieving sustainable development goals. *Sustainability* 14 (7): 4036.

Gibovic, D. and Bikfalvi, A. (2021). Incentives for plastic recycling: how to engage citizens in active collection. Empirical evidence from Spain. *Recycling* 6 (2): 29.

Golsteijn, L. and Valencia Martinez, E. (2017). The circular economy of E-waste in the Netherlands: optimizing material recycling and energy recovery. *Journal of Engineering* 2017.

Gupta, S., Modi, G., Saini, R., and Agarwala, V. (2014). A review on various electronic waste recycling techniques and hazards due to its improper handling. *International Refereed Journal of Engineering and Science (IRJES)* 3 (5): 05–17.

Han, Y., Tang, Z., Sun, J. et al. (2019). Heavy metals in soil contaminated through e-waste processing activities in a recycling area: implications for risk management. *Process Safety and Environmental Protection* 125: 189–196.

Hong, I. and Ke, J. (2011). Determining advanced recycling fees and subsidies in "E-scrap" reverse supply chains. *Journal of Environmental Management* 92 (6): 1495–1502.

Hossain, M.S., Al-Hamadani, S.M.Z.F., and Rahman, M.T. (2015). E-waste: a challenge for sustainable development. *Journal of Health and Pollution* 5 (9): 3–11.

ILO (2019). Decent work in the management of electrical and electronic waste (E-waste). Paper presented at the Global Dialogue Forum on Decent Work in the Management of Electrical and Electronic Waste (E-Waste). Geneva.

Kaya, M. (2016). Recovery of metals and nonmetals from electronic waste by physical and chemical recycling processes. *Waste Management* 57: 64–90.

Kojima, M., Yoshida, A., and Sasaki, S. (2009). Difficulties in applying extended producer responsibility policies in developing countries: case studies in e-waste recycling in China and Thailand. *Journal of Material Cycles and Waste Management* 11 (3): 263–269.

Kosior, E. and Crescenzi, I. (2020). Solutions to the plastic waste problem on land and in the oceans. In: *Plastic Waste and Recycling Environmental Impact, Societal Issues, Prevention, and Solutions* (ed. T.M. Letcher), 415–446. Elsevier.

Kumar, U. and Singh, D.N. (2014). Electronic waste: reduce, reuse, recycle & remove (R's) concept and approach. *International Journal of Engineering Research & Technology* 3 (2): 1472–1478.

Kumari, R. and Samadder, S.R. (2022). A critical review of the pre-processing and metals recovery methods from e-wastes. *Journal of Environmental Management* 320: 115887.

Lamma, O.A. (2021). The impact of recycling in preserving the environment. *IJAR* 7 (11): 297–302.

Leclerc, S.H. and Badami, M.G. (2020). Extended producer responsibility for E-waste management: policy drivers and challenges. *Journal of Cleaner Production* 251: 119657.

Lepawsky, J. (2012). Legal geographies of e-waste legislation in Canada and the US: jurisdiction, responsibility and the taboo of production. *Geoforum* 43 (6): 1194–1206.

Lindhqvist, T. (2000). Extended producer responsibility in cleaner production: policy principle to promote environmental improvements of product systems. Doctoral dissertation. Lund University, Sweden.

Madanhire, I., Mugwindiri, K., and Mbohwa, C. (2020). Evaluating the feasibility of effective E waste management: case study. Paper presented at the Proceedings of the International Conference on Industrial Engineering and Operations Management.

Murthy, V. and Ramakrishna, S. (2022). A review on global E-waste management: urban mining towards a sustainable future and circular economy. *Sustainability* 14 (2): 647.

Naik, S. and Eswari, J.S. (2022). Electrical waste management: recent advances challenges and future outlook. *Total Environment Research Themes* 1–2: 100002.

Nivedha, R. and Sutha, A.I. (2020). The challenges of electronic waste (e-waste) management in India. *European Journal of Molecular & Clinical Medicine* 7 (3): 4583–4588.

Nnorom, I.C. and Osibanjo, O. (2008). Overview of electronic waste (e-waste) management practices and legislations, and their poor applications in the developing countries. *Resources, Conservation and Recycling* 52 (6): 843–858.

Nordic Council of Ministers (2018). *Policy Brief: Recycling in the Circular Economy. How to Improve the Recycling Markets for Construction Materials, Biowaste, Plastics and Critical Metals*. Retrieved from https://norden.diva-portal.org/smash/get/diva2:1269435/FULLTEXT01.pdf (accessed 21 July 2022).

Nuwematsiko, R., Oporia, F., Nabirye, J. et al. (2021). Knowledge, perceptions, and practices of electronic waste management among consumers in Kampala, Uganda. *Journal of Environmental and Public Health* 2021.

OECD (2016). *Extended producer responsibility. Updated guidance for efficient waste management.* Retrieved from https://www.oecd.org/environment/waste/Extended-producer-responsibility-Policy-Highlights-2016-web.pdf (accessed 21 July 2022).

OECD (2018). Business models for the circular economy opportunities and challenges from a policy perspective. Retrieved from https://www.oecd.org/environment/waste/policy-highlights-business-models-for-the-circular-economy.pdf (accessed 21 July 2022).

OECD (2021). Towards a more resource-efficient and circular economy. The role of the G20. Retrieved from https://www.oecd.org/env/waste/OECD-G20-Towards-a-more-Resource-Efficient-and-Circular-Economy.pdf (accessed 21 July 2022).

Owusu-Sekyere, E., Peprah, K., and Demuyakor, R.M. (2018). Exploring the dynamics of E-waste disposal strategies in Tamale, Ghana. *Ghana Journal of Development Studies* 15 (2): 168–192.

Patwa, N., Sivarajah, U., Seetharaman, A. et al. (2021). Towards a circular economy: an emerging economies context. *Journal of Business Research* 122: 725–735.

Perkins, D.N., Drisse, B., and Mn, N. (2014). E-waste: a global hazard. annals of global health. *Annals of Global Health* 80 (4): 286–295. https://doi.org/10.1016/j.aogh.2014.10.001.

Petridis, K., Petridis, N., Stiakakis, E., and Dey, P. (2017). Investigating the factors that affect the time of maximum rejection rate of e-waste using survival analysis. *Computers & Industrial Engineering* 108: 15–26.

Roztocki, N., Soja, P., and Weistroffer, H.R. (2019). The role of information and communication technologies in socioeconomic development: towards a multi-dimensional framework. *Information Technology for Development* 25 (2): 171–183.

Saha, L., Kumar, V., Tiwari, J. et al. (2021). Electronic waste and their leachates impact on human health and environment: global ecological threat and management. *Environmental Technology & Innovation* 24: 102049.

Sajid, M., Zakkariya, K.A., and Joy, H. (2022). Determinants of E-waste recycling intention in India: the influence of environmental concern, attitude and economic incentives. *Colombo Business Journal* 13 (1).

Shahabuddin, M., Uddin, M.N., Chowdhury, J.I. et al. (2022). A review of the recent development, challenges, and opportunities of electronic waste (e-waste). *International Journal of Environmental Science and Technology* 20 (4): 4513–4520.

Shevchenko, T., Laitala, K., and Danko, Y. (2019). Understanding consumer E-waste recycling behavior: introducing a new economic incentive to increase the collection rates. *Sustainability* 11 (9): 2656.

Shevchenko, T., Saidani, M., Danko, Y. et al. (2021). Towards a smart E-waste system utilizing supply chain participants and interactive online maps. *Recycling* 6 (1): 8.

Singh, M., Thind, P.S., and John, S. (2018). An analysis on e-waste generation in Chandigarh: quantification, disposal pattern and future predictions. *Journal of Material Cycles and Waste Management* 20 (3): 1625–1637.

Sinha, R., Chauhan, G., Singh, A. et al. (2018). A novel eco-friendly hybrid approach for recovery and reuse of copper from electronic waste. *Journal of Environmental Chemical Engineering* 6 (1): 1053–1061.

Stahel, W.R. (2016). The circular economy. *Nature* 531 (7595): 435–438.

Stefanakis, A.I., Calheiros, C.S.C., and Nikolaou, I. (2021). Nature-based solutions as a tool in the new circular economic model for climate change adaptation. *Circular Economy and Sustainability* 1 (1): 303–318.

Tangaran, J., Patimo, D.M., and Dollado, R.F. (2021). Disposal management practices of unserviceable electronic resources. *Sriwijaya Journal of Environment* 6 (1): 8–12.

Thakur, P. and Kumar, S. (2021). Evaluation of e-waste status, management strategies, and legislations. *International Journal of Environmental Science and Technology* 19 (7): 6957–6966.

Thompson, S. and Oh, S. (2006). Do sustainable computers result from design for environment and extended producer responsibility? Analyzing e-waste programs in Europe and Canada. Paper presented at the Proceedings of the International Solid Waste Association's 2006 Annual Congress, Copenhagen, Denmark.

Turaga, R.M.R., Bhaskar, K., Sinha, S. et al. (2019). E-waste management in India: issues and strategies. *Vikalpa* 44 (3): 127–162.

Ugwu, C.O., Ozoegwu, C.G., Ozor, P.A. et al. (2021). Waste reduction and utilization strategies to improve municipal solid waste management on Nigerian campuses. *Fuel Communications* 9: 100025.

United Nations Environment Programme (2019). *UN report: time to seize opportunity, tackle challenge of e-waste.* Press release on 24 January 2019.

Van Yken, J., Boxall, N.J., Cheng, K.Y. et al. (2021). E-waste recycling and resource recovery: a review on technologies, barriers and enablers with a focus on oceania. *Metals* 11 (8): 1313.

Virlanuta, F.O., David, S., and Manea, L.D. (2020). The transition from linear economy to circular economy: a behavioral change. *FORCE: Focus on Research in Contemporary Economics* 1 (1): 4–18.

Wibowo, N., Piton, J.K., Nurcahyo, R. et al. (2021). Strategies for improving the e-waste management supply chain sustainability in Indonesia (Jakarta). *Sustainability* 13 (24): 13955.

Widmer, R., Oswald-Krapf, H., Sinha-Khetriwal, D. et al. (2005). Global perspectives on e-waste. *Environmental Impact Assessment Review* 25 (5): 436–458.

Wirtu, Y.D. and Tucho, G.T. (2022). E-waste: growing environmental and health problems and its management alternatives in developing countries. *Environmental Reviews* 30 (4): 524–536.

Xavier, L.H., Ottoni, M., and Lepawsky, J. (2021). Circular economy and e-waste management in the Americas: Brazilian and Canadian frameworks. *Journal of Cleaner Production* 297: 126570.

Yang, H., Ma, M., Thompson, J.R., and Flower, R.J. (2018). Waste management, informal recycling, environmental pollution and public health. *Journal of Epidemiology and Community Health* 72 (3): 237–243.

Yedla, S. (2016). Development of a methodology for electronic waste estimation: a material flow analysis-based SYE-waste model. *Waste Management & Research* 34 (1): 81–86.

Zeng, X., Gong, R., Chen, W., and Li, J. (2016). Uncovering the recycling potential of "new" WEEE in China. *Environmental Science & Technology* 50 (3): 1347–1358.

13

Effective Utilization of E-waste in Advanced Energy Technology Processes

*Barnali Bhui, Shekhar Jyoti Pathak, and Prabu Vairakannu**

Department of Chemical Engineering, Indian Institute of Technology Guwahati, Assam, 781039, INDIA
**Corresponding author*

13.1 Introduction

The intense global competition with rapid technological innovation has led to e-waste (EW) production in enormous quantity. The reduction in the lifespan of electronic products further accelerates the EW disposal rate. It is expected that the production rate will further escalate by 3.0–5.0% per year; thus, waste will be generated at a higher frequency than in the present scenario (Arya and Kumar 2020). The lifespan of electrical equipment is reduced from 8 to 2 years, while it is reduced from 4 years to 9 months for mobile phones (Zhang et al. 2012). The fate of EW materials can either be reuse, recycling, or disposal. EW cannot be disposed of like municipal waste due to the presence of heavy metals and flame retardants. A large amount of EW generation is usually accompanied by a lower recovery rate, as only 20.0% of the total EW is recycled worldwide (Yang et al. 2021). The remaining amounts are either dumped in open landfills or incinerated, thereby releasing hazardous components (Cd, Hg, Pb, Cr, Ni, Hg, etc.) into the surroundings (Islam et al. 2020). Open burning of EW emits organophosphate flame retardants in the range of 5.22×103 to 9.27×10^4 ng/g that are carcinogenic and cause land pollution (Li et al. 2019). Landfills require a large area for disposal, which is a constraint due to the limited land availability and causes environmental concerns. Thus, solid waste management researchers have shifted their focus from landfilling options to metals and energy recovery from EW materials (Jadhao et al. 2021). Discarded materials can be reprocessed to produce potential attracting products, which can be further utilized in several applications. It is estimated that the global e-waste production will reach up to 74.7 million metric tons (MMT) by 2030 since it already reached 53.6 MMT in 2019 (Jadhao et al. 2021). India is ranked as the third-largest EW producer globally. Table 13.1 shows the list of recyclers in Indian states, where Maharashtra contains the highest dismantlers when compared to other Indian states (116 metric tons per year).

Table 13.2 shows the proximate and ultimate analysis of various EW materials. It is seen that the ash content ranges from 0.5 to 70.0%. The ash content in EW is higher due to the presence of metals. These discarded waste materials have a heating value ranging from 6 to 38 MJ/kg, making them desirable energy sources that can be recovered through various thermo-chemical processes. Due to the absence of any metals in the keyboard, the calorific value is found to be the highest. PCB from discarded EW materials has attracted the main attention due to its abundant metallic contents. A PCB typically contains chips, connectors, capacitors, transistors, glass fibers, reinforcement materials, etc. (Yang et al. 2021). The metallic content is mainly Fe, Cu, Ni, Sn, Pb, etc. (Marques et al. 2013). Thus, there is an increasing need for innovative technologies to utilize the EW instead of dumping it into the open landfills.

This chapter highlights e-waste treatment techniques using conventional and advanced technologies to extract value-added products. The overview of different EW components is discussed. Further, comprehensive insight on different methods (pyrometallurgy, hydrometallurgy, biometallurgy) for extracting metals from EW is presented. The detailed mechanism of oil extraction using liquefaction is provided. An emphasis is also given to integrating the metal extraction process with power generation units using advanced technologies. Thus, novel hybrid methods are discussed for complete EW utilization (both plastic and metallic fractions). This chapter discusses the economic and environmental aspects of utilizing EW, and an exhaustive literature survey is performed based on the approaches

Electronic Waste Management: Policies, Processes, Technologies, and Impact, First Edition. Edited by Sunil Kumar and Vineet Kumar.

Table 13.1 List of dismantlers in India by state. (Data collected from Central Pollution Control Board (CPCB) as of 06.12.2021.)

State	Number of authorized recyclers	Installed capacity (metric tons per year)
Andhra Pradesh	8	32 212.5
Assam	1	120
Chhattisgarh	2	6750
Delhi	2	120
Gujarat	33	84 301.9
Goa	1	103
Haryana	42	124 281.6
Himachal Pradesh	2	1500
Jammu and Kashmir	3	705
Jharkhand	2	660
Karnataka	71	52 842
Kerala	1	1200
Maharashtra	116	106 280.5
Madhya Pradesh	2	9600
Orissa	5	5690
Punjab	7	9492
Rajasthan	23	83 254
Tamil Nadu	32	132 049
Telengana	16	91 563
Uttar Pradesh	89	494 042.7
Uttarakhand	6	147 305
West Bengal	4	1950

for recovering metals and plastic contents from different types of EW and recent developments to tackle the management of EW. Finally, the challenges and opportunities in utilizing EW are outlined.

13.2 E-waste Treatment Methods

To date, effective utilization of EW through appropriate channels is not achieved such that complete EW resources could be attained. Conventional EW recycling technologies are mainly categorized as mechanical, chemical, and thermal (Khaobang et al. 2022). Mechanical recycling includes dismantling, shredding, grinding, and separation methods (magnetic, density based, or electrical conductivity based), while chemical recycling requires certain chemicals for metals extraction (aqua regia leaching) (Debnath et al. 2018). These methods are expensive and are mainly focused on metal extraction by ignoring the fate of the plastics associated with the EW. Thermal recycling methods are superior to these methods in terms of energy and material recovery. This chapter explores the different thermal treatment pathways for treating EW as shown in Figure 13.1. Different categories of EW, such as discarded television sets, mobile phones, calculators, music systems, cameras, printed circuit boards (PCB), household electrical items, etc., are mostly available in landfills. The EWs are a good source of plastics (around 28.0%), along with metals and composite materials, as seen in Table 13.2. EW disposal through conventional treatment methods (incineration, landfill, etc.) fails to utilize the

Table 13.2 Proximate and ultimate analysis of different types of e-waste.

Analysis	Composition	Electrical cable (Anis et al. 2018)	Keyboards (Quan et al. 2013)	Printed circuit boards (Evangelopoulos et al. 2015)	Mobile phones (Annamalai and Gurumurthy 2021)	Printer cartridge plastic (Pathak et al. 2022)
Proximate analysis (wt%, wet basis)	Moisture	0.2	0.4	0.2	0.9	0.8
	Volatile matter	70.8	96.2	20.2	42.3	94.3
	Ash	21.6	0.5	68.4	46.5	1.1
	Fixed carbon	7.4	2.9	11.4	2.7	3.7
Ultimate analysis (wt%, dry and ash free basis)	C	42.6	83.7	18.9	69.3	85.5
	H	3.4	7.7	1.9	6.2	8.2
	O	53.3	2.8	5.9	21.1	5.6
	N	0.5	5.8	0.6	2.6	0.7
	S	0.2	-	0.06	0.3	-
Heating value (MJ/kg)		30.8	-	6.77	12.3	37.9

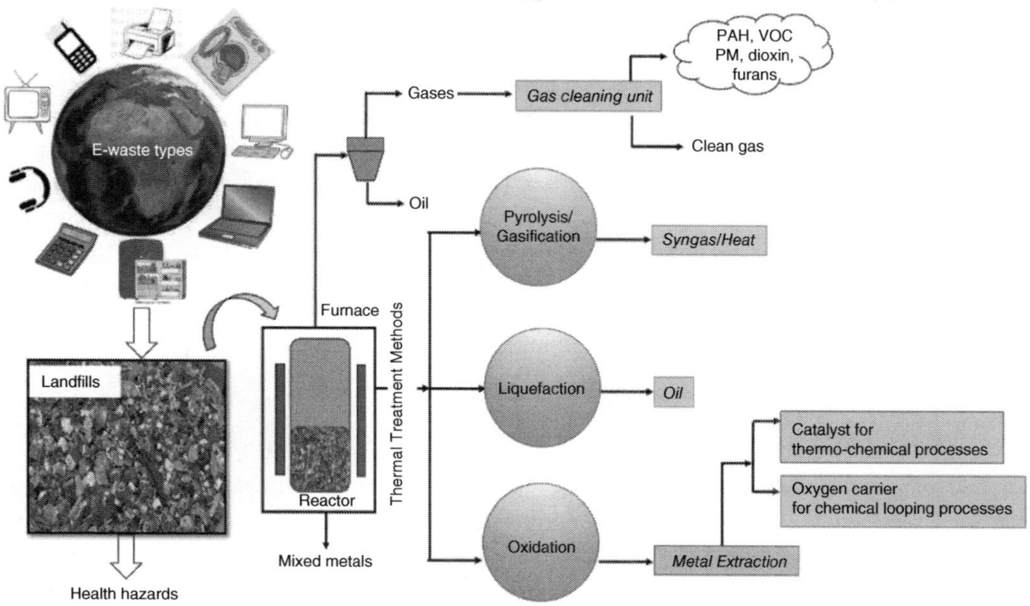

Figure 13.1 Types of thermal treatment pathways for treating e-waste to derive value-added products (PAH: polyaromatic hydrocarbons, VOC: volatile organic compounds, PM: particulate matter).

organic content for energy recovery, and the EW further pollutes the environment (Li and Xu 2019). Thus, appropriate technologies need to be implemented to treat the EW judiciously, as reduction in EW generation is nearly impossible. Plastic resins (mainly thermoplastics) existing in various electronic scraps can be utilized for energy production via thermochemical conversion methods. The thermal processes are chosen on the basis of the desirable products. Pyrolysis is usually conducted in an inert atmosphere to obtain syngas and char, while the addition of gasifying agent generates rich syngas during EW gasification. Liquefaction is an attractive technique to convert EW materials into valuable liquid products by considering milder operating conditions compared to pyrolysis. Direct liquefaction of EW solids into transportable liquid (oil) can be a potential alternative to the dependency on fossil fuels in transportation. Thus, the thermo-chemical waste treatment processes (pyrolysis, gasification, or liquefaction) have the ability to convert solid waste into value-added products such as syngas and oil after passing through a cleaning unit. Further, the metallic content in EW gets separated out easily with no additional cost. These metals then can either be used as a catalyst or as an oxygen carrier in chemical looping processes.

13.2.1 Incineration, Hydrothermal, and Biometallurgy Treatment for Metal Extraction

The existing conventional technologies, such as incineration, pyrolysis, hydrothermal, and bio-metallurgical methods, are used to recover metals (Cu, Br, Sb, etc.) from EW materials (Das et al. 2021). Hydrothermal treatment is performed in the presence of subcritical/supercritical water or certain liquid solvents as the reaction medium. The treatment is usually conducted in a closed autoclave. The pyrometallurgical process enables the segregation of metals from the e-waste with a recovery rate of over 95.0% (Hagelüken 2006). The bioleaching process includes chemolithotropic, heterotropic microorganisms to separate metals from EW. However, these methods are time consuming and have low metal extraction efficiency (Islam et al. 2020).

13.2.1.1 Biometallurgy Treatment

Certain microorganisms (bacteria and fungi) have the capability to recover metals from EW through acidolysis and complexation processes (Islam et al. 2021). These microorganisms generate organic acids, thereby enhancing the leaching rate (Hao et al. 2020). Microbes such as *Acidithiobacillus thioxidans, Acidithiobacillus, Thiobacillus ferooxidans*, etc. mainly, are used for the bioleaching processes (Pant and Dhiman 2020). This technology utilizes the leaching ability of these microorganisms to separate out metals and recover them from the solution. This method is employed due to lower environmental impacts and a lower carbon footprint on the environment when compared to other existing methods. However, maintaining optimum conditions for the growth culture microorganisms (temperature, pH, aeration rate, operation time, etc.) is required to improve the recovery efficiency. Zhu et al. (2011) observed that acidophilic bacteria extracted 96.8% Cu, 88.2% Al, and 91.6% Zn at a leaching period of 98 hours. Kaur et al. (2022)

studied the bioleaching of PCB using different microbes to extract Cu and Fe. The Cu recovery was found in the range of 10–18%, while Fe recovery was 9–13.6%. Thus, it is recommended to integrate the biological method with either the physical or chemical process in order to reduce the total operation time and achieve a high recovery yield.

13.2.1.2 Incineration

Incineration was preferred earlier due to efficient volume reduction and energy recovery since it involves open burning. However, the release of airborne contaminants has limited its usage, such as volatile organic compounds, HCl, dioxins, furans, particulate matter, etc. The electric cables are usually coated with polyvinyl chloride, which, when combusted, generates toxic chemicals (Arain et al. 2022). The furans and dioxins are liberated due to the presence of polybrominated biphenyl or polybrominated diphenyl ether in the flame retardants (Evangelopoulos et al. 2015). The brominated flame retardants are used in all EW to prevent it from fire for safety purposes. Pyrolysis is superior to incineration due to the limited release of these pollutants (Cui et al. 2021), which are liberated due to the decomposition of the brominated epoxy resins. The mechanism of furans formation from brominated epoxy resins at high temperatures is shown in Figure 13.2. The dibenzofurans gets converted into bisphenol, isopropyl phenol, etc. at lower temperature, while it gets degraded into phenol, o-cresol, p-cresol with no bromine (Das et al. 2021). A similar observation is given by Evangelopoulos et al. (2015), in which they identified that phenolic components are abundantly present in the tar after electronic waste pyrolysis, mainly in the form of styrene, bromophenol, bisphenol, etc. However, attention needs to be paid to gas purification to completely remove furans, dioxins, mainly brominated organic pollutants from the other gaseous constituents. Sahle-Demessie et al. (2021) investigated EW pyrolysis and observed the release of harmful components such as polyaromatic hydrocarbons, and heavy metals such as Cd, Cr, Pb, Hg, etc. They recovered energy of 30 MJ/kg by the thermal treatment process. Demirbas (2004) conducted pyrolysis of mixed plastics and concluded that waste materials should not contain chlorine content of more than 1.0% in order to obtain high-quality oil and char. These results indicate that high-temperature thermal processes should be favored for separating metals from EW and harness energy from non-metallic contents.

13.2.1.3 Hydrothermal Treatment

Hydrothermal treatment techniques are usually conducted at moderate temperatures (180–260°C) and saturated pressure (2–10 MPa) to produce hydrochars, while at 300–350°C and 5–20 MPa for hydrothermal liquefaction. Hydrothermal gasification is relatively operated at higher operating conditions (temperature >374°C and pressure >22 MPa) (Cui et al. 2021). Leaching is performed in the presence of acid and alkalis in order to leach out the metals. Harmful agents containing chlorine and cyanides usually get liberated (Ding et al. 2019). The process is energy intensive and generates wastewater, which results in eutrophication, acidification, etc. (Debnath et al. 2018). Brominated contaminants such as hydrobromic acid and dibenzofurans are toxic in nature (Zhan et al. 2020). This method, however, conducts dechlorination, and debromination efficiently from the char by generating HCl as Cl is replaced by the OH bond in the presence of additives (mainly NaOH, KOH) (Xiao et al. 2008). Debromination leads to the production of phenol and 4-isopropenylphenol, which could further be used in chemical industries. Addition of these alkalis makes this process more costly

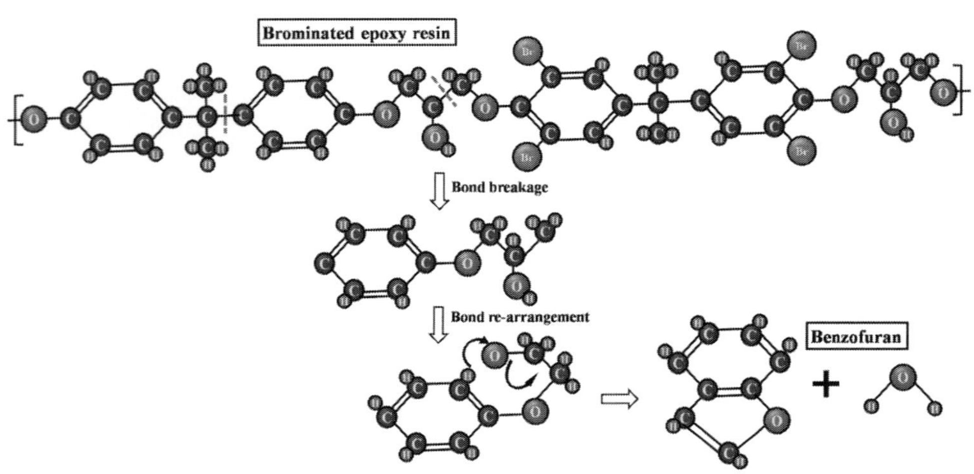

Figure 13.2 Possible reaction mechanism for dibenzofuran formation from epoxy resin (Evangelopoulos et al. 2015 / with permission of Elsevier).

and limits its usage for small-scale applications (Ghosh et al. 2015). It efficiently separates Cu and other precious metals (Ag, Au, etc.), while other metal separation remains unaffected. Thus, this method is difficult to scale up due to high operating cost and offers selective metal extraction facility. However, initiatives need to be taken to make this process feasible and economical.

13.2.2 Chemical Looping Processes for Metal Recovery and Energy Production

Treatment technologies should be targeted such that maximum resources from EW could be extracted without degrading the environment. Thus, the current study focused on utilizing the plastic content in e-waste as fuel and, further, its metallic contents as the oxygen carrier. This section intends to provide a solution to use the mixed metal oxides from EW in an advanced technology termed the chemical looping process. Figure 13.3 shows the proposed stepwise procedure for metals recovery from a discarded printed circuit board (PCB) using thermal treatment processes. The dismantled board is reduced to a particle size of 1–3 mm for thermal treatment processes using a shredder or crusher (Figures 13.3a–13.3c). Thermal methods are preferred over other methods (ion exchange, solvent extraction, leaching, etc.) due to their superior performance in metal extraction and energy recovery (Ding et al. 2019). The mixture of chipped metals and isolated board pieces is then fed to the reactor such that volatiles are liberated, while the metals and char are retained in the solid residue (Figure 13.3d). The residue containing mainly flame retardants can be used as reinforcement materials in building construction (Hao et al. 2020). Thus, thermal treatment methods facilitate the complete utilization of e-waste. Pyrolysis (500°C) followed by gasification (900°C) is performed since energy in the pyrolyzed char can be recovered by gasification (Figure 13.3e). Thus, the total energy output can either be used to produce electricity in small facilities or prepare organic components (dimethyl ether, methanol, etc.) (Condori et al. 2021). Khaobang et al. (2022) performed PCB pyrolysis at 500°C to liberate the volatiles and further conducted ultrasonication to extract the metals from the char. The char, oil, and gas yields were found to be 48.0%, 20.0%, and 32.0%, respectively. A total metal species

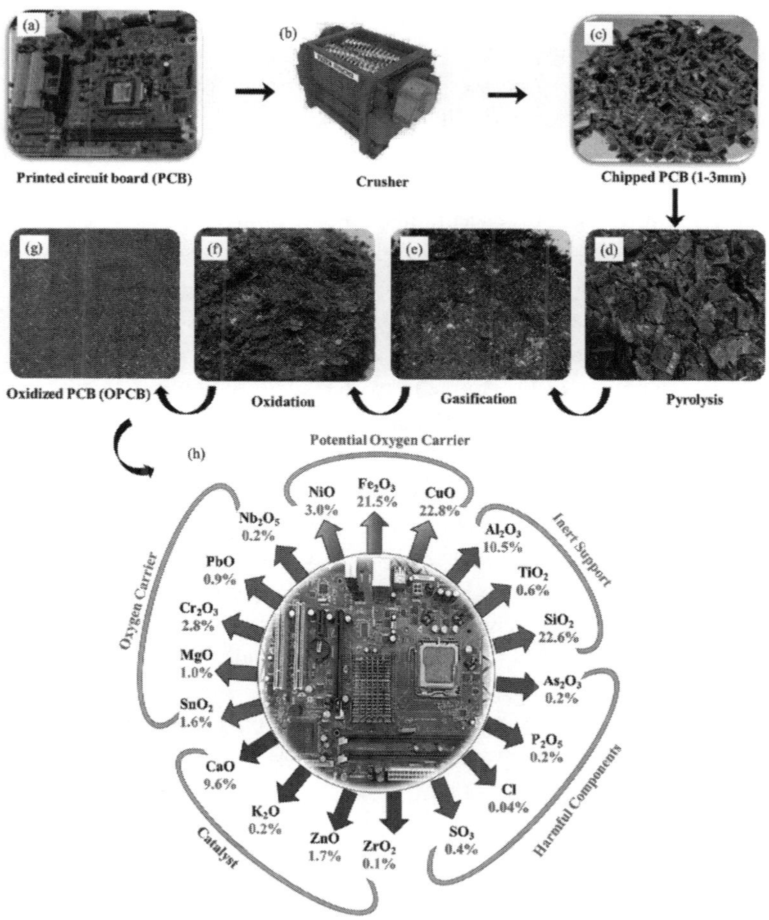

Figure 13.3 Stepwise procedure to produce mixed metal oxides from discarded e-waste (a) PCB, (b) crusher, (c) chipped PCB, (d) pyrolyzed PCB, (e) gasified PCB, (f) oxidized PCB, (g) screened OPCB, (h) OPCB oxide composition (Bhui et al. 2021 with permission from Elsevier).

of 89.8% from the PCB char could be extracted by using the ultrasonication process. They further observed that coupling pyrolysis with the gasification process (to re-utilize the char) reduced the e-waste recycling cost by 25.0–28.0% when compared to the conventional approach. Shibata et al. (2003) observed that the release rate of furans is higher than dioxins during the direct oxidation process due to the slow decomposition rate of furans under the oxygen atmosphere. Direct combustion of plastic waste materials promotes the release of these components (dioxins and furans); thus, pyrolysis and gasification processes are performed prior to the combustion (Weiland et al. 2021). However, the production rate of dioxins and furans was reduced drastically with an increase in CuO content due to its catalytic activity. Thus, the oxidation of the metals is performed after pyrolysis and gasification processes at 900°C (Figure 13.3f). The mixed metal oxides are then termed as an oxidized printed circuit board (OPCB) after screening to −250 to +212 μm, as shown in Figure 13.3(g). Figure 13.3(h) shows the metallic oxide content in the OPCB analyzed using XRF analysis. The OPCB contains 47.0% of NiO, Fe_2O_3, and CuO. Thus, this can either be used as an oxygen carrier in the chemical looping process or as a catalyst for any thermal treatment process. The presence of Al_2O_3, SiO_2, and TiO_2 are also beneficial since they will act as support materials for the oxygen carriers and will further inhibit agglomeration due to their high melting point.

Chemical looping processes are one of the emerging and versatile technologies that inherently separates CO_2 in an economical way. It has gained substantial interest since it does not require an air separation unit (ASU), unlike conventional processes. It is accomplished by employing two reactors, namely, a fuel reactor (FR) and an air reactor (AR). The solid fuel is oxidized in the presence of an oxygen carrier (OC) in the FR, while the reduced OC is re-oxidized in the AR such that multiple cycle operation can be conducted (Equation 13.1). Thus, the oxygen carrier supplies oxygen for fuel conversion. Equation (13.2) is the main reaction, which occurs inside the FR, where the OC gets reduced to its metallic form by oxidizing the fuel into CO_2 and H_2O. CO_2 as a gasifying agent is introduced into the reactor such that the char is gasified to CO and further reacts with the solid OC (Equation 13.3). However, this method suffers from a slow gasification rate since it is the rate-determining step. Thus, the concept of chemical looping oxygen uncoupling (CLOU) emerged, where certain metal oxides have the ability to release molecular oxygen at elevated temperatures (Equation 13.4). These OCs have higher reactivity than the iG-CLC OCs.

$$\textbf{AR}: \text{Me (s)} + 0.5\text{O}_2\text{(g)} \rightarrow \text{MeO (s)} \tag{13.1}$$

$$\textbf{FR}: (2\text{n}+\text{m})\ \text{MeO} + \text{C}_\text{n}\text{H}_{2\text{m}} \rightarrow (2\text{n}+\text{m})\text{Me} + \text{n}\,\text{CO}_2 + \text{m}\,\text{H}_2\text{O} \tag{13.2}$$

$$\textbf{FR}: \text{C (s)} + \text{CO}_2\text{(g)} \rightarrow \text{CO(g)} \tag{13.3}$$

$$\textbf{FR}: \text{MeO (s)} \rightarrow \text{Me (s)} + 0.5\text{O}_2\text{(g)} \tag{13.4}$$

The chemical looping operation can be categorized into various processes: (a) chemical looping combustion (CLC), (b) chemical looping reforming (CLR), and (c) chemical looping gasification (CLG). These processes follow the same basic principle; however, each process provides different desirable products. The CLC process completely oxidizes the fuel into CO_2 and H_2O; thus, pure CO_2 can directly be sent for sequestration, thereby exhibiting low energy penalty (Figure 13.4a). CLR and CLG mainly focus on producing pure H_2 and syngas by altering the process parameters compared to the CLC process. CLG produces a gas mixture (CO, H_2, CH_4, etc.), thus requiring an additional gas separation technique to separate H_2 (Figure 13.4b). However, the CLR concept employs three reactors and produces a rich stream of H_2 (Figure 13.4c). The OC is partially reduced in the CLG and CLR cases, while CLC allows a complete reduction in FR. The key features of OC are: (a) high reactivity, (b) high oxygen transport capacity, (c) low attrition rate, (d) high durability, (e) carbon deposition inhibited on the OC pores, (f) non-toxic and cheaper.

Low-cost OCs are preferred due to several reasons: (a) they make the process economical and feasible; (b) a complete separation of ash and oxygen carriers is not possible, thus, a part of OC is always disposed of along with the ash particles; and (c) ash deposition over the OC results in agglomeration and sintering. Several oxygen carriers, such as oxides of Fe, Ni, Cu, Co, Mn, etc., have already been extensively used in the chemical looping processes. Table 13.3 shows the basic properties of the oxygen carriers. NiO has a high reactivity, with an oxygen transfer capacity (OTC) of 21.0%, but its use is restricted due to its high cost and toxicity (Liu et al. 2021). Fe_2O_3 is commonly utilized in the CLC process due to its low cost and environmental friendliness. It has a lower OTC of 3.0% (Qasim et al. 2021). Cu-, Co-, and Mn-based OC are CLOU materials; thus, they have higher reactivity than Fe and Ni-based OC. However, these OC could not be used in long-run operations due to their low melting point (CuO), high attrition rate (Mn_2O_3), or toxicity (CO_3O_4).

To date, there is very limited research that has focused on oxygen carriers derived from waste materials. Hildor et al. (2019) examined the efficacy of steel converter slag (obtained from steel-producing industries) as an oxygen carrier with gaseous fuels (syngas, CH_4). It demonstrated lower agglomeration during its 60 hours of operation at 950°C. The

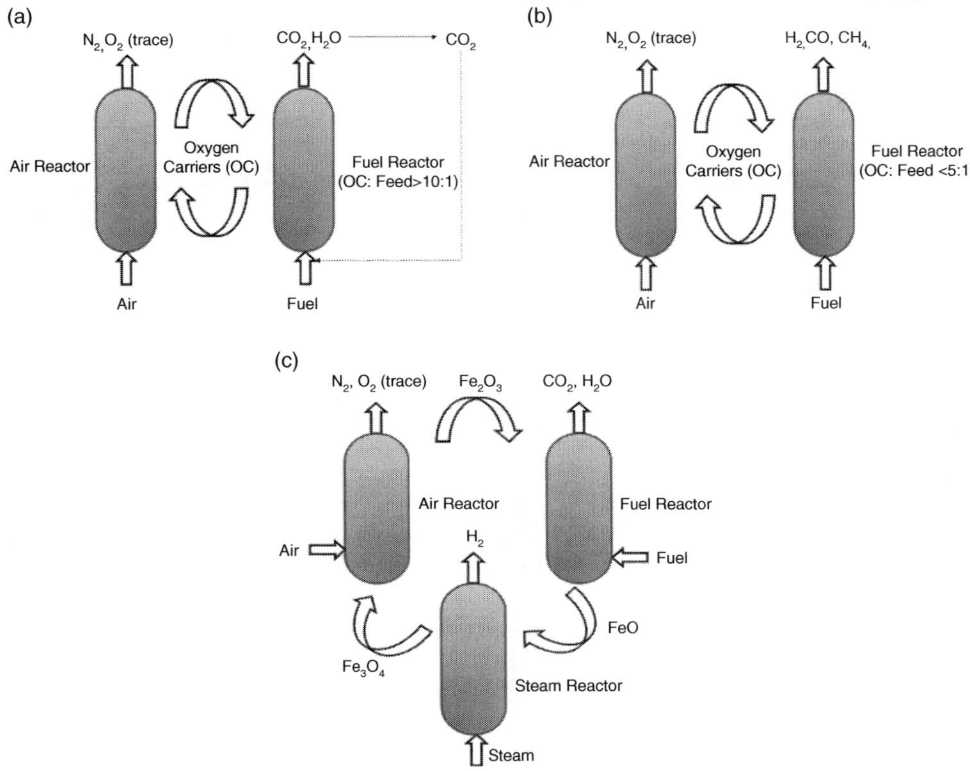

Figure 13.4 Conceptual diagram of (a) chemical looping combustion (CLC), (b) chemical looping gasification (CLG), and (c) chemical looping reforming (CLR) processes.

Table 13.3 Properties of typical oxygen carriers.

Properties	Fe_2O_3/Fe_3O_4	CuO/Cu_2O	NiO/Ni	Mn_3O_4/Mn_2O_3	Co_3O_4/CoO
Reactivity	Moderate	High	High	High	High
Oxygen transport capacity	Moderate	High	High	High	High
Cost	Low	High	High	High	High
Melting temperature	High	Low	High	High	High
Toxic	Low	Low	High	Moderate	High
Attrition rate	High	Low	High	Low	Moderate

presence of supporting materials (Al_2O_3, SiO_2, TiO_2 etc.) in slag enhanced the mechanical strength of the OC during multiple-cycle operation. Feng et al. (2022) revealed that the addition of NiO, CuO, and MnO_2 to red mud improved the CLC reactivity when compared to red mud alone. The CO_2 yield was 90.0% when operated for 10 consecutive redox cycles, even in a fixed-bed reactor. Li et al. (2022) used straw digestate as the feed to assess the performance of a perovskite ($LaCu_{0.5}Fe_{0.5}O_3$) in a thermogravimetric analyzer (TGA). The reactivity of OC escalated up to the seventh cycle due to the increase in surface oxygen content. However, its gasification efficiency decreased from 84.7% to 72.6% in the eighth cycle. Zhao et al. (2022) investigated the chemical looping reforming process by combining several oxygen carriers (Ca, Fe, and Mg) and testing it with syngas at 900°C. When compared to single-based metal oxides, the presence of Ca boosted syngas generation, while the presence of Mg increased total reactivity. Adánez-Rubio et al. (2018) tested pine sawdust with 34.0% CuO and 66.0% Mn_3O_4 oxygen carrier. They achieved a maximum CO_2 output of 97.8% and a char conversion of 96.6% at 850°C. Thus, CLC operation can be feasible even at lower temperatures in the presence of mixed metal oxides. Hammache et al. (2020) used industrial waste (waste from the fluid catalytic cracking process) to investigate the CLC performance at 1100°C in a fixed-bed reactor and used coal char as a fuel. Due to higher quantities of alumina and silica, industrial waste was discovered to be more resistant to attrition than hematite. This implies that mixed metal oxides are more reactive than single-based oxygen carriers.

The current research attempts to use EW as a viable oxygen carrier in the CLC process due to the presence of mixed metal oxides, as shown in Figure 13.3. There has been no research on the use of EW-derived oxygen carriers in this advanced carbon capture method. The interaction of the potential oxygen carriers with the volatiles is shown in Equations (13.5-13.12) (Sarafraz et al. 2017; Bhui et al. 2021). Potential oxygen carriers such as Fe_2O_3, CuO, and NiO are present in higher quantities than other oxygen carriers. Thus, it is assumed that the CLC operation is mainly conducted due to the presence of these oxides since the presence of other oxygen carriers contribute approximately 6.5%. Inert materials (Al_2O_3, SiO_2, TiO_2) contents are 33.7%. Certain harmful and toxic components such as As, P, Cl, and S are detrimental to the CLC operation since they might retard the CLC reactivity and cause corrosion during multiple-cycle operation. Thus, certain treatment methods are necessary prior to their usage. The EW also contains hazardous substances such as Pd, Cr, Ni, etc. However, these oxides can be utilized as oxygen carriers. Sarafraz et al. (2017) demonstrated the potential use of these oxygen carriers. They can be easily disposed of after several cycles since their toxicity will reduce due to agglomeration and sintering with other nontoxic components. This suggests that the recycled EW can be used to produce value-added products, thereby reducing the burden on the existing resources.

$$\text{Volatiles}(CO, H_2, CH_4) + Fe_2O_3 \rightarrow CO_2 + H_2O + Fe_3O_4 \quad (13.5)$$

$$\text{Volatiles}(CO, H_2, CH_4) + CuO \rightarrow CO_2 + H_2O + Cu_2O \quad (13.6)$$

$$\text{Volatiles}(CO, H_2, CH_4) + NiO \rightarrow CO_2 + H_2O + Ni \quad (13.7)$$

$$\text{Volatiles}(CO, H_2, CH_4) + Nb_2O_5 \rightarrow CO_2 + H_2O + NbO_2 \quad (13.8)$$

$$\text{Volatiles}(CO, H_2, CH_4) + PbO \rightarrow CO_2 + H_2O + Pb \quad (13.9)$$

$$\text{Volatiles}(CO, H_2, CH_4) + Cr_2O_3 \rightarrow CO_2 + H_2O + Cr \quad (13.10)$$

$$\text{Volatiles}(CO, H_2, CH_4) + SnO_2 \rightarrow CO_2 + H_2O + SnO \quad (13.11)$$

13.2.3 Direct Liquefaction (DL) Process for Oil Production

Material identification of e-waste plastics (EWP) on various discarded electronic devices confirms the existence of high-impact polystyrene (HIPS), polycarbonate blends, acrylonitrile butadiene styrene (ABS), polyamide (PA), polypropylene (PP), styrene acrylonitrile (SAN), polystyrene (PS), low-density polyethylene (LDPE), and polyurethane foam, etc. (Kumar et al. 2018; Zhao et al. 2019). DL technology brings a sustainable way of transforming such waste plastics into transportable liquid oil with significant heating value and fuel properties due to low oxygen content. It is usually operated under mild temperatures in the range from 300 to 450°C and at elevated pressures (40–220 bar) using different organic solvents (ethanol, methanol, tetralin, etc.) (Kameel et al. 2022; Nanda and Berruti 2021).

Generally, EWP, after size reduction, is introduced to a high-pressure autoclave reactor system to perform the liquefaction operation at a suitable temperature under pressurized conditions in H_2 or N_2 atmosphere. The products obtained are mainly in the form of gas, liquid, and solid. Gas can be collected after completion of the liquefaction experiment via a gas outlet followed by extraction of liquid and solid products using organic solvents like hexane, tetrahydrofuran, etc., in a Soxhlet extractor apparatus for a particular liquefaction operation (Figures 13.5a–13.5d). The solid residue is usually discarded (Figure 13.5e), while the liquid product can further be characterized for its quality detection (Figures 13.5g–13.5h). Liquefied oil characterization can be done through a number of analytical methods, namely Fourier Transform Infrared Spectroscopy (FTIR), Nuclear Magnetic Resonance (NMR), etc., to identify the functional groups and chemical nature of the oil. According to the FTIR study, peaks in the range 920–700 cm^{-1} correspond to C–H out of plane bending vibration in the aromatic compounds present in liquefied oil, whereas peaks at 2922 cm^{-1} and near 1455 cm^{-1} represent C–H stretch (CH_2 or CH_3) and C–H bending vibration, respectively (Figure 13.5g). These peaks are comparable with that of commercial diesel fuel. Further, a ^1H NMR study confirms high aromatic content (68.41%), followed by paraffins (20.35%) and olefins (10.89%) in the EWP liquefied oil (Pathak et al. 2022). Detailed oil analysis can be found in the literature to understand the oil characterization (Chen et al. 2019; Kumar et al. 2018). Most of the DL of waste plastic materials was conducted at high temperatures and

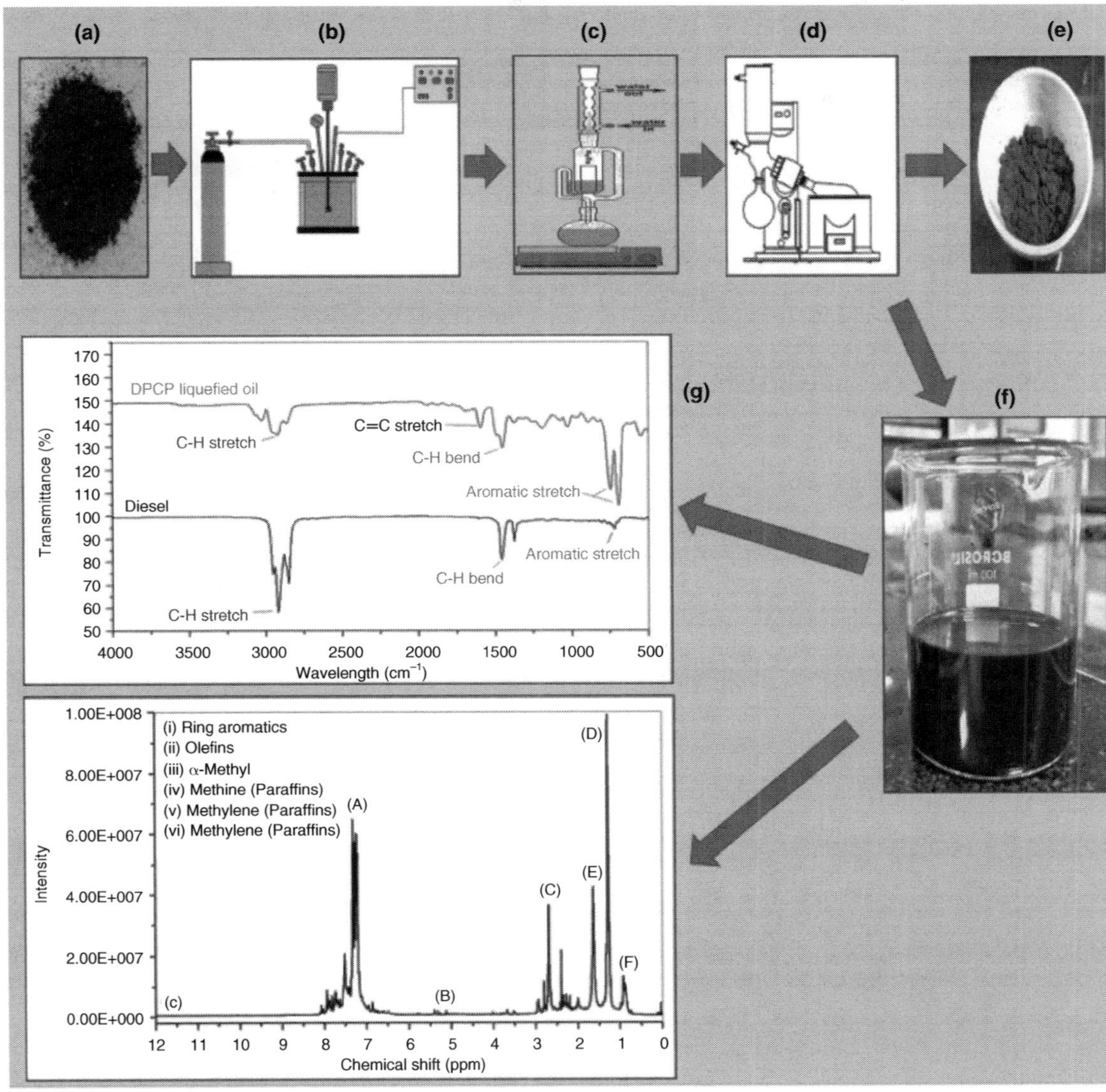

Figure 13.5 Stepwise conversion of discarded printer cartridge plastics (DPCP) into oil through the direct liquefaction process: (a) powdered DPCP, (b) high-pressure reactor system, (c) Soxhlet extraction assembly, (d) rotary evaporator, (e) solid product, (f) oil, (g) Fourier Transform Infrared Spectroscopy (FTIR) analysis, and (h) Nuclear Magnetic Resonance (NMR) analysis (Pathak et al. 2022 / with permission of Elsevier).

pressure under the hydrogen atmosphere and/or in the presence of organic solvents acting as hydrogen donors. However, in the absence of H_2 and organic solvent, DL of discarded printer cartridge plastic resulted in 48.66 wt% of oil formation at 400°C in 15 minutes of reaction time under N_2 atmosphere (Table 13.4). Thus, N_2 can potentially replace costly H_2 to perform direct liquefaction of waste plastics economically. Also, catalyst addition significantly benefits waste plastics conversion favoring higher oil yield. Zeolite Socony Mobil-5 (ZSM-5) was reported as a highly active catalyst that increased the oil yields for polyethylene plastics beyond 90.0% at 420–450°C, 55.16 bar of H_2 (cold), and 60 minute reaction time (Taghiei et al. 1994). EWP, being hydrogen-rich, can act as a hydrogen donor in the direct coal liquefaction (DCL) process. Co-liquefaction of a coal-plastic mixture resulted in higher oil yield (typically by 5.1–22.6%) than the average of the oil yields for the individual coal and plastic liquefaction with a significant reduction in hydrogen consumption by 7.7–17.9% (Wang and Chen 2004). Also, the co-liquefaction of oxygen-rich biomass with hydrogen-rich plastics was found to enhance the bio-crude oil yield and improve fuel properties (Hongthong et al. 2020).

Therefore, EWP is a cheaper hydrogen source that can be used as a co-reactant with coal and biomass to enhance the liquefied bio-oil yield and quality. Transition metal catalysts based on nickel, cobalt, and iron (Raney nickel, iron ore, nickel and iron metals, $CoMo/\gamma{-}Al_2O_3$, etc.) have been

Table 13.4 Waste plastic liquefaction under various operating parameters.

Sample Feed	Operating conditions	Solvent	Reacting atmosphere	Liquid product yield (oil)	Reference
Mixed waste plastic	430°C, 55.16 bar, 60 min	Tetralin	H_2	30.0–35.0%	Taghiei et al. (1994)
Commingled waste plastic	445°C, 55.16 bar, 20–60 min	Tetralin Waste oil	H_2	48.3%	Huffman et al. (1995)
Waste plastic mixture	500°C, 10 bar, 60 min	–	H_2	48.2%	Williams and Slaney (2007)
APC waste plastic	445°C, 55.16 bar, 60 min	–	H_2	75.0–80.0%	Feng et al. (1996)
Discarded printer cartridge plastic	400°C, 20 bar, 15 min	–	N_2	48.66%	Pathak et al. (2022)

tested in HTL and, particularly, nickel catalysts were found to enhance the bio-oil yield and quality due to its hydrogenating property (Mukundan et al. 2022). Hence, metals such as iron cobalt and nickel extracted from the EW can be an effective source to prepare low-cost catalysts for the direct liquefaction process to further improve the liquefaction efficiency. Utilizing EWP as co-reactant and the extracted metals as catalyst source in co-liquefaction not only improves the overall feed conversion and oil yields, but also mitigates the environmental problem of EW landfilling with effective waste management strategy.

13.2.3.1 Liquefaction Mechanism

Thermal liquefaction involves the scission of long polymer chains into smaller ones with chain end capping by available hydrogen, thus, converting high molecular weight plastics into low molecular weight products such as lighter hydrocarbons and valuable chemicals (Murty et al. 1996). In general, direct liquefaction of plastics subjects them to a reaction mechanism via a chain initiation, propagation, and termination process. As shown in Figure 13.6(a), the primary radicals formed due to C–C bond random homolysis combine with a hydrogen atom from the solvent phenol to produce n-paraffin (C_5–C_{32} n-paraffins) and a phenoxy radical. The phenoxy radical also combines with hydrogen atom from the high density polyethylene (HDPE) chain to regenerate a phenol molecule along with a secondary intermediate radical, which may undergo C–C bond homolysis and results in volatile α-olefin (C_2–C_{32} α-olefins) formation and primary radical regeneration. Therefore, various hydrocarbons (C_1–C_{32}) are formed during the thermal degradation of HDPE with phenol via propagation. The relatively short primary radical finally reacts with another radical to produce n-paraffin through the termination step. Similarly, thermal

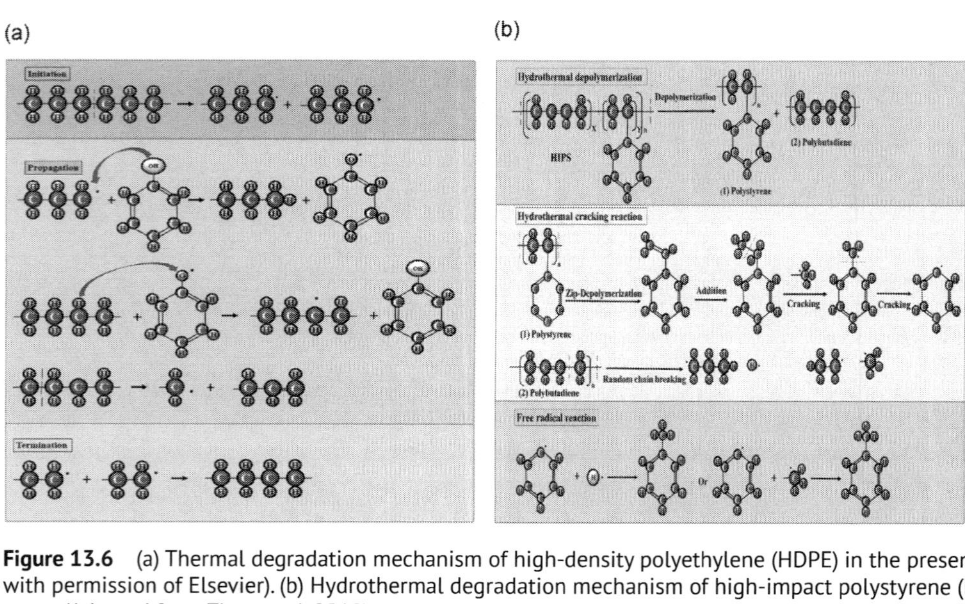

Figure 13.6 (a) Thermal degradation mechanism of high-density polyethylene (HDPE) in the presence of phenol (Vicente et al. 2009 / with permission of Elsevier). (b) Hydrothermal degradation mechanism of high-impact polystyrene (HIPS) in the presence of subcritical water (Adapted from Zhao et al. 2019).

degradation of high-impact polystyrene (HIPS) involves zip depolymerization and random chain breaking followed by combined reactions of various free radicals (Figure 13.6b). During thermal treatment, HIPS initially degrades to form polystyrene and polybutadiene. At a later stage, cracking of the polystyrene into the monomer of styrene and random chain breaking of polybutadiene to form phenyl radicals, phenylmethyl radicals, and alkyl radicals take place. The free radicals thus obtained undergo a number of reactions to form organic compounds. For example, a combination of phenylmethyl radicals and hydrogen or a combination of phenyl and alkyl radicals results in toluene formation, as shown in Figure 13.6(b). Different plastic materials found in EW items exhibit similar kinds of reaction mechanisms, and a number of organic compounds are formed, which are found in the liquefied oil composition. However, during co-thermal treatment of mixed feed of EWP and high ash coal, the impact of the inorganics and sulfur content of the coal on overall feed conversion and oil yield require broad investigation. Different reaction mechanisms should be studied to understand the chemical interaction between EWP and lower rank coals on thermal treatment.

13.2.3.2 Importance of EWP Liquefied Oil

Comparability in physicochemical properties of EWP liquefied oil with market fuels signifies the importance of DL technology as a potential EW recycling alternative (Table 13.5). A higher H/C ratio and a significant amount of aromatic content in EWP liquefied oil are crucial in reducing greenhouse gas emissions and enhancing octane quality on blending with other liquid fuels, respectively. Various hydrocarbons present in liquefied oil favor its usability as a fuel. Thermogravimetric analysis (TGA) revealed that the oil molecules obtained from hydrothermal liquefaction of various plastics (polypropylene, polyethylene terephthalate, polycarbonate, etc.) are primarily in the volatility range of gasoline, kerosene, and diesel fractions (Seshasayee and Savage 2020). EWP liquefied oil composition can be an attractive option in the industrial sector as liquid hydrocarbon products are mainly benzene, ethylbenzene, toluene, styrene, alkylbenzene, phenyl compounds, indanes, naphthalenes, and other hydrocarbons (Kumar et al. 2018; Nanda and Berruti 2021). Compounds like styrene and its derivatives, bisphenol A, and caprolactam have significance as raw materials in plastic manufacturing or chemical feedstock (Zhao et al. 2019). Oil components such as ethylbenzene and cyclohexane are the key compounds in the production of plastics, resins, fibers, surfactants, dyestuffs, and pharmaceu-ticals, whereas, long chain alkylbenzenes are used as feedstocks in surfactants manufacturing. Also, toluene, being a chemical feedstock, is used in the production of pesticides, surfactants, and solvents (Williams and Slaney 2007). As the direct liquefaction process requires a lower temperature when compared to the pyrolysis process, oil obtained via EWP liquefaction becomes more appealing than pyrolysis-based oil in terms of energy cost (Ahmad et al. 2020). Therefore, DL can be a promising technology in the energy sector to convert solid EWP material into energy-dense liquid products, which can further be processed to produce finished liquid fuels, blended with available market fuels, or used as valuable chemical intermediates. Generally, bio-oil obtained from thermal treatment of the biomass materials is of low calorific value due to high oxygen content. Its quality is improved by the addition of a large amount of hydrogen, thus making the process expensive (Abnisa and Wan Daud 2014). In this regard, the impact of adding hydrogen-rich EWP in co-thermal treatment is expected to achieve liquefied oil with quality improvements via depolymerization of the carbonaceous biomass under milder process conditions.

13.2.3.3 Drawbacks of DL of EWP

Due to the low recycling value of plastics, coupled with the addition of brominated flame retardants (BFRs), waste electrical and electronic plastics are complex feedstock and

Table 13.5 Fuel property comparison of waste plastic liquefied oil with commercial liquid fuels (Ahmad et al. 2020; Pathak et al. 2022; Wang et al. 2019).

	Waste plastic liquefied oil			Gasoline	Diesel	Crude oil
Feed source	Printer cartridge	Electronic plastic (computer display)	Waste plastic (polystyrene)			
Density (kg/m^3)	910	–	880	710–780	870–950	760–960
Calorific value (MJ/kg)	39.67	37.5	40.91 ± 0.15	42.0–46.0	42.0–45.0	42.0–47.0
Flash point (°C)	–	–	55 ± 0.11	–42.8	37.8–54.4	60.0
Carbon number distribution	C_8–C_{31}	–	–	C_4–C_{12}	C_8–C_{25}	–

require attention for processing via DL (Wang et al. 2019). High oxygen content in EWP materials is undesirable as their liquefaction mostly results in oil with a lower heating value, and the hydrogen requirement is high during oil upgradation (Seshasayee and Savage 2020). Energy requirements and capital costs in the thermal treatment of waste plastics are reported on the higher side and require major attention during liquefaction operations (Ukarde and Pawar 2021). In addition, CO and CO_2 formation in gaseous products during EWP liquefaction is undesirable considering environmental safety (Williams and Slaney 2007).

13.3 Future Aspects

A thorough investigation is necessary to explore EW thermal treatment methods to valorize waste plastics in the energy sector. Upgradation of the EW liquefied oil is expected to improve its blending with commercial liquid fuels. Apart from BFRs, plastics collected from discarded electronic devices (mobile phones, computers, PCBs, etc.) are contaminated with a variety of heavy metals (chromium, lead, mercury, cobalt, etc.) as additive compounds and require broad investigation for sustainable decontamination in the future. Also, reduction in the toxin emissions (persistent organic pollutants, dioxins, polycyclic aromatic hydrocarbons, etc.) during thermal treatment of EW needs effective strategies.

Hydrogen-rich EWP can act as a hydrogen donor in co-liquefaction studies, and therefore, a detailed investigation is recommended to study the impact of mixing EWP with a variety of biomass materials and lower-rank coals to achieve a synergistic effect on improving overall co-liquefaction feed conversion efficiency and high-quality oil yield economically and reduce the emission of toxic components. The effect of blending EWP with other feedstocks needs to be studied on the basis of oil production rate and quality. The energy efficiency of the processes should be estimated for proper understanding of their synergy. EWP liquefied oil can be a potential feedstock for producing various chemicals. Further, a detailed investigation is required for its effective utilization as an auxiliary fuel for combustion within boilers and furnaces and in modified heavy-duty engines after refining. Use of EW as a raw feedstock for thermal treatment need utmost attention keeping environmental safety in mind. Process parameters should be considered carefully to earn excellent outcomes in terms of energy recovery from EW at lower cost and contribute to waste management.

The co-utilization of EW with various feedstocks (coal, biomass, etc.) needs to be tested to study their synergy using thermal treatment methods. The impact of inorganics and sulfur content of the coal on co-thermal treatment with EW materials for oil production requires detailed investigation. Extensive research is required regarding the actual e-waste generation and consumption rate in order to quantify the exact amount of waste generation. Different processes for dismantling and plastic separation from metals should be further improved in order to enhance the recovery rate. However, these mixed metals can further be used in various applications after proper extraction due to the high content of Fe, Cu, Ni, etc. Gasification of different types of EW should be conducted in order to check the variation in syngas composition. The effect of metals on the gasification rate needs to be thoroughly examined. Finally, public awareness related to EW recycling should be reinforced.

13.4 E-waste Management Cost

13.4.1 Life Cycle Assessment (LCA)

LCA is one of the important tools that quantifies the environmental impact of a product (Carvalho and Silva 2021; Ismail and Hanafiah 2021). It is evaluated on the basis of two subsystems: (a) collection and transportation and (b) treatment and disposal, including pre-treatment and waste disposal processes. Ismail and Hanafiah (2021) evaluated the LCA of printed wire boards for different cases: (a) landfill disposal, (b) incineration with energy recovery, (c) incineration without energy recovery, and (d) materials recovery with energy recovery. They concluded that case (d) is the best management option since the waste is reduced substantially along with secondary materials and energy generation. The waste reduction rate is found to be 42.9% for case (d), while it is 0% for landfilling (case a) and 20.3% for incineration (case b). Similarly, Rocchetti and Beolchini (2014) concluded that recycling of cathode ray tubes (CRTs) and liquid crystal displays (LCDs) is more beneficial than landfilling on the basis of environmental and economic aspects. Mangmeechai (2022) observed that the highest revenue is collected from mobile phone e-waste, while CRTs produce the least revenue due to the presence of toxic components (mainly Pb). Included is the cost of the treatment. Dismantling and recycling of different e-waste generated a net profit of US$ 6.8–45 million in Thailand. He concluded that valuable metals could be recovered from e-waste in a cheaper way, provided that a high-tech process is implemented. Arain et al. (2022) performed the LCA of electrical appliances (washing machine, refrigerator, CRT, and fan). The LCA estimated the damages caused during material recovery on the basis of human health, climate change, ecosystem quality, etc. They estimated that 39.9 kg of CO_2 gets released when one refrigerator is recycled, while a fan released 5.03 kg of CO_2. They recovered 68.3% copper from electric cables. However, the maximum metal extraction is obtained from a television (CRT screen) in the form of aluminum, steel, and copper. Thus, it confirms that improper EW treatment methods (landfilling, incineration) can cause potential damage to the environment. This suggests that an advanced technology (chemical looping processes or liquefaction) needs to be imposed for e-waste handling. Chemical looping processes can utilize the metallic contents as oxygen carriers, while the liquefaction process can produce bio-oil from the plastic contents.

13.4.2 Levelized Cost of Electricity (LCOE)

LCOE is one of the economic indicators to check the economic viability of a proposed power plant. It displays the necessary price required for electricity generation in \$/MWh. The conceptual diagram for power production using OPCB as the oxygen carriers in a CLR integrated with the steam cycle is shown in Figure 13.7. The schematic diagram illustrates the complete resource utilization of PCB in an advanced technology, CLR, for multi-component generation (energy, hydrogen, syngas, and metals). The PCB is first sent to a pulverizer for size reduction. It is then transported to the gasifier to produce rich syngas and separate out the metals from the PCB. The metals are then transferred to AR for oxidizing them into their respective metal oxides. The syngas and metals are produced in an excess amount such that a part of it can be sent for commercial purposes. Solid feed is introduced directly into the FR along with the PCB syngas and OPCB. The FR outlet contains rich CO_2 and H_2O, which are then sent to the H_2S removal unit. The clean gas is then sent to the heat recovery steam generator (HRSG) unit to produce steam. It is further used in three different steam turbines to produce electricity. The detailed information is given in Rai et al. (2022). The reduced OC from FR is then dispatched to SR for further reduction to produce H_2. Thus, multi-components (CO_2, H_2, energy, syngas, and mixed metals) were obtained using this configuration. The detailed energy and economic analysis is shown in Table 13.6. The total energy input for all the cases is assumed to be 2320 MW. The PCB flow rate to the gasifier is fixed such that 50% of chemical energy is supplied from the PCB syngas. The excess syngas and metals produced are sent to storage for future purposes.

The economic analysis of each case is quantified by evaluating the levelized cost of electricity (LCOE), as shown in Equation (13.12). The levelized cost of H_2 production (LCOH) quantifies the total cost required to produce one unit of H_2 (Equation 13.13). Further, the excess amount of syngas and metals that are produced for commercial purposes is quantified on the basis of levelized cost of metals (LCOM) and levelized cost of syngas (LCOS) as shown in Equations (13.14) and (13.15).

$$\text{LCOE} = \frac{\begin{array}{c}\text{Annual capital cost (ACC)}+\\ \text{Total operating \&}\\ \text{maintenance cost (TOM)}\end{array}}{\text{Net electricity produced}} \tag{13.12}$$

$$\text{LCOH} = \frac{\text{ACC}+\text{TOM}}{\text{Annual hydrogen produced}} \tag{13.13}$$

$$\text{LCOM} = \frac{\text{ACC}+\text{TOM}}{\text{Net metal produced}} \tag{13.14}$$

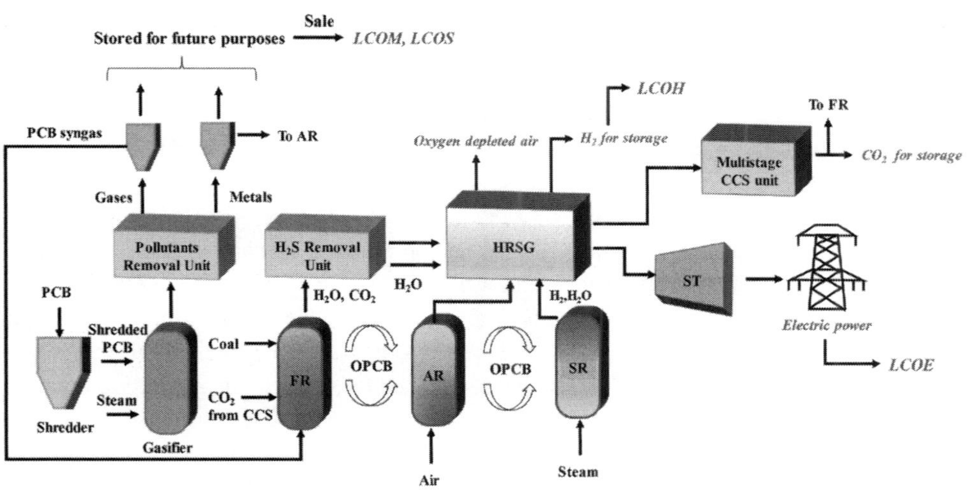

Figure 13.7 Schematic diagram of effective PCB utilization in a chemical looping reforming (CLR) system for multi-component generation using oxidized printed circuit boards (OPCB) as the oxygen carriers (Rai et al. 2022 / with permission of Elsevier). (PCB: printed circuit board; HRSG: heat recovery steam generator; FR: fuel reactor; AR: air reactor; SR: steam reactor; ST: steam turbine; LCOE: levelized cost of electricity; LCOS: levelized cost of syngas; LCOM: levelized cost of metals; LCOH: levelized cost of hydrogen production).

Table 13.6 Energy analysis of a CLR integrated steam cycle configuration (Rai et al. 2022 / with permission of Elsevier).

Parameters	PCB syngas	PCB syngas and coal
Total chemical energy input to the gasifier and fuel reactor (MW)	2320	2320
Printed circuit board to gasifier (kg/s)	200	100
Coal to fuel reactor (kg/s)		35.36
Total syngas produced from gasifier (kg/s)	1193.39	1697.90
Excess syngas sent for storage (kg/s)	1090.99	1646.70
Excess metals sent for storage (kg/s)	1594.20	2437.20
CO_2 at fuel reactor outlet (kg/s)	125.61	137.92
H_2 at steam reactor outlet (kg/s)	5.49	8.01
Energy analysis		
Total energy produced from steam turbine and H_2 (MW)	895.23	1304.82
Energy penalty (MW)		
H_2 compression (MW)	36.01	52.54
Gasifier (MW)	35.01	17.5
CO_2 compression (MW)	41.25	44.06
Pollutants removal unit (MW)	20.58	30.01
Total power consumption (MW)	135.31	146.41
Gross efficiency (%)	38.59	56.24
Net efficiency (%)	32.72	49.89
Economic analysis		
Total capital cost (M$)	6913.53	8808.01
Total operating cost (M$/ year)	56.36	61.37
Levelized cost of electricity, LCOE ($/MWh)	93.28	70.63
Levelized cost of hydrogen, LCOH ($/kg)	3.68	2.91
Levelized cost of metals, LCOM ($/ton)	24.38	15.94
Levelized cost of syngas, LCOS ($/kWh)	0.0034	0.0026

$$LCOS = \frac{ACC + TOM}{\text{Net syngas produced}} \tag{13.15}$$

The capital cost includes total installed cost, equipment cost, utilities, etc. The typical pollutants liberated from PCB gasification are HCl, H_2S, HBr, dioxins, furans, and particulate matter. Thus, the pollutants removal unit cost is included along with other equipment costs. The total operating cost incorporates the cost of e-waste segregation cost, operating and maintenance cost, slag disposal cost, and CO_2 storage and transportation cost.

The LCOE of PCB syngas alone is $93.3/MWh, which was reduced by 24.3% when a blend of coal and PCB syngas was used ($70.6/MWh). Similarly, the LCOH value was reduced from $3.7/kg to $2.9/kg for the blend. The LCOH varied from $1.66/kg to $16.9/kg by using different metal oxides (iron ore or tungsten) (Khan and Shamim 2019). Thus, it is seen that the LCOH values using OPCB are comparable with the reported literature. The levelized cost for metals and syngas production from PCB is found to be highest for PCB syngas ($24.4/ton, $0.0034/kWh). This signifies that metals obtained from discarded e-waste can be used as potential oxygen carriers in the chemical looping processes (Rai et al. 2022). Thus, blending of PCB with Indian coal proved to be beneficial due to high power production with negligible release of toxic components. However, this study should not be limited only to PCB usage. EW having either higher ash content (indicates higher metals) or plastic content (epoxy resins) should also be tested as a feedstock to estimate the net efficiency in a power plant. Damaged electric cables are usually discarded due to high content of polyvinyl chloride (Merlo et al. 2020). However, this waste has a potential to produce rich syngas by the gasification process due to its high calorific value (~31 MJ/kg) as seen in Table 13.2. This gasification step will efficiently eliminate the char from the metallic fractions. The presence of metals in cables (mainly Cu) can act as a catalyst to promote the gasification process. Further, steam gasification will result in H_2-rich syngas, thus high syngas quality can be achieved compared to using air gasification. Thus, intensive research is needed to select the potential EW that can generate high oil and syngas yield by thermal treatment methods. However, the emission of harmful toxic gases needs to be monitored for each type of EW. Thus 'zero waste with zero emissions' can be attained if steam gasification of EW is promoted.

13.5 Conclusions

It is essential to reuse the discarded e-waste due to the limited existing natural resources of metals. This chapter discussed the existing and recently developed technologies for e-waste treatment such that complete resource utilization could be achieved. Further, this chapter highlighted the metal recovery and economic value of the metals recovered and thus facilitated the advanced recovery techniques. The issues related to pollution are inevitable; however, initiatives are taken in order to minimize their liberation during future recycling practices. Pyrolysis coupled with the gasification process proved to be beneficial for obtaining multiple desirable products, i.e., syngas, oil, and mixed metal oxides. This comprehensive study will be beneficial to generate valuable products from toxic-laden e-waste materials, thereby safeguarding the environment from potential hazards. Syngas obtained from EW, when blended with coal, produced a net efficiency of 49.9% by integrating the CLR process with a steam turbine cycle system. A lower LCOE of $70.6/MWh was achieved with the EW blended system. Hence, EW utilization along with conventional feedstocks would be beneficial to overcome the energy scarcity issues.

DL shows a significant pathway for transforming EW into mixtures of lighter hydrocarbons, which, once upgraded, can be considered valuable as a fuel. Compounds present in oil have significance as chemical feedstock or intermediates in manufacturing industries. DL under inert atmosphere provides an economical way to convert EW materials into high calorific oil (~50 MJ/kg). Upgradation of the liquefied oil is expected to improve its blending with commercial liquid fuels. Further, EWP can be a good hydrogen donor in co-liquefaction with coal/biomass, and detailed investigation is necessary to observe a synergistic effect on improving overall oil yield and its quality.

References

Abnisa, F. and Wan Daud, W.M.A. (2014). A review on co-pyrolysis of biomass: an optional technique to obtain a high-grade pyrolysis oil. *Energy Conversion and Management* 87: 71–85.

Adánez-Rubio, I., Pérez-Astray, A., Mendiara, T. et al. (2018). Chemical looping combustion of biomass: CLOU experiments with a Cu-Mn mixed oxide. *Fuel Processing Technology* 172: 179–186.

Ahmad, N., Ahmad, N., Maafa, I.M. et al. (2020). Thermal conversion of polystyrene plastic waste to liquid fuel via ethanolysis. *Fuel* 279: 118498.

Anis, S. and Lestari, S.E. (2018). Production of liquid-oil from thermolysis of electrical and electronic wastes (e-waste) under microwave irradiation. *Journal of Physical Science* 29: 53–62.

Annamalai, M. and Gurumurthy, K. (2021). Characterization of end-of-life mobile phone printed circuit boards for its elemental composition and beneficiation analysis. *Journal of the Air and Waste Management Association* 71: 315–327.

Arain, A.L., Neitzel, R.L., Nambunmee, K. et al. (2022). Material flow, economic and environmental life cycle performances of informal electronic waste recycling in a Thai community. *Resources, Conservation and Recycling* 180: 106129.

Arya, S. and Kumar, S. (2020). E-waste in India at a glance: current trends, regulations, challenges and management strategies. *Journal of Cleaner Production* 271: 122707.

Bhui, B., Das, B., and Prabu, V. (2021). Experimental investigations on novel electronic waste based oxygen carriers for direct coal fuelled chemical looping combustion process. *Fuel* 305: 121535.

Carvalho, A.A.D. and Silva, E.A.D. (2021). Life cycle assessment principles in smartphone apps. *Cleaner Environmental Systems* 3: 100041.

Chen, W.T., Jin, K., and Linda Wang, N.H. (2019). Use of supercritical water for the liquefaction of polypropylene into oil. *ACS Sustainable Chemistry and Engineering* 7 (4): 3749–3758.

Condori, O., García-Labiano, F., de Diego, L.F. et al. (2021). Biomass chemical looping gasification for syngas production using LD Slag as oxygen carrier in a 1.5 kW_{th} unit. *Fuel Processing Technology* 222: 106963.

Cui, X., Zhang, J., Wang, X. et al. (2021). A review on the thermal treatment of heavy metal hyperaccumulator: fates of heavy metals and generation of products. *Journal of Hazardous Materials* 405: 123832.

Das, P., Gabriel, J.C.P., Tay, C.Y. and Lee, J.M. (2021). Value-added products from thermochemical treatments of contaminated e-waste plastics. *Chemosphere* 269: 129409.

Debnath, B., Chowdhury, R., and Ghosh, S.K. (2018). Sustainability of metal recovery from E-waste. *Frontiers of Environmental Science and Engineering* 12: 1–12.

Demirbas, A. (2004). Pyrolysis of municipal plastic wastes for recovery of gasoline-range hydrocarbons. *Journal of Analytical and Applied Pyrolysis* 72: 97–102.

Ding, Y., Zhang, S., Liu, B. et al. (2019). Recovery of precious metals from electronic waste and spent catalysts: a review. *Resources, Conservation and Recycling* 141: 284–298.

Evangelopoulos, P., Kantarelis, E., and Yang, W. (2015). Investigation of the thermal decomposition of printed circuit boards (PCBs) via thermogravimetric analysis (TGA) and analytical pyrolysis (Py-GC/MS). *Journal of Analytical and Applied Pyrolysis* 115: 337–343.

Feng, X., Li, Z., Lin, S. et al. (2022). Enhanced performance of red mud for chemical-looping combustion of coal by the modification of transition metal oxides. *Journal of the Energy Institute* 102: 22–31.

Feng, Z., Zhao, J., Rockwell, J. et al. (1996). Direct liquefaction of waste plastics and coliquefaction of coal-plastic mixtures. *Fuel Processing Technology* 49: 17–30.

Ghosh, B., Ghosh, M.K., Parhi, P. et al. (2015). Waste printed circuit boards recycling: an extensive assessment of current status. *Journal of Cleaner Production* 94: 5–19.

Hagelüken, C. (2006). Improving metal returns and eco-efficiency in electronics recycling metals smelting and refining. *Proceedings of the 2006 IEEE International Symposium on Electronics and the Environment* 218–223.

Hammache, S., Means, N., Burgess, W. et al. (2020). Investigation of low-cost oxygen carriers for chemical looping combustion at high temperature. *Fuel* 273: 117746.

Hao, J., Wang, Y., Wu, Y., and Guo, F. (2020). Metal recovery from waste printed circuit boards: a review for current status and perspectives. *Resources, Conservation and Recycling* 157: 104787.

Hildor, F., Mattisson, T., Leion, H. et al. (2019). Steel converter slag as an oxygen carrier in a 12 MW_{th} CFB boiler—ash interaction and material evolution. *International Journal of Greenhouse Gas Control* 88: 321–331.

Hongthong, S., Raikova, S., Leese, H.S., and Chuck, C.J. (2020). Co-processing of common plastics with pistachio hulls via hydrothermal liquefaction. *Waste Management* 102: 351–361.

Huffman, G.P., Feng, Z., Huggins, F.E., and Mahajan, V. (1995). Direct liquefaction of waste plastics and coliquefaction of waste plastics with coal. *Coal Science and Technology* 24: 1519–1522.

Islam, A., Ahmed, T., Awual, M.R. et al. (2020). Advances in sustainable approaches to recover metals from e-waste—a review. *Journal of Cleaner Production* 244: 118815.

Islam, A., Swaraz, A.M., Teo, S.H. et al. (2021). Advances in physiochemical and biotechnological approaches for sustainable metal recovery from e-waste: a critical review. *Journal of Cleaner Production* 323: 129015.

Ismail, H. and Hanafiah, M.M. (2021). Evaluation of e-waste management systems in Malaysia using life cycle assessment and material flow analysis. *Journal of Cleaner Production* 308: 127358.

Jadhao, P.R., Ahmad, E., Pant, K.K., and Nigam, K.D.P. (2022). Advancement in the field of electronic waste recycling: critical assessment of chemical route for generation of energy and valuable products coupled with metal recovery. *Separation and Purification Technology* 54: 120773.

Jadhao, P.R., Pandey, A., Pant, K.K., and Nigam, K.D.P. (2021). Efficient recovery of Cu and Ni from WPCB via alkali leaching approach. *Journal of Environmental Management* 296: 113154.

Kameel, N.I.A., Daud, W.M.A.W., Patah, M.F.A., and Zulkifli, N.W.M. (2022). Influence of reaction parameters on thermal liquefaction of plastic wastes into oil: a review. *Energy Conversion and Management* 14: 100196.

Kaur, P., Sharma, S., Albarakaty, F.M. et al. (2022). Biosorption and bioleaching of heavy metals from electronic waste varied with microbial genera. *Sustainability* 14 (2): 1–16.

Khan, M.N. and Shamim, T. (2019). Techno-economic assessment of a chemical looping reforming combined cycle plant with iron and tungsten based oxygen carriers. *International Journal of Hydrogen Energy* 44 (23): 11525–11534.

Khaobang, C., Sarabhorn, P., Siripaiboon, C. et al. (2022). Pilot-scale combined pyrolysis and decoupling biomass gasification for energy and metal recovery from discarded printed circuit board and waste cable. *Energy* 245: 123268.

Kumar, U., Gaikwad, V., Mayyas, M. et al. (2018). Application of high-resolution NMR and GC-MS to study hydrocarbon oils derived from noncatalytic thermal transformation of e-waste plastics. *ACS Omega* 3: 9282–9289.

Li, K. and Xu, Z. (2019). A review of current progress of supercritical fluid technologies for e-waste treatment. *Journal of Cleaner Production* 227: 794–809.

Li, T.Y., Bao, L.J., Wu, C.C. et al. (2019). Organophosphate flame retardants emitted from thermal treatment and open burning of e-waste. *Journal of Hazardous Materials* 367: 390–396.

Li, Z, Dong, X., Yan, B. et al. (2022). Chemical looping gasification of digestate: investigation on the surface and lattice oxygen of perovskite oxygen carrier. *Fuel* 318: 123663.

Liu, F., Zhao, J., Xuan, G. et al. (2021). Spatial evolution characteristics of active components of copper-iron based oxygen carrier in chemical looping combustion. *Fuel* 306: 121650.

Mangmeechai, A. (2022). The life-cycle assessment of greenhouse gas emissions and life-cycle costs of e-waste management in Thailand. *Sustainable Environment Research* 32: 1–13.

Marques, A.C., Marrero, J.M.C., and Malfatti, C.D.F. (2013). A review of the recycling of non-metallic fractions of printed circuit boards. *SpringerPlus* 2: 1–11.

Merlo, A., Lavagna, L., Suarez-Riera, D., and Pavese, M. (2020). Mechanical properties of mortar containing waste plastic (PVC) as aggregate partial replacement. *Case Studies in Construction Materials* 13: 443–467.

Mukundan, S., Wagner, J.L., Annamalai, P.K. et al. (2022). hydrothermal co-liquefaction of biomass and plastic wastes into biofuel: study on catalyst property, product distribution and synergistic effects. *Fuel Processing Technology* 238: 107523.

Murty, M.V.S., Rangarajan, P., Grulke, E.A., and Bhattacharyya, D. (1996). Thermal degradation/hydrogenation of commodity plastics and characterization of their liquefaction products. *Fuel Processing Technology* 49: 75–90.

Nanda, S. and Berruti, F. (2021). Thermochemical conversion of plastic waste to fuels: a review. *Environmental Chemistry Letters* 19: 123–148.

Pant, D. and Dhiman, V. (2020). An overview on environmental pollution caused by heavy metals released from e-waste and their bioleaching. *Advances in Environmental Pollution Management: Wastewater Impacts and Treatment Technologies* 4: 41–53.

Pathak, S.J., Gangal, A., and Prabu, V. (2022). Direct liquefaction of discarded printer cartridge plastics and its kinetic modelling. *Fuel Processing Technology* 228: 107147.

Qasim, M., Ayoub, M., Ghazali, N.A. et al. (2021). Recent advances and development of various oxygen carriers for the chemical looping combustion process: a review. *Industrial and Engineering Chemistry Research* 60: 8621–8641.

Quan, C., Li, A., and Gao, N. (2013). Combustion and pyrolysis of electronic waste: thermogravimetric analysis and kinetic model. *Procedia Environmental Sciences* 18: 776–782.

Rai, C., Bhui, B., and Prabu, V. (2022). Techno-economic analysis of e-waste based chemical looping reformer as hydrogen generator with co-generation of metals, electricity and syngas. *International Journal of Hydrogen Energy* 47: 11177–11189.

Rocchetti, L. and Beolchini, F. (2014). Environmental burdens in the management of end-of-life cathode ray tubes. *Waste Management* 34: 468–474.

Sahle-Demessie, E., Mezgebe, B., Dietrich, J. et al. (2021). Material recovery from electronic waste using pyrolysis: emissions measurements and risk assessment. *Journal of Environmental Chemical Engineering* 9: 104943.

Sarafraz, M.M., Jafarian, M., Arjomandi, M., and Nathan, G.J. (2017). The relative performance of alternative oxygen carriers for liquid chemical looping combustion and gasification. *International Journal of Hydrogen Energy* 42: 16396–16407.

Seshasayee, M.S. and Savage, P.E. (2020). Oil from plastic via hydrothermal liquefaction: production and characterization. *Applied Energy* 278: 115673.

Shibata, E., Yamamoto, S., Kasai, E., and Nakamura, T. (2003). Formation behavior of PCDD/FS in PVC pyrolysis with copper oxide. *Chemosphere* 50: 1235–1242.

Taghiei, M.M., Feng, Z., Huggins, F.E., and Huffman, G.P. (1994). Coliquefaction of waste plastics with coal. *Energy and Fuels* 8: 1228–1232.

Ukarde, T.M. and Pawar, H.S. (2021). A Cu doped TiO_2 catalyst mediated Catalytic Thermo Liquefaction (CTL) of polyolefinic plastic waste into hydrocarbon oil. *Fuel* 285: 119155.

Vicente, G., Aguado, J., Serrano, D.P., and Sánchez, N. (2009). HDPE chemical recycling promoted by phenol solvent. *Journal of Analytical and Applied Pyrolysis* 85: 366–371.

Wang, L. and Chen, P. (2004). Development of first-stage co-liquefaction of Chinese coal with waste plastics. *Chemical Engineering and Processing: Process Intensification* 43: 145–148.

Wang, X., Sun, M., and Wang, Q. (2019). Liquefaction characteristics of waste electronic plastics in supercritical isopropyl alcohol. *Ekoloji* 28: 3235–3247.

Weiland, F., Lundin, L., Celebi, M. et al. (2021). Aspects of chemical recycling of complex plastic waste via the gasification route. *Waste Management* 126: 65–77.

Williams, P.T. and Slaney, E. (2007). Analysis of products from the pyrolysis and liquefaction of single plastics and waste plastic mixtures. *Resources, Conservation and Recycling* 51: 754–769.

Xiao, Y., Oorsprong, M., Yang, Y., and Voncken, J.H.L. . (2008). Vitrification of bottom ash from a municipal solid waste incinerator. *Waste Management* 28: 1020–1026.

Yang, W.D., Sun, Q., and Ni, H.G. (2021). Cost-benefit analysis of metal recovery from e-waste: implications for international policy. *Waste Management* 123: 42–47.

Zhan, L., Jiang, L., Zhang, Y. et al. (2020). Reduction, detoxification and recycling of solid waste by hydrothermal technology: a review. *Chemical Engineering Journal* 390: 124651.

Zhang, Y., Liu, S., Xie, H. et al. (2012). Current status on leaching precious metals from waste printed circuit boards. *Procedia Environmental Sciences* 16: 560–568.

Zhao, X., Xia, Y., Zhan, L. et al. (2019). Hydrothermal treatment of e-waste plastics for tertiary recycling: product slate and decomposition mechanisms. *ACS Sustainable Chemistry and Engineering* 7: 1464–1473.

Zhao, Y., Li, Y., Jin, B., and Liang, Z. (2022). Layered double hydroxide derived bifunctional Ca-Fe-Mg material for integrated CO_2 capture and utilization via chemical looping strategy. *Chemical Engineering Journal* 431: 133826.

Zhu, N., Xiang, Y., Zhang, T. et al. (2011). Bioleaching of metal concentrates of waste printed circuit boards by mixed culture of acidophilic bacteria. *Journal of Hazardous Materials* 192 (2): 614–619.

14

Mechanical Effects of Recycling Plastics from Electronic Waste in Concrete: A Detailed View

Ana Laura De la Colina Martínez[1,*] *and David Joaquín Delgado Hernández*[2]

[1] *Doctora en Ciencia de Materiales por la Facultad de Química, Universidad Autónoma del Estado de México, Paseo Colón Esquina Paseo Tollocan S/N, 50000, Toluca, México, México*
[2] *Facultad de Ingeniería, Universidad Autónoma del Estado de México, Avenida Universidad S/N, Cerro de Coatepec, Ciudad Universitaria, 50110, Toluca, México*
* *Corresponding author*

Introduction

This review is divided into three sections. The first one corresponds to the actual researchers on recycling electronic waste polymers as fillers and reinforcers for construction materials. In the second section, the relevance of using gamma radiation as a technology that improves the mechanical performance of polymers and composites is discussed, especially in the line of the morphological, mechanical, and physicochemical modifications that it induces. Finally, the last section presents the gamma radiation applied as a technology that improves the recycling of e-plastic waste, particularly on electronic waste polycarbonate.

14.1 Electronic Waste Polymers

14.1.1 Global Production of Electronic Waste

In 2019, humanity emitted almost 54 million metric tons (Mt) of Waste Electrical and Electronic Equipment (WEEE), widely known as electronic waste. By 2030, this number will reach almost 75 Mt. Recent statistics reveal that China was the most important electronic waste generator, having produced approximately 10 Mt of electronic waste (7.2 kg per capita). In the Americas, the top producer of electronic waste is the United States of America (USA), having generated almost 7.0 Mt (21 kg per capita). After the USA, the three countries that generate the most electronic waste are India (3.2 kg), Japan (2.6 kg), and Brazil (2.1 kg). Together, these five countries generated almost 50% of electronic waste.

The electronic waste includes an extensive variety of manufactured goods, which can be classified into the following groups that match directly to their waste management characteristics: temperature interchange apparatus, lamps, large machines, small machines, and small IT and telecommunication apparatus (United Nations Environment Programme 2017).

14.1.2 Recycling Electronic Waste

Almost 18% of 2019's electronic waste was collected and recycled. The fate of the other 82.6% (44.3 Mt) is in doubt, and its situation and the ecological effect contrasts through the diverse zones (Figure 14.1). Recycling actions are not keeping pace with the macro augmentation of electronic waste (Forti et al. 2020). For example, in 2015 the smartphone life cycle was 19.5 months in China and 21.6 months in the USA (Balde et al. 2017). Continuing in this way, the style in which people produce, consume, and dispose of electrical and electronic equipment is unsustainable (Forti et al. 2020).

14.1.3 Recycling Electronic Waste Plastics

A large variety of valuable materials as plastics (27–30%) are in electric and electronic waste (Balde et al. 2017; Yang et al. 2013). In general, the plastics existent in the electronic waste are: acrylonitrile-butadiene-styrene (ABS) (30%), high-impact polystyrene (HIPS) (25%), polycarbonate (PC) (10%), polycarbonate/acrylonitrile-butadiene-styrene (PC/ABS) blends (9%), polypropylene (PP) (8%), polyphenylene ether/high-impact polystyrene (PPE/HIPS) blends (7%), and others such as polyvinyl chloride (PVC), polystyrene (PS), polyamide (PA), and polybutylene terephthalate (PBT) (Yang et al. 2013). However, it is possible to find other plastics.

There are three common methods of plastic recycling: chemical, mechanical, and thermal. In the chemical

Electronic Waste Management: Policies, Processes, Technologies, and Impact, First Edition. Edited by Sunil Kumar and Vineet Kumar.

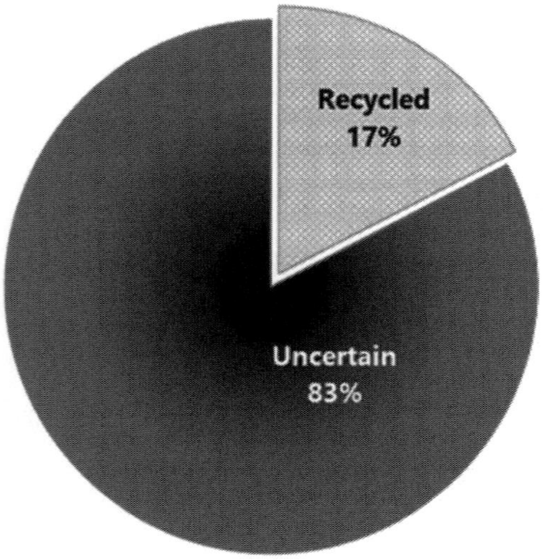

Figure 14.1 Electronic waste recycled and non-recycled in 2019.

method, after identification and sorting of different waste plastics, they can be de-polymerized, and they can be used as raw materials. In the mechanical technique, waste plastics are pelletized, to create different plastic merchandise; and in the thermal way, waste plastics are burned as alternative fuel (United Nations Environment Programme 2017). However, the main problems concerning recycling electronic waste plastics are the polymers mixtures, the variability of their composition and the presence of additives, which are the essential ingredients that convert them into commercially useful plastics. They perform three essential functions: the process of fabrication, the plastic properties, and to correct the flaws of the first two classes of additives. They are known as fillers, pigments, plasticizers, stabilizers, and others (Coleman 2011). In addition, the separation of the electronic waste plastics contains a series of technical challenges, since they can contain flame retarders (De la Colina-Martínez et al. 2017).

The slow adoption of recycling of electronic waste, and particularly of plastics inside it, generates aftereffects such as the emission of toxic substances, the production of greenhouse gases, and the spending of resources. Subsequently, numerous countries are confronted by the substantial ecological and social risks of ineffectively controlled electronic waste. In this sense, it is especially important to appropriately manage e-plastic waste, adjust recycling and collection tax, and decrease the amount of e-plastic waste that finishes in landfills (Forti et al. 2020). The incorporation of electronic waste into the civil engineering field seems to be one of the finest keys, due to its monetary and ecological benefits.

14.1.4 Recycling Electronic Waste Plastics in Concrete

Use of electronic waste in the civil engineering field is divided in two areas, which correspond to the principal binding raw material used: cement and asphalt. When using cement, it is possible to obtain two products, concrete and mortar. This chapter will focus specifically on concrete.

Many studies are being carried out on the use of plastic waste in concrete. In general it is possible to recognize that the decrease in density is clearly equivalent to the plastic aggregates content (Choi et al. 2005). Besides, the inclusion of plastic waste in concrete produces the reduction of air content (Soroushian et al. 2003), slump (Hama and Hilal 2019; Soroushian et al. 2003) and compact factor (Hama and Hilal 2019). Compressive strength of concrete decreases with the addition of plastic waste (Choi et al. 2005; Marzouk et al. 2007; Soroushian et al. 2003). Both splitting tensile strength and modulus of elasticity of concrete decrease with the increase of plastic waste aggregates (Choi et al. 2005). Significant benefits in the impact resistance are due to the inclusion of plastic waste. However, there are important differences in this improvement, caused by diverse plastic waste reinforcement systems. Moreover, the presence of plastic waste in concrete decreases air permeability (reduced air permeability supports long-term stability of concrete). The inclusion of plastic waste in concrete also causes a certain decrease in the abrasion resistance of concrete (Siddique et al. 2008; Soroushian et al. 2003).

There have been some reviews that summarize the utilization of plastic waste in concrete, for example, those by Gu and Ozbakkaloglu (2016), Saikia and De Brito (2012), Pacheco-Torgal et al. (2012), and Siddique et al. (2008). In this last review, plastic wastes (debris and fibers) are classified. Concretes containing plastic waste such as HDPE, PA, polyethylene terephthalate (PET), polystyrene (PS), PVC, expanded polystyrene (EPS), melamine, polyurethane (PUR), phenol formaldehyde (PF), and PP were reviewed. Virgin plastic materials such as EPS, PP, and nylon were equally reviewed. On the other hand, Sharma and Bansal (2016) reviewed various forms of plastic waste such as flakes (PET and HDPE), shredded fibers of bags of polyethylene (PE), bottle fibers (PET), and granulates and pipes (PVC) utilized to substitute diverse components of concrete.

However, all these articles and reviews do not cover the role of electronic waste plastics in concrete. The improvement of innovative materials of construction consuming electronic waste plastics is also very significant to the environment and to the construction industry. The use of them in construction material with concrete could help save energy, reduce disposal problems, and minimize environmental contamination.

Electronic waste plastics, as a complementary material in concrete, show encouraging results in different studies. The aims of this use are to prevent the leaching of toxic elements in oceans, rivers, and soils and to save the valuable natural resources that are constantly extracted to acquire aggregates. Researchers have tested different kinds of electronic waste plastics (bricks and fibers) and have included distinct types of them into hydraulic and polymeric concrete. However, it is necessary to validate the properties of the resulting composites (Luhar and Luhar 2020).

For these reasons, the move toward the utility of electronic waste plastics in civil engineering offers new and exciting opportunities. A detailed analysis on the influence of electronic waste plastics on the characteristics of concrete, in particular workability, density, chloride and sulfate resistance, compressive strength, tensile strength, splitting tensile strength, flexural strength, modulus of elasticity, toughness, and water penetration depth of concrete are discussed below.

14.1.5 Investigations of Electronic Waste Plastics in Concrete

In this section, electronic waste plastics (applied by researchers) are catalogued without specifying their composition. In addition, the main way in which these electronic waste plastics are added to concrete is in the form of debris and in substitution of concrete aggregates as coarse or fine particles.

14.1.5.1 Coarse Aggregate Replacement

The following assessment included a series of studies in which electronic waste plastic replaces part of the coarse aggregate of concrete. The physical and mechanical properties that concrete acquires, once waste plastic is added, are discussed below.

- **Compressive strength of concrete**

The document presents the experiment of concrete prepared with e-plastic waste as part of coarse aggregate. The e-plastic waste particles with a specific gravity 1.0 and angular shape were utilized as limited coarse aggregate. Electronic waste in the character of imprecisely scrapped, residual, obsolescent, fragmented, electronic or electrical appliances from market informal recyclers have been accumulated, broken, and milled to the scrap dimension. The fragmented particle size is expected to be between 1.18 mm and 2.36 mm (Lakshmi and Nagan 2011).

The test results (Lakshmi and Nagan 2011) show that the 28-day concrete compressive strength goes from 28.8 MPa (conventional concrete) up to 18.8 MPa and 8.2 MPa for 8% and 20% of e-plastic, respectively. However, the incorporation of fly ash (10%) slightly decreases this compressive strength to 27.4 MPa for 8% of e-plastic, and to 22.0 MPa for 20% of e-plastic. The incorporation of fly ash leads to a loss in the resistance of almost 5% and 24%, respectively, avoiding in that order, a loss of 35% and 71% in the absence of fly ash. See Figure 14.2.

In the same way, another study shows the performance of electronic waste plastic fragments in concrete and analyzes the viability of making use of crushed e-plastic particles as limited replacements for coarse aggregate. The results show that when e-plastic unaccompanied was utilized, there was a reduction in compressive strength, but once 10% fly ash was included, effects equivalent to the reference sample were achieved. It is for this reason that use of this electronic waste in concrete will diminish the percentage of traditional coarse aggregates (Akram et al. 2015). All fly ashes exhibit cementitious characteristics that modify the chemical and physical behavior of cement (Basham et al. 2007).

Other investigation shows a decrease in the compressive strength when the proportion of e-plastic waste increases, the greatest reduction being 44% in reference to the standard mixture (Sabău and Vargas 2018).

For its part, an experimental analysis shows that the 28-day compressive strength decreases slightly from 47.18 MPa (conventional concrete) to 44.07 MPa when 10% of e-plastic has been added (indicating a loss in resistance of 7%). See Figure 14.3. However, for higher values of e-plastic such as 20%, it decreases to 24.69 MPa, indicating

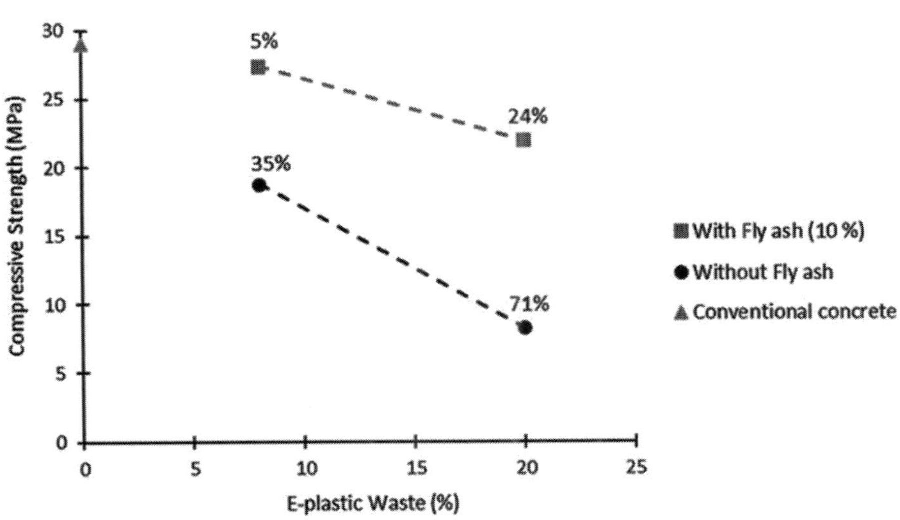

Figure 14.2 28-days compressive strength of concrete with e-plastic waste.

a loss in resistance of almost 48%. In contrast, this last value could be acceptable, if what is desired is to incorporate 20% e-plastic and compressive strengths at 28 days close to 20 MPa (Manjunath 2016).

This last study can be compared with other investigation that covers the revision of concrete mixes with e-plastic waste. The mix proportion was M 20, where M alludes to the blend and 20 to the 28-day compressive strength (Bureau of Indian Standards 2000) in MPa. The compressive strength was compared for diverse proportions of e-plastic waste with conventional concrete. The 28-day compressive strength decreases from 21.2 MPa to 17.14 MPa (for 10% of e-plastic) and 9.62 MPa (for 20% of e-plastic). That is a loss in resistance of 19% and 55%, respectively. See Figure 14.3 (Nadhim et al. 2018).

In Figure 14.3, the incorporation of 10% e-plastic waste is favored by using a cement that gives greater resistance (50 MPa) (Manjunath 2016); then the compressive strength does not fall below 7%. However, when using a cement that provides a lower resistance (20 MPa) (Nadhim et al. 2018), the possibility of this e-plastic incorporation (10%) is lost, since the compressive strength falls as much as 19%.

In a further study, the electronic plastic waste was cut and divided into three sizes, 4.7 mm, 10 mm, and 20 mm. This plastic was replaced by coarse aggregate in the concrete in various proportions (from 5% to 25%). M 20 and M 25 grade concretes were used. Testing was done on 150 mm x 150 mm x 150 mm-sized blocks. The 28-day compressive strength results for M 25 concrete show a loss of 12% and 40%, and for M 20 concrete show a loss of 10% and 37.5%, when the e-plastic reached 10% and 20%, in that order (Manatkar and Deshmukh 2015).

It can be established that concrete with lower strengths (20 MPa or 25 MPa) supports major the incorporation of 10% or 20% of e-plastic, when it is crushed and used in various sizes, 4.75 mm, 10 mm, and 20 mm, compared with the results of Nadhim et al. (2018) (Figure 14.3), resulting in a strength loss of 12% and 40%, respectively.

Another research study shows an investigation on structural concrete with electronic plastic waste. In this investigation the losses in 28-day compressive strength were 8%, 16%, 31%, 41%, and 48%, respectively. The 28-day compressive strength of concrete with this plastic was demonstrably lower than that of the standard concrete (40 MPa) (Senthil Kumar and Baskar 2015).

- **Flexural strength of concrete**

In an experimental study, the flexural strength of concrete at 28 days when 10% and 20% e-plastic waste were added was the same as for conventional concrete (4.4 MPa). However, for the highest value of e-plastic waste (30%), this value decreases to 2.5 MPa (Manjunath 2016). See Figure 14.4.

Another research study shows that the flexural strength of treated concrete was relatively less than that of the standard concrete (Senthil Kumar and Baskar 2015). In this analysis the flexural strength of concrete at 28 days with 10% and 20% of e-plastic waste was 6.4 MPa and 5.8 MPa, respectively. However, for the highest value of e-plastic waste (50%), this value decreases to 4.5 MPa. See Figure 14.4.

Figure 14.4 shows the flexural strength decrease for two diverse mixes of concrete with e-plastic waste; the first mix reached a 28-day compressive strength of 50 MPa (Manjunath 2016), and the second reached 40 MPa (Senthil Kumar and Baskar 2015). The losses in percentage also are shown.

With adherence to IS 456 (2000), the flexural strength of concrete can be calculated from the compressive strength with Equation (14.1), where ***fcr*** is flexural strength and ***fck*** is compressive strength (each one in N/mm^2).

$$fcr = 0.7\sqrt{fck} \tag{14.1}$$

With this in mind, the flexural strength expected for a conventional concrete (calculated) and the flexural strength of concrete with electronic waste plastic from the above researches (Manjunath 2016; Senthil Kumar and Baskar 2015) are presented. The results show that there is a

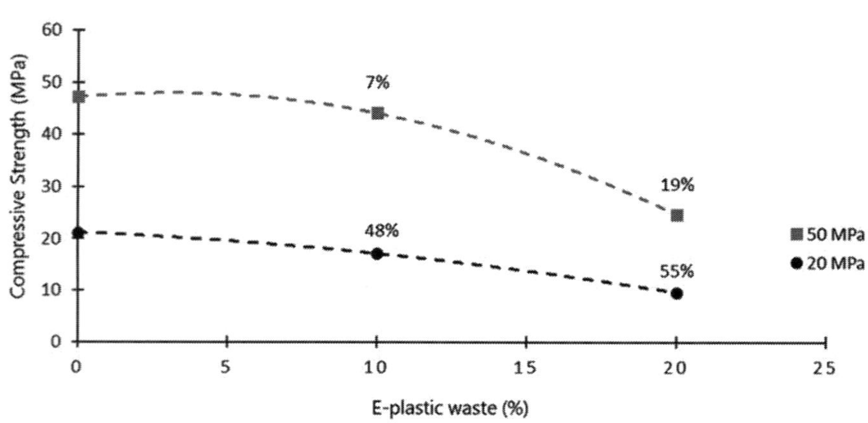

Figure 14.3 Compressive strength of concrete at 28 days with e-plastic waste.

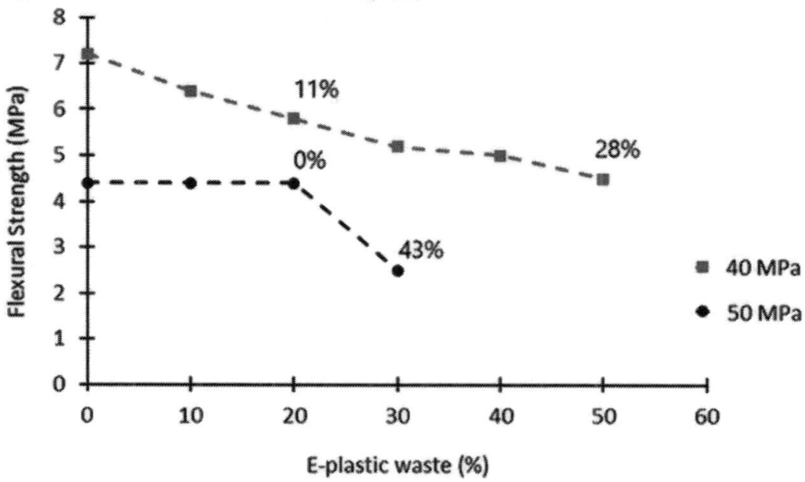

Figure 14.4 Flexural strength of concrete at 28 days with e-plastic waste.

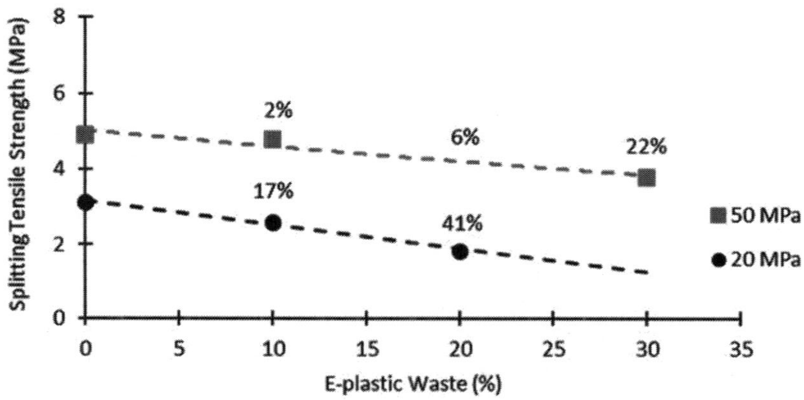

Figure 14.5 Splitting tensile strength of concrete at 28 days with e-plastic waste.

significant gap in the middle of the flexural strength expected and the flexural strength of concrete with electronic waste plastic. For instance, the calculated flexural strength for a concrete with a compressive strength of 40 MPa was comparatively less than that of the concrete with electronic waste plastic (30 MPa). In this report the flexural strength of concrete at 28 days, with 10% and 20% e-plastic waste, was 6.4 MPa and 5.8 MPa, respectively. However, the flexural strength calculated when 10% and 20% e-plastic waste were added was 4.5 MPa and 3.2 MPa, respectively. It shows that the expected values for conventional concrete (calculated) are much lower than those obtained experimentally for concrete with electronic waste plastic.

- **Splitting tensile strength**

In this experimental investigation, the splitting tensile strength at 28 days, in which a standard test cylinder is loaded in compression on its side, was almost the same as the conventional concrete (4.9 MPa) when 10% and 20% e-plastic waste were added. However, for the highest value of e-plastic waste (30%), this strength decreases to 3.8 MPa (Manjunath 2016). See Figure 14.5.

The research above can be compared with the investigation made by Nadhim et al. (2018), which covers the study of concrete mixes with e-plastic waste. The mix proportion was obtained with M 20 grade concrete for a compressive strength of 20 N/mm^2 at 28 days (Bureau of Indian Standards 2000). The results show that splitting tensile strength at 28 days falls from 3.12 (conventional concrete) up to 2.60 and 1.85 MPa, for 10% and 20% of e-plastic waste, respectively. See Figure 14.5.

In Figure 14.5 the inclusion of 10% and 20% e-plastic waste is favored by using a cement that gives greater resistance to concrete (50 MPa), since the resistance does not fall above 6%. However, when using a cement that provides lower resistances such as M 20 grade concrete (20 MPa) (Nadhim et al. 2018), the possibility of these incorporations (10% and 20% e-plastic waste) is deficient, since the resistances fall to 17% and 41%, respectively.

Another research study (Senthil Kumar and Baskar 2015) demonstrates that the splitting tensile strength of limitedly replaced concrete was (6, 19, 24, 37, and 49%, respectively) less than that of the target concrete (3.1 MPa).

In another investigation, it was found that splitting tensile strength of e-plastic waste aggregates-replaced concrete in the presence of fly ash (10%) is more (2.7 MPa) than the standard concrete (2.5). It was also perceived that the standard concrete samples had a sudden rupture complemented by noise. However, for the experiments with the plastic waste, the rupture was softer and without any sound. It shows that e-plastic will develop plasticity (Akram et al. 2015). The splitting tensile strength at 28 days is lower in the samples that only incorporate e-plastic. It is also true that this mixture seems to withstand better the incorporation of more e-plastic, exhibiting lower percentage losses (0, 13, and 15%) than those presented by the mixture that incorporates fly ash (9, 16, and 33%).

- **Tensile strength of concrete**

In the article by Lakshmi and Nagan (2010), concrete with addition of 10% fly ash also was investigated. The strength criteria was M 20 grade of concrete with compressive strength of 20 N/mm^2 at 28 days (Bureau of Indian Standards 2000). The compressive strength, splitting tensile strength, and flexural strength of the concrete were evaluated.

A laboratory comparison explains that the splitting tensile strength of conventional concrete is maybe as much as 150% of the direct tensile strength (Gerald and Neville 2015). In this sense, the tensile strength of concrete mix from the above search (Lakshmi and Nagan 2010) could be calculated. See Figure 14.6.

- **Concrete resistance to chloride and sulfate attack**

A certain search shows the effects of chloride and sulfate attack on concrete with e-plastic waste incorporating fly ash. For the chloride attack test the tester cubes were immersed in sulfuric acid solution (5%) and hydrochloric acid solution (5%). After 105 days, measurement of the weight and the compressive strength of the cubes was completed. For the sulfate attack test the cubes (3375 mm^3) after 28 days of curing were immersed in sodium sulfate solution (10%). After 105 days, measurement of the weight and the compressive strength of the cubes was completed. The findings show a higher weight loss in grams for the conventional concrete (not including fly ash) than for the concrete with e-plastic waste and fly ash (10%). In the chloride attack test, for example, the weight loss of the conventional concrete was 265 g compared to 180 g from the concrete with e-plastic waste and fly ash. Similarly, in the sulfate resistance test, the weight loss was 265 g contrasted to 178 g. The results also show a higher compressive strength loss for the conventional concrete than for the concrete with e-plastic waste and fly ash (10%). For instance, in the chloride attack the compressive strength loss was 2.65% for the conventional concrete compared to 2.46% for the concrete with e-plastic waste and fly ash, and in the sulfate attack it was 3.63% compared to 3.52% (Lakshmi and Nagan 2011). Concrete prepared with e-plastic waste and fly ash (10%) shows less erosion and more compressive strength than conventional concrete (Figure 14.7).

Other experiments also confirm greater stability to chlorides and sulfates in concrete samples incorporating fly ash. Figure 14.8 presents the effects of fly ash on both chloride penetration (ASTM standard, C1202-10) and sulfate resistance of concrete (ASTM standard, C1012-02). The inclusion of fly ash in concrete origins a notable diminution in chloride ion penetration and a favorable decrease in length change at 180 days (Dissanayake and Priyadarshana 2016).

- **Density of concrete**

The results of another paper show an important reduction (22%) in the density of the concrete combinations with e-plastic waste, which suggests that more weightless concrete can be created with this e-plastic waste (Sabău and Vargas 2018). Based on the findings obtained, the

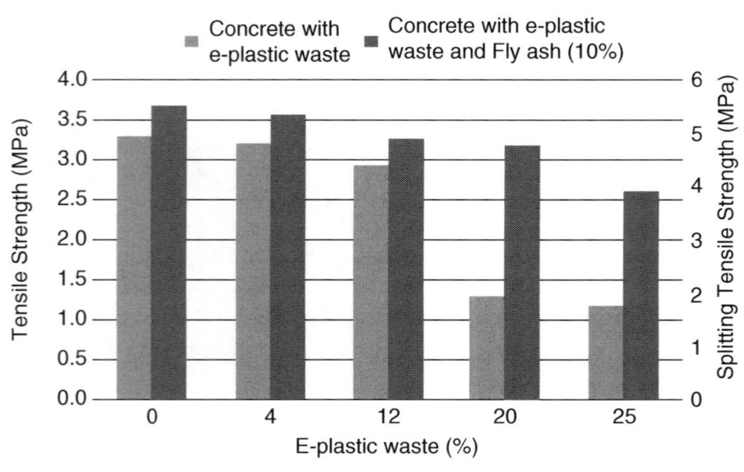

Figure 14.6 Tensile strength of concrete mixes, calculated from splitting tensile strength.

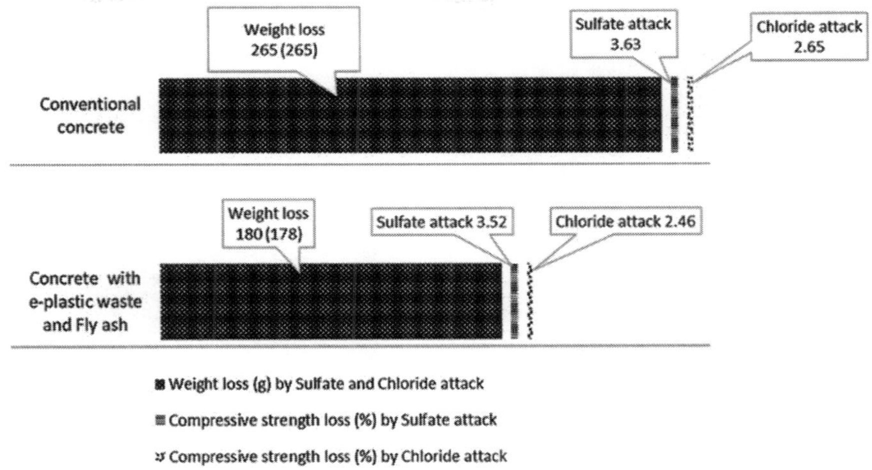

Figure 14.7 Concrete with e-plastic resistance to chloride and sulfate attack.

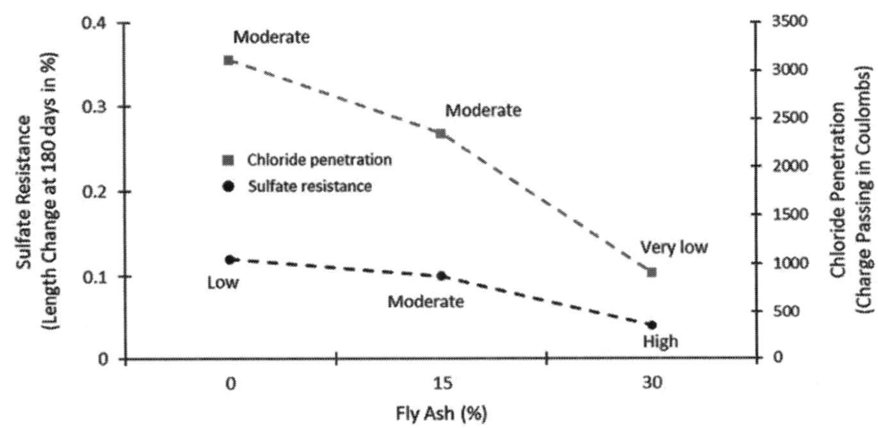

Figure 14.8 Resistance of concrete with fly ash to chloride and sulfate attack.

concrete mix without e-plastic waste can be categorized as standard weight concrete, with a density of 2426 kg/m^3. Predictably, for the concrete mixtures with e-plastic waste lower densities have been recorded. The lowest value (1892 kg/m^3) was recorded for the concrete mix with the highest percentage of e-plastic waste (60%), which means a 22% weight reduction.

Another research study shows that the addition of e-plastic waste decreased the fresh and dry density of concrete (Senthil Kumar and Baskar 2015). Figure 14.9 shows how densities diminish to a loss of 14% and 18%, respectively. When a maximum of 50% of e-plastic waste was added, the minimum dry density of 2139 kg/m^3 was recorded. However, obtaining low-density concretes (1120–1920 kg/m^3), according to the American Concrete Institute (ACI 2003), is still far from the results obtained in this probe.

- **Workability of concrete**

There are diverse ways of classifying workability, depending on the slump obtained. As a general rule, a small slump level implies low workability and a big value signifies high workability. For instance, according to Koehler and Fowler (2003) and Wilby (1991), slump can be ordered as follows: Very Low (0–25 mm), Low (25–50 mm), Medium (50–100 mm), High (100–180 mm), and Very High (>180 mm). On the other hand, European Standard EN 206-1 2000 (EN 206-1 2000), grades it in five categories: S1 (10–40 mm), S2 (50–90 mm), S3 (100–150 mm), S4 (160–210 mm), and S5 (≥220 mm). Corresponding to the standard IS 456 (2000) (Bureau of Indian Standards 2000), slump can be categorized as: Very Low (compacting factor determination will be more applicable than slump), Low (25–75 mm), Medium (50–100 mm), High (100–150 mm), and Very High (determination of flow will be more appropriate than slump).

An experimental paper shows that although the workability of the concrete mix (slump) remains high or S3 when adding 10% e-plastic waste (114 mm), contrasted to typical concrete (128 mm), it is also true that higher percentages of e-plastic waste (30%) decrease the slump to values close to 75 mm, which is equivalent to medium or S2 workability (Figure 14.10).

Another research study (Senthil Kumar and Baskar 2015) presents that although the workability of the concrete mix (slump) remains Low or S1 when adding 10% e-plastic waste (24 mm) in comparison with regular concrete (32 mm), it is also true that higher percentages of e-plastic waste (20% to 50%) decrease the slump to Very Low or S1 workability (Figure 14.10).

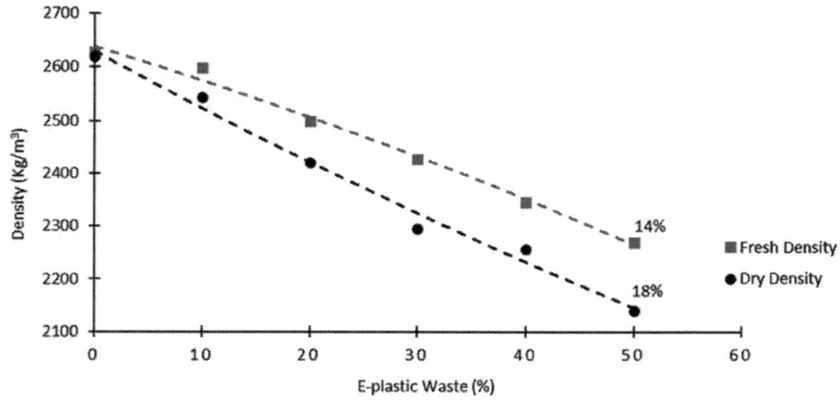

Figure 14.9 Effect of e-plastic waste on the fresh and dry density of concrete.

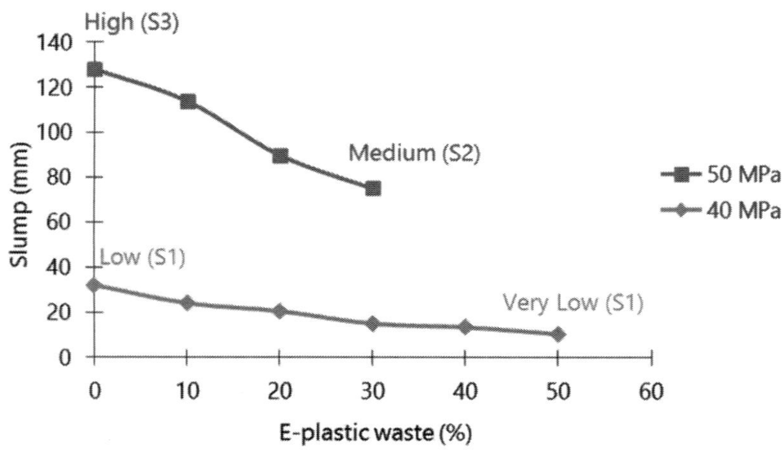

Figure 14.10 Slump of concrete with electronic waste plastic.

14.1.5.2 Fine Aggregate Replacement

The following analysis included a series of studies in which electronic waste plastic replaces part of the fine aggregate of concrete. The physical and mechanical properties of these concretes are discussed below.

- **Compressive strength of concrete**

In another paper, the compressive strength was analyzed and related with reference concrete (Arora et al. 2013). The results show a slight decrease in 28-day compressive strength of 8, 12, and 18% when using 2, 6, and 8% electronic waste plastic, respectively.

On the other hand, the results of Damal et al. (2015) show a considerable decrease in 28-day compressive strength of 7, 48, and 53% when using 7.5, 15.0, and 21.5% electronic waste plastic, respectively.

The document in which various percentages of e-plastic aggregates (0, 5, 10, 15, and 20%) and silica fume (0, 5, and 10%) were considered (Farooq 2019) shows that the increment in the electronic waste plastic produces a reduction in compressive strength values (at 28 days). However, if silica fume is added to the concrete, it culminates in advantageous values slightly superior to the values for the standard concrete.

- **Flexural strength of concrete**

The results of Farooq (2019) show that an increase in the electronic waste plastic produces a reduction in flexural strength values (at 28 days). However, if silica fume is added to the concrete it finishes in strong values slightly superior to those for the reference concrete.

- **Splitting tensile strength of concrete**

The research (Farooq 2019) shows that an increase in the electronic waste plastic produces a reduction in splitting tensile strength values (at 28 days). However, when silica is added to the concrete, improvement values slightly superior to those for the standard concrete are obtained.

- **Density of concrete**

The reference study by Farooq (2019) shows that an increase in the electronic waste plastic produces a reduction in 28-days dry density values. However, if silica is added to the concrete, it amplifies the dry density values.

- **Workability of concrete**

The results of the article by Farooq (2019) show that an increase in the electronic waste plastic produces a slight increase in slump values.

14.1.5.3 Fine and Coarse Aggregate Replacement

The following review included a report in which electronic waste plastic replaces part of fine and coarse aggregates of concrete. The physical and mechanical properties of concrete are discussed below.

In a research paper by Evram et al. (2020), the specific gravity of electronic waste plastic was 1.07 and its size ranged between 0 and 14 mm. The e-plastic waste was utilized to substitute fine and coarse aggregates from 0 to 40% by volume.

- **Compressive strength of concrete**

In an investigation, the integration of e-plastic decreased the 28-day compressive strength compared to that of the standard mixture. However, with the assimilation of MD, there was a minor augment in the compressive strength of the blends (Evram et al. 2020).

- **Toughness of concrete**

In general, in the research article, the e-plastic addition decreased the concrete toughness; but, with the inclusion of MD, there was a minor augment in the toughness of the blends (Evram et al. 2020).

- **Splitting tensile strength of concrete**

In the analysis (Evram et al. 2020) the 28-day splitting tensile strength of the test combinations with e-plastic were inferior to those with the standard concrete.

- **Modulus of elasticity of concrete**

In the research (Evram et al. 2020), the modulus of elasticity of the standard concrete was 41 GPa and declined by 7, 15, 24, and 29% at 10, 20, 30, and 40% e-plastic substitute percentages, respectively. MD inclusion did not indicate a substantial influence on the modulus of elasticity.

- **Water penetration depth of concrete**

An investigation shows that the combinations including 10% e-plastic at 28 days, either with or without MD, had somewhat lower water penetration depths than the indicator blend (Evram et al. 2020).

- **Density of concrete**

The research article shows that the dry unit weight of 28-day concrete containing e-plastic decreased as the replacement ratio increased (Evram et al. 2020).

- **Workability of concrete**

In the investigation, the slump of concrete mixtures with e-plastics and without MD was around 220 mm at all replacement levels. However, addition of MD up to 15% at each e-plastic replacement level decreased slump value up to 180 mm (Evram et al. 2020).

14.2 Gamma Radiation

In this section, the relevance of using gamma radiation as a technology that improves the mechanical performance of polymers and composites is discussed, especially in the line of the morphological, mechanical, and physicochemical modifications that it induces.

14.2.1 Properties of Gamma Radiation

Gamma radiation, usually produced by radioactive elements, is a kind of electromagnetic radiation that is emitted by unstable atomic nuclei when they go from a high to a low energy state (Peña Pichardo 2018). Gamma radiation, which is an ionizing radiation, consists of photons that have sufficient energy to remove the electrons bound to the atoms. In the case of Cobalt-60 (^{60}Co), the unstable nucleus releases a beta particle with a maximum energy of 0.31 MeV (β-) and two gamma rays (ɣ) of 1.17 and 1.33 MeV (Centro Nacional de Aceleradores 2016; Silva Aquino 2012) (Figure 14.11).

The advantages over similar methods are that irradiated products can be used immediately, samples are minimally heated, it is a very accurate method, and it is reproducible. In addition, it has a high penetrability, subsequently, it is often used on products that are not uniform and have a

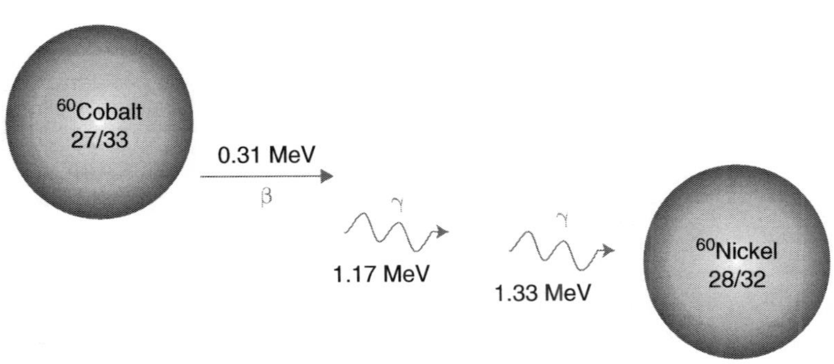

Figure 14.11 Cobalt-60 decays to Nickel-60.

high density. It is also easy to check, as only the dosage needs to be monitored (Chmielewski 2011).

Gamma rays impact the material substances in three diverse routes: the photoelectric effect, the Compton effect, and pair production.

14.2.1.1 Photoelectric Effect

This effect happens when photons, with the lowest energy, collide with matter. The electrons absorb all the energy of the photons. These electrons are released from the atom with a kinetic energy (Cruz Bastida and García Hernández 2015; Peña Pichardo 2018). In this case, the gamma ray, when interacting with an electron, transfers its energy completely.

14.2.1.2 Compton Effect

In this effect, photons with a medium energy, interacting with matter, collide with the electrons of the atom and release them. After colliding with the electrons, the incident electromagnetic radiation (such as gamma radiation) acquires a longer wavelength, decreases in energy, and takes a new direction. (Cruz Bastida and García Hernández 2015; Peña Pichardo 2018).

Although there are some similarities between the photoelectric effect and the Compton effect, since in both cases electrons are emitted due to the impact of photons against a material, the differences are particularly important. In the first case, the photon gives up all its energy to the electron and disappears, while in the second case, it is deflected with a lower energy (Cruz Bastida and García Hernández 2015; Peña Pichardo 2018).

14.2.1.3 Pair Production

When the photon (with an energy superior to 1.02 MeV) interacts with the atomic nucleus, it transforms into (materializes) an electron-positron pair (Cruz Bastida and García Hernández 2015; Peña Pichardo 2018).

14.2.2 Gamma Irradiation Effects on Plastics

14.2.2.1 Surface Morphology

Evaluation of SEM (scanning electron microscopy) images of some plastics such as PP and PI (from 0 to 230.4 kGy, at 27°C and in air) shows that the roughness of these particles increases with increasing dose. The exterior of gamma-irradiated PP is found to consist of imperfections, holes/hollows, cracks, fractures, and bubbles of different dimensions. This roughness can be attributed to various elements associated with the fracture of the bonds. Moreover, this external roughness agrees well with the Fourier transform infrared spectroscopy (FTIR) of PP, where chain cleavage and oxidation are evident. On the other hand, the roughness of PI is significantly lower than that of PP. This is still constant with the permanency of bonds as well as oxidation resistance of PI compared to radiation results as detected in FTIR and UV-visible investigations (Mathakaria et al. 2014).

The SEM study of the PE-electrolyte layers offers the physical evidence of their morphology. It can be appreciated that the exterior of the layer is soft and the spherical semi-crystalline regions inside polymers (spherulites) are visibly clear as well as are slight superficial fissures. In PE irradiated layers, the spherical semi-crystalline regions practically disappear. The external morphology appears to be more consistent after electron beam than gamma radiation, where more accumulation is seen. These modifications reveal movement of the molecules due to the influence of the radiation. These consequences suggest the polymer degradation due to chain scission, which relates well with the X-ray diffraction and dielectric results (Raghu et al. 2016).

14.2.2.2 Contact Angle

Particularly, it is noted that the contact angle of PP falls from 85.2° (0 kGy) to 62.2° (230.4 kGy). This reduction is approximately 27% and can be assigned to the oxidation and the roughening of the polymer surface. It has been examined in the FTIR and EDS (energy dispersive spectroscopy) analyses that PP is considerably oxidized later than radiation. It is detected that the contact angle of PI diminishes from the original result of 70.8° (0 kGy) to 67.6° (230.4 kGy). This reduction was 5%. Consequently, the PI remains to reveal its radiation stability (Mathakaria et al. 2014).

14.2.2.3 Evolved Gases

The yields of total evolved gas (containing CO_2, CO, N_2, H_2, and CH_4) were investigated from films of polyimides (as Kapton, Upilex-R, and Upilex-S). The evolved gas escalates consistently with gamma dose, but it has a tendency to decline at greater doses (Hegazy et al. 1992).

14.2.2.4 Tensile Strength

The tensile strength increases consistently with dose, but the yield per dose tends to diminish at higher doses (DuPont 2020).

An other review presents the effect of gamma radiation on the tensile strength of different Tyvek® (packaging of HDPE filaments) formulations (1073B, 1059B, and 2FS) (DuPont 2017). In this case, the tensile strength decreases proportionally with dose.

14.2.2.5 Elongation

Elongation (which is measured in percent) from films of polyimide by gamma irradiation decreases abruptly with dose (DuPont 2020). The effect of gamma radiation on elongation of some different Tyvek®(HDPE filament packing) formulations (1073B, 1059B, and 2FS) was researched (DuPont 2017). As the gamma dose increases, the elongation decreases. The higher the percentage, the better the performance.

14.2.2.6 Penetration

ASTM standard, F2638 is a real-time assessment that reports a pMax, which is the highest penetration for a specified material. This measurement be subject to some physical characteristics of the matter (mass, density, and diameter) (DuPont 2018).

Tyvek® is a packaging material made of continuous, randomly oriented, HDPE filaments. Some different Tyvek® formulations (1073B, 1059B, and 2FS) have a pMax of less than 0.5%, since other porous barrier materials (coated paper or synthetic fiber-reinforced) range from 8% to 18% (DuPont 2018).

The influence of gamma radiation on particle penetration of diverse Tyvek® (packaging of HDPE filaments) formulations (1073B, 1059B, and 2FS) was studied (DuPont 2017). The lesser the percentage of pMax, the greater the effectiveness. As the gamma dose increases, the particle penetration also increases (DuPont 2017).

14.2.2.7 Dielectric Strength

Dielectric strength is any of the main attributes of insulators. The dielectric strength of PI was studied, and it decreases regularly with dose, although the yield per dose tends to expand at greater doses (DuPont 2020).

14.2.2.8 AC Conductivity

The structural modifications in the irradiated PE-electrolyte of polyethylene are completely an effect on physical characteristics. AC conductivity was discovered to be augmented with increasing the dose (Raghu et al. 2016).

14.2.3 Gamma Radiation Effects on Concretes with Waste Plastics

The use of gamma radiation is becoming more important from an ecological and economic point of view, as it is seen as an underused alternative technology for the recycling of plastics, due to the fact that it modifies the chemical structure and properties of most polymers (Ávila Córdoba et al. 2015).

Some studies attest to the effects of gamma radiation as an alternative that has proven to be efficient in increasing the mechanical strength values of concrete when the gamma dose is applied to concrete containing waste plastics inside or when concrete is elaborated with irradiated waste plastic.

The results of irradiated (by 7 days during solidification) and non-irradiated specimens of concrete show that the compressive strengths of samples have been improved by gamma irradiation (Rezaei Ochbelagh et al. 2011).

The effects of the gamma dose on concrete can be explained as follows. When the concrete solidifies, the water (distributed in the concrete) vaporizes and voids are created in the concrete. These voids are usually trapped in the concrete. Due to these pores (macroscopic and microscopic) the strength of the concrete decreases considerably. Although macropores can be reduced by vibration, it does not affect the micropores. However, micropores can be reduced by irradiating the concrete during the drying process. This radiation can be gamma ray, neutron, or electron beam (Rezaei Ochbelagh et al. 2011).

On the other hand, the gamma radiation may change the texture of irradiated waste plastic. For instance, experimental results show that if the gamma irradiation augments, irregular roughness on the exterior of polycarbonate particles can be found (De la Colina Martínez et al. 2019). The same comportment is observed in PP (Mathakaria et al. 2014).

14.2.3.1 Compressive Strength

The following is an investigation with PET from waste bottles and its use as a sand substitute in concrete. The particle size and the percentage of post-consumer PET were variables. The gamma radiation dose applied to the processed concrete after 28 days of curing was 100 kGy. This investigation reveals that concrete with waste PET (2.5% by volume and 0.5 mm in size) exhibits a higher compressive strength when is irradiated (at 100 kGy) than when not (Ávila Córdoba et al. 2015).

By using gamma radiation, in another experiment, it was possible to improve the mechanical characteristics of hydraulic concrete blends developed with polypropylene fibers (PP-fibers) and marble. In this case, concrete irradiated at 50 kGy including 1.5% by volume of PP-fibers (10 mm long and 30–40 μm diameter), exhibits a higher compressive strength than concrete non-irradiated with the same percentage and kind of PP fibers (Martínez-Barrera et al. 2011).

In another research study, the compressive strength results, for concrete with waste tire debris, decrease gradually when the quantity of tire debris increases. Furthermore, all these blends of concrete exhibited results inferior to target concrete (24.1 MPa). Additionally, only concretes with 10% of tire debris irradiated at 200 and 300 kGy showed superior results, up to 23% of improvement, over those for standard concrete (Herrera-Sosa et al. 2015).

14.2.3.2 Modulus of Elasticity

PET from waste bottles was used as a sand substitute in concrete. The particle size and the percentage of PET were variables. The gamma radiation dose applied to the processed 28-day concrete was 100 kGy. The results show that concrete with PET (5% by volume and 0.5 mm in size) exhibits a higher modulus of elasticity when irradiated at 100 kGy (6 GPa) than when not irradiated (4 GPa) (Ávila Córdoba et al. 2015).

Other innovative investigation shows that the modulus of elasticity values vary as a function of concentration of waste tire debris and radiation dose. The values decrease when increasing the concentration of non-irradiated particles. The values are higher with respect to reference concrete for concrete with particles gamma irradiated to 300 kGy. However, the values are lower compared to those for standard concrete when the tire debris is irradiated to 200 kGy (Herrera-Sosa et al. 2015). Furthermore, only concretes with 10% of tire debris irradiated at 300 kGy showed higher values, up to 21% of improvement, than those for standard concrete.

14.2.3.3 Splitting Tensile Strength

In original research, the splitting tensile strength results decay when increasing the quantity of tire debris non-irradiated from waste. The results are higher for blends with tire debris gamma irradiated to 300 kGy. However, these strengths are lower with respect to target concrete, particularly when the concrete has tire debris irradiated to 200 kGy (Herrera-Sosa et al. 2015). A dose of 300 kGy allows an increase of these strength up to 4% with respect to the reference concrete.

14.2.3.4 Flexural Strength

In a recent article, the results of flexural strength change due to the amount of tire debris from waste and gamma irradiation dose. The flexural strengths decline when increasing the quantity of non-irradiated tire debris. The flexural strengths are lower with respect to standard concrete for concrete with particles irradiated to 200 and 300 kGy (Martínez-Barrera et al. 2020). However, concretes with 10% of tire debris irradiated at 300 kGy showed the lowest loss of strength, which corresponded to 2%.

As can be seen from the results of mechanical tests shown in this section, including compressive strength, modulus of elasticity, and splitting tensile strength, the use of gamma radiation improves the properties of plastics and promotes the performance of concrete, even when it contains significant percentages of plastic waste such as PP, PET, or tire debris. Furthermore, losses in flexural strength do not go beyond 2% in concrete with 10% of waste tire debris.

14.3 Recycling Electronic Waste Polycarbonate in Concrete by Gamma Radiation

In this section gamma radiation as a technology that improves the recycling of electronic waste plastics such as polycarbonate is discussed.

14.3.1 Gamma Radiation Effects on Polycarbonate

When a plastic material is exposed to ionizing radiation, physical and chemical changes occur. This kind of radiation affects mechanical properties, crystallinity, density, insulating capacity, the coefficient of thermal expansion, as well as its resistance to solvents and abrasion (Buttafava et al. 2005; Martínez-Barrera et al. 2011). Gamma radiation can modify the chemical structure of plastics through three different processes: cross-linking, scission, and grafting (Ávila Córdoba et al. 2015).

In the special case of polycarbonate, when exposed to gamma radiation, its main chain can break (chain scission). However, its high structural rigidity allows many of the bonds that are broken to recombine. Besides, its aromatic structure absorbs photon energy, and this prevents the polymer chain from cracking. This contributes to it being one of the most gamma radiation-resistant polymers (Massey 2005). Some research on the behavior of commercial polycarbonate irradiated with gamma rays report that it becomes yellowish and the molecular weight decreases (Durolon®, 10–300 kGy) (Araújo et al. 1999). The molecular weight decreases and the Young's modulus as the stiffness increases (PC-Light®, 340 kGy) (de Melo et al. 2007). Furthermore, it can become structurally compacted (Makrofol® BL 2–4, 20–200 kGy) (Nouh and Abutalib 2014) or Vicker's hardness can increase (Makrofol® DE 7–2, 1000 kGy) (Tayel et al. 2015).

14.3.1.1 Chemical-Structural Changes

The analysis of the variations produced in the structure of polycarbonate by gamma radiation was carried out by FTIR. The representative bands of carbonate group vibration, C = 0 (1795 cm^{-1}), and the bands of methyl group vibrational modes, CH_3 (2969 and 1230 cm^{-1}), were the most modified bands after irradiation. The area under these bands gradually decreases with increasing dose. In addition, a new band emerges at 3548 cm^{-1}. This band arrives still at small irradiation doses and successively augments in magnitude. This shows the appearance of phenoxy arms (OH). The broadband section (3650–3440 cm^{-1}) offers the information of bonded (3523 cm^{-1}) and free (3548 cm^{-1}) hydroxyl stretching (Kalkar et al. 1992).

The difference of relative area under the infrared absorption bands 1795 cm^{-1} (C = 0), 1230 cm^{-1} (CH_3), and the broadband profile 3650–3440 cm^{-1} (OH) as a function of gamma dose (from 0 to 700kGy) varies from 10 up to 2 AU to C = 0, from 8 up to 5.8 AU to CH_3, and from 0 up to 5 AU to OH group (Kalkar et al. 1992).

14.3.1.2 Molecular Weight

In this research, molar mass was obtained by intrinsic viscosity through the Mark–Houwink equation. It shows that when gamma dose increases (from 0 up to 125 kGy) the molecular weight decreases (from 25 000 up to 22 000). The FTIR conclusions ratify that chain rupture happens mainly in the carbonyl group (de Melo et al. 2007).

On the other hand, (Weber et al. 2017) obtained a fall in molecular weight of 34 000 to 14 000 g/mol, when polycarbonate was irradiated to 340 kGy.

In other research, the molecular weight of the polycarbonate samples was established by means of High-Performance Liquid Chromatography (HPLC). Chloroform was the solvent and polystyrene was the standard. The absence of cross-linking was indicated by the unaltered solubility of the irradiated polycarbonate samples. The reduction in molecular weight due to irradiation dose illustrates a substantial degradation across the principal chain (Kalkar et al. 1992).

14.3.1.3 Glass Transition

In this investigation, the glass transition temperature decreases extremely fast when polycarbonate is gamma irradiated at 150 KGy. Tg of polycarbonate decreases from 149°C for non-irradiated experiments to 134°C for irradiated to the maximum dose (680 kGy). The increase in chain scission amplifies mobility and leads to an expansion in free volume and decrease in Tg (Kalkar et al. 1992).

14.3.1.4 Intrinsic Viscosity

This research shows that when gamma dose increases (from 0 to 125 kGy) the intrinsic viscosity of polycarbonate slightly decreases (from 49 to 44 dL/g). The degradation appears to go on as long as the polymer is subjected to gamma radiation (de Melo et al. 2007).

14.3.1.5 Tensile Strength

This experiment shows that when gamma dose increases (from 0 to 125 kGy) the tensile strength of polycarbonate decreases (from 63.5 to 61 MPa). These results confirm that polycarbonate has appropriate stability at these doses (de Melo et al. 2007). Alternatively, another study obtained a reduction in ultimate strength of 68 to 54 MPa when polycarbonate was irradiated to 340 kGy (Weber et al. 2017).

14.3.1.6 Elongation at Break

This investigation shows that when gamma dose increases (from 0 to 125 kGy) the elongation at break of polycarbonate decreases (from 104 to 94%) (de Melo et al. 2007). On the other hand, Weber et al. (2017) obtained a decrease in elongation at break of polycarbonate (PC light) of 77 to 55% when it was irradiated to 340 kGy.

14.3.1.7 Apparent Fracture Toughness

In this analysis the fracture toughness of polycarbonate decreases from 3.72 to 3.45 MPa $m^{1/2}$ when the samples are irradiated at 75, 100, and 125 kGy. These results confirm that gamma radiation generates a diminution in the toughness of this plastic (de Melo et al. 2007).

14.3.1.8 Rockwell Hardness

A decrease in Rockwell hardness (M) of 55 to 37 when polycarbonate was gamma irradiated at 340 kGy was obtained (Weber et al. 2017).

14.3.1.9 Fracture Morphology

An evaluation of the recycled polycarbonate samples (without irradiation) shows that the fracture of the particles is characteristic of a ductile fracture process. However, as the gamma irradiation dose increases and reaches the highest dose (125 kGy), the breakage is uniform with a brittle fracture pattern (de Melo et al. 2007).

14.3.1.10 Yellowness Index

The yellowness index is a numerical quantity that refers to the change in color of a sample from transparent to yellow. Yellowness is in the wavelength range 570–580 nm. Irradiation on the polymer generates phenyl and phenoxy radicals, and these produce yellowing of the polycarbonate. This research shows that gamma irradiation generates an increase in the yellow color of this plastic (from 0 to 90, when it is irradiated from 0 up to 100 kGy) (Araújo et al. 1999).

14.3.2 Gamma Radiation Effects on Recycled Electronic Waste Polycarbonate

Recycled electronic waste polycarbonate means that electronic waste polycarbonate was heat and mechanically processed. This process includes separation, washing, and pelletizing. In the pelletizing process the resin is heat, passing through the warm filter unit, and then it is extruded. The cooling unit is a strand bath where the cold water circulates to harden viscous resin. Finally, the hardened resin is cut in the form of pellets. The polycarbonate used may be aromatic or aliphatic. However, due to its higher durability, aromatic polycarbonate is preferred (Watanabe and Tanaka 2014).

14.3.2.1 Surface Morphology

In the case of recycled plastic particles a shiny and lustrous aspect is perceived; however, after irradiating at the highest gamma dose (1200 kGy), an irregular finish with little separated clusters is acquired as shown in Figure 14.12 (De la Colina Martínez et al. 2019).

14.3.2.2 Chemical-Structural Changes

- **Amorphous state**

The XRD results from De la Colina Martínez et al. (2019) show that the amorphous state is the same for both types of recycled polycarbonate: non-irradiated and irradiated (1200 kGy). It is also possible to detect in both cases the presence of rutile (TiO_2), which is commonly attributed to a white pigment widely used in the plastic industry.

- **Chemical elemental composition**

The EDS spectra of the recycled polycarbonate particles irradiated (to 600 and 1200 kGy) and not irradiated show the presence of chemical elements such as carbon, oxygen, and titanium (De la Colina Martínez et al. 2019). There is a slight decrease of carbon and a vague increase of oxygen due to the increase of gamma dose.

14.3.2.3 Plastic to Brittle Behavior

In the case of non-irradiated recycled polycarbonate it was simple to cut the plastic particles with the aid of a nail clipper. However, with the highest radiation dose (1200 kGy), the plastic behavior of particles changed from supple to rigid and they were fragmented erratically. From these results it is easy to see a plastic to brittle behavior of recycled polycarbonate when gamma dose increases (De la Colina Martínez et al. 2019).

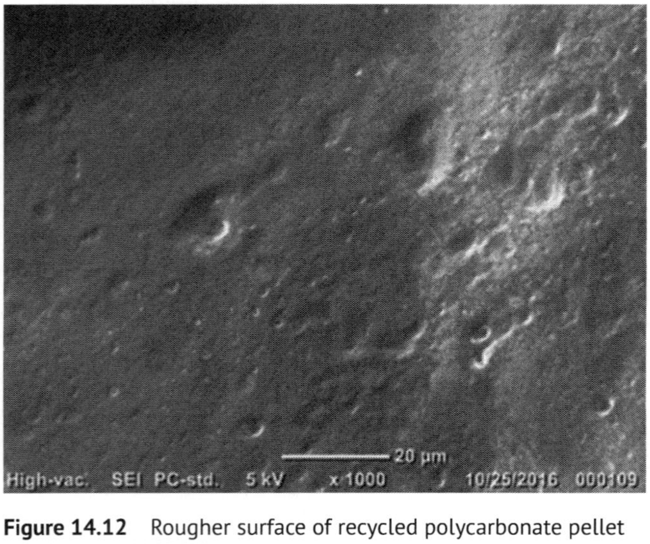

Figure 14.12 Rougher surface of recycled polycarbonate pellet irradiated at 1200 kGy.

14.3.3 Concrete with Non-irradiated and Irradiated Recycled Electronic Waste Polycarbonate

14.3.3.1 Compressive Strength

The recycled polycarbonate was developed by extrusion from waste electrical sockets. The shape of the recycled polycarbonate was cylindrical with a size of 1 mm in diameter and 3 mm in length (1 × 3 mm). The concrete was made from recycled polycarbonate, Portland cement, and aggregates. The recycled polycarbonate replaced the sand in percentages of 1, 3, and 6% by volume (De la Colina-Martínez et al. 2017).

The 28-day compressive strength of concrete with this polycarbonate shows that the maximum strength was for the blend with 1% of recycled polycarbonate (25 MPa). However, the concretes with 3% (24.4 MPa) and 6% (24 MPa) of recycled polycarbonate do not have lower values than those for reference concrete (24 MPa) (De la Colina-Martínez et al. 2017). In this sense, the sand used for the concrete could be replaced with the recycled polycarbonate by 6% without affecting the strength of the concrete.

Since the concrete can include up to 6% of recycled polycarbonate (1 × 3 mm) without loss of compressive strength, recycled polycarbonate was gamma irradiated to different doses. The highest values were for concretes gamma irradiated to 1000 and 1200 kGy (24 MPa) (De la Colina Martínez et al. 2019). For these concretes, the performance in strength was 12% with respect to standard concrete (21.5 MPa). However, the polycarbonate gamma irradiated from 0 up to 800 kGy did not bring any improvement to the strength, because it decreases (19 MPa) (De la Colina Martínez et al. 2019).

In order to find out if concrete could withstand a higher percentage of irradiated polycarbonate, it was irradiated at 1000 kGy and cut, obtaining cylindrical recycled polycarbonate pellets with a size of 1 × 1.5 mm (De la Colina Martínez et al. 2019). Although in this test, the compressive strengths have decreased with respect to the previous experiment, there is an increase in compressive strength when 15% of polycarbonate is included. Furthermore, the compressive strength of this blend is slightly superior to that of standard concrete. In this sense, the sand used for the concrete could be replaced by the irradiated recycled polycarbonate (15%) without affecting the strength of the concrete.

14.3.3.2 Density

Cylindrical recycled polycarbonate obtained by extrusion from waste electrical sockets was used to formulate concrete with Portland cement. The polycarbonate (1 × 3 mm) was used as substitute of the sand in percentages of 1, 3, and 6%

by volume. Findings show that increments of the recycled polycarbonate concentration provoke diminution of the calculated equilibrium density. For concrete without polycarbonate a density of 1851 kg/m^3 is obtained, while for the highest polycarbonate concentration (6%) the lowest density is 1825 kg/m^3 (De la Colina-Martínez et al. 2017). Concrete, structural lightweight has an equilibrium density in between 1120 and 1920 kg/m^3 and a minimal 28-day compressive strength of 17 MPa. It contains exclusively lightweight material or a combination of normal and lightweight ingredients (ACI 2003). For these reasons, the concrete obtained above could be an additive from concrete structural lightweight.

Cylindrical recycled polycarbonate, extruded from used electrical sockets, was mixed with Portland cement, sand, and water to formulate the concrete. The polycarbonate (1 × 1.5 mm) was gamma irradiated to 1000 kGy and recycled as a limited substitute of the sand in percentages of 3, 6, and 15% by volume. Results show that increments of the recycled polycarbonate concentration provoke diminution of the calculated equilibrium density. For concrete without polycarbonate a density of 1849 kg/m^3 is obtained, while for the highest polycarbonate concentration (15%) the lowest density is 1814 kg/m^3 (De la Colina Martínez et al. 2019).

14.4 Conclusion

From this review it is concluded that a large number of plastics from electronic waste could be recycled as aggregates (fine or coarse) in concrete in percentages of less than 10%. The traditional loss of mechanical strength, when higher percentages of plastics are added, can be improved by using, for example, fly ash, waste marble dust, and high-performance cements (with high compressive strengths). It should be noted that the mechanical properties of concrete can also be significantly improved if plastics are added in pellet form or if they are modified by gamma radiation.

References

ACI (2003). *213R-03 Guide for Structural Lightweight-Aggregate Concret, Concretee*. ACI Committee Report. http://civilwares.free.fr/ACI/MCP04/213r_03.PDF.

Akram, A., Sasidhar, C., and Pasha, K.M. (2015). E-waste management by utilization of e-plastics in concrete mixture as coarse aggregate replacement. *International Journal of Innovative Research in Science, Engineering and Technology* 4 (7): 5087–5095. https://doi.org/10.15680/ijirset.2015.0407008.

Araújo, E.S., Khoury, H.J., and Silveira, S.V. (1999). Effects of gamma-irradiation on some properties of durolon polycarbonate. *Radiation Physics and Chemistry* 53 (1): 79–84. https://doi.org/10.1016/S0969-806X(97)00300-9.

Arora, A., Urmil, D., and Dave, V. (2013). Utilization of E-waste and plastic bottle waste in concrete. *International Journal of Students Research in Technology & Management* 1 (4): 398–406.

Ávila Córdoba, L.I., Martnez-Barrera, G., Barrera-Díaz, C.E. et al. (2015). PET de desecho y su uso en concreto. In: *Materiales Sustentables Y Reciclados En la Construcción* (ed. G. Martnez-Barrera, J.B. Hernández Zaragoza, T. López Lara, and C. Menchaca Campos), 95–122. OmniaScience. https://doi.org/10.3926/oms.246.

Balde, C.P., Forti, V., Gray, V. et al. (2017). *The Global E-waste Monitor 2017*. United Nations University. https://doi.org/10.1016/j.proci.2014.05.148.

Basham, K.D., Clark, M., France, T., and Harrison, P. (2007). What is fly ash? *Concrete Construction* 1. https://www.concreteconstruction.net/how-to/materials/what-is-fly-ash_o.

Bureau of Indian Standards (2000). IS 456 (2000)—Plain and reinforced concrete—code and practice, Bureau of Indian Standards. https://law.resource.org/pub/in/bis/S03/is.456.2000.pdf.

Buttafava, A., Consolati, G., Mariani, M. et al. (2005). Effects induced by gamma irradiation of different polyesters studied by viscometry, thermal analysis and positron annihilation spectroscopy. *Polymer Degradation and Stability* 89 (1). https://doi.org/10.1016/j.polymdegradstab.2005.01.009.

Centro Nacional de Aceleradores (2016). *Research Report 2015–2016*. https://cna.us.es/images/memorias/MEMORIA_CNA_2015-2016.pdf.

Chmielewski, A. (2011). *Gamma Irradiators for Radiation Processing*. Vienna, Austria: International Atomic Energy Agency.

Choi, Y.-W., Moon, D.-J., Chung, J.-S., and Co, S.-K. (2005). Effects of waste PET bottles aggregate on the properties of concrete. *Cement and Concrete Research* 35 (4). https://doi.org/10.1016/j.cemconres.2004.05.014.

Coleman, E.A. (2011). Plastics Additives. In: *Applied Plastics Engineering Handbook* (ed. Myer Kutz), 419–428. Elsevier. https://doi.org/10.1016/B978-1-4377-3514-7.10023-6.

Cruz Bastida, J. and García Hernández, D. (2015). *El Fotón de Asclepio*. México: CONACYT.

Damal, V.S., Londhe, S.S., and Mane, A.B. (2015). Utilization of electronic waste plastic in concrete. *International Journal of Engineering Research and Applications* 5 (4): 35–38. https://www.ijera.com/papers/Vol5_issue4/Part%20-%202/F504023538.pdf.

De la Colina-Martínez, A.L., Martínez-Barrera, G., Barrera-Díaz, C.E. et al. (2017). Characterization of recycled polycarbonate from electronic waste and its use in hydraulic concrete: improvement of compressive performance. *Advances in Concrete Construction* 5 (6): 563–573. https://doi.org/10.12989/acc.2017.5.6.563.

De la Colina Martínez, A.L., Martínez Barrera, G., Barrera Díaz, C.E. et al. (2019). Recycled polycarbonate from electronic waste and its use in concrete: effect of irradiation. *Construction and Building Materials* 201: 778–785. https://doi.org/10.1016/j.conbuildmat.2018.12.147.

de Melo, N.S., Pondé Weber, R., and Miguez Suarez, J.C. (2007). Toughness behavior of gamma-irradiated polycarbonate. *Polymer Testing* 23 (3): 315–322. https://doi.org/10.1016/j.polymertesting.2006.10.014.

Dissanayake, R. and Priyadarshana, T. (2016). Chloride penetration and sulfate resistance of concrete incorporating nano silica nano SiO2 micro silica micro SiO2 and fly ash'. *Proceedings of Fourth International Conference on Advances in Civil, Structural and Mechanical Engineering, ACSM 2016*. https://doi.org/10.15224/978-1-63248-096-5-14.

DuPont (2017). *Technical Reference Guide for DuPont™ Transition Tyvek® 1073B, Transition Tyvek® 1059B and Tyvek® 2FS™—metric units.*

DuPont (2018). *Tyvek® for Microbial Barrier Packaging of Medical Devices & Pharmaceuticals.*

DuPont (2020). *DuPont ™ Kapton ® Summary of Properties*, http://www.dupont.com/content/dam/dupont/products-and-services/membranes-and-films/polyimde-films/documents/DEC-Kapton-summary-of-properties.pdf.

EN 206-1 (2000). *Concrete—Part 1: Specification, Performance, Production and Conformity*. European Committee for Standardization (CEN).

Evram, A., Akçaoğlu, T., Ramyar, K. et al. (2020). Effects of waste electronic plastic and marble dust on hardened properties of high strength concrete. *Construction and Building Materials* 263 (6). https://doi.org/10.1016/j.conbuildmat.2020.120928.

Farooq, A. (2019). Impact on concrete properties using e-plastic waste fine aggregates and silica fum. *Gospodarka Surowcami Mineralnymi / Mineral Resources Management* [Preprint]. https://doi.org/10.24425/gsm.2019.128516.

Forti, V., Baldé, C.P., Kuehr, R., and Bel, G. (2020). *The Global E-waste Monitor 2020: Quantities, Flows, and the Circular Economy Potential, United Nations University (UNU)/ United Nations Institute for Training and Research (UNITAR)–co-hosted SCYCLE Programme, International Telecommunication Union (ITU) & International Solid Waste Association (ISWA), Bonn/Geneva/Rotterdam.*

Gerald, B. and Neville, P.E. (2015). *Concrete Manual Based on the 2015 IBC® and ACI 318–14.*

Gu, L. and Ozbakkaloglu, T. (2016). Use of recycled plastics in concrete: a critical review. *Waste Management* [Preprint]. https://doi.org/10.1016/j.wasman.2016.03.005.

Hama, S.M. and Hilal, N.N. (2019). Fresh properties of concrete containing plastic aggregate. *Use of Recycled Plastics in Eco-efficient Concrete*. https://doi.org/10.1016/b978-0-08-102676-2.00005-0.

Hegazy, E.S.A., Sasuga, T., Nishii, M., and Seguchi, T. (1992). Irradiation effects on aromatic polymers: 1. Gas evolution by gamma irradiation. *Polymer* [Preprint]. https://doi.org/10.1016/0032-3861(92)90074-7.

Herrera-Sosa, E.S., Martínez-Barrera, G., Barrera-Díaz, C. et al. (2015). Recovery and modification of waste tire particles and their use as reinforcements of concrete. *International Journal of Polymer Science* [Preprint]. https://doi.org/10.1155/2015/234690.

Kalkar, A.K., Kundagol, S., Chand, S. et al. (1992). Effect of gamma-irradiation on structural and electrical properties of poly (bisphenol-A carbonate) films. *International Journal of Radiation Applications and Instrumentation. Part C*. 39 (5): 435–442. https://doi.org/10.1016/1359-0197(92)90056-L.

Koehler, E.P. and Fowler, D.W. (2003). *Summary of Concrete Workability Test Methods*. International Center for Aggregates Research.

Lakshmi, R. and Nagan, S. (2010). Studies on concrete containing E plastic waste. *International Journal of Environmental Sciences* 1 (3).

Lakshmi, R. and Nagan, S. (2011). Investigations on durability characteristics of E-plastic waste incorporated concrete. *Asian Journal of Civil Engineering* 12 (6).

Luhar, S. and Luhar, I. (2020). A green conception in the construction sector: incorporation of E-waste into concrete. *Handbook of Environmental Materials Management*. https://doi.org/10.1007/978-3-319-58538-3_193-1.

Manatkar, P.A. and Deshmukh, G.P. (2015). Use of non-metallic e-waste as a coarse aggregate in a concrete. *International Journal of Research in Engineering and Technology* [Preprint]. https://doi.org/10.15623/ijret.2015.0403040.

Manjunath, B.T.A. (2016). Partial replacement of E-plastic waste as coarse-aggregate in concrete. *Procedia Environmental Sciences* 30: 731–739. https://doi.org/10.1016/j.proenv.2016.07.079.

Martínez-Barrera, G., Ureña-Nuñez, F., Gencel, O., and Brostow, W. (2011). Mechanical properties of polypropylene-fiber reinforced concrete after gamma irradiation. *Composites Part A: Applied Science and Manufacturing* 42 (5): 567–572. https://doi.org/10.1016/j.compositesa.2011.01.016.

Martínez-Barrera, G., del Coz-Díaz, J.J., Álvarez-Rabanal, F.P. et al. (2020). Waste tire rubber particles modified by gamma radiation and their use as modifiers of concrete. *Case Studies in Construction Materials* 12 (2020): e00321. https://doi.org/10.1016/j.cscm.2019.e00321.

Marzouk, O.Y., Dheilly, R.M., and Queneudec, M. (2007). Valorization of post-consumer waste plastic in cementitious concrete composites. *Waste Management* [Preprint]. https://doi.org/10.1016/j.wasman.2006.03.012.

Massey, L.K. (2005). Chapter 11—Polycarbonate (PC). In: *The Effect of Sterilization Methods on Plastics and Elastomers*, 2e. 81–109. https://doi.org/10.1016/B978-081551505-0.50013-5.

Mathakaria, N.L., Bhoraskar, V.N., and Dhole, S.D. (2014). A comparative study on the effects of Co-60 gamma radiation on polypropylene and polyimide. *Radiation Effects and Defects in Solids* [Preprint]. https://doi.org/10.1080/10420150.2014.939176.

Nadhim, S., Shree, P.N., and Kumar, G.P. (2018). A comparative study on concrete containing E- plastic waste and fly ash concrete with conventional concrete. *PARISHODHANA In-house Journal of Science and Technology* 6 (1).

Nouh, S.A. and Abutalib, M.M. (2014). Effect of gamma irradiation on the structural, thermal and optical properties of Makrofol BL 2-4′. *Radiation Effects and Defects in Solids* 169 (10): 817–827. https://doi.org/10.1080/10420150.2014.956320.

Pacheco-Torgal, F., Ding, Y., and Jalali, S. (2012). Properties and durability of concrete containing polymeric wastes (tyre rubber and polyethylene terephthalate bottles): an overview.*Construction and Building Materials* [Preprint]. https://doi.org/10.1016/j.conbuildmat.2011.11.047.

Peña Pichardo, P. (2018). *Elaboración y caracterización de materiales compuestos elaborados con resina poliéster, fibras textiles de desecho y agregados minerales, para su uso como bloque de construcción*. Doctoral thesis. Universidad Autónoma del Estado de México.

Raghu, S., Archana, K., Sharanappa, C. et al. (2016). Electron beam and gamma ray irradiated polymer electrolyte films: dielectric properties. *Journal of Radiation Research and Applied Sciences* 9 (2): 117–124. https://doi.org/10.1016/j.jrras.2015.10.007.

Rezaei Ochbelagh, D., Azimkhani, S., and Gasemzadeh Mosavinejad, H. (2011). Effect of gamma and lead as an additive material on the resistance and strength of concrete. *Nuclear Engineering and Design* [Preprint]. https://doi.org/10.1016/j.nucengdes.2011.03.001.

Sabău, M. and Vargas, J.R. (2018). Use of e-plastic waste in concrete as a partial replacement of coarse mineral aggregate. *Computers and Concrete* [Preprint]. https://doi.org/10.12989/cac.2018.21.4.377.

Saikia, N. and De Brito, J. (2012). Use of plastic waste as aggregate in cement mortar and concrete preparation: a review. *Construction and Building Materials* 34: 385–401.

Senthil Kumar, K. and Baskar, K. (2015). Recycling of E-plastic waste as a construction material in developing countries. *Journal of Material Cycles and Waste Management* 17: 718–724. https://doi.org/10.1007/s10163-014-0303-5.

Sharma, R. and Bansal, P.P. (2016). Use of different forms of waste plastic in concrete—A review. *Journal of Cleaner Production* [Preprint]. https://doi.org/10.1016/j.jclepro.2015.08.042.

Siddique, R., Khatib, J., and Kaur, I. (2008). Use of recycled plastic in concrete: a review. *Waste Management* [Preprint]. https://doi.org/10.1016/j.wasman.2007.09.011.

Silva Aquino, K.A.D. (2012). Sterilization by gamma irradiation. In: *Gamma Radiation* (ed. F. Androvic). London, UK: InTechOpen Ltd. https://doi.org/10.5772/34901..

Soroushian, P., Plasencia, J., and Ravanbakhsh, S. (2003). Assessment of reinforcing effects of recycled plastic and paper in concrete. *ACI Materials Journal* [Preprint]. https://doi.org/10.14359/12620.

Tayel, A., Zaki, M.F., El Basaty, A.B., and Hegazy, T.M. (2015). Modifications induced by gamma irradiation to Makrofol polymer nuclear track detector. *Journal of Advanced Research* 6 (2): 219–224. https://doi.org/10.1016/j.jare.2014.01.005.

United Nations Environment Programme (2017). *Compendium of Technologies for the Recovery of Materials from WEEE/E-waste*.

Watanabe, N. and Tanaka, T. (2014). Extrusion pelletizer for polycarbonate resin and method for extrusion-pelletizing polycarbonate resin. United States Patent Application Pub. No. US 2015/0360393 A1. https://patentimages.storage.googleapis.com/c9/9d/4a/7aa7bdd79b2f7a/US20150360393A1.pdf.

Weber, R.P., Monteiro, S.N., Suarez, J.C.M. et al. (2017). Fracture toughness of gamma irradiated polycarbonate sheet using the essential work of fracture. *Polymer Testing* 57: 115–118. https://doi.org/10.1016/j.polymertesting.2016.11.021.

Wilby, C.B. (1991). *Concrete Materials and Structures: A University Civil Engineering Text*. Cambridge University Press.

Yang, X., Sun, L., Xiang, J. et al. (2013). Pyrolysis and dehalogenation of plastics from waste electrical and electronic equipment (WEEE): a review. *Waste Management* 33 (2): 462–473. https://doi.org/10.1016/j.wasman.2012.07.025.

15

Recovery of Rare Earth Elements and Critical Metals from Electronic Waste

*Sedevino Sophia and Vidya Shetty K**

Department of Chemical Engineering, National Institute of Technology Karnataka, Surathkal, Srinivasnagar. Surathkal, Mangalore - 575025. India
** Corresponding author*

15.1 Introduction

15.1.1 Genesis of Electronic Waste

With the advancements in digital technology, the demand for newer electronic devices and the replacement rate of the electronic products have escalated dramatically. This has led to massive electronic waste generation (Ghulam and Abushammala 2023; Martins et al. 2019; Zhu et al. 2021). Electronic waste (e-waste) refers to equipment for which the economic life has ended and which is no longer used by consumers (Işıldar et al. 2018). E-waste classically includes mobile phones, televisions, computers, monitors, cathode ray tubes, printers, and liquid crystal display (LCD) units (Veit and Bernardes 2015).

The present scenario of e-waste has attracted interest and concerns among researchers and legislators on finding ways to manage the generated e-waste (Li et al. 2019). As a result of technological progress, waste electrical and electronic equipment (WEEE) are growing at a rate three times faster than that of municipal waste (Otsuki et al. 2020). The global generation of e-waste was assessed as 53.6 million metric tons (Mt) in 2019 and is expected to reach nearly 120 Mt by 2050 (Forti et al. 2020; Ghulam and Abushammala 2023; Mtibe et al. 2023). In the year 2019, Asia generated a maximum amount of e-waste amounting to 24.9 Mt, and America followed it with generation of 13.1 Mt and then Europe with 12 Mt (Nithya et al. 2021). Meanwhile, a study by the United States Environmental Protection Agency (US EPA) reported that some fraction of e-waste ends up in the landfill; nevertheless it may represent with at least 70% hazardous waste affecting water bodies, agricultural land, and open landfills (Islam et al. 2020). Several heavy metals and toxic substances seep from the e-waste into the soil, which can perhaps retard the growth of plants in the nearby area and decrease the land productivity (Bhardwaj et al. 2023; Cayumil et al. 2016; Paranjape and Yadav 2023). Thus, countries must have a proper strategy to manage e-waste in order to preserve the environment and public health (Sahle-Demessie et al. 2021).

15.1.2 Metals in Electronics

In electronic equipment, metals play a predominant role due to their unique physical and chemical properties that make the equipment more functional and intelligent (Zhang et al. 2017). E-waste is considered to be highly heterogeneous, containing mixed materials with about 60 different elements. Critical metals (CMs), precious metals (Cucchiella et al. 2016; Isildar 2018; Sharma et al. 2023), and rare earth elements (REEs) (Swain and Mishra 2019; Tunali et al. 2021) are utilized based on their precise features and specifications in an electronic device (Zhang et al. 2017). For instance, indium (In) is used in touch screens' conductive coatings; arsenic (As) is used in radio frequency amplifiers; tungsten (W) in vibration and heat sinks; germanium (Ge) is integrated in displays. REEs such as dysprosium (Dy), praseodymium (Pr), and neodymium (Nd) are used in computer hard disk drives' magnets because of their magnetic properties (Işıldar et al. 2018). CMs are technologically important in terms of their necessity for the functionality of various emerging technologies, but have supply constraints due to geological scarcity (Watari et al. 2020). CMs are identified based on unstable supply, huge demand, imports, and other market factors (Can Sener et al. 2021). Apart from REEs and precious metals, platinum group metals and other valuable metals such as lithium (Li), cobalt (Co), nickel (Ni), and uranium (U) are also included under CMs (Can Sener et al. 2021).

Electronic Waste Management: Policies, Processes, Technologies, and Impact, First Edition. Edited by Sunil Kumar and Vineet Kumar.

The REEs are a group of 17 elements that include scandium (Sc), yttrium (Y), and the elements from lanthanum through lutetium (La–Lu), which are called lanthanoids (Wall 2021). They possess magnetic and luminescent properties due to which they serve as essential components of digital and low-carbon technologies.

Though they are found in few quantities in electronic equipment, the application of such REEs and precious metals for the production of electronic devices is crucial (Lister et al. 2016). This has led European countries through the European Commission and the United States through their Department of Energy to explore urban mining where e-waste can be a secondary resource of REE metals of great economic importance, at higher risk of supply and of environmental aspects (Lukowiak et al. 2020; Morf et al. 2013). The urban mining of e-waste has secured importance and recognition, as the recovery of valuable metals from this source is ten times greater than in natural ores. Natural ore mining requires geological study and surveys, and the inability to predict the number of metals to be found make it uneconomical for extraction of metals (Annamalai and Gurumurthy 2021). Recovery of metals from e-waste can serve as a strategy to end the identification of new or unexplored mines and also to reduce the risk associated with WEEE (Amato et al. 2020). An innovative approach was initiated in Japan in which discarded e-waste mobile phones were used for producing the gold (Au), silver (Ag), and bronze medals for the 2020 Olympic Games (Buechler et al. 2020). Gold medals were made with 32 kg of recovered Au, silver medals with 3500 kg of recovered Ag, and bronze medals from 2200 kg of recovered bronze (Nithya et al. 2021).

15.1.3 Exigence of Rare Earth Elements and Critical Metals

The utilization of the REEs in electronic equipment has increased because of their distinct and eccentric properties. However, there are limited sources for such exploitable mineral reserves for REEs, which makes extraction uneconomical (Ganguli and Cook 2018). Some fragment of the mining sector also ends up as waste at the end of the electronic equipment life cycle, causing a long-term effect on environmental sustainability (Xavier et al. 2019). In fact, REEs and CMs are not major mining products, but by-products that are generated from the extraction of abundant base metals (Efstratiadis and Michailidis 2022; Royen and Fortkamp 2016). REEs are mostly found to be paramagnetic in nature, due to which they are found in combination with transition metals such as Fe (iron) and Co (cobalt) (Wall 2021). With respect to the recycling process and the high demand for REEs and CMs, the supply-demand interrelation becomes imbalanced. Considering these conditions, the REEs and CMs are predestined to be at a higher risk of supply (Sikander et al. 2018).

CMs can be characterized when their supply rates are relatively lower than their increasing demand. This also includes those metals that have low dispersity, a low recycling rate, and a low potential for substitution (Bakas et al. 2016). They are identified as CMs based on the access to sustainability and resources (Girtan et al. 2021). In order to maintain such resources, it becomes necessary to reduce the dependencies, diversify the supply from both the primary as well as secondary sources, and improve the resource efficiency and circularity. One of the solutions regarding critical materials is the adoption of the circular economy wherein the critical raw materials are reused in multiple product life cycles, and there is an efficient plan to recycle for maximizing the economic productivity and maintaining the global demand as well (Charles et al. 2020). Circular economy has already been accepted by the European Union (EU) where the secondary raw materials are in use. The latest updated 2020 EU report listed 30 materials that are considered to be the CMs (European Commission 2020), presented in Table 15.1. The United States Geological Survey (USGS) has presented the final list of critical minerals of 2022, which includes 50 critical minerals (Nassar and Fortier, 2021). Following the commitment toward the transition to non-fossil fuel energy, several types of clean energy technologies are being adopted in India. The critical mineral needs in India depend on the various technologies being adopted for the energy transition. Solar power, wind power, and electric vehicles are playing a major role in India's clean energy transition (Chadha and Sivamani 2022). In India, a scheme called FAME (Faster Adoption and Manufacturing of Hybrid and Electric Vehicles) that promotes electric vehicles (EVs) is under execution (Mint 2021). There is an increase in demand for several CMs and REEs as the country is moving toward energy transition. However, India lacks nickel (Ni), neodymium (Nd), cobalt (Co), molybdenum (Mo), REEs, and indium (In) reserves. India's current reserves of Cu and Ag are not sufficient to meet the country's requirement (Chadha and Sivamani 2022).

Oxygen (O), silica (Si), aluminum (Al), iron (Fe), calcium (Ca), and other abundant elements make up 99.23% of the continental crust's mass. Scarce metals constitute less than 0.1% of the crust, and they can be viewed as rare metals (Zepf 2013). However, most of the REEs in the earth's crust can be found more abundantly than platinum group elements (Smith et al. 2017). They can be found in as many as 200 minerals and deposits. The only downside is their low concentrations in lower grade ores (Deboer and Lammertsma 2013; Khanna et al. 2018). Furthermore, the milling and processing of REEs can be a complex operation in which ore-specific processes can lead to potential

Table 15.1 2020 list of critical raw materials by EU (European Commission 2020).

2020 Critical raw materials listed by EU			
Antimony	Gallium	Natural rubber	Vanadium
Baryte	Germanium	Niobium	Tungsten
Beryllium	Hafnium	Platinum group metals	Bauxite
Bismuth	Heavy rare earth elements	Phosphate rock	Lithium
Borate	Light rare earth elements	Phosphorus	Titanium
Cobalt	Indium	Scandium	Strontium
Coking coal	Magnesium	Silicon metals	
Fluorspar	Natural graphite	Tantalum	

environmental pollution if not handled professionally (Smith et al. 2017).

REEs include the metallic lanthanide elements yttrium (Y) and scandium (Sc). Y and Sc are regarded as REEs due to their propensity to exist in the same deposits and similar chemical characteristics to that of lanthanoids (Balaram 2019; Dushyantha et al. 2020; Jowitt et al. 2018). REEs are categorized on the basis of their atomic weight (Fernandez 2017) or atomic numbers (Ganguli and Cook 2018). Light REEs are those having atomic numbers ranging from 57 to 62 such as La, Ce, Pr, Nd, and Sm, and the heavy REEs are those with atomic number 63–71 such as Ho, Er, Tm, Yb, and Lu (Ambaye et al. 2020; Dev et al. 2020; Eggert et al. 2016; Goodenough et al. 2018; Wall 2021). Y has similar chemical and physical properties as lanthanoids, so it is also considered a heavy REE (Fernandez 2017). Promethium (Pm) is a radioactive element that is present in natural materials but in trace quantities, and it is the rarest occurring as it does not have stable isotopes (Balaram 2019; Dushyantha et al. 2020; Gwenzi et al. 2018).

15.1.4 Key Role of REEs and CMs in Electronics

The demand for REEs has reached its pinnacle in electronics industry because they are required in the minute components for the Information Technology (IT) sector and in batteries. For instance, REEs are used in fluorescent lamps, permanent magnets (Ganguli and Cook 2018), batteries, accumulators, semiconductors, and capacitors. Critical elements such as phosphorus can be found in CRTs (cathode ray tubes), LEDs (light emitting diodes), plasma display panels, fluorescent lamps, etc., neodymium-iron-boron (NdFeB) magnets in headphones, mobile phones, hard disks, speakers, and nickel-metal-hydride (NiMH) batteries (Marra et al. 2018b). Figure 15.1 presents the critical materials and REEs present in different electronic waste (Sethurajan et al. 2019).

This chapter presents the different methods of recovery of CMs and REEs from e-waste, with due emphasis on the significant role of the pretreatment processes for increasing

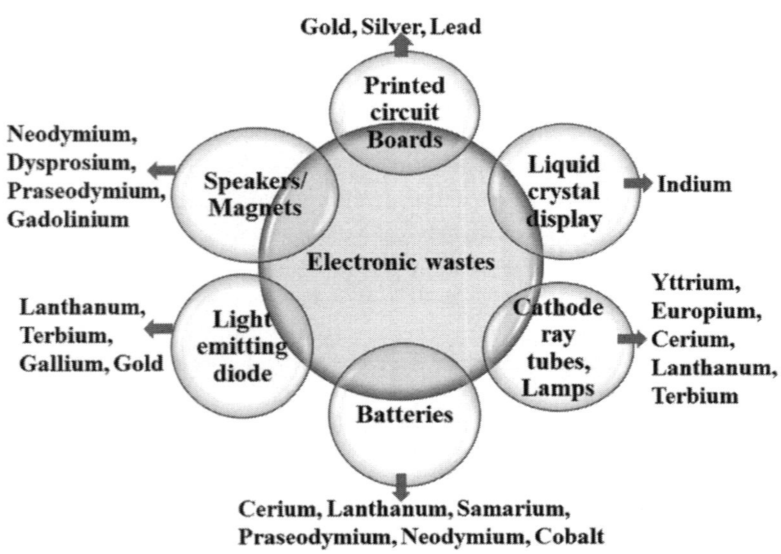

Figure 15.1 Critical metals and REEs present in various electronic waste materials (Sethurajan et al. 2019 / Taylor & Francis/ CC BY 4.0).

the efficiency of recovery of critical metals. The pyrometallurgical, hydrometallurgical, and biological processes of recovery are described, highlighting the drawbacks of the current recovery systems toward the preservation of the environment. This chapter critically reviews the recent studies to provide a better understanding of the recovery of CMs and REEs and to help in building a support system for current policymakers and future research.

15.2 Recovery of CMs and REEs from Electronic Waste

To proceed with recovery processes, understanding the metal composition in an item of electronic equipment is a crucial part so as to optimize the operating conditions in the pre-processing stage (Batinic et al. 2018). The categories of an electronic waste include general electronic waste, waste hard disk drives (HDDs), spent batteries, magnet scraps, spent catalyst, and spent LEDs (Sethurajan et al. 2019). Table 15.2 presents the critical materials present in an electronic device.

The recycling process for e-waste includes pretreatment such as disassembling (Nithya et al. 2021), upgrading (metal contents are improved), and refining, which is the final extraction of the metal product (Thethwayo 2021).

The collection and consolidation of e-waste is a crucial step before the recycling chain proceeds to any phase of the recovery system (Cecere and Martinelli 2017; Xavier et al. 2023). Though this basically depends on the logistics, it also requires consumers' awareness on how and to whom to return the end-of-life e-waste for proper channeling of the collection and recovery processes. In urban mining, municipal waste collection plays a vital role (Tesfaye et al. 2017). A formal collection system by designated government bodies has to be recognized, or authorized commercial collection points and pick-up services can facilitate in the recycling process through advanced technology for the safe and efficient recovery of the valuable materials (Ahirwar and Tripathi 2021).

15.2.1 Pretreatment Process

The physical pretreatment process refers to the separation of the metallic and non-metallic fractions of the e-waste (Chakraborty et al. 2022). The pretreatment process requires manual dismantling and disassembling and is a significant stage for most of the e-waste (Ashiq et al. 2019). The disassembly allows the separation of valuable components and decreases the impurities resulting in a higher input stream to the main recycling feed process (Sethurajan et al. 2019; Hsu et al. 2019). In particular, semi-automatic and automatic disassembly have also been reported, such as done by robots. Daisy and Liam, developed by Apple, mechanically disassemble a great number of iPhones at a time and also recycle the components for reuse (Kumar et

Table 15.2 Critical materials present in electronics (Chancerel et al. 2015; Manhart et al. 2012; Prasad et al. 2019; Zhang et al. 2017).

Critical materials in electronics	Role and relevance in electronics
Antimony	Printed wiring boards, flame retardants, and cathode ray tubes
Beryllium	Electric/electronic connectors, power supply boxes
Borates	Glass of LCD, also flame retardants
Chromium	Stainless steel, decorative hardener
Cobalt	Li-ion and NiMH batteries, cathode ray tubes, Ni-Cd batteries
Germanium	LEDs and electronic components
Indium	LCD panels, to a minor extent also LEDs, solder, and semiconductors
Magnesium	Casings
Natural graphite	Li-ion batteries
Niobium	Magnets, resistors, conductive, plasma
Platinum group metals	Palladium in electronic components, PCB, platinum and ruthenium in hard disk drive platter, iridium in LEDs
REE (light and heavy)	Magnets in motors, drive, loudspeakers, NiMH batteries, phosphors of CCFL, LED backlighting systems, voice coil accelerators
Silicon metal	Silicon, semiconductors in chips
Tantalum	Capacitors

al. 2022). Disassembling is followed by mechanical shredding using crushers or a pre-granulator. Non-metallic elements are then segregated from metallic elements and used for concentrating them using magnetic separators, eddy current, gravity separation, and density separation techniques (Debnath et al. 2018; Willner et al. 2021). Magnetic field-producing eddy currents are those that induce a magnetic field around the conductive metal particles, which on exposure to the magnetic field of the rotor leads to formation of a repulsive force called the Lorenz force that can repel those metal conducting particles apart from the material flow. Thereby, non-conductive materials such as plastics which are unaffected by the repulsive force fall under the effect of gravity, centrifugation, and frictional force (Paranjape and Yadav 2023). The gravity separation method, which is based on difference in density of different materials, can also aid in the pretreatment processes. The density separation technique can be exploited to separate light materials such as aluminum and glass, which have a density of around 2.7 g/cm^3, or plastics less than 2 g/cm^3 and ferromagnets with more than 7 g/cm^3 (Kaya 2016). Processes like shredding and cutting allow the materials' size to be reduced. Ball milling can be employed to pulverize the e-waste printed circuit boards (PCBs) to powder (Hsu et al. 2019). Once the shredded materials/powder form is achieved, it can be employed to recover the metals by pyrometallurgical, hydrometallurgical, or other methods to procure acceptable secondary resources (Vermesan et al. 2020). In the study by Baez et al. (2022), the correlation between the elemental concentrations and the particle size of the pretreated WPCBs was investigated for the recovery of the REEs present in e-waste from three different companies. The heterogeneous e-waste matrix was initially dry-sieved where different size fractions were studied. According to their findings, there was an increase in the concentrations of REEs (Y, La, and Gd) with the decrease in the particle size (fractions <0.25 mm); higher concentrations of these REEs were found in smaller particle sizes than in the coarse particles. They proposed that to enrich the recovery of REE content, a cost-effective sieving step is employed that can eliminate the loss of the REEs and promote higher recovery from the e-waste. Figure 15.2 illustrates a typical pretreatment stage involved in processing waste electrical and electronic equipment (WEEE).

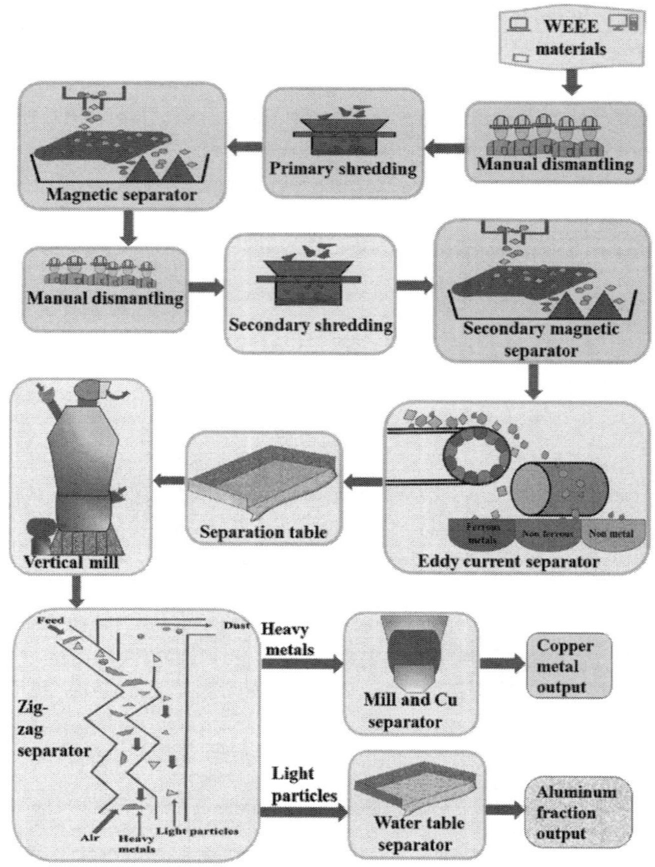

Figure 15.2 Pretreatment stage involved in typical waste electrical and electronic equipment (WEEE) processing.

Chancerel et al. (2009) reported the recovery of precious metals by utilizing the pre-shredded and shredded PCBs. Precious metals were found to be spread over all the output fractions. In another study by Sprecher et al. (2014), it was reported that the recovery of metals using shreded material led to a lower recovery rate. Marra et al. (2018a) also observed that 68% of the precious metals in appreciable numbers could be found in the output fractions. They also found that about 80% of the REEs in the WEEE feed were easily converted into dust particles and attached to ferrous surfaces due to their magnetic properties. To ensure a higher level of concentration with standard quality of the REEs in the waste streams, it poses an important question as to whether the sorting processes/physical pretreatment can be removed so as to be economically viable (Tsamis and Coyne 2015). Nevertheless, with regard to the loss of CMs, setting up a proper mechanical treatment upstream or tail-end processes can be employed. Recently, recycling methods have been upgraded using automated sensor-based sorting. Materials can be separated according to their optics, color, and chemical composition through the use of detectors coupled with advanced visible light, x-ray, and infrared light sources (Tabelin et al. 2021).

Other technologies and innovations for pretreatment steps are also being designed for the liberation of compounds and thus, increasing the input feeds for the recovery of valuable metals. Yang et al. (2021) used abrasive water jet cutting technology in which the size of the WPCBs were reduced from 14 mm to <1 mm. The solder mask acts as a protective layer for the metals preventing them from oxidation, and it can be removed by chemical treatment methods (Kang et al. 2021). Its complete liberation can be carried by the addition of sodium hydroxide (NaOH) (Moyo et al. 2020). Organic swelling process-based delamination (Ghosh et al. 2015) using organic solvents like dimethylformamide (DMF) and dimethyl sulfoxide (DMSO), etc. with ultrasonic bath or microwave treatment can remove the organic materials in waste PCBs (Yousef et al. 2018; Zhu et al. 2013). Pretreatment processes have also been found to result in material loss in the form of shredding dust (Paranjape and Yadav 2023). Nonetheless, one of the major challenges found in the recovery methods of e-waste is the presence of only a small quantity of critical elements in electronic equipment. This may misrepresent the presentation of recovery of such metals as nonprofitable. However, the low recovery efficiency rate would be because of lack of knowledge of a particular design (Ambaye et al. 2020). It is essential to understand the design of WEEE units and then choose the relevant method for pretreatment. The other option is to design specialized systems to enhance the recovery of CMs and to minimize their loss in the preprocessing stage (Batinic et al. 2018).

15.3 Types of Metal Recovery Operations

15.3.1 Pyrometallurgical Method

The methods for metal recovery from e-waste include pyrometallurgical methods such as smelting/incineration in a plasma and blast furnace with the involvement of high temperature in a gas phase, dressing operation, and sintering and melting processes (Chakraborty et al. 2022; Efstratiadis and Michailidis 2022). Figure 15.3 shows that e-waste scraps are treated in a blast furnace that is operated under high temperature, and then the refractory oxides are removed to obtain the slag for further recovery of metals (Nithya et al. 2021). Currently, Pb and Cu are recovered from e-waste in Cu and Pb smelters (Khaliq et al. 2014). Table 15.3 lists various industries using pyrometallurgical methods for recovery of metals from e-waste.

The Umicore smelter in Belgium (Kaya 2016) is the largest pyrometallurgical plant processing e-waste material. The first step is the smelting process wherein the e-waste is smelted in a furnace to extract precious metals. The process is known as precious metals operations (PMO), wherein the plastic fraction and other organic substances present in the feed e-waste material act as the reducing agents and energy source that substitute for coke (Hagelüken 2005). Precious metals and Cu are separated from other base metals into molten metal in the smelter, meanwhile the majority are separated as a lead slag. The molten metal is then electrowinned and precious metals are recovered (Khaliq et al. 2014). The lead slag is processed in an integrated smelter and a refining plant that includes a blast furnace, lead refinery (converter) that is utilized for recovering the base metals and precious metals. Further, in the special metal recovery plants, In, selenium (Se), and tellurium (Te) are recovered (Cui and Zhang 2008; Hagelüken 2005) under the base metal operations

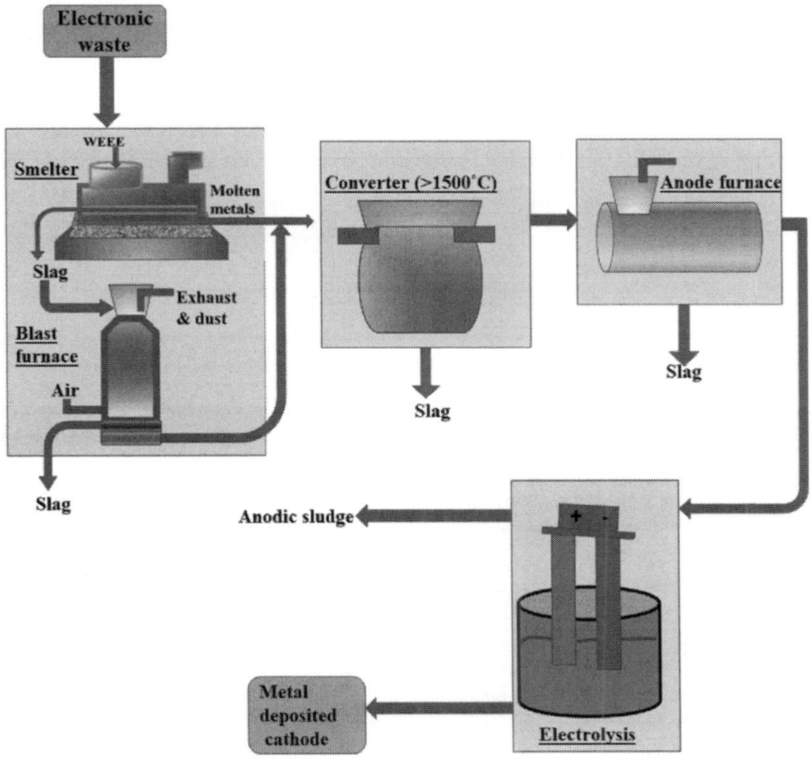

Figure 15.3 Metal extraction operations by the pyrometallurgy method. (Adapted from Chakraborty et al. 2022).

Table 15.3 Various industrial facilities using pyrometallurgical methods for recovery of metals from e-waste (Khaliq et al. 2014; Tesfaye et al. 2017).

Facility	Location	Process features	Metals recovered	Capacity (kt/year)
Aurubis smelter	Lünen, Germany	Black Cu processing, electrofining and precious metal refining	Ag, Au, Cu, Pb, Sn, and Zn	300 (only a fraction is e-waste)
Umicore smelter	Hoboken, Belgium	Smelting, Cu leaching, electrowinning, precious metal refining	Ag, As, Au, Bi, Cu, In, Ir, Ni, Pb, Pd, Pt, Rh, Ru, Sb, Se, Sn	350
Noranda smelter	Quebec, Canada	Feeding of e-waste smelting. Upgrading in converter and anode furnace	Ag, Au, Cu, Ni, Pd, Pt, Se, and Te	100
DOWA Group smelter	Kosaka, Japan	Smelting, Cu refining, and precious metal refining	Ag, Au, Bi, Cu, Ni, Pb, Sb, Sn, Te	150

(BMO) (Cui and Zhang 2008). The metals such as Ag, Au, Pt group metals, In, Bi (bismuth), Sn (tin), Se, Te, Sb (antimony), As (arsenic), and the base metals Cu, Ni, and Pb are recovered. As REEs have a high affinity for oxygen, they end up in the slag of the smelter and the blast furnace (Binnemans et al. 2013a). Among the REEs present in these slags, cerium (Ce) is a majority, but it is not categorized as a critical metal (Falconnet 1985). Further, Rhodia (Solvay) and Umicore collaborated on the development of a method for recycling REEs from NiMH (nickel-metal-hydride) rechargeable batteries. In a shaft furnace, the metals are converted into a Ni-Co-Cu-Fe alloy and a slag that includes mainly the oxides of Ca, Al, Si, and Fe, along with Li and REEs. Further, rare earth concentrates are produced from the oxide slags after Li recovery, and these concentrates could be further fed to a rare earth separation plant. This process has been adopted by Rhodia in La Rochelle, France (Verhaeghe et al. 2011).

Park et al. (2019) reported a three-step massive treatment involving the pyrolysis process, physical treatment methods, and pyrometallurgical processing for mobile phone scrap. They processed 1000 kg of scrap with pyrolysis including the physical classification method. Smelting of the insoluble residue was carried out so that oxide-complicated materials were acid-leached effectively. Metals like Ni, Au, Ag, and Sn were recovered in a molten state. Nonetheless, pyrometallurgy is widely applicable for high-grade ore and requires high energy input, making the treatment of smaller or low-grade ore more difficult and uneconomical (Tunsu et al. 2015). Further, additional processes are required for the production of pure REEs as they tend to form mixed products (Tipre et al. 2021). A large amount of precious metals and REEs end up in the output from the pyrometallurgical refining treatment as ferrous materials or dust, without undergoing the refining process (Ganguli and Cook 2018). In the slag phase, the REEs get entrapped because of their affinity toward oxygen and are easily lost during the pyrometallurgical processes (Binnemans et al. 2013b). Avarmaa et al. (2019) studied the behavior of the distribution coefficients of In, Ga, and Ge in the base metal recoveries of metallic Cu and those of silica-saturated iron-silicate slags where the secondary Cu smelting and the refining process conditions were executed under the range of partial pressure of oxygen (pO_2) of 10^{-9} to 10^{-5} atm (atmospheric pressure) at 1300°C. They reported that the distribution coefficient of In increased as a function of decreasing pO_2, whereas that of Ga decreased as the pO_2 increased and Ge was almost completely vaporized from the sample. There would be a lower risk of In losses by direct smelting of LCDs material in an integrated smelter. However, most of the LCD panels are made of glasses that can end up in the slag, similar to other precious metals. Eventually, the requirement of a disproportionately huge amount of energy input to retrace and recover a minute quantity of metals becomes uneconomical for large-scale investment, and the extraction of the CMs from the slags becomes a limitation (Gotze and Rotter 2012).

Since the year 2002, Dowa's Kosaka smelting unit in Japan treats e-waste as a secondary feed-in to its Oukokumpu flash smelter to recover Cu, and approximately 10% of the total Cu throughput from the facility resulted from e-waste (Tabelin et al. 2021). Umicore, Aurubis, Outotec, and TSL recycling are some of the companies that employ a pyrometallurgical method to recover metals from e-waste (Debnath et al. 2018). For instance, Umicore recycles 30 tons of In annually from e-waste (Zhang et al. 2017). Nonetheless, one of the challenges to recover REEs is that even though they have high melting points, they are soft, Ag-colored metals that can be tarnished easily in the atmosphere (Wall 2021). Drawbacks of pyrometallurgy are the large slag generation, loss of precious metals, generation of air pollutants such as dioxins and fluoride products, and difficulty in recovering selective metals (Cardoso et al. 2019; Sahu et al. 2022). The pyrometallurgical processes for recovery of CMs and REEs require high thermal energy. It has been reported that a large quantity of slag is generated by pyrometallurgical operations, and it is difficult to recover metals from low grade ores (Cardoso et al. 2019). The recovery of REEs from the e-waste becomes difficult as they are present in low concentrations.

15.3.2 Hydrometallurgical Processes

An alternate extraction method of metals is the hydrometallurgical process that utilizes lixiviants for cyanide, thiosulfate, halide, and thiourea leaching (Cui and Zhang 2008; Huang et al. 2022; Nithya et al. 2021). The hydrometallurgical method includes leaching using various chemicals, selective extraction using precipitation, crystallization, electrorefining, cementation, solvent extraction, ion exchange, etc. (Ashiq et al. 2019; Efstratiadis and Michailidis 2022) that results in a cost-effective method for the extraction of REEs. The rate of leaching and the leaching efficiency depend on various parameters, which include particle size, type and concentration of the solvent used for leaching, pH, solid to liquid ratio, agitation, temperature, and redox potential, etc. (Sethurajan et al. 2019). Oxidants and reductants can be added to improve the leaching efficiency. Precious metals like Au, Ag, Pd, Pt, etc. are leached by halide leaching, thiourea leaching, cyanide leaching, and thiosulfate leaching. Chemical leaching can also be performed by using ligands that get complexed with metals (Pant et al. 2012). Ethylenediaminetetraacetic acid (Cheikh et al. 2010), diethylene triamine pentaacetate (DTPA), and NTA along with citrate, oxalate, and tartrate are used as chelating agents for the extraction of metals (Elliott and Shastri 1998). Strong acids and basic solutions are used for selective dissolution and precipitation of different metals of interest (Jha et al. 2016; Tipre et al. 2021). Various inorganic acids such as hydrochloric acid (HCl), sulfuric acid (H_2SO_4), and nitric acid (HNO_3) have been reported to leach more than 90% of the metals from PCBs, which include Cu, Zn, and Ni (Kim et al. 2003; Kinoshita et al. 2003). The metal of interest is made to leach into the acid solution. Hydrometallurgical etching is a chemical leaching in which chemicals like ferrous chloride ($FeCl_2$), cupric chloride ($CuCl_2$), and HCl (Barbieri et al. 2010) or organic solvents (Lee et al. 2009) are used to extract metals. The next step is that the metals are isolated and can be concentrated for recovery of REEs (Bindschedler et al. 2017). Figure 15.4 presents different types of hydrometallurgical treatment operations.

The REEs are generally found in the SmCo (samerium cobalt) and NdFeB magnet scraps. Since the metallic REEs are highly reactive, leaching is required to recover them. Generally, if sulfuric acid is used for the leaching purpose, then a double salt of Sm is obtained, whereas if HCl/ HNO_3 is used, solvent extraction can be applied directly. Furthermore, the double salt precipitation methods can be employed to achieve the selective precipitation procedures for Sm. As the Sm-containing salts like $Sm_2(SO_4)_3.2Na_2SO_4.2H_2O$ have low solubility, Sm can be separated from the leaching solution (Sun et al. 2017). Stein et al. (2022) studied the efficiency of leaching using organic acid such as citric and acetic acid for the recovery of REE from NdFeB from permanent magnets. The microwave leaching with 0.5 M citric acid resulted in leaching 57% of Nd and 58% of Pr, whereas by using acetic acid at 0.5 M, 48% of Nd and 65% of Pr were leached within 15 minutes.

Yang et al. (2017) performed a complete dissolution of NdFeB magnet and recovery of REEs by their selective leaching and conversion to a new solid phase. Resende and Morais (2010) have reported leaching of Eu and Y from computer monitors, where sulfuric acid is used as a leaching agent, which showed optimum recovery of 97% of REE

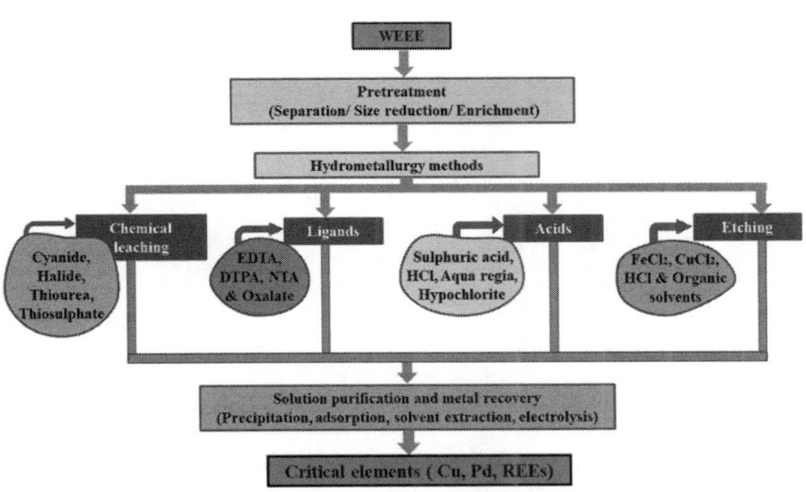

Figure 15.4 Hydrometallurgical treatment operations (Pant et al. 2012 / with permission of Elsevier).

by dissolution. Purification through a solvent extraction technique was carried out followed by selective precipitation, and further, calcination of the precipitate to obtain their oxides. Another study by Deshmane et al. (2019) developed a novel membrane assisted extraction (MSX) technique for selective extraction of more than 95% of Nd, Pr, and Dy from different scrap magnets. The magnets were crushed and dissolved in nitric acid to prepare the feed for recovering the REEs. In MSX, an organic phase containing tetraoctyl diglycolamide (TODGA) as extractant and an organic solvent were immobilized in a hydrophobic membrane. Since REEs have a higher affinity toward TODGA due to the ionic size of the REEs that fits perfectly in the cavity of the TODGA geometry, it results in selective extraction of REEs over non-REEs. Constantine et al. (2022) explored using spent NiMH batteries for the recovery of Ni, Co, and REEs. In their work, four different acids, citric acid, acetic acid, ascorbic acid, and 1-hydroxyethane-1,1-diphosphonic acid (etidronic acid), were used as lixiviants. Citric acid was considered, as it provided enough acidity to be able to form metal-citrate complexes that assisted in the leaching of REEs, Ni, and Co. With further addition of dilute H_3PO_4 (phosphoric acid) to the supercritical water extraction leaching solution, they were able to achieve 99.31% of La recovery, 98.75% of Ce recovery, and 99.05% of Nd. Barragan et al. (2020) implemented a combination of hydrometallurgical and electrochemical processes for recovery of Cu of 96 wt% (weight percent) and Sb of 81 wt% pure precipitate from e-waste. Cu metals were recovered using ferric chloride, followed by an electrowinning process that uses a rotating cylinder electrode for its efficient predictable fluid flow pattern. Sb precipitates were recovered by altering the pH of the solution. Ionic liquid (IL) salts have been found to be utilized in the recovery of critical metal. The extraction of In from waste LCDs using IL was performed with ascorbic acid. The separation of In was achieved resulting into two phases: In in an IL organic-rich phase and Fe in ascorbic acid aqueous solution (Paiva and Nogueira 2021). Similar leaching using HCl and HNO_3 acids have been reported by Hammache et al. (2021) to leach Nd and Pr. Deng et al. (2021) have used an ultrafast electrothermal process in which flash joule heating (FJH) activates the waste by thermally degrading it and increases the solubility of REE species to improve the extractability. The waste activated by FJH when treated with dilute HCl doubled the leachability and resulted in high recovery. This shows that the efficiency of hydrometllurgical methods can be improvised by pre-activation of the waste. Table 15.4 presents the critical metals that were recovered by hydrometallurgical methods from e-waste.

The use of pyrometallurgical and hydrometallurgical methods for recovery of critical metals can be energy

Table 15.4 Critical metals recovered by hydrometallurgical method from e-waste.

Critical metals recovered	Source	Leachants	Conditions: temperature and time	Recovery efficiency	References
In	LCD panel	H_2SO_4 acid	80 °C for 10 min	>90%	Rocchetti et al. (2015)
In	LCD screens	H_2SO_4 acid	90 °C, 60 h	96.4% In	Silveira et al. (2015)
Co	Spent LIBs	Glycine	360 min at 80 °C	>95% Co	Nayaka et al. (2016)
Co and REEs	NiMH batteries	H_2SO_4	90 °C for 6 h	98% Co and REEs	Becker et al. (2016)
Mn, Li, Co, and Ni	Spent LIBs	L-tartaric acid ($C_4H_6O_6$) and hydrogen peroxide (H_2O_2)	70 °C and 30 min	99.31% Mn, 99.07% Li, 98.64% Co, and 99.31% Ni	He et al. (2017)
La, Ce, Nd, Pr, and Sm	Spent Ni-MH batteries	First leaching with water. Second leaching: $NaHSO_3$ in H_2SO_4 (at 95 °C)	Pretreatment: Baking with 2 ml H_2SO_4 at 300 °C for 90 min. First leaching with water at 75 °C. Second leaching: $NaHSO_3$ in H_2SO_4 (at 95 °C)	First leaching step: 91% of Ni, 94% of Zn, and 91% of REEs and 20–30% of Co, Fe, and Mn. Second leaching step: residual of Mn and Co.	Meshram et al. (2017)
La, Nd, Sm, Pr, and Ce	NiMH batteries	Baking followed by H_2O and H_2SO_4 leaching	Pretreatment of baking with 2 mL H_2SO_4 at 300 °C Leaching H_2O, at 75 °C in 1 h	2.7% La, 4.5% Nd, 6.2% Sm, 2.29% Pr, and 2.58% Ce	Marra et al. (2018a)

(Continued)

Table 15.4 (Continued)

Critical metals recovered	Source	Leachants	Conditions: temperature and time	Recovery efficiency	References
Li and Co	Spent LIBs	Lactic acid and ascorbic acid	360 min and 90 °C	99% Li and 98% Co	Singh and Tandon (2014)
Nd, Fe, and B	NdFeB magnets	H_2SO_4	27 °C and 15 min	99.8% Nd and >91.3% Dy	Erust et al. (2021)
Ge	Semiconductor diodes	Acetic acid	240 min at 90 °C	70% Ge	Rafiee et al. (2021)
Y, La, and Gd		Microwave-assisted acid digestion using aqua regia with addition of hydrogen peroxide	10 min to reach 200 ± 5 C, then temp was held for another 20 min at 1000 W max power in a CEM MARS Xpress microwave	80%	Baez et al. (2022)
Nd, Fe, and B	NdFeB magnets	H_2SO_4 acid	~25 °C, 30 min	70.21% Fe and 20.46% Nd	Uysal et al. (2022)
La, Ce, Nd, Ni, and Co	Spent NiMH batteries	Citric acid	150 °C in 5 min	90.85% La, 88.84% Ce, 90.85% Nd, 90.70% Ni, and 91.62% Co	Constantine et al. (2022)

Table 15.5 Comparison of pyrometallurgy, hydrometallurgy, and bio-hydrometallurgy in metal extraction and refining techniques (Işıldar et al. 2018; Zhou et al. 2021).

Parameters	Pyrometallurgy	Hydrometallurgy	Bio-hydrometallurgy
Process conditions	High thermal conditions with high energy consumption	Use of corrosive acids	Safe conditions with low energy consumption, low toxicity or nontoxic chemicals
Selectivity	Low selectivity, only fraction of metals, high adaptability for different raw materials	High	High
Final residues	Generation of huge slags	Low and production of secondary pollution of waste liquids	Low to none, circulated water
Scale of operations	Suitable for large scale industrial production	Feasible and flexible to smaller scales	Flexible to smaller scales and few industrial requirements
Operational and energy costs	Higher	Lower	Lower
Environmental impact	Emission of gaseous chemicals such as dioxins	Moderate due to the toxic chemicals	Low
Social acceptance	Low	Medium due to the toxic reagents used and end products	High with cleaner processes and auto-pollution control

consuming and produce unwanted toxic secondary wastes including thorium (Th), uranium (U), etc., making it not environmentally friendly. For instance, the practices of separation, purification, pre-concentration, and also the use of strong acids and bases can become cost inefficient (Yu et al. 2020). Moreover, pyrometallurgical or hydrometallurgical methods used in the mineral sector are not as effective when targeted in lower concentrations on the extraction of REEs and precious metals (Bachér and Kaartinen 2017). Consequently, the need for alternative technologies that qualify the logical and coherent recovery of REEs and CMs is prudent (Yu et al. 2020). Biological processes may help in reducing the energy consumption and negative impact on the environment for the recovery of metals (Islam et al. 2020). A comparison of various recovery methods for metal extraction and refining techniques is shown in Table 15.5.

15.4 Biological Processes

The biological processes used for the recovery of REEs and CMs involve bioleaching, which is used for the extraction of metals from the e-waste to liquid media using microorganisms. Further, other biological processes aided by microorganisms such as biosorption and bioaccumulation can selectively recover REEs and CMs from the leaching solution. Bioprecipitation and biomineralization can also be used to recover REEs and CMs from the leaching solution. Figure 15.5 presents the different mechanisms of biosorption, bioaccumulation, and bioleaching (Dev et al. 2020; Nancharaiah et al. 2015).

15.4.1 Bioleaching

Bioleaching is an emerging and promising process that can operate at atmospheric pressure without depending on extreme temperature and gas emissions found in pyrometallurgical processes and without employment of harmful reagents used in hydrometallurgical processes (Debnath et al. 2018). The bioleaching process aids in metal ion mobilization from solid sources by biologically catalyzed reactions. Microorganisms that excrete organic acids (gluconic acid, citric acids, etc.) and inorganic acids (cyanide or sulfuric acid) are employed to aid in the dissolution of metals or to facilitate oxidation and reduction of metals (Opare et al. 2021; Rasoulnia et al. 2020).

Generally, depending on the metabolism requirements, the microorganisms can be classified as chemolithoautotrophs and chemoorganotrophs. Acidophiles are under the chemolithoautotrophs classification that grow at acidic pH below 2 by feeding on the atmospheric CO_2 (carbon dioxide), which serves as a carbon source, and inorganic compounds containing Fe^{2+} and S^0 serve as the energy source. In addition, based on the growth of the microbial environment temperature, they can be further subdivided into three categories, wherein microorganisms such as extremophiles, moderate thermophiles, and mesophiles are employed in the recovery of metals because of their higher tolerance for metal toxicity. On the contrary, chemoorganotrophs are those microorganisms that feed on the organic carbon as the energy source and are pH tolerant (alkaline) (Srivastava et al. 2020). For metal recovery from WEEE, bioleaching can be achieved by employing autotrophic microorganisms like S (sulfur) and Fe oxidizer and heterotrophic microorganisms that are cyanide-producing telluric microorganisms (Gavrilescu 2022; Işıldar et al. 2019). Heterotrophic bacteria such as *Acinetobacter* sp. Cr B2 have also been reported for bioleaching of Cu from electronic waste (Jagannath et al. 2017).

The conduct of bioleaching of REEs can follow either a contact bioleaching that includes a one-step method and two-step method or an indirect bioleaching process called spent-medium bioleaching, which is a non-contact process. In one-step bioleaching, both the REE containing material and the microbial medium are added simultaneously for the reaction to occur. In a two-step process, the REE-containing material is added to a previously cultured microbial medium. Non-contact bioleaching employs microorganism to produce or secrete essential metabolites, which act as leaching agents. Then the cell free growth media containing these metabolites can be used with the material that contains REE (Rasoulnia et al. 2020). Mineralytic effects of bacteria and fungi on minerals in the indirect method mainly depend on three principles, namely

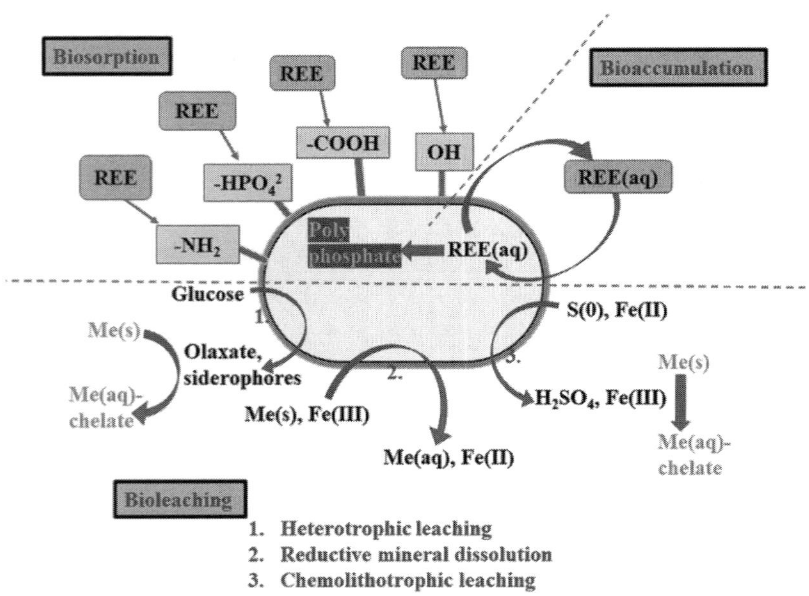

Figure 15.5 Mechanisms involved in biosorption, bioaccumulation, and bioleaching (Nancharaiah et al. 2015 / with permission of Elsevier).

acidolysis, complexolysis, and redoxolysis (Krebs et al. 1997). Metal mobilization is carried out by microorganisms by: (i) the organic or inorganic acids (protons) such as malic, gluconic, or sulfuric acids (H_2SO_4) formed by them; (ii) the excretion of complexing agents by the microorganisms; (iii) redoxolysis, where microbial growth occurs from metal ions by electron transfer resulting in the dissolution of metals (Opare et al. 2021). Table 15.6 presents an overview of bioleaching of critical metals by various microorganisms from e-waste.

Two-step bioleaching was carried out by Marra et al. (2018a) for the recovery of REEs from WEEE. In the initial step, *Acidithiobacillus thioxidans* was utilized, which aided in lowering the pH of the leaching solution, and Ce, Eu, and Nd were mobilized more than 95% while the extraction yield of La, Y were found to be 80%. In the next step, 48% of Au was mobilized by *Pseudomonas putida* producing cyanide, within 3 hours from the residue of the first step.

Pourhossein et al. (2021b) used a hybrid acidic-cyanide for the bioleaching of precious metals and CMs from spent LED lamps. Biogenic ferric was produced by the *Acidithiobacillus ferrooxidans* cells, and these were employed for pretreatment of the spent LEDs. The pretreated LED powder was separated from the solution by centrifugation and further dried, which allowed the base metals and heavy metals such as Sn, Ni, Cu, Al, Ga, Pb, Cr, and Fe to leach out. Further, *Bacillus megaterium* was used for the production of cyanide to increase the leaching of precious metals like Au and Ag from the bio-pretreated residue. It was shown that the hybrid direct bioleaching resulted in bioleaching of 93% Au, 91% Ag, 98% Ni, 87% Cu, and 84% Ga, which were better results than those obtained

Table 15.6 Bioleaching of critical metals by various microorganisms from e-waste.

Microorganism used	Electronic waste	Metals recovered	Operating conditions	References
Acidithiobacillus ferrooxidans	Spent Li-ion battery	98.4% Co	35°C, 160 rpm, 1% pulp density, and 0.02 gL^{-1} of Ag concentration	Zeng et al. (2013)
Acidithiobacillus thiooxidans	WEEE shredding dust	Ce, Eu, Nd (>99%); La, Y (80%)	5% inoculum size, 1% pulp density, 30°C, pH 4	Kelly and Wood (2015)
A. ferrooxidans, A. thiooxidans, and *L. ferrooxidans*	CRT fluorescent powder	70% Y	30°C, S/L: 1.5% w/v, 150 rpm	Reed et al. (2016)
Acidithiobacillus thiooxidans (A.t) and *Leptospirillum ferriphilum (L.f)*	Spent electric vehicle, Li-ion battery	95% Li, 96% Co, 97% Ni, and 96% Mn	1% pulp density, pH adjustment had effect on Ni and Co. Release of Li due to biogenic H_2SO_4	Xin et al. (2016)
Acidithiobacillus thiooxidans	Discarded LCDs	100% In	1.6% w/v pulp density, pH 2.6, sulfur 8.6 gL^{-1}	Jowkar et al. (2018)
Mixed culture of *A. ferrooxidans* and *A. thiooxidans*	Spent Li-ion batteries	99.2% Li, 89.4% Ni, and 50.4% Co	*A.f/A.t* ratio of 3:2, pH 1.5, $FeSO_4$ of 36.7 gL^{-1}, and sulfur of 5 gL^{-1}	Heydarian et al. (2018)
Acidithiobacillus thiooxidans	WEEE shredding dust	>99% Ce, Eu, and Nd	30°C, 1.2–1.5 pH	(Marra et al. 2018b)
Aspergillus niger strains MM1 and SG1 and	Spent Li-ion batteries	82% Co and 100% Li	Supernatant of *A. niger* strain, 0.25% w/v	Biswal et al. (2018)
Acidithiobacillus ferrooxidans and *L. ferrooxidans*	Waste magnets	86% Dy, 91% Nd, and 100% Pr	10% inoculum size, 1% w/v, 85 rpm for 14 d	Auerbach et al. (2019)
Acidithiobacillus caldus, Leptospirillum ferriphilum, Sulfobacillus spp, and *Ferroplasma* spp.	Li-ion batteries	99.9% Co, 99.7% Ni, and 84% Li	9 K medium with 10% inoculation, 45°C, and 130 rpm. $FeSO_4.7H_2O$ (44.2 g/l) and elemental sulfur (10 g/l) were used as nutrients.	Ghassa et al. (2020)
Pretreatment with *Acidithiobacillus thiooxidans* and bioleaching by *Bacillus megaterium*	Light-emitting diodes	93% Au, 91% Ag, 87% Cu, 98% Ni, and 84% Ga	9 K medium, pH 2, 140 rpm at 29°C for 10 d	Pourhossein et al. (2022)

by unpretreated LEDs. Another study by Pourhossein et al. (2021a) reported the extraction of In and Sr (strontium) from organic LED-type smart phone touch screens. The interaction of recovery of Sr and In using *Acidithiobacillus ferroxidans* with several influential variables such as pH, ferrous sulfate, elemental S, and solid content was evaluated by using the statistical approach of response surface methodology. The optimum condition showed the complete extraction of In, whereas Sr showed only 5% extraction efficiency.

Mishra et al. (2008) investigated bioleaching of spent lithium-ion secondary batteries using *Acidithiobacillus ferrooxidans*. The metabolites such as H_2SO_4 and ferric ion are produced by the microorganism in the bacterial leaching medium. The dissolution of Co could be achieved faster than that of Li. A decrease in dissolution was observed when the initial Fe(II) concentration was increased. At higher solid-to-liquid ratio metal dissolution was lower, owing to the arrest of cell growth. Roy et al. (2021) also reported on recovery of Ni, Mn, Co, and Li by bioleaching from spent LIBs (lithium-ion batteries) using an autotrophic bacterium, *Acidithiobacillus ferroxidans*. They reported that biogenic H_2SO_4 and Fe^{3+} were produced at higher concentration in the culture medium on higher dosage of $FeSO_4$ in the nutrient medium. Higher leaching efficiency of 92% Mn, 90% Ni, 89% Li, and 82% Co using 100 g/L solid content in 72 h could be achieved by using three cycles of bacterial cultures. Alavi et al. (2021) employed two different fungi, *Aspergillus niger* and *Aspergillus tubingensis*, in bioleaching where one-step, two-step, and spent medium bioleaching using different energy source of carbohydrate material-sucrose, impure sucrose, and vinasse was conducted. The presence of vinasse illustrated the higher bioleaching among the three types.

15.4.2 Biosorption

As a part of purification process of REE wherein the REEs or CMs dissolved in leaching solution are to be recovered, ion exchange and crystallization methods are employed. However, these processes suffer the drawback of the generation of secondary waste effects and they become expensive. Biosorption is one of the affordable methods for adsorption of metals, and since the last decade, several contributions have been reported using microorganisms as biosorbents (Gupta et al. 2019). It is a promising purification step for sequestering metallic cations from aqueous solution for continuous metal recovery methods (Andrès and Gérente 2011). When biosorption is not regulated metabolically, the rate of metal adsorption can be faster. Microbial biomass of bacteria, fungi, algae, etc., proteins and other bio-related materials can drive as biosorbents (Lo et al. 2014). Living cells can also lead to a passive metabolism where metals can adhere to the surface biomolecules like S-layer protein. When a saturation level is reached, no new metallic ion would be able to adhere to the biosorbents (Velásquez and Dussan 2009). The physicochemical interaction by different mechanisms such as ion exchange, adsorption, precipitation, and electrostatic reactions between the metal ions in solution and the functional groups on the surface of biological material aid in biosorption (Wang et al. 2017). In simple terms, it is a biological approach that allows the binding of metal ions in the aqueous solution with the biomass or their metabolites (Awasthi et al. 2016).

Giese (2020) studied the mechanism of biosorption chemistry of lanthanide and proposed that some species of lanthanide might form in the presence of hydroxyl ligands such as $Ln(OH)_{2+}$. The REEs occur dominantly in 3^+ oxidation state in aqueous phase unless the concentration of the OH group is high, making them strong acids. They are expected to have an electrostatic nature of bonding instead of covalent bonding (Giese 2020). Once the desired metal gets biosorbed onto the biomass/biological component, the recovery can be performed using solutions of elution that contain EDTA or Na salts (Bindschedler et al. 2017). Introduction of functional groups like -COOH, $-NH_2$, -OH, and -C=O to the biosorbents can improve on the activity for the metal sorption, which can also increase the adsorption capacity (Yu et al. 2019). Dolker and Pant (2019) extracted Li from Li-ion battery waste by a chemical-biological hybrid system. The proteinaceous S-layer in the identified *Lysinibacillus* sp. bacterial surface was found to bind metals in the solution using specialized proteins and organic acids as a chelator. The bacteria-citric acid hybrid showed a higher recovery of Li and Co in one pot process rather than performing individually. It was found that the formation of new hydroxyl sites in citric acid improved the leaching process, which may be attributed to increased pH that may have increased the hydroxyl sites. The treatment resulted in increased Li leaching by 25% and Co biosorption by 98%.

Factors that affect the biosorption process (Debnath et al. 2018) are:

- Type and availability of biological ligands to facilitate metal binding.
- Type of biosorbents (living or nonliving).
- Chemical, stereochemical, and the coordination properties of targeted metals.
- Metal solution characteristics (e.g., pH and competing ions).

Brewer et al. (2019) developed a novel flow-through process for biosorption. They encapsulated an engineered *Escherichia coli* strain in a permeable PEG (polyethylene glycol) diacrylate hydrogel by an emulsion process. The microbe bead adsorbent surface contained a functional group that were selective for REE adsorption. Nd was effectively adsorbed at a flow velocity of 3 m/h at pH 4–6 as demonstrated by a breakthrough experiment. A bench-scale breakthrough curve showed REE purity of 97% in the adsorbed fraction. Seven different microbial cultures with a potential to biosorb and bioleach were examined by Kaur et al. (2022). The pieces of PCB were incubated along with the microbial culture in liquid broth. *Pleurotus florida* showed biosorption of Cu and *Pseudomonas* spp. exhibited bioleaching of iron. Laccase enzyme was shown to be attributed to the biocatalysis.

15.4.3 Bioaccumulation

Unlike biosorption, which deals with adsorption on the surface of the cell, bioaccumulation is driven by ATP active transport or precipitation and metabolic chemical reactions that occur intracellularly (Lo et al. 2014). The adsorbed REEs on the cell surface of the microorganisms are translocated by the importer complex, which are present in the lipid bilayer of the cell membrane. Eventually, the REEs are seized in the intracellular space by the active metabolic proteins and peptide ligand (Dev et al. 2020). At a certain metal concentration, the efflux mechanism can be activated preventing excess metal concentration to enter the cell (Velásquez and Dussan 2009). The metals accumulated inside the cells can be easily harvested by disruption of cells (Lo et al. 2014). Sometimes, the toxic metals can replace or substitute significant elements from the cell, and PolyP metabolism may mediate the intracellular accumulation of REE. Scanning electron microscope (SEM) and transmission electron microscope (TEM) images presented by Maleke et al. (2019) in their study showed that Eu was accumulated mostly on the surface of the bacterial envelope, membrane, and in the cytoplasm. Čížková et al. (2021) studied the bioaccumulation of REEs from waste luminophores of energy-saving bulbs and fluorescent lamps using *Galdieria phlegrea*. The luminophores contained yttrium abundantly at 90% w/w along with other REES. Yttrium was found to accumulate abundantly in the algal biomass, followed by Eu and La. Di Piazza et al. (2017) isolated *Penicillium expansum*, which were able to bioaccumulate REEs and aided in the recovery of lanthanum (390 ppm) and terbium (1520 ppm) from WEEE. Horiike and Yamashita (2015) were able to successfully bioaccumulate Dy using an acidophilic fungal strain, *Penidiella* sp. T9, which was able to decrease the high concentration of Dy from the environment. Although bioaccumulation and bioreduction have not been studied thoroughly, they may be a viable option for REE and CM recovery (Rene et al. 2021).

15.4.4 Bioprecipitation and Biomineralization

Precipitation is widely used for the recovery or removal of valuable metals from process effluents or solution containing precious metals. The most commonly used precipitating agents are sulfides, hydroxides, or carbonates. Addition of such additives shift the ionic equilibrium of the systems and aid in achieving the precipitation of any targeted metals (Sethurajan and Gaydardzhiev 2021). Bioprecipitation processes have been found to depend on microorganisms that produce biogenic sulfide-reducing microorganisms that form insoluble metal sulfidic precipitates (Işıldar et al. 2019). Calvert et al. (2019) demonstrated a 99% precipitation efficiency of Al, Ni, Co, and Cu from LIBs using hydrogen sulfide. A sulfate-reducing bacterial consortia produced the hydrogen sulfide, which was predominantly rich with *Desulfovibrio* spp. Janyasuthiwong et al. (2016) illustrated the recovery of Cu, Ni, and Zn from the desktop PCBs, in which leaching was achieved using HNO_3 as lixiviant. The precipitation was accomplished by using sulfate-reducing bacteria in an inverse fluidized bed reactor yielding >90% Cu recovery.

Biomineralization is a process by which microbes produce a variety of minerals both intracellularly and extracellularly. The dissolved metals or REEs from the leaching solution may be recovered by biomineralization to form nanoparticles. Studies on the synthesis of nanoparticles using the cell free extract and the cell biomass that utilizes the intracellular enzymes have been reported by Mukherjee et al. (2018). Biomineralization finds its application in wastewater treatment, nanotechnology, and metal recovery (Nancharaiah et al. 2015). In some ways, the chemical method of synthesizing nanoparticles (NPs) tends to have low stability, and controlling the crystallization in resulting NPs is difficult. Biologically derived NPs are least toxic because of the presence of capping agents or surfactants that are produced by microorganisms resulting in higher solubility and biocompatibility (Rienzie et al. 2019). Extracellular synthesis in the presence and absence of cells have also been reported. In the absence of cells, it utilizes the cell free supernatant of the microbial cells (Shivaji et al. 2011; Thakkar et al. 2010). Maleke et al. (2019) investigated the interaction of Eu that was bioaccumulated in *Thermus scotoductus* SA-01, thermophilic bacterium. The bacterial strain seemed to be stimulated by the bioaccumulated Eu at low concentration, and Eu could also be biomineralized by the *T. scotoductus* strain as $Eu(CO_3)_3$. This put forward the idea that the particular strain can be employed in the recovery of rare earth elements. Das and Ting (2017) used supernatant of the cyanogenic bacterium *Chromobacterium violaceum* pBAD strain to recover gold from electronic scrap. They adopted pretreatment of the waste with nitric acid followed by ultrasound using H_2SO_4 and H_2O_2 (hydrogen peroxide) to reduce the competition for cyanide ion from base metals (which is predominantly Cu) for metal-cyanide complexation. Further, they reported the biomineralization of Au to form crystalline gold nanoparticles from the leachate using *C. metallidurans* and *Delftia acidovorans* via extracellular precipitation (Das et al. 2015; Das and Ting 2017).

Chandane et al. (2020) used the cell-free supernatant of *Acidiphilium acidophilum* to leach Cu, and the leached Cu was recovered from Cu liquor by synthesizing an organ-inorganic hybrid (OIH) containing Cu as the inorganic part by using crude egg white as a polyphenol oxidase (PPO) enzyme source. Creamer et al. (2006) demonstrated the precipitation of Au(III) to Au(0) using biomass of *Desulfovibrio desulfuricans* using waste electronic scrap leachate. The pre-palladization of the cells in the absence of Cu^{2+} influenced Pd(II) removal from the leachate containing Cu^{2+} and Pd(II), and 95% Pd(II) removal could be achieved.

Biometallurgical methods for the recovery of CMs or REEs is an effective way to recover the resources from e-waste without producing any secondary pollutants. Yet, more pathways and mechanisms of the biological methods have to be elucidated for clearer understanding of the mechanisms and to achieve selective recovery.

15.5 Conclusion

The progress and advancement in the electronic industries versus the level of the recycling rate has confounded in terms of environmental management and sustainability. The utilization of the REEs and CMs in electronic equipments has increased because of their distinct and eccentric properties. Metals fall in the category of CMs when their supply rates are relatively lower than their increasing demand. This also applies to those metals that have low dispersity, a low recycling rate, and lower potential for substitution. To proceed with a recovery process, understanding the metal composition in a piece of electronic waste is a crucial element so as to optimize the operating conditions in each phase of the recovery system. As WEEE is a heterogeneous mixture of various metallic and nonmetallic components with varying composition and concentration, the recovery becomes a challenging task. The choice of a recovery method depends on the e-waste source, its chemical composition and REE content, the overall cost, energy requirement, and the environmental footprint. The process operating cost has to be low. When fossil fuel or biomass are used as a source of energy, the environment footprint of the process becomes larger. The drive toward greener energy must be emphasized. It is essential to either understand the design of WEEE units and then choose the relevant method for pre-treatment, or to design specialized systems to enhance the recovery of CMs and to minimize their loss.

The conventional methods of recovery of metals such as pyrometallurgy require high thermal energy to process. A large amount of slag is generated by pyrometallurgical operations, and it is difficult to recover when the fraction of REEs or CMs is very low in the WEEE. Hydrometallurgical methods for recovery employ different chemical solvents and harmful reagents that can be effective and selective but can raise environmental concern in the future. Biological methods of recovery of metals have been found to be an alternative way to process low-grade ore and under mild conditions. The use of microbes for recovery has found to be less harmful and does not require high temperatures, resulting in a safer and more reliable method. There are biological options available for leaching out REEs and CMs from electronic scrap material like bioleaching as well as for recovery of the leached metals and REEs from the leachate by using biosorption, bioprecipitation, and biomineralization. However, the research in this area is still primitive. Identification of specific microorganisms with a high degree of selectivity is essential for the recovery of REEs and other CMs to provide economically and technologically feasible alternatives. Further, process optimization in terms of the media composition, environmental factors such as pH, temperature, etc. is required. The design and development of suitable bioreactors or contactors through lab-scale and pilot-scale studies followed by appropriate scale-up strategies is highly essential for commercial application of these processes. There is a need for both the fundamental and applied research in this area. However, there are several challenges associated as the electronic waste is highly heterogeneous, and it may not be possible to arrive at a 'one size fits all' strategy. Further, these bio-based processes are time consuming and low-rate processes with an associated risk of contamination leading to operational challenges. Therefore, appropriate decisions have to be made by the stakeholders through adequate studies and research on which of the REEs or CMs are to be targeted for recovery, based on how much reduction in consumption of the precious primary raw material resources could be achieved by increasing the recycling rate, as well as based on the associated environmental risks if not recycled.

Nomenclature

Atm	atmospheric
BMO	base metal operations
CM	critical metals
CO_2	carbon dioxide
DTPA	diethylene triamine pentaacetate
DoE	Department of Energy
EC	European Commission
EDTA	ethylenediaminetetraacetic acid
EEDs	electrical and electronic devices
EVs	electric vehicles
e-waste	electronic waste
FAME	Faster Adoption and Manufacturing of Hybrid
IL	ionic liquid
IUPAC	International Union of Pure and Applied Chemistry
LCD	liquid crystal display
LED	light emitting diode
LIBs	lithium-ion batteries
MSX	membrane-assisted extraction
Mt	million metric tons
NP	nanoparticle
NdFeB	neodymium-iron-boron
NiMH	nickel metal hydride
NTA	nitrilotriacetic acid
PCB	printed circuit board
PEG	polyethylene glycol
pO_2	partial pressure of oxygen
PMO	precious metal operations
REE	rare earth elements (REEs)
SmCo	samerium cobalt
TODGA	tetraoctyl diglycolamide
USGS	United States Geological Survey
US EPA	US Environmental Protection Agency
WEEE	waste electrical and electronic equipment

References

Ahirwar, R. and Tripathi, A.K. (2021). E-waste management: a review of recycling process, environmental and occupational health hazards, and potential solutions. *Environmental Nanotechnology, Monitoring and Management* 15: 100409. https://doi.org/10.1016/j.enmm.2020.100409.

Alavi, N., Partovi, K., Majlessi, M. et al. (2021). Bioleaching of metals from cellphones batteries by a co-fungus medium in presence of carbon materials. *Bioresource Technology Reports* 15: 100768. https://doi.org/10.1016/j.biteb.2021.100768.

Amato, A., Becci, A., and Beolchini, F. (2020). Sustainable recovery of Cu, Fe and Zn from end-of-life printed circuit boards. *Resources, Conservation and Recycling* 158: 104792. https://doi.org/10.1016/j.resconrec.2020.104792.

Ambaye, T.G., Vaccari, M., Castro, F.D. et al. (2020). Emerging technologies for the recovery of rare earth elements (REEs) from the end-of-life electronic wastes: a review on progress, challenges, and perspectives. *Environmental Science and Pollution Research* 27 (29): 36052–36074. https://doi.org/10.1007/s11356-020-09630-2.

Andrès, Y. and Gérente, C. (2011). Removal of rare earth elements and precious metal species by biosorption. In: *Microbial Biosorption of Metals* (eds. P. Kotrba, M. Mackova, and T. Macek). Dordrecht: Springer. https://doi.org/10.1007/978-94-007-0443-5_8.

Annamalai, M. and Gurumurthy, K. (2021). Characterization of end-of-life mobile phone printed circuit boards for its elemental composition and beneficiation analysis. *Journal of the Air and Waste Management Association* 71 (3): 315–327. https://doi.org/10.1080/10962247.2020.1813836.

Ashiq, A., Kulkarni, J., and Vithanage, M. (2019). Hydrometallurgical recovery of metals from E-waste. In: *Electronic Waste Management and Treatment Technology*, 225–246. Elsevier. https://doi.org/10.1016/B978-0-12-816190-6.00010-8.

Auerbach, R., Bokelmann, K., Stauber, R. et al. (2019). Critical raw materials—advanced recycling technologies and processes: recycling of rare earth metals out of end of life magnets by bioleaching with various bacteria as an example of an intelligent recycling strategy. *Minerals Engineering* 134: 104–117. https://doi.org/10.1016/j.mineng.2018.12.022.

Avarmaa, K., Klemettinen, L., O'Brien, H. et al. (2019). Critical metals Ga, Ge and In: experimental evidence for smelter recovery improvements. *Minerals* 9 (6): 1–13. https://doi.org/10.3390/min9060367.

Awasthi, A.K., Zeng, X., and Li, J. (2015). Environmental pollution of electronic waste recycling in India: a critical review. *Environmental Pollution* 211: 259–270. https://doi.org/10.1016/j.envpol.2015.11.027.

Bachér, J. and Kaartinen, T. (2017). Liberation of printed circuit assembly (PCA) and dust generation in relation to mobile phone design in a size reduction process. *Waste Management* 60: 609–617. https://doi.org/10.1016/j.wasman.2016.09.037.

Baez, A.G., Muñoz, L.P., Garelick, H., and Purchase, D. (2022). Environmental technology and innovation characterization of industrially pre-treated waste printed circuit boards for the potential recovery of rare earth elements. *Environmental Technology & Innovation* 27: 102481. https://doi.org/10.1016/j.eti.2022.102481.

Bakas, I.I., Herczeg, M., Vea, E.B. et al. (2016). Critical metals in discarded electronics: mapping recycling potentials from selected waste electronics in the Nordic region. Nordic Council of Ministers. TemaNord 2016: 526. http://dx.doi.org/10.6027/TN2016-526.

Balaram, V. (2019). Rare earth elements: a review of applications, occurrence, exploration, analysis, recycling, and environmental impact. *Geoscience Frontiers* 10 (4): 1285–1303. https://doi.org/10.1016/j.gsf.2018.12.005.

Barbieri, L., Giovanardi, R., Lancellotti, I., and Michelazzi, M. (2010). A new environmentally friendly process for the recovery of gold from electronic waste. *Environmental Chemistry Letters* 8 (2): 171–178. https://doi.org/10.1007/s10311-009-0205-2.

Barragan, J.A., Ponce De León, C., Alemán Castro, J.R. et al. (2020). Copper and antimony recovery from electronic waste by hydrometallurgical and electrochemical techniques. *ACS Omega* 5 (21): 12355–12363. https://doi.org/10.1021/acsomega.0c01100.

Batinic, B., Vaccari, M., Savvilotidou, V. et al. (2018). Applied WEEE pre-treatment methods: maximizing the recovery of critical metals opportunities to. *Global Nest Journal* 20 (4): 706–711. https://doi.org/10.30955/gnj.002589.

Becker, K., Chmielarz, A., Szołomicki, Z. et al. (2016). Hydrometalurgiczny recykling akumulatorów Ni-MH i Li-ion. *Rudy i Metale Nieżelazne Recykling* 61 (6): 235–243. https://doi.org/10.15199/67.2016.6.1.

Bhardwaj, S., Khurana, S., and Kumar, A. (2023). Electronic (E-waste) conduct: chemical assessment and treatment methods. In: *Waste Management and Resource Recycling in the Developing World*. 143–161. https://doi.org/10.1016/B978-0-323-90463-6.00028-2.

Bindschedler, S., Vu Bouquet, T.Q.T., Job, D. et al. (2017). Fungal biorecovery of gold from E-waste. *Advances in Applied Microbiology* 99: 53–81. https://doi.org/10.1016/bs.aambs.2017.02.002.

Binnemans, K., Jones, P.T., Blanpain, B. et al. (2013a). Recycling of rare earths: a critical review. *Journal of Cleaner Production* 51: 1–22. https://doi.org/10.1016/j.jclepro.2012.12.037.

Binnemans, K., Pontikes, Y., Jones, P.T. et al. (2013b). Recovery of rare earths from industrial waste residues: a concise review. In: *3rd International Slag Valorisation Symposium*, 191–205.

Biswal, B.K., Jadhav, U.U., Madhaiyan, M. et al. (2018). Biological leaching and chemical precipitation methods for recovery of Co and Li from spent lithium-ion batteries. *ACS Sustainable Chemistry and Engineering* 6 (9): 12343–12352. https://doi.org/10.1021/acssuschemeng.8b02810.

Brewer, A., Dohnalkova, A., Shutthanandan, V. et al. (2019). Microbe encapsulation for selective rare-earth recovery from electronic waste leachates. *Environmental Science and Technology* 53 (23): 13888–13897. https://doi.org/10.1021/acs.est.9b04608.

Buechler, D.T., Zyaykina, N.N., Spencer, C.A. et al. (2020). Comprehensive elemental analysis of consumer electronic devices : rare earth, precious, and critical elements. *Waste Management* 103: 67–75. https://doi.org/10.1016/j.wasman.2019.12.014.

Calvert, G., Kaksonen, A.H., Cheng, K.Y. et al. (2019). Recovery of metals from waste lithium ion battery leachates using biogenic hydrogen sulfide. *Minerals* 9 (9): 1–19. https://doi.org/10.3390/min9090563.

Can Sener, S.E., Thomas, V.M., Hogan, D.E. et al. (2021). Recovery of critical metals from aqueous sources. *ACS Sustainable Chemistry and Engineering* 9 (35): 11616–11634. https://doi.org/10.1021/acssuschemeng.1c03005.

Cardoso, C.E.D., Almeida, J.C., Lopes, B. et al. (2019). Recovery of rare earth elements by carbon-based nanomaterials—a review. *Nanomaterials* 9 (814): 1–35.

Cayumil, R., Khanna, R., Rajarao, R. et al. (2016). Concentration of precious metals during their recovery from electronic waste. *Waste Management* 57: 121–130. https://doi.org/10.1016/j.wasman.2015.12.004.

Cecere, G. and Martinelli, A. (2017). Drivers of knowledge accumulation in electronic waste management: an analysis of publication data. *Research Policy* 46 (5): 925–938. https://doi.org/10.1016/j.respol.2017.03.005.

Chadha, R. and Sivamani, G. (2022). Critical minerals for India: assessing their criticality and projecting their needs for green technologies. Centre for Social and Economic Progress. https://csep.org/policy-brief/critical-minerals-for-india-assessing-their-criticality-and-projecting-their-needs-for-green-technologies-2 (accessed 06 August 2022).

Chakraborty, S.C., Zaman, M.W.U., Hoque, M. et al. (2022). Metals extraction processes from electronic waste: constraints and opportunities. *Environmental Science and Pollution Research* 29 (22): 32651–32669. https://doi.org/10.1007/s11356-022-19322-8.

Chancerel, P., Marwede, M., Nissen, N.F., and Lang, K. (2015). Resources, conservation and recycling estimating the quantities of critical metals embedded in ICT and consumer equipment. *Resources, Conservation & Recycling* 98: 9–18. https://doi.org/10.1016/j.resconrec.2015.03.003.

Chancerel, P., Meskers, C.E.M., Hagelüken, C., and Rotter, V.S. (2009). Assessment of precious metal flows during preprocessing of waste electrical and electronic equipment. *Journal of Industrial Ecology* 13 (5): 791–810. https://doi.org/10.1111/j.1530-9290.2009.00171.x.

Chandane, P., Jori, C., Chaudhari, H. et al. (2020). Bioleaching of copper from large printed circuit boards for synthesis of organic-inorganic hybrid. *Environmental Science and Pollution Resources* 27: 5797–5808.

Charles, R.G., Douglas, P., Dowling, M. et al. (2020). Towards increased recovery of critical raw materials from WEEE—evaluation of CRMs at a component level and pre-processing methods for interface optimisation with recovery processes. *Resources, Conservation and Recycling* 161 (April): 104923. https://doi.org/10.1016/j.resconrec.2020.104923.

Cheikh, M., Magnin, J.P., Gondrexon, N. et al. (2010). Zinc and lead leaching from contaminated industrial waste sludges using coupled processes. *Environmental Technology* 31 (14): 1577–1585. https://doi.org/10.1080/09593331003801548.

Čížková, M., Mezricky, P., Mezricky, D. et al. (2021). Bioaccumulation of rare earth elements from waste luminophores in the red algae, Galdieria phlegrea. *Waste and Biomass Valorization* 12 (6): 3137–3146. https://doi.org/10.1007/s12649-020-01182-3.

Constantine, J., Lie, J., and Liu, J.-C. (2022). Recovery of rare earth elements from spent NiMH batteries using subcritical water extraction with citric acid. *Journal of Environmental Chemical Engineering* 10 (3): 108000. https://doi.org/10.1016/j.jece.2022.108000.

Creamer, N.J., Baxter-Plant, V.S., Henderson, J. et al. (2006). Palladium and gold removal and recovery from precious metal solutions and electronic scrap leachates by Desulfovibrio desulfuricans. *Biotechnology Letters* 28 (18): 1475–1484. https://doi.org/10.1007/s10529-006-9120-9.

Cucchiella, F., Adamo, I.D., Rosa, P., and Terzi, S. (2016). Automotive printed circuit boards recycling: an economic analysis. *Journal of Cleaner Production* 121: 130–141. https://doi.org/10.1016/j.jclepro.2015.09.122.

Cui, J. and Zhang, L. (2008). Metallurgical recovery of metals from electronic waste: a review. *Journal of Hazardous Materials* 158 (2–3): 228–256. https://doi.org/10.1016/j.jhazmat.2008.02.001.

Das, S., Dong, B., and Ting, Y. (2015). Gold biodissolution from electronic scrap and biomineralization of bacterial gold nanoparticles. *Advanced Materials Research* 1130: 668–672. https://doi.org/10.4028/www.scientific.net/AMR.1130.668.

Das, S. and Ting, Y.P. (2017 October). In-vitro synthesis of gold nanoparticles using thiourea leachate of electronic waste. In: *Proceedings Sardinia 2017/ Sixteenth International Waste Management and Landfill Symposium, 2–6 October 2017*, Italy.

Debnath, B., Chowdhury, R., and Ghosh, S.K. (2018). Sustainability of metal recovery from E-waste. *Frontiers of Environmental Science and Engineering* 12 (6): 1–12. https://doi.org/10.1007/s11783-018-1044-9.

Deboer, M.A. and Lammertsma, K. (2013). Scarcity of rare earth elements. *ChemSusChem* 6 (11): 2045–2055. https://doi.org/10.1002/cssc.201200794.

Deng, B., Luong, D.X., Wang, Z., Kittrell, C., McHugh, E.A., and Tour, J.M. (2021). Urban mining by flash Joule heating. *Nature Communications* 12 (1): 5794. https://doi.org/10.1038/s41467-021-26038-9.

Deshmane, V.G., Islam, S.Z., and Bhave, R.R. (2019). Selective recovery of rare earth elements from a wide range of e-waste and process scalability of membrane solvent extraction. *Environmental Science and Technology* 54: 550–558. https://doi.org/10.1021/acs.est.9b05695.

Dev, S., Sachan, A., Dehghani, F. et al. (2020). Mechanisms of biological recovery of rare-earth elements from industrial and electronic wastes: a review. *Chemical Engineering Journal* 397: 124596. https://doi.org/10.1016/j.cej.2020.124596.

Di Piazza, S., Cecchi, G., Cardinale, A.M. et al. (2017). Penicillium expansum link strain for a biometallurgical method to recover REEs from WEEE. *Waste Management* 60: 596–600. https://doi.org/10.1016/j.wasman.2016.07.029.

Dolker, T. and Pant, D. (2019). Chemical-biological hybrid systems for the metal recovery from waste lithium ion battery. *Journal of Environmental Management* 248: 109270. https://doi.org/10.1016/j.jenvman.2019.109270.

Dushyantha, N., Batapola, N., Ilankoon, I.M.S.K., and Rohitha, S. (2020). The story of rare earth elements (REEs): occurrences, global distribution, genesis, geology, mineralogy and global production. *Ore Geology Reviews* 122: 103521. https://doi.org/10.1016/j.oregeorev.2020.103521.

Efstratiadis, V.S. and Michailidis, N. (2022). Sustainable recovery, recycle of critical metals and rare earth elements from waste electric and electronic equipment (circuits, solar, wind) and their reusability in additive manufacturing applications: a review. *Metals* 12 (5): 794. https://doi.org/10.3390/met12050794.

Eggert, R., Wadia, C., Anderson, C. et al. (2016). Rare earths: market disruption, innovation, and global supply chains. *Annual Review of Environment and Resources* 41 (August): 199–222. https://doi.org/10.1146/annurev-environ-110615-085700.

Elliott, H.A. and Shastri, N.L. (1998). Extractive decontamination of metal-polluted soils using oxalate. *Water, Air, and Soil Pollution* 110: 335–346.

Erust, C., Akcil, A., Tuncuk, A. et al. (2021). A multi-stage process for recovery of neodymium (Nd) and dysprosium (Dy) from spent hard disc drives (HDDs). *Mineral Processing and Extractive Metallurgy Review* 42 (2): 90–101. https://doi.org/10.1080/08827508.2019.1692010.

European Commission (2020). Communication from the Commission to the European Parliament, the Council, the European Economic and Social Committee and the Committee of the Regions.

Falconnet, P. (1985). The economics of rare earths. *Journal of the Less-Common Metals* 111 (1–2): 9–15. https://doi.org/10.1016/0022-5088(85)90163-8.

Fernandez, V. (2017). Rare-earth elements market: a historical and financial perspective. *Research Policy* 53: 26–45. https://doi.org/10.1016/j.resourpol.2017.05.010.

Forti, V., Baldé, C.P., Kuehr, R., and Bel, G. (2020). *The Global E-waste Monitor 2020: Quantities, Flows and the Circular Economy Potential.* United Nations University (UNU)/ United Nations Institute for Training and Research (UNITAR)–co-hosted SCYCLE Programme. In: *International Telecommunication Union* (ITU) and (July Issue). http://ewastemonitor.info.

Ganguli, R. and Cook, D.R. (2018). Rare earths: a review of the landscape. *MRS Energy & Sustainability* 5 (1): 1–16. https://doi.org/10.1557/mre.2018.7.

Gavrilescu, M. (2022). Microbial recovery of critical metals from secondary sources. *Bioresource Technology* 344: 126208. https://doi.org/10.1016/j.biortech.2021.126208.

Ghassa, S., Farzanegan, A., Gharabaghi, M., and Abdollahi, H. (2020). Novel bioleaching of waste lithium ion batteries by mixed moderate thermophilic microorganisms, using iron scrap as energy source and reducing agent. *Hydrometallurgy* 197: 105465. https://doi.org/10.1016/j.hydromet.2020.105465.

Ghosh, B., Ghosh, M.K., Parhi, P. et al. (2015). Waste printed circuit boards recycling: an extensive assessment of current status. *Journal of Cleaner Production* 94: 5–19. https://doi.org/10.1016/j.jclepro.2015.02.024.

Ghulam, S.T. and Abushammala, H. (2023). Challenges and opportunities in the management of electronic waste and its impact on human health and environment. *Sustainability* 15 (3): 1837. https://doi.org/10.3390/su15031837.

Giese, E.C. (2020). Biosorption as green technology for the recovery and separation of rare earth elements. *World Journal of Microbiology and Biotechnology* 36 (4): 1–11. https://doi.org/10.1007/s11274-020-02821-6.

Girtan, M., Wittenberg, A., Grilli, M.L. et al. (2021). The critical raw materials issue between scarcity, supply risk, and unique properties. *Materials* 14: 1–18.

Goodenough, K.M., Wall, F., and Merriman, D. (2018). The rare earth elements: demand, global resources, and challenges for resourcing future generations. *Natural Resources Research* 27 (2): 201–216. https://doi.org/10.1007/s11053-017-9336-5.

Gotze, R. and Rotter, V.S. (2012). Challenges for the recovery of critical metals from waste electronic equipment—a case study of indium in LCD panels. In: *2012 Electronics Goes Green 2012+* 1–8. IEEE.

Gupta, N.K., Gupta, A., Ramteke, P. et al. (2019). Biosorption—a green method for the preconcentration of rare earth elements (REEs) from waste solutions: a review. *Journal of Molecular Liquids* 274: 148–164. https://doi.org/10.1016/j.molliq.2018.10.134.

Gwenzi, W., Mangori, L., Danha, C. et al. (2018). Science of the total environment sources, behaviour, and environmental and human health risks of high-technology rare earth elements as emerging contaminants. *Science of the Total Environment* 636: 299–313. https://doi.org/10.1016/j.scitotenv.2018.04.235.

Hagelüken, C. (2005). Recycling of electronic scrap at Umicore's integrated metals smelter and refinery. *Proceedings—European Metallurgical Conference* EMC 2005 59 (3): 152–161.

Hammache, Z., Bensaadi, S., Berbar, Y. et al. (2021). Recovery of rare earth elements from electronic waste by diffusion dialysis. *Separation and Purification Technology* 254: 117641. https://doi.org/10.1016/j.seppur.2020.117641.

He, L.P., Sun, S.Y., Mu, Y.Y. et al. (2017). Recovery of lithium, nickel, cobalt, and manganese from spent lithium-ion batteries using L-tartaric acid as a leachant. *ACS Sustainable Chemistry & Engineering* 5 (1): 714–721. https://doi.org/10.1021/acssuschemeng.6b02056.

Heydarian, A., Mousavi, S.M., Vakilchap, F., and Baniasadi, M. (2018). Application of a mixed culture of adapted acidophilic bacteria in two-step bioleaching of spent lithium-ion laptop batteries. *Journal of Power Sources* 378: 19–30. https://doi.org/10.1016/j.jpowsour.2017.12.009.

Horiike, T. and Yamashita, M. (2015). A new fungal isolate, Penidiella sp. strain T9, accumulates the rare earth element dysprosium. *Applied and Environmental Microbiology* 81 (9): 3062–3068. https://doi.org/10.1128/AEM.00300-15.

Hsu, E., Barmak, K., West, A.C., and Park, A.H.A. (2019). Advancements in the treatment and processing of electronic waste with sustainability: a review of metal extraction and recovery technologies. *Green Chemistry* 21 (5): 919–936. https://doi.org/10.1039/c8gc03688h.

Huang, Y.F., Chou, S.L., and Lo, S.L. (2022). Gold recovery from waste printed circuit boards of mobile phones by using microwave pyrolysis and hydrometallurgical methods. *Sustainable Environment Research* 32 (1): 1–11. https://doi.org/10.1186/s42834-022-00118-x.

Isildar, A. (2016). Biological versus chemical leaching of electronic waste for copper and gold recovery (Doctoral dissertation). Universite Paris-Est, Paris, France. Retrieved from https://tel.archives-ouvertes.fr/tel-01738056.

Işıldar, A., Rene, E.R., van Hullebusch, E.D., and Lens, P.N.L. (2018). Electronic waste as a secondary source of critical metals: management and recovery technologies. *Resources, Conservation and Recycling* 135 (July 2017): 296–312. https://doi.org/10.1016/j.resconrec.2017.07.031.

Işıldar, A., van Hullebusch, E.D., Lenz, M. et al. (2019). Biotechnological strategies for the recovery of valuable and critical raw materials from waste electrical and electronic equipment (WEEE)—a review. *Journal of Hazardous Materials* 362 (January 2018): 467–481. https://doi.org/10.1016/j.jhazmat.2018.08.050.

Islam, A., Ahmed, T., Awual, R. et al. (2020). Advances in sustainable approaches to recover metals from e-waste—a review. *Journal of Cleaner Production* 244: 118815. https://doi.org/10.1016/j.jclepro.2019.118815.

Jagannath, A., Shetty, K.V., and Saidutta, M.B. (2017). Bioleaching of copper from electronic waste using Acinetobacter sp. Cr B2 in a pulsed plate column operated in batch and sequential batch mode. *Biochemical Pharmacology* https://doi.org/10.1016/j.jece.2017.02.023.

Janyasuthiwong, S., Ugas, R., Rene, E.R. et al. (2016). Effect of operational parameters on the leaching efficiency and recovery of heavy metals from computer printed circuit boards. *Journal of Chemical Technology and Biotechnology* 91 (7): 2038–2046. https://doi.org/10.1002/jctb.4798.

Roy, J.J., Srinivasan, M., and Cao, B. (2021). Bioleaching as an eco-friendly approach for metal recovery from spent NMC-based lithium-ion batteries at a high pulp density. *ACS Sustainable Chemistry and Engineering* 9 (8): 3060–3069. https://doi.org/10.1021/acssuschemeng.0c06573.

Jha, M.K., Kumari, A., Panda, R. et al. (2016). Hydrometallurgy review on hydrometallurgical recovery of rare earth metals. *Hydrometallurgy* 165: 2–26. https://doi.org/10.1016/j.hydromet.2016.01.035.

Jowitt, S.M., Werner, T.T., Weng, Z., and Mudd, G.M. (2018). Recycling of the rare earth elements. *Current Opinion in Green and Sustainable Chemistry* 13: 1–7. https://doi.org/10.1016/j.cogsc.2018.02.008.

Jowkar, M.J., Bahaloo-Horeh, N., Mousavi, S.M., and Pourhossein, F. (2018). Bioleaching of indium from discarded liquid crystal displays. *Journal of Cleaner Production* 180: 417–429. https://doi.org/10.1016/j.jclepro.2018.01.136.

Kang, K.D., Ilankoon, I.M.S.K., Dushyantha, N., and Chong, M.N. (2021). Assessment of pre-treatment techniques for coarse printed circuit boards (PCBs) recycling. *Minerals* 11 (10): 1–18. https://doi.org/10.3390/min11101134.

Kaur, P., Sharma, S., Albarakaty, F.M. et al. (2022). Biosorption and bioleaching of heavy metals from electronic waste varied with microbial genera. *Sustainability (Switzerland)* 14 (2): 935. https://doi.org/10.3390/su14020935.

Kaya, M. (2016). Recovery of metals and nonmetals from electronic waste by physical and chemical recycling processes. *Waste Management* 57: 64–90. https://doi.org/10.1016/j.wasman.2016.08.004.

Kelly, D.P. and Wood, A.P. (2015). Acidithiobacillus. *Bergey's Manual of Systematics of Archaea and Bacteria* 1–5. https://doi.org/10.1002/9781118960608.gbm01079.

Khaliq, A., Rhamdhani, M.A., Brooks, G., and Masood, S. (2014). Metal extraction processes for electronic waste and existing industrial routes: a review and Australian perspective. *Resources* 3 (1): 152–179. https://doi.org/10.3390/resources3010152.

Khanna, R., Ellamparuthy, G., Cayumil, R. et al. (2018). Concentration of rare earth elements during high temperature pyrolysis of waste printed circuit boards. *Waste Management* 78: 602–610. https://doi.org/10.1016/j.wasman.2018.06.041.

Kim, C., Lee, Y., and Ong, S.K. (2003). Factors affecting EDTA extraction of lead from lead-contaminated soils. *Chemosphere* 51 (9): 845–853. https://doi.org/10.1016/S0045-6535(03)00155-3.

Kinoshita, T., Akita, S., Kobayashi, N. et al. (2003). Metal recovery from non-mounted printed wiring boards via hydrometallurgical processing. *Hydrometallurgy* 69 (1–3): 73–79. https://doi.org/10.1016/S0304-386X(03)00031-8.

Krebs, W., Brombacher, C., Bosshard, P.P. et al. (1997). Microbial recovery of metal from solids. *FEMS Microbiology Reviews* 20 (December): 605–617.

Kumar, A., Holuszko, M.E., and Song, S. (2022). Recycling technologies—physical separation. In: *Electronic Waste: Recycling and Reprocessing for a Sustainable Future* (eds. M.E. Holuszko, A. Kumar, and D.C.R. Espinosa), 95–134. Wiley. https://doi.org/10.1002/9783527816392.ch6.

Lee, J.C., Yoo, J.M., Jeong, J.K., and Jha, M.K. (2009). Novel pre-treatment process for liberation of metals from waste printed circuit boards using organic solution. International patent application. Publication of WO2009064063A1.

Li, Z., Diaz, L.A., Yang, Z. et al. (2019). Comparative life cycle analysis for value recovery of precious metals and rare earth elements from electronic waste. *Resources, Conservation and Recycling* 149: 20–30. https://doi.org/10.1016/j.resconrec.2019.05.025.

Lister, T.E., Diaz, L.A., Clark, G.G., and Keller, P. (2016). Process development for the recovery of critical materials from electronic waste. Idaho Falls, Idaho, US: Idaho National Lab (INL).

Lo, Y.C., Cheng, C.L., Han, Y.L. et al. (2014). Recovery of high-value metals from geothermal sites by biosorption and bioaccumulation. *Bioresource Technology* 160: 182–190. https://doi.org/10.1016/j.biortech.2014.02.008.

Lukowiak, A., Zur, L., Tomala, R. et al. (2020). Rare earth elements and urban mines: critical strategies for sustainable development. *Ceramics International* 46 (16): 26247–26250. https://doi.org/10.1016/j.ceramint.2020.03.067.

Maleke, M., Valverde, A., Vermeulen, J.G. et al. (2019). Biomineralization and bioaccumulation of europium by a thermophilic metal resistant bacterium. *Frontiers in Microbiology* 10 (JAN): 1–10. https://doi.org/10.3389/fmicb.2019.00081.

Manhart, A., Buchert, M., Bleher, D., and Pingel, D. (2012). Recycling of critical metals from end-of-life electronics. Introduction & methodology. *2012 Electronics Goes Green* IEEE Conference, 1–5.

Marra, A., Cesaro, A., and Belgiorno, V. (2018a). Separation efficiency of valuable and critical metals in WEEE mechanical treatments. *Journal of Cleaner Production* 186: 490–498. https://doi.org/10.1016/j.jclepro.2018.03.112.

Marra, A., Cesaro, A., Rene, E.R. et al. (2018b). Bioleaching of metals from WEEE shredding dust. *Journal of Environmental Management* 210: 180–190. https://doi.org/10.1016/j.jenvman.2017.12.066.

Martins, T.A.G., Gomes, K.E., Rosario, C.G.A. et al. (2019). Characterization of printed circuit boards of obsolete (PCBs) aimed at the production of copper nanoparticles. In: *Characterization of Minerals, Metals, and Materials* 2019, January, 425–432. https://doi.org/10.1007/978-3-030-05749-7.

Meshram, P., Somani, H., Pandey, B.D. et al. (2017). Two stage leaching process for selective metal extraction from spent nickel metal hydride batteries. *Journal of Cleaner Production* 157: 322–332. https://doi.org/10.1016/j.jclepro.2017.04.144.

Mint (2021). India extends FAME scheme by two years to incentivise purchase of electric vehicles. Retrieved from https://www.livemint.com/auto-news/india-extends-fame-scheme-bytwo-years-to-incentivise-purchase-of-electricvehicles-11624631393528.html (accessed 31 July 2022).

Mishra, D., Kim, D.J., Ralph, D.E. et al. (2008). Bioleaching of metals from spent lithium ion secondary batteries using Acidithiobacillus ferrooxidans. *Waste Management* 28 (2): 333–338. https://doi.org/10.1016/j.wasman.2007.01.010.

Morf, L.S., Gloor, R., Haag, O. et al. (2013). Precious metals and rare earth elements in municipal solid waste—sources and fate in a Swiss incineration plant. *Waste Management* 33 (3): 634–644. https://doi.org/10.1016/j.wasman.2012.09.010.

Moyo, T., Chirume, B.H., and Petersen, J. (2020). Assessing alternative pre-treatment methods to promote metal recovery in the leaching of printed circuit boards. *Resources, Conservation & Recycling* 152 (October 2019): 104545. https://doi.org/10.1016/j.resconrec.2019.104545.

Mtibe, A., Mokhena, T.C., and John, M.J. (2023). Sustainable valorization and conversion of e-waste plastics into value-added products. *Current Opinion in Green and Sustainable Chemistry* 40: 100762. https://doi.org/10.1016/j.cogsc.2023.100762.

Mukherjee, K., Gupta, R., Kumar, G. et al. (2018). Synthesis of silver nanoparticles by Bacillus clausii and computational profiling of nitrate reductase enzyme involved in production. *Journal of Genetic Engineering and Biotechnology* 16: 527–536. https://doi.org/10.1016/j.jgeb.2018.04.004.

Nancharaiah, Y.V., Mohan, S.V., and Lens, P.N.L. (2015). Biological and bioelectrochemical recovery of critical and scarce metals. *Trends in Biotechnology* 34 (2): 137–155. https://doi.org/10.1016/j.tibtech.2015.11.003.

Nassar, N.T., and Fortier, S.M. (2021). Methodology and technical input for the 2021 review and revision of the U.S. critical minerals list. U.S. Geological Survey Open-File *Report 2021–1045*, https://doi.org/10.3133/ofr20211045.

Nayaka, G.P., Pai, K.V., Santhosh, G., and Manjanna, J. (2016). Recovery of cobalt as cobalt oxalate from spent lithium ion batteries by using glycine as leaching agent. *Journal of Environmental Chemical Engineering* 4 (2): 2378–2383. https://doi.org/10.1016/j.jece.2016.04.016.

Nithya, R., Sivasankari, C., and Thirunavukkarasu, A. (2021). Electronic waste generation, regulation and metal recovery: a review. *Environmental Chemistry Letters* 19 (2): 1347–1368. https://doi.org/10.1007/s10311-020-01111-9.

Opare, E.O., Struhs, E., and Mirkouei, A. (2021). A comparative state-of-technology review and future directions for rare earth element separation. *Renewable and Sustainable Energy Reviews* 143: 110917. https://doi.org/10.1016/j.rser.2021.110917.

Otsuki, A., De La, L., King, A. et al. (2020). Non-destructive characterization of mechanically processed waste printed circuit boards—particle liberation analysis. *Waste Management* 102: 510–519. https://doi.org/10.1016/j.wasman.2019.11.006.

Paiva, A.P. and Nogueira, C.A. (2021). Ionic liquids in the extraction and recycling of critical metals from urban mines. *Waste and Biomass Valorization* 12 (4): 1725–1747. https://doi.org/10.1007/s12649-020-01115-0.

Pant, D., Joshi, D., Upreti, M.K., and Kotnala, R.K. (2012). Chemical and biological extraction of metals present in E waste: a hybrid technology. *Waste Management* 32 (5): 979–990. https://doi.org/10.1016/j.wasman.2011.12.002.

Paranjape, P. and Yadav, M.D. (2023). Recent advances in the approaches to recover rare earths and precious metals from E-waste: a mini-review. *Canadian Journal of Chemical Engineering* 101 (2): 1043–1054. https://doi.org/10.1002/cjce.24435.

Park, H.S., Han, Y.S., and Park, J.H. (2019). Massive recycling of waste mobile phones: pyrolysis, physical treatment, and pyrometallurgical processing of insoluble residue. *ACS Sustainable Chemistry and Engineering* 7 (16): 14119–14125. https://doi.org/10.1021/acssuschemeng.9b02725.

Pourhossein, F., Mousavi, S.M., and Beolchini, F. (2022). Innovative bio-acid leaching method for high recovery of critical metals from end-of-life light emitting diodes. *Resources, Conservation and Recycling* 182: 106306. https://doi.org/10.1016/j.resconrec.2022.106306.

Pourhossein, F., Mousavi, S.M., Beolchini, F., and Lo Martire, M. (2021a). Novel green hybrid acidic-cyanide bioleaching applied for high recovery of precious and critical metals from spent light emitting diode lamps. *Journal of Cleaner Production* 298: 126714. https://doi.org/10.1016/j.jclepro.2021.126714.

Pourhossein, F., Rezaei, O., Mousavi, S.M., and Beolchini, F. (2021b). Bioleaching of critical metals from waste OLED touch screens using adapted acidophilic bacteria. *Journal of Environmental Health Science and Engineering* 19 (1): 893–906. https://doi.org/10.1007/s40201-021-00657-2.

Prasad, M.N.V., Vithanage, M., and Vithanage, M., eds. (2019). *Handbook of Electronic Waste Management: International Best Practices and Case Studies*. Elsevier.

Rafiee, P., Ghassa, S., Moosakazemi, F. et al. (2021). Recovery of a critical metal from electronic wastes: germanium extraction with organic acid. *Journal of Cleaner Production* 315: 128223. https://doi.org/10.1016/j.jclepro.2021.128223.

Rasoulnia, P., Barthen, R., and Lakaniemi, A.M. (2020). A critical review of bioleaching of rare earth elements: the mechanisms and effect of process parameters. *Critical Reviews in Environmental Science and Technology* 51 (4): 378–427. https://doi.org/10.1080/10643389.2020.1727718.

Reed, D.W., Fujita, Y., Daubaras, D.L. et al. (2016). Bioleaching of rare earth elements from waste phosphors and cracking catalysts. *Hydrometallurgy* 166: 34–40. https://doi.org/10.1016/j.hydromet.2016.08.006.

Rene, E.R., Sethurajan, M., Kumar Ponnusamy, V. et al. (2021). Electronic waste generation, recycling and resource recovery: technological perspectives and trends. *Journal of Hazardous Materials* 416: 125664. https://doi.org/10.1016/j.jhazmat.2021.125664.

Resende, L.V. and Morais, C.A. (2010). Study of the recovery of rare earth elements from computer monitor scraps—leaching experiments. *Minerals Engineering* 23 (3): 277–280. https://doi.org/10.1016/j.mineng.2009.12.012.

Rienzie, R., Perera, A.T.D., and Adassooriya, N.M. (2019). Biorecovery of precious metal nanoparticles from waste electrical and electronic equipments. *Electronic Waste Management and Treatment Technology*, 133–152. https://doi.org/10.1016/B978-0-12-816190-6.00006-6.

Rocchetti, L., Amato, A., Fonti, V. et al. (2015). Cross-current leaching of indium from end-of-life LCD panels. *Waste Management* 42: 180–187. https://doi.org/10.1016/j.wasman.2015.04.035.

Royen, H. and Fortkamp, U. (2016). Rare earth elements—purification, separation and recycling. IVL Swedish Environmental Research Institute.

Sahle-Demessie, E., Mezgebe, B., Dietrich, J. et al. (2021). Material recovery from electronic waste using pyrolysis: emissions measurements and risk assessment. *Journal of Environmental Chemical Engineering* 9 (1): 104943. https://doi.org/10.1016/j.jece.2020.104943.

Sahu, S., Mohapatra, M., and Devi, N. (2022). Effective hydrometallurgical route for recovery of energy critical elements from E-wastes and future aspects. *Materials Today: Proceedings*. https://doi.org/10.1016/j.matpr.2022.05.491.

Sethurajan, M. and Gaydardzhiev, S. (2021). Bioprocessing of spent lithium ion batteries for critical metals recovery—a review. *Resources, Conservation and Recycling* 165: 105225. https://doi.org/10.1016/j.resconrec.2020.105225.

Sethurajan, M., van Hullebusch, E.D., Fontana, D. et al. (2019). Recent advances on hydrometallurgical recovery of critical and precious elements from end of life electronic wastes—a review. *Critical Reviews in Environmental Science and Technology* 49 (3): 212–275. https://doi.org/10.1080/10643389.2018.1540760.

Sharma, R., Suhendra, N.F., Jung, S.H., and Lee, H.I. (2023). Gold recovery at ultra-high purity from electronic waste using selective polymeric film. *Chemical Engineering Journal* 451 (P1): 138506. https://doi.org/10.1016/j.cej.2022.138506.

Shivaji, S., Madhu, S., and Singh, S. (2011). Extracellular synthesis of antibacterial silver nanoparticles using psychrophilic bacteria. *Process Biochemistry* 46 (9): 1800–1807. https://doi.org/10.1016/j.procbio.2011.06.008.

Sikander, A., Savvilotidou, V., Jia, X., and Nicomel, N. (2018). The presence of rare earth elements and critical metals in waste electric and electronic equipment: challenges for recovery. *Global Nest Journal* 20 (4): 773–777. https://doi.org/10.30955/GNJ.002582.

Silveira, A.V.M., Fuchs, M.S., Pinheiro, D.K. et al. (2015). Recovery of indium from LCD screens of discarded cell phones. *Waste Management* 45: 334–342. https://doi.org/10.1016/j.wasman.2015.04.007.

Singh, S.B. and Tandon, P.K. (2014). Catalysis: a brief review on nano-catalyst. *Journal of Energy and Chemical Engineering* 2 (3): 106–115.

Smith, Y., Kumar, P., and McLennan, J. (2017). On the extraction of rare earth elements from geothermal brines. *Resources* 6 (3): 39. https://doi.org/10.3390/resources6030039.

Sprecher, B., Xiao, Y., Walton, A. et al. (2014). Life cycle inventory of the production of rare earths and the subsequent production of NdFeB rare earth permanent magnets. *Environmental Science and Technology* 48 (7): 3951–3958. https://doi.org/10.1021/es404596q.

Srivastava, R.R., Ilyas, S., Kim, H., and Choi, S. (2020). Biotechnological recycling of critical metals from waste printed circuit boards. *Journal of Chemical Technology and Biotechnology* 95: 2796–2810. https://doi.org/10.1002/jctb.6469.

Stein, R.T., Kasper, A.C., and Veit, H.M. (2022). Recovery of rare earth elements present in mobile phone magnets with the use of organic acids. *Minerals* 12: 668. https://doi.org/10.3390/min12060668.

Sun, Z., Cao, H., Xiao, Y. et al. (2017). Toward sustainability for recovery of critical metals from electronic waste: the hydrochemistry processes. *ACS Sustainable Chemistry & Engineering* 5: 21–40. https://doi.org/10.1021/acssuschemeng.6b00841.

Swain, N. and Mishra, S. (2019). A review on the recovery and separation of rare earths and transition metals from secondary resources. *Journal of Cleaner Production* 220: 884–898. https://doi.org/10.1016/j.jclepro.2019.02.094.

Tabelin, C.B., Park, I., Phengsaart, T. et al. (2021). Copper and critical metals production from porphyry ores and E-wastes: a review of resource availability, processing/recycling challenges, socio-environmental aspects, and sustainability issues. *Resources, Conservation and Recycling* 170: 105610. https://doi.org/10.1016/j.resconrec.2021.105610.

Tesfaye, F., Lindberg, D., Hamuyuni, J. et al. (2017). Improving urban mining practices for optimal recovery of resources from e-waste. *Minerals Engineering* 111: 209–221. https://doi.org/10.1016/j.mineng.2017.06.018.

Thakkar, K.N., Mhatre, S.S., and Parikh, R.Y. (2010). Biological synthesis of metallic nanoparticles. *Nanomedicine: Nanotechnology, Biology, and Medicine* 6 (2): 257–262. https://doi.org/10.1016/j.nano.2009.07.002.

Thethwayo, B. (2021). The recovery of metals from electronic waste: a review. *Metalurgija* 60 (3–4): 444–446. https://doi.org/10.1201/9780429023903-11.

Tipre, D.R., Khatri, B.R., Thacker, S.C., and Dave, S.R. (2021). The brighter side of e-waste—a rich secondary source of metal. *Environmental Science and Pollution Research* 28 (9): 10503–10518. https://doi.org/10.1007/s11356-020-12022-1.

Tunali, M., Tunali, M., and Yenigun, O. (2021). Characterization of different types of electronic waste: heavy metal, precious metal and rare earth element content by comparing different digestıon methods. *Journal of Material Cycles and Waste Management* 23 (1): 149–157. https://doi.org/10.1007/s10163-020-01108-0.

Tunsu, C., Petranikova, M., Gergorić, M. et al. (2015). Reclaiming rare earth elements from end-of-life products: a review of the perspectives for urban mining using hydrometallurgical unit operations. *Hydrometallurgy* 156: 239–258. https://doi.org/10.1016/j.hydromet.2015.06.007.

Tsamis, A., and Coyne, M. (2015). *Recovery of Rare Earths from Electronic Wastes: An Opportunity for High-Tech SMEs*. European Parliament, Directorate-General for Internal Policies, Policy Department A: Economic and Scientific Policy, 1–44. https://doi.org/10.1007/978-94-017-5344-9_136.

Uysal, E., Al, S., Emil-Kaya, E. et al. (2022). Hydrometallurgical recycling of waste NdFeB magnets: design of experiment, optimisation of low concentrations of sulphuric acid leaching and process analysis. *Canadian Metallurgical Quarterly* 1–12. https://doi.org/10.1080/00084433.2022.2058152.

Veit, H.M. and Bernardes, A.M. (2015). *Electronic Waste: Generation and Management*. Cham, Switzerland: Springer. https://doi.org/10.1007/978-3-319-15714-6.

Velásquez, L. and Dussan, J. (2009). Biosorption and bioaccumulation of heavy metals on dead and living biomass of Bacillus sphaericus. *Journal of Hazardous Materials* 167 (1–3): 713–716. https://doi.org/10.1016/j.jhazmat.2009.01.044.

Verhaeghe, F., Goubin, F., Yazicioglu, B. et al. (2011). Valorisation of battery recycling slags. *2nd International Slag Valorisation Symposium; the Transition to Sustainable Materials Management, Katholieke Universiteit Leuven* 18 (20.04): 2011.

Vermesan, H., Tiuc, A.-E., and Purcar, M. (2020). Advanced recovery techniques for waste materials from it and telecommunication equipment printed circuit boards. *Sustainability* 12: 74.

Wall, F. (2021). Rare earth elements. In: *Encyclopedia of Geology* https://doi.org/10.1016/b978-0-08-102908-4.00101-6.

Wang, M., Tan, Q., Chiang, J.F., and Li, J. (2017). Recovery of rare and precious metals from urban mines—a review. *Frontiers of Environmental Science & Engineering* 11 (5): 1. https://doi.org/10.1007/s11783-017-0963-1.

Watari, T., Nansai, K., and Nakajima, K. (2020). Review of critical metal dynamics to 2050 for 48 elements. *Resources, Conservation and Recycling* 155 (January): 104669. https://doi.org/10.1016/j.resconrec.2019.104669.

Willner, J., Fornalczyk, A., Jablonska-Czapla, M. et al. (2021). Germanium, tellurium and thallium in electronic waste. *Materials* 14: 3722.

Xavier, L.H., Giese, E.C., Ribeiro-Duthie, A.C., and Lins, F.A.F. (2019). Sustainability and the circular economy: a theoretical approach focused on e-waste urban mining. *Resources Policy* 101467. https://doi.org/10.1016/j.resourpol.2019.101467.

Xavier, L.H., Ottoni, M., and Abreu, L.P.P. (2023). A comprehensive review of urban mining and the value recovery from e-waste materials. *Resources, Conservation & Recycling* 190: 106840. https://doi.org/10.1016/j.resconrec.2022.106840.

Xin, Y., Guo, X., Chen, S. et al. (2016). Bioleaching of valuable metals Li, Co, Ni and Mn from spent electric vehicle Li-ion batteries for the purpose of recovery. *Journal of Cleaner Production* 116: 249–258. https://doi.org/10.1016/j.jclepro.2016.01.001.

Yang, M., Liu, H., Ye, B., and Qian, W. (2021). Recycling of printed circuit boards by abrasive waterjet cutting. *Process Safety and Environmental Protection* 148: 805–812. https://doi.org/10.1016/j.psep.2021.01.052.

Yang, Y., Walton, A., Sheridan, R. et al. (2017). REE recovery from end-of-life NdFeB permanent magnet scrap: a critical review. *Journal of Sustainable Metallurgy* 3 (1): 122–149. https://doi.org/10.1007/s40831-016-0090-4.

Yousef, S., Tatariants, M., Tichonovas, M. et al. (2018). Recycling of bare waste printed circuit boards as received using an organic solvent technique at a low temperature. *Journal of Cleaner Production* 187: 780–788. https://doi.org/10.1016/j.jclepro.2018.03.227.

Yu, Z., Han, H., Feng, P. et al. (2019). Recent advances in the recovery of metals from waste through biological processes. *Bioresource Technology* 297: 122416. https://doi.org/10.1016/j.biortech.2019.122416.

Yu, Z., Han, H., Feng, P. et al. (2020). Recent advances in the recovery of metals from waste through biological processes. *Bioresource Technology* 297: 122416. https://doi.org/10.1016/j.biortech.2019.122416.

Zeng, G., Luo, S., Deng, X. et al. (2013). Influence of silver ions on bioleaching of cobalt from spent lithium batteries. *Minerals Engineering* 49: 40–44. https://doi.org/10.1016/j.mineng.2013.04.021.

Zepf, V. (2013). Rare earth elements: what and where they are.In: *Rare Earth Elements*. Berlin, Heidelberg: Springer. 11–39. https://doi.org/10.1007/978-3-642-35458-8_2.

Zhang, S., Ding, Y., Liu, B., and Chang, C. (2017). Supply and demand of some critical metals and present status of their recycling in WEEE. *Waste Management* 65: 113–127. https://doi.org/10.1016/j.wasman.2017.04.003.

Zhou, M., Li, B., Li, J., and Xu, Z. (2021). Pyrometallurgical technology in the recycling of a spent lithium ion battery: evolution and the challenge. *ACS ES&T Engineering* 1 (10): 1369–1382. https://doi.org/10.1021/acsestengg.1c00067.

Zhu, P., Chen, Y., Wang, L.Y. et al. (2013). The separation of waste printed circuit board by dissolving bromine epoxy resin using organic solvent. *Waste Management* 33 (2): 484–488. https://doi.org/10.1016/j.wasman.2012.10.003.

Zhu, X., Nie, C., Ni, Y. et al. (2021). Advanced utilization of copper in waste printed circuit boards : synthesis of nano-copper assisted by physical enrichment. *Journal of Hazardous Materials* 401: 123294.https://doi.org/10.1016/j.jhazmat.2020.123294I.

16

Finite Element Analysis of Mortars with Recycled PC from Electronic Waste

Ana Laura De la Colina Martínez[1,], David Joaquín Delgado Hernández[2], Boris Miguel López Rebollar[3], Gonzalo Martínez Barrera[4], Liliana Ivette Ávila Córdoba[5], Carlos Eduardo Barrera Díaz[6], and Fernando López Gayarre[7]*

[1] *Doctora en Ciencia de Materiales por la Facultad de Química, Universidad Autónoma del Estado de México, Paseo Colón esquina Paseo Tollocan S/N, 50000, Toluca, México, México*
[2] *Facultad de Ingeniería, Universidad Autónoma del Estado de México, Avenida Universidad S/N, Cerro de Coatepec, Ciudad Universitaria, 50110, Toluca, México*
[3] *Instituto Interamericano de Tecnología y Ciencias del Agua, Universidad Autónoma del Estado de México, Carretera Toluca-Atlacomulco, km 14.5, 50200, San Cayetano, México*
[4] *Laboratorio de Investigación y Desarrollo de Materiales Avanzados (LIDMA), Facultad de Química. Universidad Autónoma del Estado de México Km. 12 de la Carretera Toluca-Atlacomulco, 50200, San Cayetano, México*
[5] *Facultad de Ingeniería, Universidad Autónoma del Estado de México, Avenida Universidad S/N, Cerro de Coatepec, Ciudad Universitaria, 50110, Toluca, México*
[6] *Facultad de Química. Universidad Autónoma del Estado de México, Carretera Toluca-Atlacomulco Km. 12, 50200, San Cayetano, México*
[7] *Departamento de Construcción e Ingeniería de Fabricación, Universidad de Oviedo, Campus de Gijón, 33203, Gijón, Asturias*
* *Corresponding author*

16.1 Introduction

Global demand for PC reached 4.6 million tons (MT) in 2019, 4.5 MT in 2020, and 4.3 MT in 2021 (Mordor Intelligence 2023). The global demand for PC is growing at a CAGR (compound annual growth rate) of 5.7% from 2022 to 2030 (Spherical Insights 2023). In March 2022, Covestro AG started two polycarbonate compounding production lines in India, particularly for the automotive and electrical and electronics industries. In May 2022, SABIC (Saudi Basic Industries Corporation) announced the start-up of pre-commissioning activities at a polycarbonate production complex in Tianjin, China (Mordor Intelligence 2023). A rapid development of PC, combined with a short life cycle, has resulted in a continuous increase of waste from electrical and electronic equipment (e-waste or WEEE). In 2019, the world generated 53.6 million metric tons (Mt), and only 17.4% of this was officially documented as properly collected and recycled. Most of the e-waste was generated in Asia (24.9 Mt), while Europe generates the most in kilograms per capita (16.2 kg per capita). Europe is the continent with the highest documented formal e-waste collection and recycling rate (42.5%). Asia ranked second at 11.7%, the Americas and Oceania stood at 9.4% and 8.8%, respectively, while Africa ranked last at 0.9%. However, statistics can vary substantially and depend on a number of factors (e.g. income level, policy in place, structure of the waste management system, etc.) (Forti et al. 2020). The unscientific and crude methods adopted for resource recovery from E-waste are called informal recycling or backyard recycling. In this case, the E-waste management activities (e.g., transportation, processing, segregation, repairing, refurbishing, and dismantling) are carried out by unskilled people in an improper manner. Approximately 82.6% of E-waste generated globally is informally recycled in countries such as India, Ghana, and China (Rautela et al. 2021).

Using plastic waste in cement mortar could be one of the best ecological solutions to decrease its environmental impact, thanks to its economic and eco-friendly advantages. For instance, using HDPE and PET as aggregate in concrete maintains mechanical properties and minimizes global warming potential (Gravina et al. 2021). In addition, by utilizing fly ash and recycled plastic waste in the concrete the carbon emissions were lowered by 13% compared to conventional emissions (Ersan et al. 2022).

Despite these previous results, PC from electronic waste used as aggregate in cement mortar has been scantily studied. In 2010 a study reported the addition of sharpened particles of PC directly obtained from industrial waste (size 3.15 mm in almost 85%) as a partial replacement for sand. Portland cement with a resistance of 52.5 MPa to 28 days was used. The experimental results show a decrease in

Electronic Waste Management: Policies, Processes, Technologies, and Impact, First Edition. Edited by Sunil Kumar and Vineet Kumar.

compressive strength from 53 to 35 MPa with the increase of plastic content (0 to 15%). Even though the mixture was made with a cement of 52.5 MPa, the margin of loss of compressive strength is very high. Consequently, the use of this PC would be very limited. Besides, the elastic modulus decreases when plastic aggregates content increases. Its value varies between 37.3 to 29 GPa (0 to 15% of PC). This drop in the elastic modulus was probably due to the low values of the elastic modulus of PC (2700 MPa) and to the poor bond between the matrix and plastic aggregates (Hannawi et al. 2010).

In 2020 a study reported the addition of waste PC (a significant fraction of sub-rounded particles and a small fraction of elongated particles, size 2–5 mm in almost 72%, and size 5–10 mm in almost 18%) that did not come from e-waste as partial replacement for sand. Portland cement with a resistance of 42.5 MPa to 28 days was used. The experimental results show a decrease in compressive strength from 62 to 60 MPa with the increase of plastic content (0 to 15%). In addition, the flexural strength shows a decrease from 8.0 to 6.9 MPa with the increase of plastic content (0 to 15%). However, the water absorption coefficient shows a decrease from 0.8 to 0.24 ($mg/cm^2s^{1/2}$) with the increase of plastic content (0 to 20%). This means that the possibility of incorporating 15% of PC (or more) is limited due to an augment of capillary water infiltration (Kaur and Pavia 2020). Although the mechanical results seem promising, it should be noted that the strength obtained for the mortar control (60 MPa) is too high for a 42.5 MPa cement. The results seem to be more typical of a 52.5 MPa cement, as can be seen in the study by Ekolu and Murugan (2012).

On the other hand, an examination was made of the compressive strength and modulus of elasticity of concrete (not mortar) with different amounts of cylindrical particles of PC from plastic waste (not from electronic waste) (size 3.15 mm) used as a sand replacement. Portland cement with a resistance of 42.5 MPa to 28 days was used. The results show a systematic reduction in concrete compressive strength from 43.7 MPa to 27.6 MPa with the increase of plastic content (0 to 10% of PC). The margin of loss of compressive strength is very high, despite the low percentage of PC added, if we consider that the replacement was only carried out on the sand and not on the gravel that is present in the concrete mix. Therefore, the use of this PC would be very limited. In addition, the elastic modulus reduces from 31.1 to 24.7 GPa (0 to 10% of PC) (Al-Tayeb 2020).

In terms of the mathematical modelling, Finite Element Analysis (FEA) has mainly been used to estimate, in a numerical way, the stresses and strains of an element or prototype subjected to different loading scenarios. The physical and mechanical properties of the prototype will depend directly on the characteristics of the material selected for the simulation. Although FEA offers approximate results, it is widely used for the optimization of products or materials at a low cost. Here, since the analysis was carried out through computational means and in a very short period, the use of FEA was believed to be adequate. Note that this optimization at the laboratory level would have required specialized equipment and more time, resources, and materials (Chandrakant Baadkar 2010; Evangelos 2018; Lee 2018).

The use of recycled PC from electronic waste, with uniform cylindrical shape and size, unlike the studies cited above, could improve the mechanical resistance and then the possibility of using it in mortar, especially when it has already been tested in concrete and has given very good results (De la Colina-Martínez et al. 2017; 2019). Therefore, using plastic waste in cement mortar could be one of the best ecological solutions for this type of plastic, due to its economic and environmental advantages. Combining this fact with the possibility of evaluating the performance of these new materials at the structural level opens the possibility of new insights and applications.

To show the effects of the recycled PC from electronic waste in two types of cement mortar, and on the numerical elastic behavior of two prototypes (cylindrical model and rectangular prism), the aims of this work are: (1) to characterize sands and recycled PC pellets, (2) to obtain both mechanical and physical properties of cement mortars, prepared with Portland cement, sand, water, and recycled PC pellets from electronic waste, (3) to obtain the stresses and strains of two prototypes using Finite Element Analysis (FEA) by means of the ANSYS Mechanical© software, and (4) to compare the elastic behavior of both prototypes.

16.2 Material and Methods

16.2.1 Materials

Mortar mixtures to be examined were made in the laboratory using the following materials: cement, sand, water, and recycled PC pellets from electronic waste. Portland cement Type I, sulfate resistant, with a resistance of 42.5 MPa to 28 days was used.

Sand size 0/4 mm was used, and it was provided by the local company MONCA, located in Gijón, Asturias, Spain (Figure 16.1a).

Recycled PC pellets from electronic waste (e-PC) size 1 mm diameter and 3 mm length (1 × 3 mm), were obtained by extrusion from electrical outlets wall switch waste. PC pellets (Figure 16.1b) were provided by the Polyvima Company (located at San Lorenzo Tepaltitlán, Toluca, México). The physical and mechanical properties of the PC

Figure 16.1 Materials: (a) sand size 0/4 mm, and (b) recycled PC pellets size 1 mm diameter and 3 mm length. Scale is given in cm.

were provided for the company according to Resin Pellets and/or Makrolon® PC sheet, which are in accordance with Covestro and Professional Plastics (Covestro 2015; Professional Plastics 2017).

16.2.2 Methods

16.2.2.1 Cement Mortar Elaboration

Cement mortars were prepared by mixing Portland cement Type I (sulfate resistant, with a resistance of 42.5 MPa to 28 days), sand, water, and recycled PC pellets from electronic waste. PC pellets were used as a partial replacement for sand in cement mortar as shown in Table 16.1. The idea was to examine how PC affects the physical and mechanical properties of mortars when it substitutes different-size sand particles.

First, the sand used in GA mortars, as "acquired," was replaced by PC in the percentage indicated in the table. Then, the sand used in GG mortars was sieved, and only the sand particles that were "retained" in the 2 mm mesh size were replaced by PC in the percentage indicated in Table 16.1.

Table 16.1 Sand and PC constituents: Control mortar (0% of PC); GA mortars (replacement of acquired sand); and GG mortars (replacement of retained sand in the 2 mm mesh size).

PC (volume %)	Mortars evaluated (GA and GG)		PC (g)	Sand (g)
0	✓	✓	0	1416
9	✓	✓	57	1289
15	✓	✓	96	1204
30	✓	✓	192	990

Three different kinds of mortars were elaborated and tested: (a) Cement mortar with cement, sand, and water, which was denominated as control cement mortar (without PC); (b) cement mortar, namely GA; and (c) cement mortar, namely GG. For GA and GG mortars the sand was replaced by recycled PC pellets at concentrations of 0, 9, 15, and 30% by volume. In addition, for all samples the water/cement ratio (w/c) was 0.5 (236 g of water and 472 g of cement).

The mix of mortar components was made in an automatic mixer. Mortar prismatic specimens (40 × 40 × 160 mm) were elaborated by compaction in a steel mold according to the UNE-EN 196–1:2018 standard (AENOR 2018).

16.2.2.2 Granulometric Characterization of Sands and PC Pellets

Sand and PC granulometric characterizations were made according to the UNE-EN 933–1:2012 standard (AENOR 2012). The sieve analysis was made with the mesh series according to the UNE-EN 933–2:2022 standard (AENOR 2022).

16.2.2.3 Compressive Strength of Mortar

Three test specimens for each type of cement mortar were obtained according to the UNE-EN 1015–11:2020 standard (AENOR 2020). The test specimens were cured in climatic chambers according to the standard (AENOR 2020). With the intention of knowing the behavior of the cement mortar over time, the test specimens were maintained at 20°C ± 2°C and a 65% ± 5% RH prior to the mechanical and physical tests, which were performed to 36 days. The compressive strengths of the mortar specimens were measured in a Universal Testing Machine brand MEM 101.

16.2.2.4 Flexural Strength of Mortar

Three test specimens for each type of cement mortar were obtained according to the UNE-EN 1015–11:2020 standard (AENOR 2020). The test specimens were cured in climatic chambers according to the standard (AENOR 2020). With the intention of knowing the behavior of the cement mortar over time, the test specimens were maintained at 20°C ± 2°C and a 65% ± 5% RH prior to the mechanical and physical tests, which were performed to 36 days. The flexural strengths of the mortar specimens were measured in a Universal Testing Machine brand MEM 101.

16.2.2.5 Water Absorption Coefficient of Mortar

The determination of the water absorption coefficient due to capillary action was measured according to the UNE-EN 1015–18:2003 standard (AENOR 2003). For this determination, three specimens of each trial formulation were tested at 36 days.

16.2.2.6 Density of Mortar

The test specimens were cured according to the UNE-EN 1015–11:2020 standard (AENOR 2020) and were maintained at 20°C ± 2°C and a 65% ± 5% RH before testing. For each trial formulation, three specimens (40 × 40 × 160 mm) were tested at 36 days.

16.2.2.7 Structural Analysis and Simulation Procedure

ANSYS Mechanical© software was used to calculate and to compare stresses and strains by means of the FEA of two prototypes: a cylindrical model (50 × 100 mm), which is employed to calculate the modulus of elasticity, according to ASTM C469 standard (ASTM C469-14e1: Standard Test Method for Static Modulus of Elasticity and Poisson's Ratio of Concrete in Compression (definitive 2023) 2014); and a rectangular prism (40 × 40 × 160 mm), which is utilized to calculate compressive and flexural strengths, according to the UNE-EN 1015–11: 2020 standard (AENOR 2020).

16.2.2.7.1 Establishment of the Expected Elastic Limits of the Mortars

First, it was required to establish the expected elastic limits of the mortars obtained in the laboratory (GA and GG). For that, the performances of these mortars were compared with those found in the literature. Figure 16.2 shows the mix most representative, e.g., the cement mortar elaborated with sharpened particles of PC from industrial waste (curve H) (Hannawi et al. 2010), the concrete elaborated with waste cylindrical pellets of PC (curve M) (Al-Tayeb 2020), and the concrete standard (line C) (AENOR 2013). As can be seen, the compressive strength values of the GA and GG mortars are between the values of concrete (line C) and those of cement mortar made with sharp PC particles (curve H).

After choosing C and H materials, the compressive strength values of the mortar (GA and GG) were then utilized to calculate both the expected modulus of elasticity and the expected deformation values, according to Equations (16.1), (16.2), and (16.3).

Equation (16.1) was applied to calculate the modulus of elasticity, assuming that the prototypes were made of mortar and PC from industrial waste, according to Hannawi et al. (2010). Equation (16.2) was employed to calculate the modulus of elasticity, assuming that the prototypes were made of concrete, according to the EN 1992–1–1:2013 standard (AENOR 2013). The expected compressive strains at mean compressive strength were calculated using Equation (16.3) (AENOR 2013).

$$E_c = 0.96\left(f_c\right)^{0.92} \qquad (16.1)$$

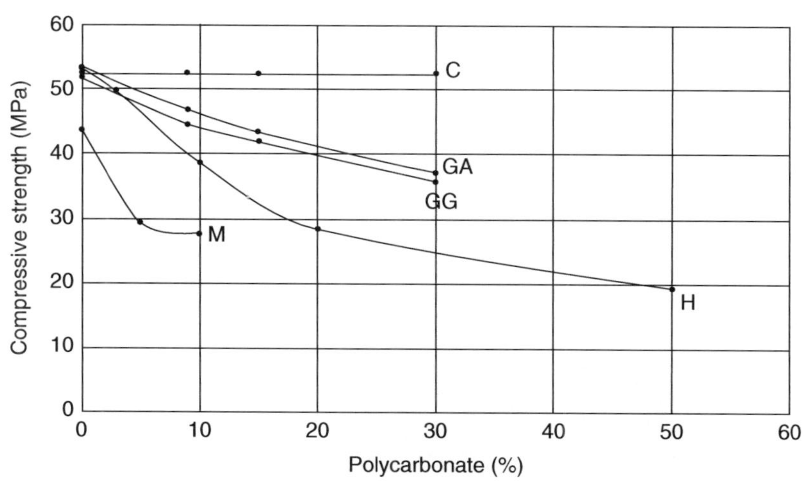

Figure 16.2 The compressive strengths of mortars obtained in the laboratory (curves GA and GG), compared with those found in the literature. Curve H shows the cement mortar elaborated with sharpened particles of PC from industrial waste (Hannawi et al. 2010); curve M shows the concrete elaborated with waste cylindrical pellets of PC (Al-Tayeb 2020); and line C shows the concrete standard (AENOR 2013).

$$E_c = 22\left(\frac{f_c}{10}\right)^{0.30} \quad (16.2)$$

$$\varepsilon_{c1}\left(\frac{0}{00}\right) = 0.76\left(f_{cm}\right)^{0.31} \quad (16.3)$$

where E_c is the modulus of elasticity, f_c is the compressive strength of mortar obtained experimentally, and ε_{c1} is the compressive strain at mean compressive strength or maximum stress (f_{cm}), that is, the compressive strength of mortar obtained experimentally. See Figure 16.3.

The resultant values were used as the expected elastic limits. These limits were named "C elastic behavior" (C-GA and C-GG) and "H elastic behavior" (H-GA and H-GG). They were utilized to calculate stresses and strains by means of FEA.

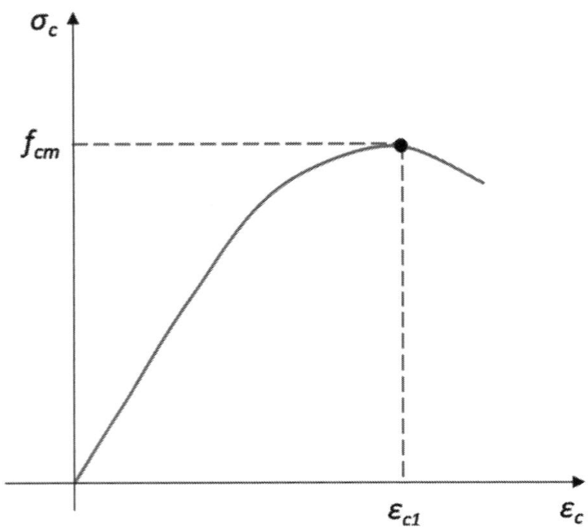

Figure 16.3 Compressive strain at mean compressive strength, where E_c is the modulus of elasticity, f_c is the compressive strength of mortar obtained experimentally, and ε_{c1} is the compressive strain at mean compressive strength or maximum stress (f_{cm}), that is, the compressive strength of mortar obtained experimentally.

16.2.2.7.2 Application of ANSYS Mechanical© Software

ANSYS Mechanical© software was applied. First, the geometrical models (rectangular prism and cylindrical model) were drawn. The spatial discretization of the two models was carried out through a mesh with 0.5 mm hexahedral elements, to facilitate the accuracy of the strains (Figure 16.4).

The assignment of the physical and mechanical properties of the material obtained in the laboratory was performed. The expected values (modulus of elasticity and deformation) and loading conditions (the compressive strengths obtained experimentally) were established. The three-dimensional (3D) prototypes with the values of the equivalent deformation, corresponding to each PC percentage, were acquired (Figure 16.5). The results were then analyzed and graphically represented.

Figure 16.5 shows the 3D prototypes with the values of the equivalent deformation corresponding to the GG mortar with 30% PC.

It is important to mention that one of the parameters obtained with the software was Poisson's ratio (υ), which measures the deformation in the material in a direction perpendicular to the direction of the applied force (Belyadi et al. 2019) and depends mainly on the strength of the material and its geometry. In the case of an entirely brittle material, Poisson's ratio is zero, whereas for a completely elastic material, $\upsilon = 0.7$. For example, for quartz glass, Poisson's ratio is small (0.17), whereas for rubber, it is large (0.6); most steels have values close to 0.3 (Poplavko 2019).

16.3 Results and Discussion

16.3.1 Experimental Stage

16.3.1.1 Granulometric Characterization of Sands and PC Pellets

The particle-size distribution of sand is shown in Figure 16.6. The sieve analysis made with the mesh series (UNE-EN 933–2:2022 standard) (AENOR 2022) reveals a distinctive Fineness Modulus (FM) of 4.4. For Youcai and

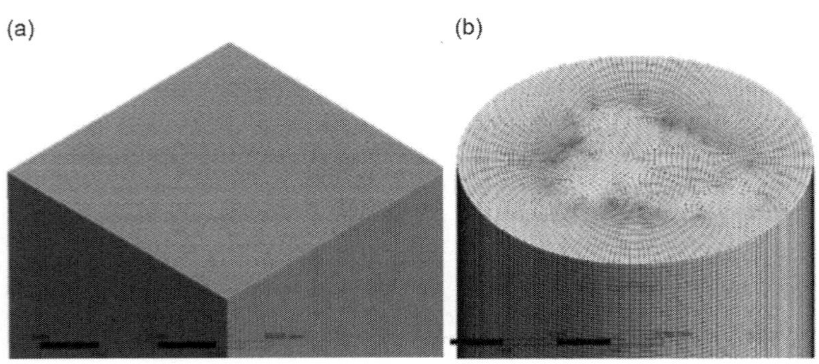

Figure 16.4 Geometrical models: (a) rectangular prism and (b) cylindrical model. The spatial discretization of the two models was carried out through a mesh with 0.5 mm hexahedral elements, to facilitate the accuracy of the strains.

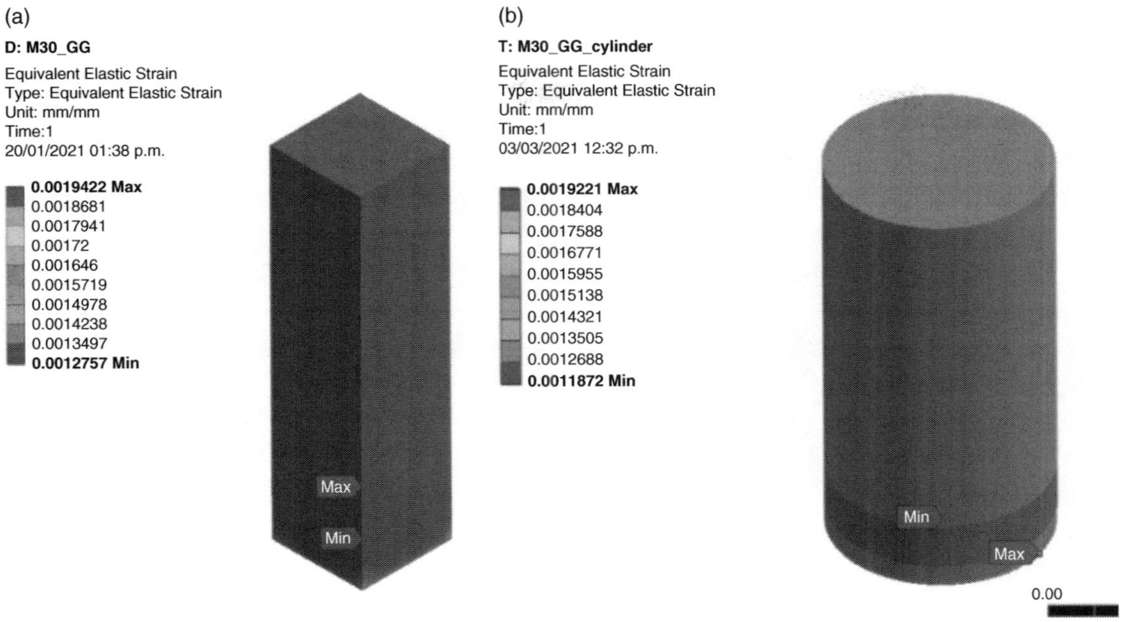

Figure 16.5 3D prototypes: (a) rectangular prism and (b) cylindrical model, corresponding to the GG mortar with 30% PC.

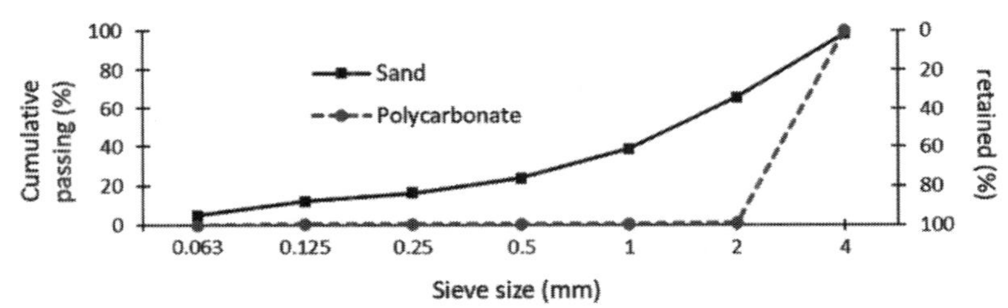

Figure 16.6 Granulometric distribution of sand and recycled PC pellets. It reveals a concentration of 99% of PC pellets retained in the 2 mm mesh size.

Sheng (2017), the bigger the fineness modulus, the more coarse the sand. The sand with the same fineness modulus can have very different grading regions. The particle-size distribution of recycled PC (Figure 16.6) revealed a concentration of 99% of PC pellets retained in the 2 mm mesh size.

16.3.1.2 Compressive and Flexural Strength of Mortar

Different concentrations of recycled PC pellets were added to the cement mortar. A representative sample of the PC integration into the mortars is shown in Figure 16.7. PC pellets are the round dark gray points, which are well joined into the mortar.

In Figure 16.8a compressive strength values of GA and GG mortars are shown. In general terms, as the PC concentration increases the compressive strength decreases. However, GA mortars exhibit slightly higher strengths than GG. Such behaviors can be explained since GA mortars have a higher amount of sand particles (2 mm) than GG. Seeing that the replacement of these particles affects the resistance, it is more convenient to replace the sand as acquired.

This decrease in compressive strength contrasts with other investigations. For example, GA and GG mortars show a less abrupt strength decrease than mortars with sharp PC particles, reported by Hannawi et al. (2010). However, the decrease in strength of GA and GG mortars was more abrupt than that of concrete with cylindrical PC pellets, reported by Al-Tayeb (2020), although the initial strength (without plastic) of this concrete was 20% lower than those obtained for the GA and GG mortars. The drop in compressive strengths due to the addition of PC aggregates can be attributed mainly to the moderated bond between the matrix and plastic aggregate. However, pelletized PC (in GA and GG mortars), seems to give better results than sharpened particles of PC.

In Figure 16.8b the flexural strengths of GA and GG mortars are shown. In this case, as PC concentration increases, flexural strength decreases. However, the addition of PC motivates the reduction of the values more sharply when PC replaces the 2 mm sand particles (GG mortars), showing that PC does not keep the same adhesion properties as the 2 mm sand particles. Afterward, it is more convenient to replace the sand as acquired.

Figure 16.7 Digital image of crosscut mortar. The round dark gray points are the PC pellets, which are well integrated into the mortar.

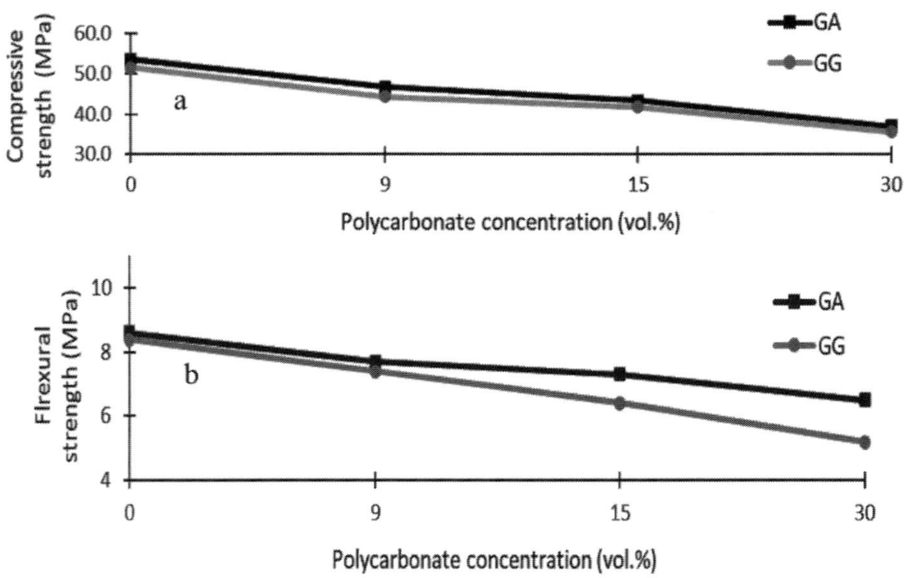

Figure 16.8 (a) Compressive strength of GA and GG mortars. (b) Flexural strength of GA and GG mortars. GA mixtures have better mechanical resistances, and up to 15% of recycled PC can be used with corresponding mechanical benefits.

16.3.1.3 Water Absorption Coefficient of Mortar

Recycled PC pellets were added to the mortar. The water absorption coefficients were calculated, and the values are constants for all GA and GG mortars (0.1 $mg/cm^2s^{1/2}$). This behavior shows that the PC addition has no influence on the mortar's water absorption, even when the samples contain 30% of PC. These water absorption values do not affect the mortar durability to sodium chloride and sulfates, because they do not increase. These results contrast with the values reported by Hannawi et al. (2010) and Kaur and Pavia (2020), who found an increase in the water absorption values when the plastic aggregates increase.

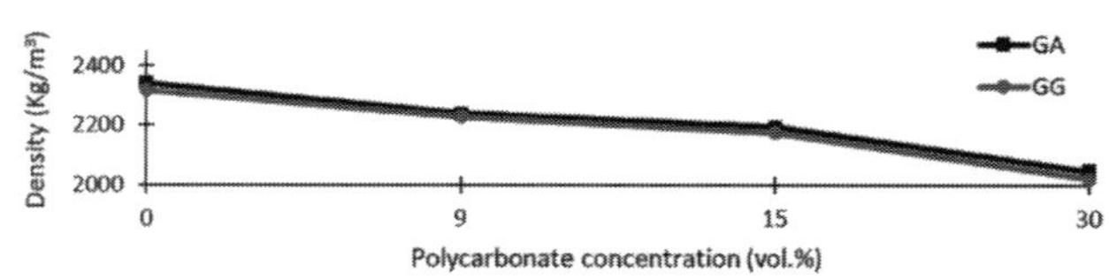

Figure 16.9 Densities of GA and GG mortars. The increase in PC favored the decrease in density of the mortars.

16.3.1.4 Density of Mortar

Recycled PC pellets were added to the cement mortar. The densities at 36 days from GA and GG mortars are shown in Figure 16.9. In general, as the PC concentration increases the density decreases since the PC density is 1.2 g/cm^3. However, the addition of PC motivates the reduction of the density values more sharply when PC replaces the 2 mm sand particles (GG mortars), showing that PC, in these circumstances, can make an interesting contribution to a further decrease in density. Even though the decrease in mortar density has been reported by Hannawi et al. (2010) this supplementary drop has not been reported before.

16.3.2 Structural Analysis and Simulation Stage

The ANSYS Mechanical© software was used to obtain the stresses and strains of the two prototypes by means of FEA. Table 16.2 shows the properties obtained experimentally from the mortars used for the FEA simulation. Tables 16.3 and 16.4 show the elastic behaviors expected for H and C materials, respectively, and the Poisson's ratio (υ) obtained from prototypes, calculated with the software.

Table 16.2 Density and compressive strengths of GA and GG mortars experimentally obtained and used for the FEA simulation.

	Mortars		
	Experimentally obtained		
Properties	Recycled polycarbonate	Density	Compressive strength
GG mortars	(%)	(Kg/m^3)	(MPa)
	0	2317	51.6
	9	2226	44.4
	15	2173	41.8
	30	2019	35.6
GA mortars	(%)	(Kg/m^3)	(MPa)
	0	2342	53.5
	9	2237	46.8
	15	2192	43.4
	30	2046	37.0

The influence of the percentage of PC on the stress and strain of GG and GA mortars is shown in Figure 16.10. Although GG mortars show lower strength, it is also observed that in both mortars, as the percentage of PC increases, equivalent elastic strain and stress decrease. This indicates a direct relationship between the amount of PC and the stress-strain function.

The results show that in both prototypes (rectangular prism and cylindrical model) there is a significant variation of Poisson's ratio with respect to the percentage of PC. For instance, prisms with H elastic behavior (Figure 16.11a) show that as the percentage of PC increases, the values of Poisson's ratio become smaller (0.05 for GG and 0.06 for GA) representing a stiffer behavior compared to prisms with C elastic behavior (Table 16.2), which show a slight increase of Poisson's ratio, reaching values of 0.1.

In relation to the cylindrical models with H elastic behavior (Figure 16.11b), these show that, as the percentage of PC increases, the values of Poisson's ratio decrease to values close to 0.1, which represents a behavior that tends towards stiffness, compared to the cylindrical models with C elastic behavior (Table 16.2) that present increasing values of Poisson's ratio, reaching values of 0.2 with the highest percentage of PC (30%).

With these results, it is established that the use of PC in the mortar mix promotes the modification of stiffness or flexibility, properties closely related to Poisson's ratio.

To examine the elastic behavior of the mortars, Figure 16.12 shows the relationship between the equivalent deformation and the percentage of PC used. Specimens

Table 16.3 Elastic behaviors expected for H material and the Poisson's ratio (υ) obtained from prototypes, calculated with the software.

	H material			
	Expected		Calculated by FEA	
Properties	Modulus of elasticity	Strain	Rectangular prism (40 × 40 × 160 mm)	Cylinder (50 × 100 mm)
GG mortars	H-GG(GPa)	(mm/mm)	Poisson's ratio	Poisson's ratio
	36.1	0.0025	0.090	0.1700
	31.4	0.0022	0.075	0.1300
	29.8	0.0021	0.065	0.1100
	25.7	0.0019	0.050	0.0900
GA mortars	H-GG(GPa)	(mm/mm)	Poisson's ratio	Poisson's ratio
	37.3	0.0026	0.1100	0.1900
	33.0	0.0023	0.0900	0.1600
	30.8	0.0022	0.0800	0.1500
	26.6	0.0020	0.0600	0.1200

Table 16.4 Elastic behaviors expected for C material and the Poisson's ratio (υ) obtained from prototypes, calculated with the software.

	C material			
	Expected		**Calculated by FEA**	
Properties	Modulus of elasticity	Strain	Rectangular prism (40 × 40 × 160 mm)	Cylinder (50 × 100 mm)
GG mortars	C-GG(GPa)	(mm/mm)	Poisson's ratio	Poisson's ratio
	36.0	0.0024	0.0850	0.1500
	34.4	0.0023	0.1000	0.1700
	33.8	0.0022	0.1000	0.1800
	32.2	0.0021	0.1000	0.2000
GA mortars	C-GG(GPa)	(mm/mm)	Poisson's ratio	Poisson's ratio
	36.4	0.0024	0.0800	0.1500
	34.9	0.0023	0.0900	0.1700
	34.2	0.0023	0.1000	0.1800
	32.6	0.0021	0.1100	0.2000

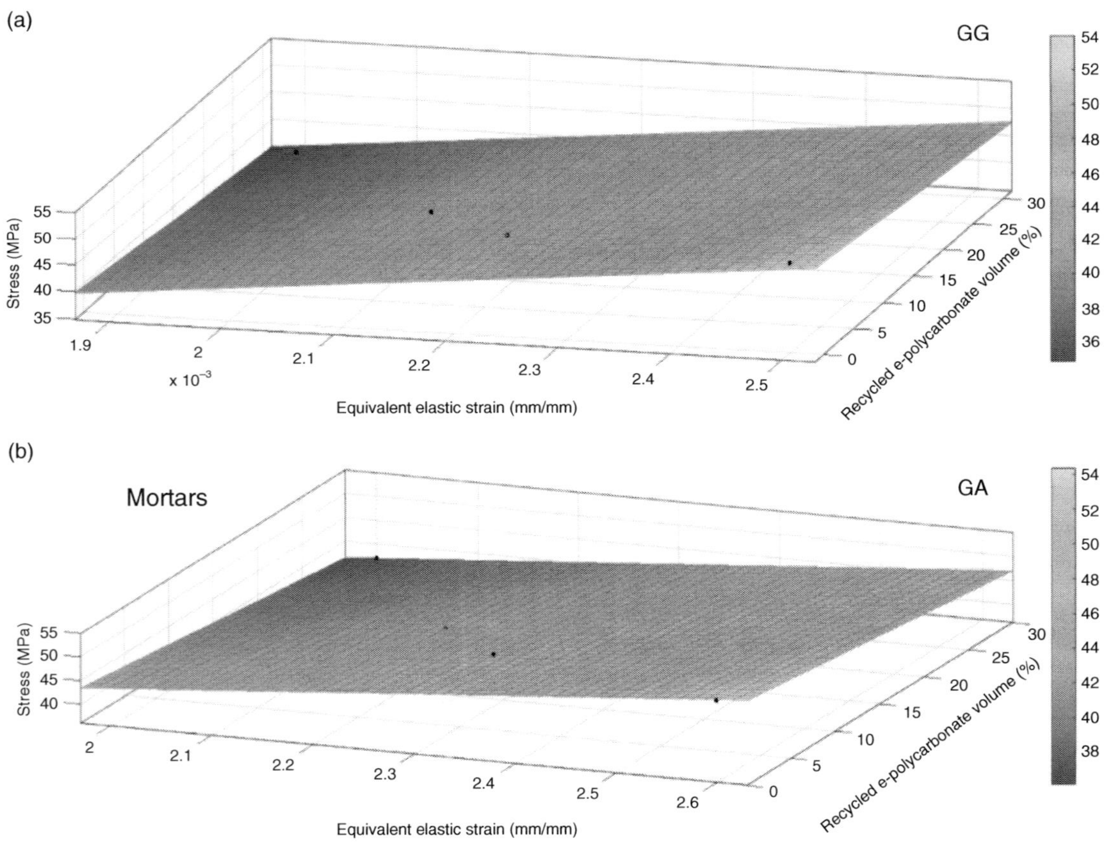

Figure 16.10 In both mortars (GA and GG), as the percentage of PC increases, equivalent elastic strain and stress decrease.

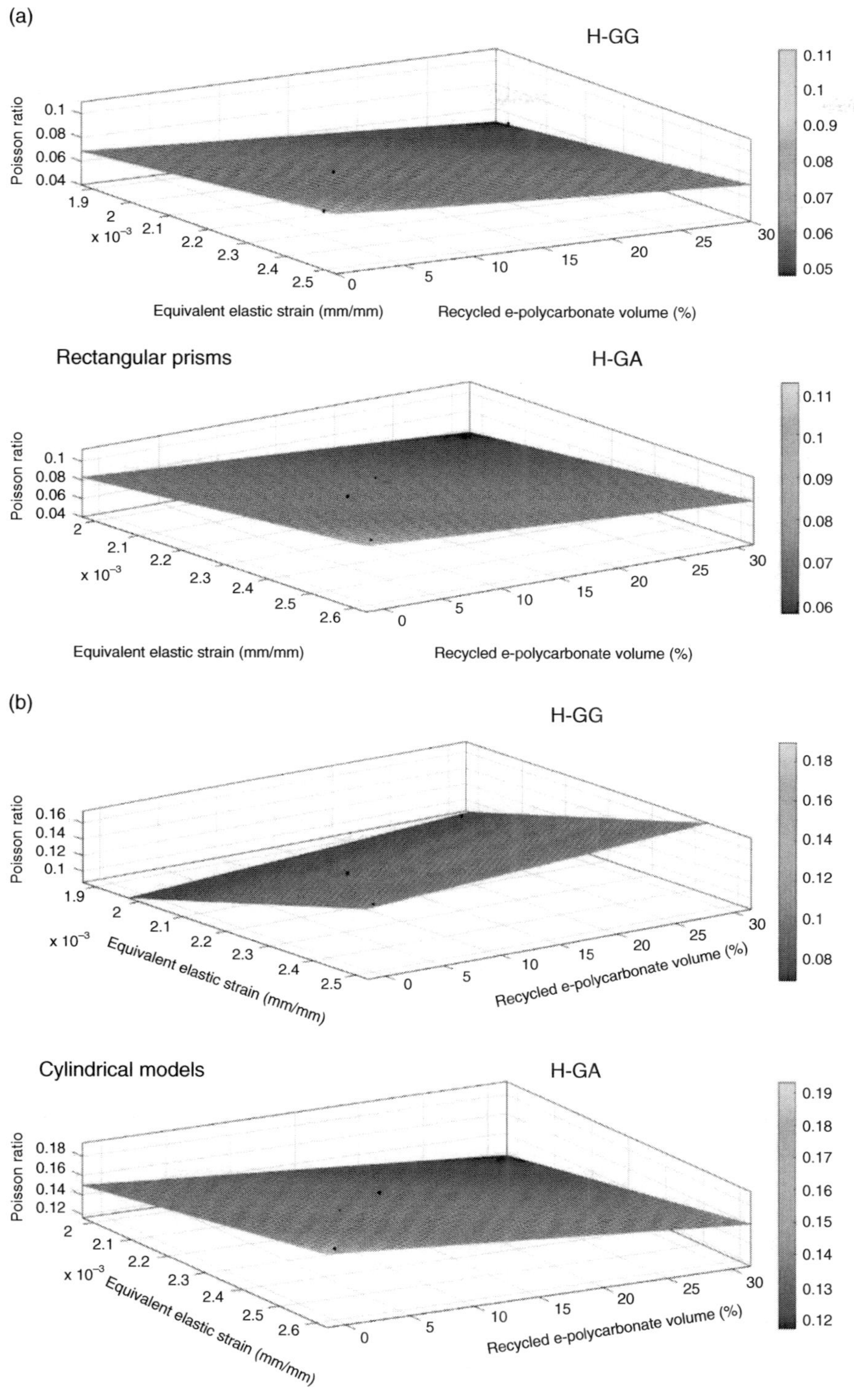

Figure 16.11 Variation of Poisson's ratio for (a) rectangular prisms and (b) cylindrical models. PC in the mortar mix promotes the modification of stiffness or flexibility, which is closely related to Poisson's ratio.

with high percentages of PC resist less deformation at breakage. Also, for the same percentage of PC, mortars with unscreened sand (GA) show greater flexibility than mortars with screened sand (GG). This indicates that by modifying these variables it is possible to produce mortars within the elastic limits suggested by the GA and GG curves.

To explore the values of the modulus of elasticity and deformation, Figure 16.13 shows a comparison of them. For example, curves C-GG and C-GA show a narrower range than curves H-GG and H-GA, which have a wider range of modulus of elasticity. Thus, the inclusion of PC allows one to establish a wider range of elasticity for the design of mortars.

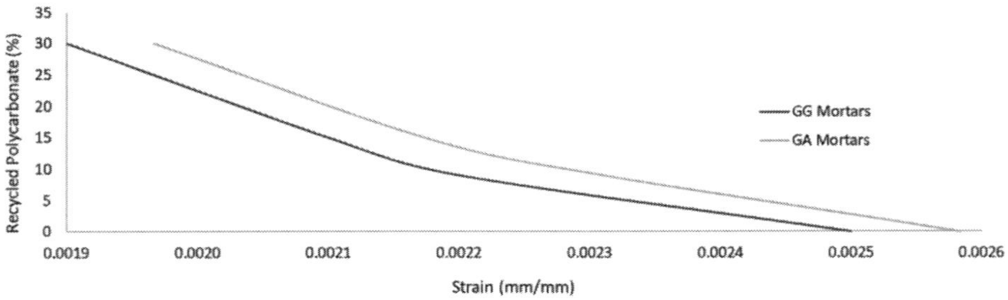

Figure 16.12 Equivalent deformation as a function of PC percentage. Specimens with high percentages of PC resist less deformation at breakage. For the same percentage of PC, mortars with unscreened sand (GA) show greater flexibility than mortars with screened sand (GG).

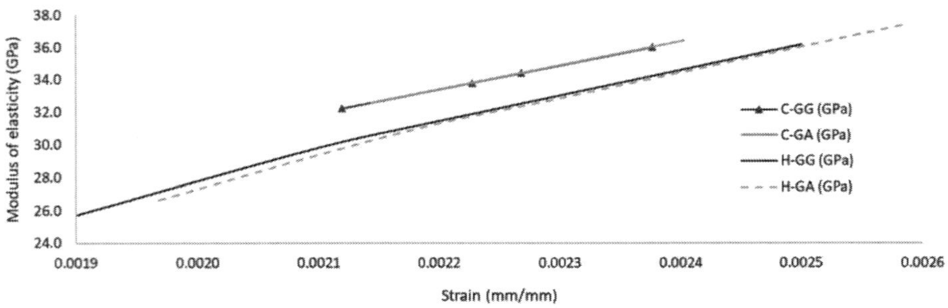

Figure 16.13 Comparison of the values of the modulus of elasticity and deformation. Curves H-GG and H-GA show a wider range of modulus of elasticity than curves C-GG and C-GA. The inclusion of PC allows one to design the range of elasticity for the new mortars.

Finally, Figure 16.14 shows a comparison between Poisson's ratio and deformation for the rectangular prisms and cylindrical models according to the type of mortar used and the elastic behavior analyzed (C-GG, C-GA or H-GG, H-GA). According to the results obtained by FEA analysis from rectangular prisms, Figure 16.14a shows that the C-GG and C-GA mortars maintain a Poisson's ratio between 0.08 and 0.11 within a small equivalent strain range of 2.1–2.4 × 10^{-3} mm/mm. In Figure 16.14b the same tendency is observed with the results obtained by FEA analysis from the cylindrical models, where the C-GG and C-GA mortars maintain a Poisson's ratio between 0.15 and 0.20 within the same equivalent strain range (2.1–2.4 × 10^{-3} mm/mm). Moreover, in both prototypes, when the amount of PC increases, the Poisson's ratio increases and the deformation decreases (Table HD), indicating that if the PC concentration increases, transversal deformation is bigger, and the longitudinal deformation is lower. Thus, the mix is more flexible when the amount of PC increases.

On the other hand, specimens H-GG and H-GA present a totally different behavior. In both prototypes (Figure 16.14), there is a decrease in the deformation as Poisson's ratio decreases, indicating that if the PC concentration increases, transversal deformation decreases as well as the longitudinal deformation. Thus, the mix is much less flexible when the amount of PC increases.

As can be seen, a taller and more slender model, such as a rectangular prism (4:1 ratio), will behave more rigidly (as it will fail before it deforms too much) than a cylindrical model (2:1 ratio), which will deform more before failure. For this reason, the Poisson ratio obtained from mortar with C elastic behavior is higher for the cylindrical model (0.15–0.20) than for the rectangular prism (0.08–0.11). In the same way, the Poisson ratio achieved from mortar with H elastic behavior is higher for the cylindrical model (0.19–0.09) than for the rectangular prism (0.11 – 0.05) even when they present a margin that is interposed. With these ideas in mind, in the next section the main conclusions of the study will be drawn.

16.4 Conclusions

This article shows that it is possible to obtain acceptable compressive and flexural strengths with up to 15% PC from e-waste, recycled in the form of cylindrical pellets of uniform size. This percentage of PC also favored the

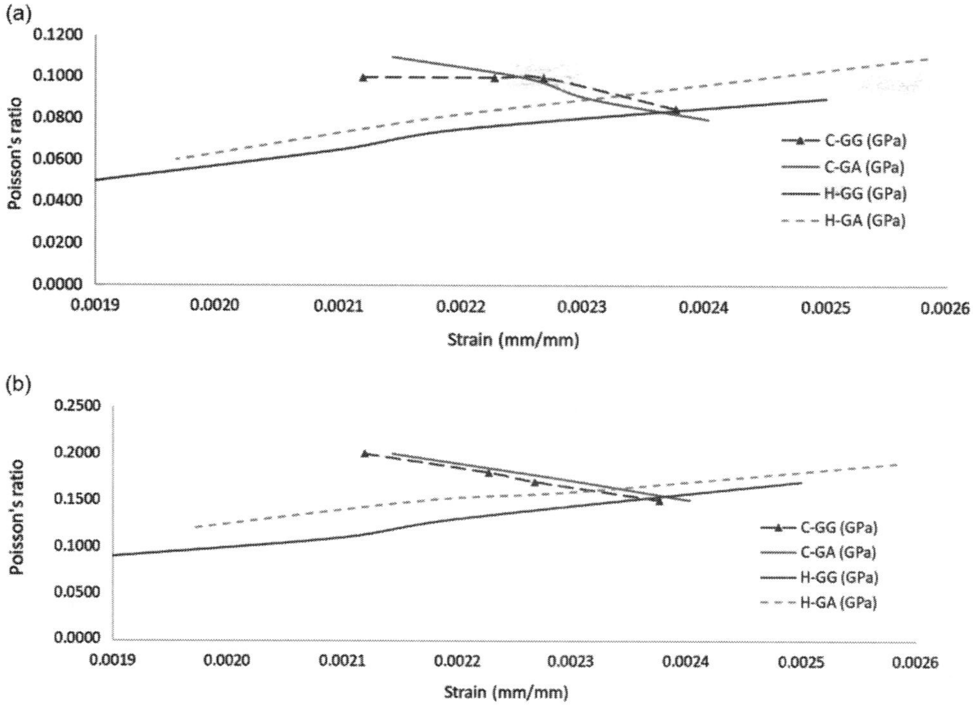

Figure 16.14 Comparison between Poisson's ratio and deformation, for (a) rectangular prisms and (b) cylindrical models. In the case of the C-GG and C-GA mortars, the rectangular prisms and the cylindrical models maintain a Poisson's ratio between 0.08 and 0.11 and between 0.15 and 0.20, respectively. However, H-GG and H-GA mortars present a totally different behavior. In both prototypes there is a decrease in the deformation as the Poisson's ratio decreases.

decrease in density of the mortars and did not affect the water absorption coefficient ($0.1\ mg/cm^2s^{1/2}$) of theme, even when the PC reached a level of 30%. In addition, specimens with high percentages of PC resist less deformation at breakage. For the same percentage of PC, mortars with unscreened sand (GA) show greater flexibility than mortars with screened sand (GG).

The use of FEA analysis also facilitated the comparison of two different prototypes with different elastic behaviors. For instance, the comparison between Poisson's ratio and deformation from both prototypes shows that in the case of the C-GG and C-GA mortars, the rectangular prisms and the cylindrical models maintain a Poisson's ratio between 0.08 and 0.11 and between 0.15 and 0.20, respectively. However, H-GG and H-GA mortars present a totally different behavior. In both prototypes there is a decrease in the deformation as the Poisson's ratio decreases. In a comparison of the values of the modulus of elasticity and deformation, curves H-GG and H-GA show a wider range of modulus of elasticity than curves C-GG and C-GA. The inclusion of PC allows one to design the range of elasticity for the new mortars. Therefore, the Poisson ratio obtained was higher for the cylindrical model than for the rectangular prism. It means that a taller, more slender model, such as a rectangular prism (4:1 ratio), was more rigid than a cylindrical model (2:1 ratio), which will deform more before failure.

This article shows how to go beyond laboratory results. By employing the FEA study, it was possible, in a short time, to obtain a more detailed analysis of the mechanical and structural behavior of new mortars that aim to be environmentally friendly.

Acknowledgments

The authors thank the National Council of Science and Technology (CONACYT) of México, for the scholarship granted to Ana Laura De la Colina Martínez. They also thank Iñigo López Boadella and Rubén Fernández Díaz for their technical support at Department of Construction and Manufacturing Engineering, University of Oviedo, Asturias, Spain.

Conflicts of Interest

Authors declare that they have no conflicts of interest.

Funding

The authors thank the National Council of Science and Technology (CONACYT) for the scholarship granted to Ana Laura De la Colina Martínez (CVU: 265754).

References

AENOR (2003). UNE-EN 1015–18:2003. Methods of test for mortar for masonry—Part 18: determination of water absorption coefficient due to capillary action of hardened mortar (definitive 2023).

AENOR (2012). UNE-EN 933–1:2012. Tests for geometrical properties of aggregates—Part 1: determination of particle size distribution—Sieving method (definitive 2023).

AENOR (2013). UNE-EN 1992–1-1: 2013. Eurocode 2: design of concrete structures—Part 1–1: general rules and rules for buildings (definitive 2023).

AENOR (2018). UNE-EN 196–1:2018. Methods of testing cement—Part 1: determination of strength (definitive 2023).

AENOR (2020). UNE-EN 1015–11:2020. Methods of test for mortar for masonry—Part 11: determination of flexural and compressive strength of hardened mortar.

AENOR (2022). UNE-EN 933–2:2022 Tests for geometrical properties of aggregates—Part 2: determination of size distribution—Test sieves, nominal size of apertures.

Al-Tayeb, M.M. (2020). Effect of partial replacements of sand by polycarbonate plastic waste on the first crack impact resistance of concrete beam. *Civil and Environmental Research* 12 (2). https://doi.org/10.7176/cer/12-2-05.

ASTM C469-14e1: Standard Test Method for Static Modulus of Elasticity and Poisson's Ratio of Concrete in Compression (definitive 2023) (2014). *American Society for Testing and Materials.*

Belyadi, H., Fathi, E., and Belyadi, F. (2019). Rock mechanical properties and in situ stresses. *Hydraulic Fracturing in Unconventional Reservoirs.* https://doi.org/10.1016/b978-0-12-817665-8.00013-8.

Chandrakant Baadkar, C. (2010). Semi-trailer structural failure analysis using finite element method. Master thesis. University of Canterbury, Christchurch, New Zealand.

Covestro (2015). *Makrolon® polycarbonate sheet. Fabrication Guide, Research Report BRO0000915.*

De la Colina Martínez, A.L., Barrera, G.M., Barrera Diaz, C.E. et al. (2019). Recycled polycarbonate from electronic waste and its use in concrete: effect of irradiation. *Construction and Building Materials* 201: 778–785. https://doi.org/10.1016/j.conbuildmat.2018.12.147.

De la Colina-Martínez, A.L., Martinez-Barrera, G., Barrera-Diaz, C.E., and Avila-Cordoba, L.I. (2017). Characterization of recycled polycarbonate from electronic waste and its use in hydraulic concrete: improvement of compressive performance. *Advances in Concrete Construction* 5 (6): 563–573. https://doi.org/10.12989/acc.2017.5.6.563.

Ekolu, S.O. and Murugan, S. (2012). Durability index performance of high strength concretes made based on different standard portland cements. *Advances in Materials Science and Engineering.* https://doi.org/10.1155/2012/410909.

Ersan, Y.C., Gulcimen, S., Imis, T.N. et al. (2022). Life cycle assessment of lightweight concrete containing recycled plastics and fly ash. *European Journal of Environmental and Civil Engineering* 26 (7). https://doi.org/10.1080/19648189.2020.1767216.

Evangelos, G. (2018). Experimental and FE analysis of 3D printed fiber reinforced honeycomb structured composite materials. Master thesis. International Hellenic University, Thessaloniki, Greece.

Forti, V., Baldé, C.P., Kuehr, R., and Garam Bel, G. (2020). *The Global E-waste Monitor 2020: Quantities, Flows, and the Circular Economy Potential.* United Nations University (UNU)/United Nations Institute for Training and Research (UNITAR)—co-hosted SCYCLE Programme, International Telecommunication Union (ITU), and International Solid Waste Association (ISWA), Bonn/Geneva/Rotterdam.

Gravina, R.J., Xie, T., Bennett, B., and Visintin, P. (2021). HDPE and PET as aggregate replacement in concrete: life-cycle assessment, material development and a case study. *Journal of Building Engineering* 44. https://doi.org/10.1016/j.jobe.2021.103329.

Hannawi, K., Kamali-Bernard, S., and Prince, W. (2010). Physical and mechanical properties of mortars containing PET and PC waste aggregates. *Waste Management* 30 (11): 2312–2320.

Kaur, G. and Pavia, S. (2020). Physical properties and microstructure of plastic aggregate mortars made with acrylonitrile-butadiene-styrene (ABS), polycarbonate (PC), polyoxymethylene (POM) and ABS/PC blend waste. *Journal of Building Engineering* 31. https://doi.org/10.1016/j.jobe.2020.101341.

Lee, H.-H. (2018). *Finite Element Simulations with ANSYS Workbench 18*. SDC Publications.

Mordor Intelligence (2023). *Bisphenol-A (BPA) Market—Growth, Trends, Covid-19 Impact, and Forecasts (2023–2028)*.

Poplavko, Y.M. (2019). Mechanical properties of solids. *Electronic Materials*. https://doi.org/10.1016/b978-0-12-815780-0.00002-5.

Professional Plastics (2017). *Polycarbonate Material Data Sheet General Purpose, Glazing Grade*.

Rautela, R., Arya, S., Vishwakarma, S. et al. (2021). E-waste management and its effects on the environment and human health. *Science of the Total Environment*. https://doi.org/10.1016/j.scitotenv.2021.145623

Spherical Insights (2023). *Global Polycarbonate Market*.

Youcai, Z. and Sheng, H. (2017). *Pollution Control and Resource Recovery: Industrial Construction and Demolition Wastes*. Chapter Eight—Recycling Technologies and Pollution Potential for Contaminated Construction and Demolition Waste in Recycling Processes. https://doi.org/10.1016/B978-0-12-811754-5.00008-7.

17

Gamma Radiation Effects on Recycled e-PC: Bisphenol A Leaching

Ana Laura De la Colina Martínez[1,*], *David Joaquín Delgado Hernández*[2], *Gonzalo Martínez Barrera*[3], *Liliana Ivette Ávila Córdoba*[4], *Carlos Eduardo Barrera Díaz*[5], *Fernando Ureña Núñez*[6], *and María Magdalena García Fabila*[7]

[1] *Doctora en Ciencia de Materiales por la Facultad de Química, Universidad Autónoma del Estado de México, Paseo Colón Esquina Paseo Tollocan S/N, 50000, Toluca, México, México*
[2] *Facultad de Ingeniería, Universidad Autónoma del Estado de México, Avenida Universidad S/N, Cerro de Coatepec, Ciudad Universitaria, 50110, Toluca, México*
[3] *Laboratorio de Investigación y Desarrollo de Materiales Avanzados (LIDMA), Facultad de Química. Universidad Autónoma del Estado de México, Km. 12 de la Carretera Toluca-Atlacomulco, 50200, San Cayetano, México*
[4] *Facultad de Ingeniería, Universidad Autónoma del Estado de México, Avenida Universidad S/N, Cerro de Coatepec, Ciudad Universitaria, 50110, Toluca, México*
[5] *Facultad de Química. Universidad Autónoma del Estado de México, Carretera Toluca-Atlacomulco Km. 12, 50200, San Cayetano, México*
[6] *Instituto Nacional de Investigaciones Nucleares, Carretera México-Toluca S/N, 52750 La Marquesa Ocoyoacac, MEX, México*
[7] *Laboratorio de Análisis Instrumental, Facultad de Química, Universidad Autónoma del Estado de México, Paseo Colon esquina Paseo Tollocan S/N, 50180Toluca, México,*
[*] *Corresponding author*

17.1 Introduction

Polycarbonate (PC) is a plastic commonly produced with bisphenol A (BPA) and phosgene (Allard 2014), and is one of the most used engineering thermoplastics. The electrical and electronics sector dominates the global PC scene, with a world market share of 46% in 2017 (IHS Markit 2018; Mordor Intelligence 2018). The global volume consumption of BPA, calculated at 8.0 million metric tons (Mt) in 2016, would reach 10.6 Mt by 2022 (Cision Pr Newswire 2016; Statista 2016). The rapid development in the electrical/electronic technology, combined with a short life cycle and a variety of uses, has resulted in a continuous increase of electronic waste with some illustrative numbers. For example, in the US alone, 9.5 million tons (Mt) were disposed of in 2008, and this was expected to increase to 12.3 Mt in 2020 (Achilias and Antonakou 2015).

PC has a great disadvantage; it leaches BPA to the environment. This may happen due to the residual BPA (derived from an incomplete polymerization) or due to the hydrolysis of the PC itself (Metz 2016; Welshons et al. 2006; Yamamoto and Yasuhara 1999). BPA is an endocrine disruptor, the estrogenic effects of which are observed at extremely low concentrations (Vandenberg et al. 2012; Vom Saal et al. 2005). Industrials have reported that there is a maximum of 100 μg/g residual BPA in the PC polymer (Aschberger et al. 2010; EPA 2015; Mercea 2009).

One alternative to reduce the BPA that the PC releases to the environment when it is present in electronic waste could be to use it as a concrete reinforcer. In fact, previous research has shown that recycled PC from electronic waste improves the compressive strength of hydraulic concrete (De la Colina-Martínez et al. 2017). In addition, the resistance increases when recycled PC from electronic waste is gamma irradiated (De la Colina Martínez et al. 2019). This behavior occurs when gamma radiation (over 1000 kGy) extensively texturizes the pellets' surface, which provides a greater grip between the pellets and the hardened concrete mix. The results show that the composite material obtained could be used in the construction industry. Although gamma radiation would raise costs considerably, it is also true that this research shows the relevance of alternative techniques, which improve the texture of plastic surfaces, as well as the repercussions and challenges of this.

Due to the toxicity of BPA, it is essential to specify the levels generated by recycled PC from electronic waste irradiated and non-irradiated, to determine the amount of BPA that could be present in the concrete mixtures. Although some research has reported that no BPA was detected in PC pellets, PC dialyzers, and a PC-containing trocar after

Electronic Waste Management: Policies, Processes, Technologies, and Impact, First Edition. Edited by Sunil Kumar and Vineet Kumar.

being gamma irradiated at 25 kGy (Scientific Committee on Emerging and Newly Identified Health Risks 2015; Shintani 2001), the effects of gamma radiation on the migration of BPA from recycled PC from electronic waste has not been studied.

To expose the effects of gamma radiation on the amount of BPA leaching from recycled PC from electronic waste, the aims of this work are: (a) to characterize the recycled PC from electronic waste, both irradiated and non-irradiated; (b) to obtain the BPA lixiviated from the irradiated and non-irradiated recycled PC in electronic waste by two methods, namely, BPA dissolution and BPA extraction; (c) to quantify by HPLC the BPA acquired from each method; (d) to compare the degradation products revealed in the HPLC chromatograms with some similar results reported in the literature; and (e) to determine the amount of BPA that could be released by concrete, prepared with different percentages of recycled PC from electronic waste, both non-irradiated and irradiated.

17.2 Materials and Methods

17.2.1 Recycled PC from Electronic Waste

The pellets of recycled PC from electronic waste (namely, recycled e-PC pellets) have an average size of 1 mm in diameter and 3 mm in length (1 × 3 mm). For this research, they were obtained by extrusion from discarded wall electrical switches and were provided by the Polyvima Company (located at Toluca, México).

17.2.1.1 Gamma Radiation of Recycled e-PC Pellets

Recycled e-PC pellets were gamma irradiated at doses of 25, 50, 75, 100, 300, 600, 1200, and 1600 kGy, at room temperature and in closed packages containing a limited amount of air, by using a calibrated industrial cobalt-60 source, model JS-6500, which has a certification ISO-9001: 2015. Determination of dose rate was made by using the well-known method of alanine. Each gamma dose was calculated with the dose rate of 14.6 kGy/h. (For instance, 100 kGy required an irradiation time of 6.84 h).

17.2.1.2 Physical and Chemical Characterization of Recycled e-PC Pellets, Gamma Irradiated and Non-irradiated

The morphological characterization of recycled e-PC pellets was evaluated by means of a scanning electron microscopy (SEM), in a JEOL-JSM-6510 LV. Samples were observed in secondary electron mode at 5 kV. Additionally, digital photographs were obtained.

The recycled e-PC pellets were characterized by X-ray diffraction (XDR), using a Siemens D-5000 diffractometer coupled to a copper-anode X-ray tube. The diffraction patterns were scanned from 5° to 70° 2θ angles.

The elemental quantification at microscopic level of recycled e-PC pellets was analyzed by using the SEM at 20 kV, and energy dispersive spectroscopy (EDS), in a JEOL-JSM-5900 LV.

The identification of functional groups of recycled e-PC pellets was estimated using the Fourier transform infrared spectroscopy (FTIR) in an IR Prestige-21.

The thermal analysis of recycled e-PC pellets was evaluated by using differential scanning calorimetry and thermogravimetric analysis (DSC-TGA) in a SDT Q-600. The heating profile was run from 0°C to 600°C. The purge was under nitrogen atmosphere with a flow rate of 100 ml/min. The heating rate was 10°C/min.

17.2.2 Preparation of BPA Standard Solution

A stock solution (500 ppm) of BPA was made with BPA (≥99.9% purity, Sigma-Aldrich®) and methanol HPLC grade (J.T. Baker®). For the calibration curve, a standard solution of BPA (10 ppm) was prepared utilizing the stock solution diluted with bi-distilled water at pH 7. Both methanol and bi-distilled water were vacuum filtered through a membrane filter with a pore size of 0.45 μm (Millipore®).

17.2.2.1 BPA Dissolution and BPA Extraction Methods

Two different methods were used to acquire BPA from recycled e-PC pellets. In the first method, BPA dissolution, the recycled e-PC pellets (10 g) were placed in a transparent glass tube containing deionized water at pH 7 (10 ml). The temperature was maintained at 20°C for 10 minutes. In the second method, BPA extraction, in accordance with the UNE-EN 14372:2005 standard (AENOR 2005) and with Troiano and Goodman (2009), the recycled e-PC pellets (10 g) were placed in a transparent glass tube containing deionized water at pH 7 (10 ml). The temperature was maintained at 40ºC for 24 hours through a hot water bath, which remained in a closed laboratory fume hood, the door of which had 5 mm tempered glass. The BPA extraction and dissolution methods were performed in April 2018, in Toluca, México (latitude 19.272828, longitude −99.659696, and altitude 2,667 m above sea level). On these days, the maximum direct illumination intensity was approximately 1050 W/m^2 (Honsberg and Bowde 2020). The maximum UV indexes were 5 (UVA) and 4 (UVB) (Gobierno de la Ciudad de México 2020). Approximately 5 ml of water from the BPA extraction method was left behind and transferred into amber-colored 5 ml vials and kept in refrigeration until the HPLC (High Performance Liquid Chromatography) analysis.

17.2.2.2 Identification and Quantification of BPA by HPLC Analysis

To know the concentration of BPA leached from the recycled e-PC pellets, the BPA samples, acquired by both methods (BPA dissolution and BPA extraction), were analyzed by HPLC. To do so, a Waters-1525 binary HPLC pump and a Waters 2487 dual λ absorbance detector ultraviolet/visible (UV/Vis) were used. An analytical column C_{18} (XTerra MS, 5 μm, 4.6 × 250 mm) was utilized. The isocratic elution was carried out with a mixture of methanol:water (v/v) in the ratio of 50:50 at a flow rate of 1 ml/m with an equilibration time of 10 minutes. The injection volume was 80 μl. The column temperature and the UV detection wavelength were set at 25ºC and 230 nm, respectively. Data were recorded and analyzed with the Brezze© software.

The BPA calibration curve was obtained with a BPA dilution of 10 ppm at six levels in the range of 1–10 ppm, with two replicates per concentration. The correlation coefficient of the calibration curve was ≥0.999. Having described the methodology followed to carry out the research, in the next section the results obtained will be discussed.

17.3 Results and Discussion

17.3.1 SEM and Digital Images: Morphological Characterization of Recycled e-PC Pellets

The morphological characteristics of the recycled e-PC pellets are shown in Figure 17.1. The recycled e-PC pellets exhibit an almost cylindrical shape (Figure 17.1a). They have an average size of 1 mm in diameter and 3 mm in length. In most cases, the recycled e-PC pellets are hollow. The tiny holes have a diameter of about 0.1 mm (Figure 17.1b, where some recycled e-PC pellets were amplified to illustrate these characteristics).

The surfaces of irradiated and non-irradiated recycled e-PC pellets are shown in Figure 17.2. In all cases, the white powder dispersed corresponds to rutile (TiO_2), which is imbedded in the non-irradiated (Figure 17.2a) and irradiated (Figures 17.2b and 17.2c) recycled e-PC pellets.

More detailed images, shown in (De la Colina Martínez et al. 2019), reveal that non-irradiated particles have a smooth and glossy surface, while the irradiated particles (1200 kGy) have a rougher surface with small, detached particles. When the dose of this electromagnetic radiation increases, rough textures on the surface of these particles can be obtained.

17.3.2 EDS: Elemental Quantification of Recycled e-PC

The elemental composition spectra of irradiated and non-irradiated recycled e-PC pellets are shown in Figure 17.3. In all cases, EDS analysis of the non-irradiated (Figure 17.3a) and irradiated recycled (Figures 17.3b and 17.3c) e-PC pellets indicate the presence of carbon (C), oxygen (O), titanium (Ti), and aluminum (Al) elements.

The relative proportion of elements in terms of atomic % and weight % of irradiated and non-irradiated recycled e-PC pellets are shown in Table 17.1. It is possible to

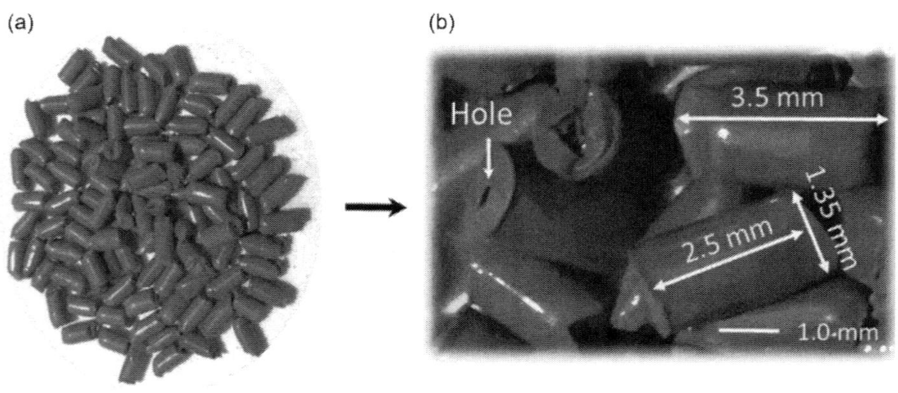

Figure 17.1 Digital images of recycled e-PC pellets: (a) recycled e-PC pellets that exhibit an almost cylindrical shape and (b) recycled e-PC pellets amplified.

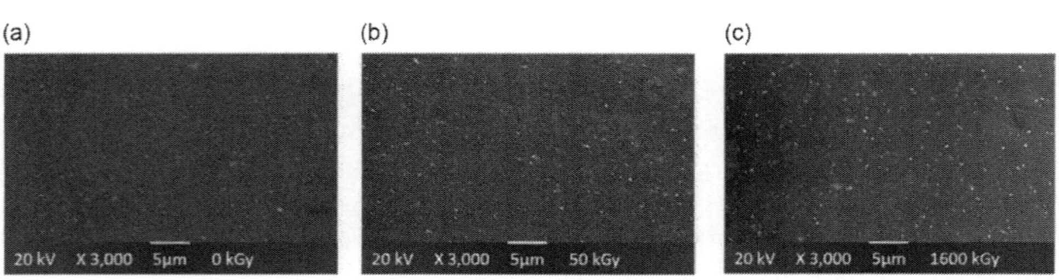

Figure 17.2 SEM images of irradiated and non-irradiated recycled e-PC pellets: (a) 0 kGy, (b) 50 kGy, and (c) 1600 kGy.

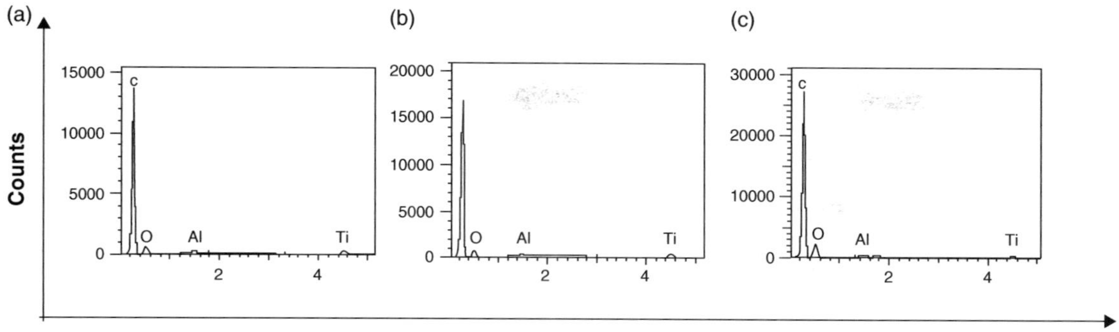

Figure 17.3 EDS spectra of irradiated and non-irradiated recycled e-PC pellets: (a) 0 kGy, (b) 50 kGy, and (c) 1600 kGy.

Table 17.1 Relative proportion of elements (atomic % and weight %) of irradiated and non-irradiated recycled e-PC pellets.

	Atomic %			Weight %		
Element	**0 kGy**	**50 kGy**	**1600 kGy**	**0 kGy**	**50 kGy**	**1600 kGy**
C	81.49	80.92	76.66	75.85	75.26	70.77
O	17.95	18.55	23.06	22.26	22.98	28.36
Al	0.12	0.13	0.08	0.26	0.27	0.17
Ti	0.44	0.40	0.19	1.63	1.49	0.70

observe the absence of hazardous substances such as mercury (Hg), cadmium (Cd), lead (Pb), and hexavalent chromium (Cr^{6+}), which are regulated by both European Union Directives 2012/19/EU and 2011/65/UE (European Parliament 2011 2012). The ratio oxygen/carbon (O/C) increases from 0.22 to 0.30, while increasing the gamma radiation from 0 kGy to 1600 kGy. It is interesting that the same trend is also observed in (De la Colina Martínez et al. 2019). It could be due to the oxygen attack to the polymer matrix, which produces a superficial absorption of oxygen during the radiation in air (Hacioğlu et al. 2016; Krasnansky et al. 1961). However, it may be said that the irradiated and non-irradiated recycled e-PC are indeed PC. To confirm this, XRD and FTIR analyses were performed.

17.3.3 XRD Analysis of Recycled e-PC

Recycled e-PC pellets were evaluated by X-ray diffraction (XRD). Figure 17.4 shows the XRD spectra of irradiated (50 and 1600 kGy) and non-irradiated (0 kGy) recycled e-PC pellets. The results show that all XRD spectra present the same diffraction peaks. One peak (17.2°) corresponds to amorphous PC and the remaining six peaks correspond to Rutile (TiO_2), which is a pigment widely used in the plastics industry. No new peaks appear in the profile after irradiation, indicating that the material does not undergo any transformation (Pondé Weber et al. 2010).

According to Sinha et al. (2004), the peak at 17.2° indicates the presence of one type of nano-crystalline zone that is dispersed in an amorphous matrix. Their results reflect that with increasing gamma dose (from 0 to 1000 kGy), the intensity of this peak increases; it means that the amorphous character of the PC decreases, and crystallinity increases with increase in dose. However, in this study and in (De la Colina Martínez et al. 2019; Kumar et al. 2017), the intensity of this peak decreases, when recycled e-PC is irradiated at 600, 1200, and 1600 kGy. It means (even though a dose of 50 kGy could lead to a type of crystalline zone) that the amorphous character of the recycled e-PC increases when it is gamma irradiated.

17.3.4 FTIR Analysis of Recycled e-PC Pellets

Figure 17.5 shows the FTIR spectra of irradiated and non-irradiated recycled e-PC pellets. All FTIR spectra show the following characteristic vibrational bands, present in the common PC chemical structure (Universidad de Barcelona 2019). The peak at 2966 cm^{-1} is attributed to the stretching vibration of the C–H bond from the CH_3 group (sp3 C–H) (Universidad de Barcelona 2019). The peak located at 1770 cm^{-1} is assigned to the carbonate group (C=O) vibration, which was also reported by Sweileh, Al-Hiari, and Kailani (2010). The peak at 1504 cm^{-1} corresponds to the ring (C=C) vibration from the phenol (Mayo et al. 2004). The

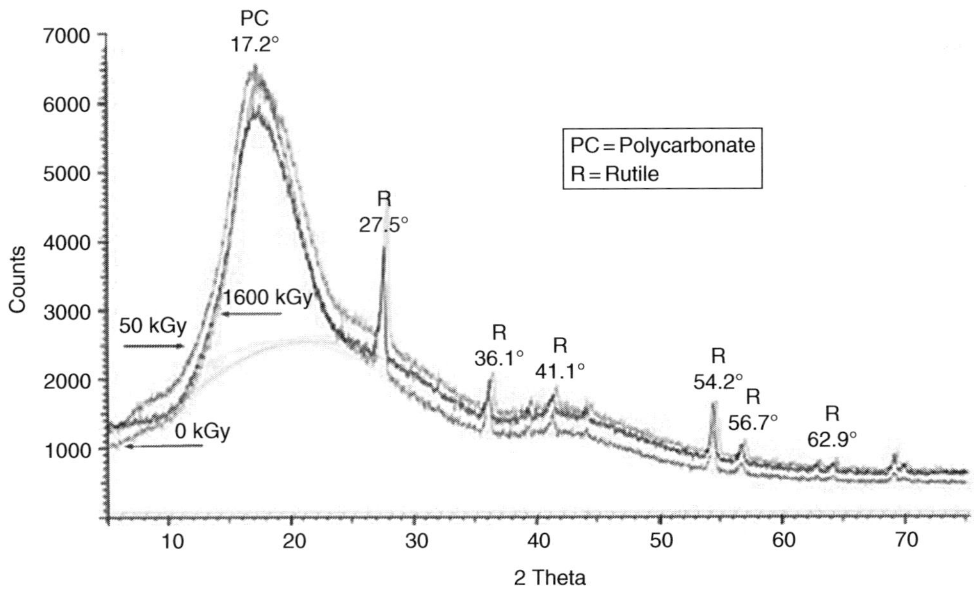

Figure 17.4 XRD spectra of recycled e-PC pellets irradiated and non-irradiated: line gray, 0 kGy; line blue, 50 kGy; and line red, 1600 kGy.

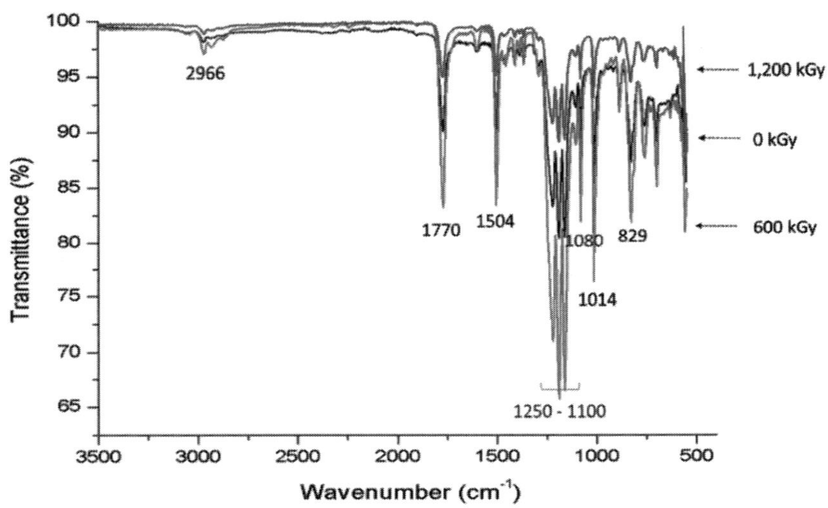

Figure 17.5 FTIR spectra of irradiated and non-irradiated recycled e-PC pellets.

bands within the range of 1250–1100 cm^{-1} are associated with stretching deformations of the asymmetric carbonate group (O–C–O). The peak at 1080 cm^{-1} is attributed to the vibration of the methyl (–CH_3) group. The band at 1014 cm^{-1}, which corresponds to the symmetric carbonate group (O–C–O) in stretching vibration mode, was also reported by Parshin, Gunyakov, and Zyryanov (2013). Finally, the vibrational band at 829 cm^{-1} is assigned to the para-substituted phenol ring (Hokenek 2009). These results show that the recycled e-PC molecules are very stable to gamma radiation doses of 600 and 1200 kGy.

The high stability of recycled e-PC to gamma dose (1200 kGy) contrasts significantly with the low or medium stability of some commercial PC (non-recycled and without TiO_2) as Lexan Margard® (de Melo et al. 2007), Durolon® (Araújo et al. 1999), PCLight® (Pondé Weber et al. 2010), Makrofol® (Abdul-Kader et al. 2018), and FRG® (Sinha et al. 2004), the degradation of which begins at 125, 300, 340, 450, and 1000 kGy, respectively. In the degradation of PC, gamma radiation is responsible for the decrease of the carbonyl group C=O (Araújo et al. 1999; Sinha et al. 2004), the breaking of O–C–O bonds (Abdul-Kader et al. 2018), and the formation of phenolic hydroxyl groups that has been attributed to changes in the OH stretch region (3650–3350 cm^{-1}) (Kalkar et al. 1992). The ESR (Electron-Spin Resonance) spectra of Durolon (Araújo et al. 1999), irradiated with 100 kGy, presented two peaks of absorption (singlets) that have been ascribed to polymeric phenoxyl and phenyl radical produced by radiolysis. The phenyl radicals decay rapidly, while the phenoxyl radicals decay more slowly (Araújo et al. 1999; Chung 1996).

17.3.5 DSC-TGA Analysis of Recycled e-PC Pellets

Figures 17.6–17.8 show the DSC-TGA thermograms of non-irradiated and irradiated recycled e-PC pellets. The glass transition temperature (*Tg*) was evaluated by the DSC analysis. The extrapolated onset temperature, which indicates the temperature at which the weight loss begins (T_{onset}) (PerkinElmer 2010), and the weight loss percentage of gamma irradiated recycled e-PC pellets were evaluated by the TGA analysis.

The DSC results show a diminution in the *Tg* (from 158 to 145°C) when the gamma dose increases (from 0 to 1200 kGy). It reveals that the recycled e-PC scissions increase when the radiation dose increases. For the *Tg* determination, a zoom in each curve was made, within the expected zone (based on the literature) of inflection. Owing to scissioning of the polymeric chains, the mobility of the chains increases, and as a result, the *Tg* of the polymer decreases (Sinha et al. 2004). In addition, the TGA results show the diminution in the T_{onset} (from 404 to 392°C) when the radiation dose increases (from 0 to 1200 kGy). This behavior is

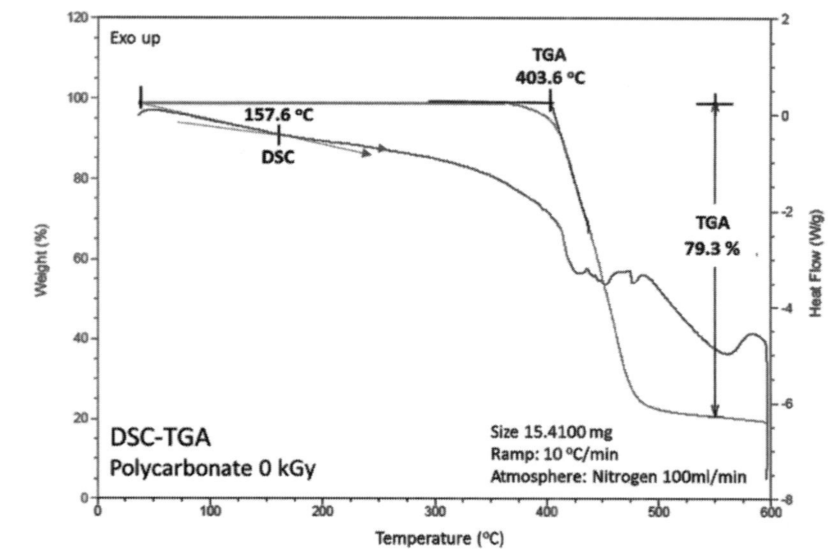

Figure 17.6 DSC-TGA thermogram of recycled e-PC pellets non-irradiated.

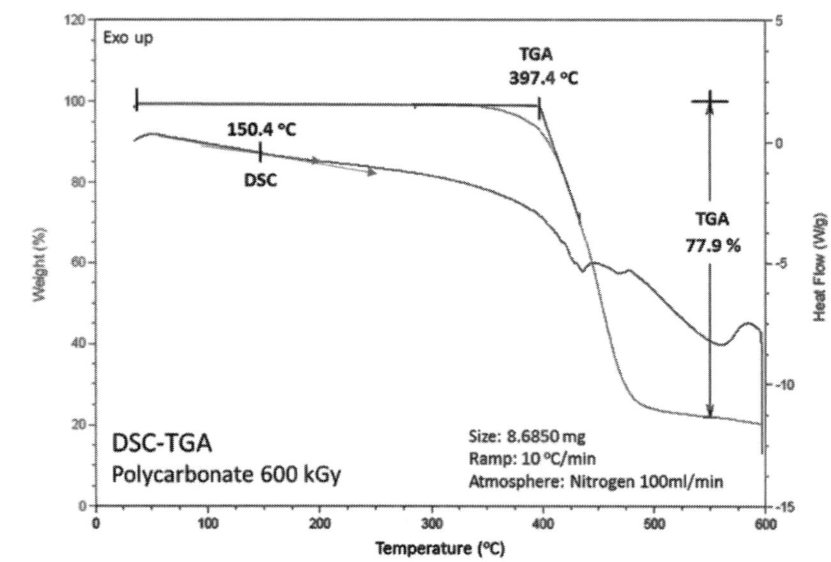

Figure 17.7 DSC-TGA thermogram of recycled e-PC pellets irradiated at 600 kGy.

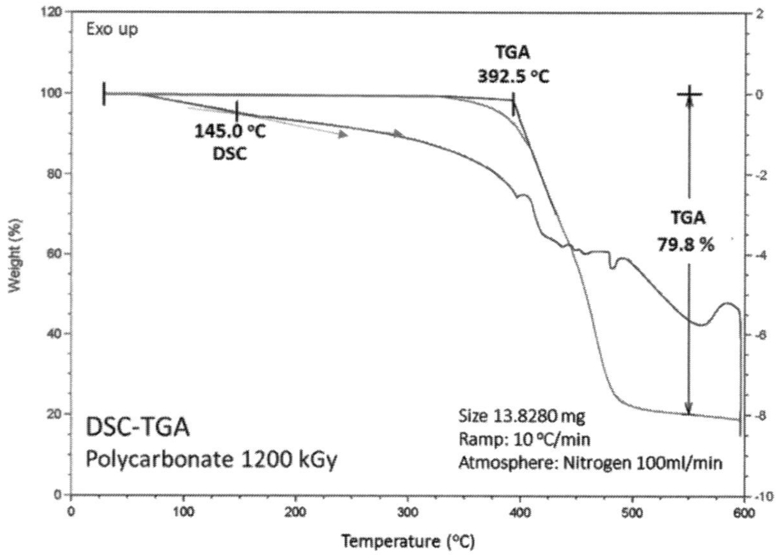

Figure 17.8 DSC-TGA thermogram of recycled e-PC pellets irradiated at 1200 kGy.

consistent with the polymer scission. These results show that the gamma radiation promotes thermal degradation (Araújo et al. 1999; de Melo et al. 2007; Hacioğlu et al. 2016).

On the other hand, the *Tg* of non-irradiated recycled e-PC (158°C) contrasts with the *Tg* of both clear resins, Makrolon® (141–148°C) (Covestro 2018) and Lexan®143R (141°C) (Prospector 2016). Moreover, the T_{onset} of non-irradiated recycled e-PC (404°C) contrasts with the T_{onset} of clear resin Lexan®143R (336°C) (Katiyar and Balasubramanian 2014; Yadav et al. 2017). This increase in both *Tg* and T_{onset} shows that the TiO_2 improves the thermal stability of the recycled e-PC (da Silva et al. 2018; Jahantigh and Nazirzadeh 2017).

Furthermore, the TGA results show that the weight loss percentages (close to 79%) are similar for the irradiated and non-irradiated recycled e-PC pellets. According to Katiyar and Balasubramanian (2014) and Méndez-Liñán et al. (2010), the char residue produced in the pyrolysis of PC constitutes a quarter (22–25%) of the weight of the PC pyrolyzed. Apaydin-Varol, Polat, and Putun (2014) revealed that the main gases released during the PC pyrolysis were CO_2, CH_4, CO, and H_2O. Organic species, including aldehydes, ketones, and carbonyls were detected also. At the end of the pyrolysis process, the char formation occurred.

17.3.6 HPLC Analysis of BPA Leached from Recycled e-PC Pellets

The calibration curve of the BPA standard is shown in Figure 17.9. The linear equation from the BPA standard solution was as follows: $y = 178472x - 15912$, where *x* is the resulting BPA concentration and *y* is the peak area. The calibration curve for BPA is linear with a correlation coefficient of ≥ 0.999 in the range of 1 to 10 ppm.

Detectable levels of BPA leached from irradiated and non-irradiated recycled e-PC pellets were quantified. Figure 17.10

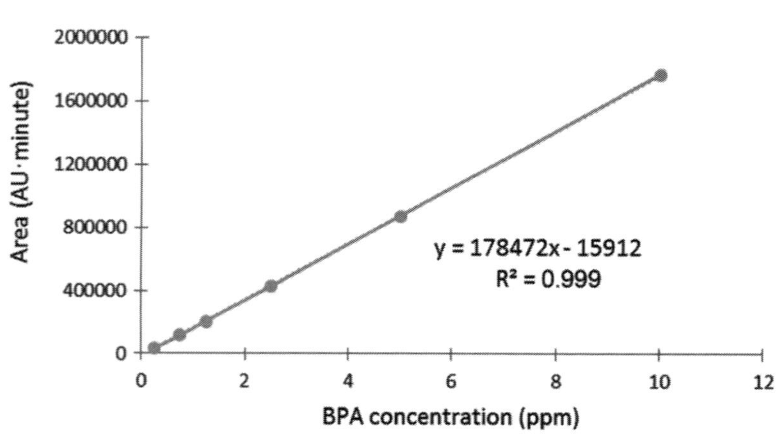

Figure 17.9 BPA calibration curve.

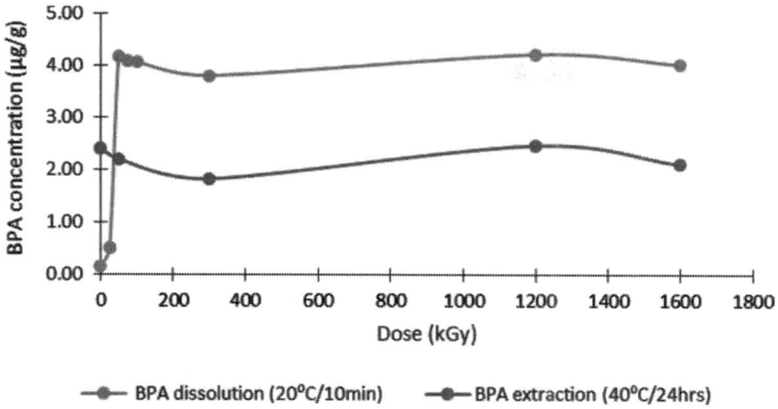

Figure 17.10 BPA concentration vs gamma radiation dose.

shows the BPA concentration identified in both methods, BPA dissolution (20°C/10 min) and BPA extraction (40°C/24 h). By using the BPA dissolution method (20°C/10 min), the BPA concentration from the non-irradiated recycled e-PC was 0.15 µg/g. The BPA leached from the irradiated recycled e-PC at 25 kGy was 0.51, and from the irradiated recycled e-PC within the range of 50 to 1600 kGy was 4.0 µg/g. On the other hand, the results of the BPA extraction method (40°C/24 h) show that the BPA concentration in all cases, for both non-irradiated and irradiated recycled e-PC pellets, was approximately 2.2 µg/g.

The results of both the BPA extraction method (40°C/24 h) and the BPA dissolution method (20°C/10 min), can be explained considering the following aspects:

(A) The BPA leaching from the PC products has two possible origins, migration of residual BPA and hydrolysis of the PC itself (Metz 2016; Welshons et al. 2006; Yamamoto and Yasuhara 1999). The BPA is the main hydrolysis product of PC (Si et al. 2014; Yamamoto and Yasuhara 1999). In water, at 40°C, PC does not decompose within 8 hours under neutral or alkaline conditions (Si et al. 2014). Therefore, it is to be expected that in the standard (UNE-EN 14372:2005) conditions (24 h, 40°C, and pH 7), the residual BPA is the one that will be released. Since all the samples come from the same recycled e-PC, it is possible to consider that the total amount of residual BPA is the same in all cases (recycled e-PC irradiated and non-irradiated).

(B) Several studies show that BPA was not degraded when it was kept in water at 25°C, exposed to 8:16 hour fluorescent light:dark cycles for 8 days (Dorn et al. 1987). Furthermore, BPA was not degraded when it was exposed to UV radiation (10 mW/cm^2) by 20 hours in water at room temperature (Ohko et al. 2001). However, more than 99% of the BPA was degraded in water at room temperature, in the presence of TiO_2, after 15 hours of UV irradiation (10 mW/cm^2) (Ohko et al. 2001). Also, BPA in the presence of TiO_2 was degraded up to 97% in water (21.2 to 30.5°C) after 6 hours of UV irradiation (18 mW/cm^2) (Wang et al. 2009).

(C) TiO_2 is a wide band gap semiconductor, which can only be activated with light energy larger than its band gap (~3.2 eV), and this limits its application within the UV region of the solar spectrum (Bello Lamo et al. 2014). In the presence of UV, TiO_2, and O_2, the photocatalytic reaction generates oxidizing HO• radicals for the degradation of organic compounds such as BPA (Chiang et al. 2004).

In the BPA extraction method (40°C/24 h), the recycled e-PC lasted 24 hours in water at 40°C in the presence of TiO_2 (1.6%, estimated by EDS) under a minimal amount of UV light at 16:8 hours of the dark:light period. During the first 16 hours (dark period), the BPA lixiviation process takes place without BPA degradation. However, during the remaining 8 hours (light period), a UV-photocatalytic reaction occurred. At the end of the 8 hours, in all cases, there was a remaining amount of BPA that reached 2.2 µg/g and some degradation products of BPA (Figure 17.13).

The results of the BPA dissolution method (20°C/10 min), which do not show any degradation products (Figure 17.12), can be related with the Ra (roughness average) values obtained from PC that is gamma irradiated. These results show an increase in the surface roughness of PC with increasing radiation dose. The increase in surface roughness showed a sudden increase up to 50 kGy dose and remained almost unchanged thereafter (up to 500 kGy) (Güven et al. 1997). These results show a trend like that shown by the amount of BPA during the dissolution method.

The major topological effect of gamma irradiation in the PC is the formation of hills instead of wells. This can be explained by the formation of gases (carbon monoxide, hydrogen, carbon dioxide, and methane) (Navarro-González and Aliev 2001) inside the polymer during gamma radiolysis. These gases most probably force themselves out from the surfaces causing the formation of irregular protrusions (Güven et al. 1997). Also, the contact angle of PC decreases with increasing gamma dose, from 77.6 (0 kGy) to 47.8 (1000 kGy). This decrease in contact angle means that the wettability of irradiated PC increases with increasing gamma doses (Tayel et al. 2015). It is known that the surface roughness affects the wettability by the formation of micro-pores (Abdul-Kader 2013; Wróbel et al. 1978). The presence of these micro-pores on the plastic surface is reflected by a lower wetting time (Wróbel et al. 1978).

The factors mentioned above increase the contact between residual BPA and water. By immersing the pellets in water at 20°C, the residual BPA dissolves almost immediately. When the gamma dose increases, the micro-pore on the PC increases too, and the amount of BPA dissolved in the water increases. Eventually, more of the residual BPA will be dissolved, until there is no more BPA to leach out, even when the PC surface is completely rough (1600 kGy).

Figure 17.11 shows the HPLC chromatogram of BPA present in the standard solution at 10 ppm. This HPLC chromatogram is very similar to those shown by Troiano and Goodman (2009).

Figure 17.12 shows the HPLC chromatograms of BPA leached from the non-irradiated (Figure 17.12a) and irradiated (Figures 17.12b and 17.12c) recycled e-PC pellets, detected in the BPA dissolution method (20°C/10 min). The results show that there are no degradation products in the dissolution of BPA.

Figure 17.13 shows the HPLC chromatograms of BPA leached from non-irradiated (Figure 17.13a) and irradiated (Figures 17.13b and 17.13c) recycled e-PC pellets detected in the BPA extraction method (40°C/24 h). By comparing these HPLC chromatograms with the HPLC chromatograms of Yang (2015) and consulting the degradation products of UV-photocatalytic degradation of BPA in aqueous

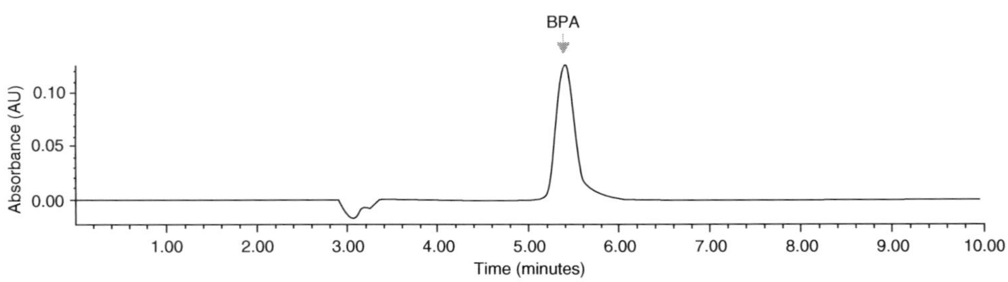

Figure 17.11 HPLC chromatogram of BPA present in the standard solution at 10 ppm.

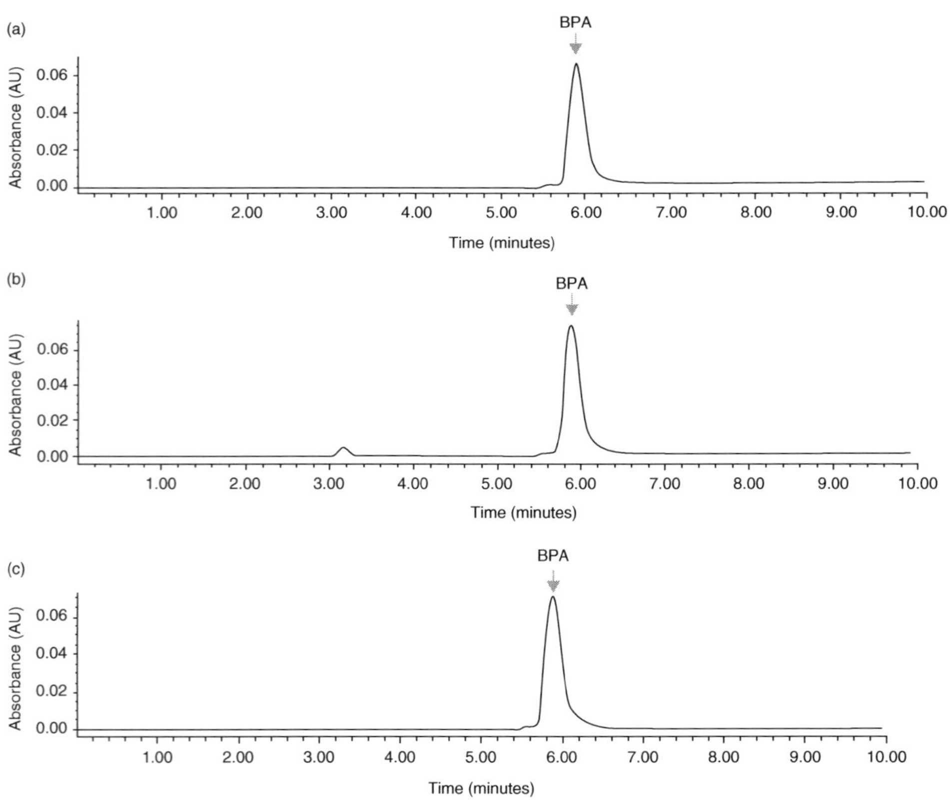

Figure 17.12 HPLC chromatogram of BPA present in non-irradiated recycled e-PC pellets (BPA dissolution method: 20°C/10 min): (a) 0 kGy, (b) 50 kGy, and (c) 1600 kGy.

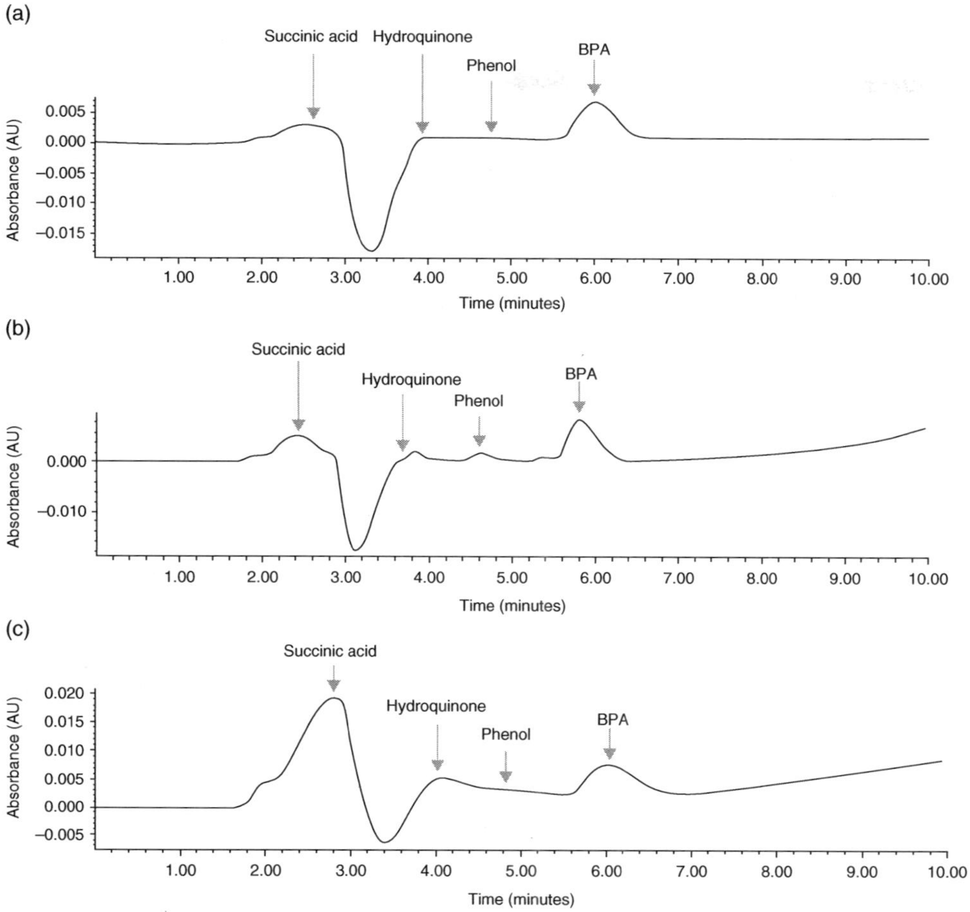

Figure 17.13 HPLC chromatograms of BPA present in recycled e-PC pellets (BPA extraction method: 40°C/24 h): (a) 0 kGy, (b) 50 kGy, and (c) 1600 kGy.

solution with TiO_2, where BPA is mineralized via HO• radicals, forming hydroquinone, phenol, and low molecular weight organic acids (Garay-Rodríguez et al. 2018) such as succinic acid (Mokhtari et al. 2016), therefore, it can be inferred that hydroquinone, phenol, and succinic acid may be present in the samples of the BPA extraction method, due to the oxidation of BPA via OH• radicals (Garay-Rodríguez et al. 2018; Mokhtari et al. 2016).

The previous sequence of chromatograms (Figure 17.12) shows that the dissolution method does not induce the degradation of bisphenol A, which does occur in the extraction method (Figure 17.13). It would be important that the standard (UNE-EN 14372:2005) should warn about the presence of TiO_2 in the samples to be analyzed, since it can alter the results, leading to errors and lower BPA values.

These results show that concrete composite, once hardened and ready to be used, contains a small amount of BPA. It is very likely that the leached BPA is trapped in the hardened concrete, and only the leached BPA could gradually seep through the pores of the concrete, if the concrete absorbs water. This filtration process should be related with both the BPA solubility in water, which is 300, 257, and 253 mg/l at 25, 24, and 22°C respectively, in the pH range of 4 to 10 (Aschberger et al. 2010; Shareef et al. 2006), and the water permeability coefficient of concrete, which varies between 0.02×10^{-11} and 1.2×10^{-11} m/s (Ahmad and Hossain 2017).

Table 17.2 shows the amount of BPA in a cubic meter of air after 8 hours/day, 5 days/week. It was calculated with the amount of water drained from the surface of one slab (size 100 cm × 100 cm × 15 cm) made with this composite material. The water drained was calculated with the equation of the permeability coefficient $k = QL/HAt$, where k is the water permeability coefficient of concrete (1.2×10^{-9} cm/s), Q is the amount of water drained in cm^3 throughout time t in s, L is the sample length in cm (15 cm), H is the difference between the level of the upper h2 (15 cm) and the lower h1 (0 cm) water column, and A is the sample cross-section area in cm^2 (100 cm × 100 cm = 10,000 cm^2) (Ribeiro et al. 2018). The amount of BPA in the drained water was calculated with the BPA solubility in water at 25°C (300 mg/l or 0.3 mg/cm^3).

Table 17.2 BPA in one cubic meter of air by 8 hours/day, 5 days/week.

Slab *	Drained water from the slab		BPA in the drained water **	BPA in a cubic meter of air
Built with concrete and recycled e-PC	At a rate of 1.2×10^{-9} cm/s ***	8 hours/day, 5 days/week	8 hours/day, 5 days/week	8 hours/day, 5 days/week
(number of slabs)	(cm^3/seg)	(cm^3)	(mg)	(mg/m^3)
1	1.20E-05	1.73	0.52	0.52

* Slab size: 100 cm × 100 cm × 15 cm.
** BPA solubility in water at 25°C: 300 mg/l or 0.3 mg/cm^3.
*** Water permeability coefficient of concrete: 1.2×10^{-11} m/s (1.2×10^{-9} cm/s).

Table 17.3 Total BPA present in one slab according to the dissolution method.

Concrete	Recycled e-PC				Total BPA in one slab*		
Total amount	In the concrete		In a cubic meter of concrete**	In one slab***	0 kGy (0.15 mg/Kg)	25 kGy (0.51 mg/Kg)	50–1600 kGy (4 mg/Kg)
(Kg)	(Vol. %)	(Kg)	(Kg)	(Kg)	(mg)	(mg)	(mg)
1.496	1	1.94E-03	2.6	0.39	0.06	0.20	1.56
1.490	3	5.81E-03	7.8	1.18	0.18	0.60	4.70
1.482	6	1.16E-02	15.5	2.33	0.35	1.19	9.31
1.458	15	2.90E-02	38.8	5.82	0.87	2.97	23.26

* According to the dissolution method.
** Density of concrete: 2000 kg/m^3.
*** Slab size: (100 cm × 100 cm × 15 cm).

In 2014, SCOEL (Scientific Committee on Occupational Exposure Limits) published its recommendation to lower the occupational exposure limit (OEL) for BPA inhalation (concentration of BPA in the air by 8 hours/day, 5 days/week), from 10 mg/m^3 to 2 mg/m^3. This new limit is identified only for workers in the manufacture of BPA (i.e. product sampling and bag filling) (National Institute for Public Health and the Environment 2016).

The results of Table 17.2 show that the BPA concentration (0.52 mg/m^3), is below the limit (2 mg/m^3) imposed by SCOEL.

Table 17.3 shows the total BPA that could be present in one slab, depending on the dose of gamma radiation applied, in accordance with the concrete formulations cited in (De la Colina Martínez et al. 2019), and with the BPA obtained from the dissolution method.

The results in Table 17.3 show that if the recycled e-PC is used without irradiation (0 kGy) and irradiated at 25 kGy, the total BPA that could be permeated from the polymer does not exceed 2 mg, except in the case of 15% (2.97 mg). The results also show that if the recycled e-PC is irradiated between 50 and 1600 kGy, the total BPA that could be permeated, in the cases of 3, 6, and 15% of recycled e-PC, exceeds 2 mg, except in the case of 1% (1.56 mg). Considering the results of Table 17.2, although the total BPA in the slab was 2.97, 4.7, 9.3, and 23.26 mg, the amount of BPA that could be permeated would only be 0.5 mg/m^3, since BPA has a solubility in water of 300 mg/l at 25°C. Therefore, it is possible to suggest that, based on the norm (OEL = 2 mg/m^3) imposed by SCOEL, even if the recycled e-PC was irradiated at 1600 kGy, it could be used up to 15% in the concrete without causing damage to human health.

17.4 Conclusions

This study aimed to evaluate the effects of gamma radiation on the amount of BPA leaching from recycled PC from electronic waste (recycled e-PC). The results show that the BPA released from recycled e-PC increases (from 0.15 μg/g to 4.0 μg/g) when the gamma radiation increases (from 0 to 1600 kGy). The leached BPA, which could migrate to the air in the most extreme conditions (1600 kGy), would be in amounts lower (0.5 mg/m^3) than those allowed by SCOEL (2.0 mg/m^3), in terms of occupational exposure limit. In addition, there are certain factors that could facilitate the recycling of polycarbonate into concrete; for instance, it is very stable at a gamma radiation dose of 1200 kGy, and also it does not have any hazardous substances such as mercury (Hg), cadmium (Cd), lead (Pb), and hexavalent chromium (Cr^{6+}), which are regulated by the two European Union Directives 2012/19/EU and 2011/65/EU.

There is the probability of reducing the BPA that the e-PC waste releases to the environment, if recycled e-PC could be used (irradiated or non-irradiated) as a concrete reinforcer.

Additionally, it would be important that the UNE-EN 14372: 2005 standard considers the influence of rutile (TiO_2) on PC samples, since it could falsify the BPA levels determined by the procedure that this standard uses.

References

Abdul-Kader, A.M. (2013). The optical band gap and surface free energy of polyethylene modified by electron beam irradiations. *Journal of Nuclear Materials* 435 (1–3): 231–235. https://doi.org/10.1016/j.jnucmat.2013.01.287.

Abdul-Kader, A.M., Zaki, M.F., Radwan, R.M. et al. (2018). Influence of gamma irradiation on physical and chemical properties of Makrofol (NTD) material. *Radiation Physics and Chemistry* 151: 12–18. https://doi.org/10.1016/j.radphyschem.2018.05.010.

Achilias, D.S. and Antonakou, E.V. (2015). Chemical and thermochemical recycling of polymers from waste electrical and electronic equipment. In: *Recycling Materials Based on Environmentally Friendly Techniques* (ed. D.S. Achilias), 37–61. London, United Kingdom: IntechOpen.

AENOR (2005). *UNE-EN 14372:2005 Child use and care articles—Cutlery and feeding utensils—Safety requirements and tests.*

Ahmad, S.I. and Hossain, M.A. (2017). Water permeability characteristics of normal strength concrete made from crushed clay bricks as coarse aggregate. *Advances in Materials Science and Engineering* [Preprint]. https://doi.org/10.1155/2017/7279138.

Allard, P. (2014). Chapter 27, Bisphenol A. In: *Biomarkers in Toxicology* (ed. R.C. Gupta), 459–474. https://doi.org/10.1016/B978-0-12-404630-6.00027-0.

Apaydin-Varol, E., Polat, S., and Putun, A.E. (2014). Pyrolysis kinetics and thermal decomposition behavior of polycarbonate—a TGA-FTIR study. *Thermal Science* 18 (3): 833–842. https://doi.org/10.2298/TSCI1403833A.

Araújo, E.S., Khoury, H.J., and Silveira, S.V. (1999). Effects of gamma-irradiation on some properties of durolon polycarbonate. *Radiation Physics and Chemistry* 53 (1): 79–84. https://doi.org/10.1016/S0969-806X(97)00300-9.

Aschberger, K., Munn, H., Olsson, S. et al. (eds.). (2010). *European Union Risk Assessment Report—BPA (4,4′-Isopropylidenediphenol (Bisphenol-A)), Assessment.* https://doi.org/10.2788/40195.

Bello Lamo, M.P., Williams, P., Reece, P. et al. (2014). Study of gamma irradiation effect on commercial TiO2 photocatalyst. *Applied Radiation and Isotopes* 89: 25–29. https://doi.org/10.1016/j.apradiso.2014.02.001.

Chiang, K., Lim, T.M., Tsen, L., and Lee, C.C. (2004). Photocatalytic degradation and mineralization of bisphenol A by TiO2 and platinized TiO2. *Applied Catalysis A: General* 261 (2): 225–237.

Chung, J.Y.J. (1996). Stabilization of polycarbonate against gamma-irradiation. *Annual Technical Conference–ANTEC, Conference Proceedings.*

Cision Pr Newswire (2016). *Bisphenol-A—A Global Market Overview.*

Covestro (2018). *Makrolon Product Range: Typical Values.*

da Silva, D.J., Tsuyama Escote, M., Cruz, S.A. et al. (2018). Polycarbonate/TiO2 nanofibers nanocomposite: preparation and properties. *Polymer Composites* 39 (S2): 780–790. https://doi.org/10.1002/pc.24232.

De la Colina Martínez, A.L., Martínez-Barrera, G., Barrera-Díaz, C.E. et al. (2019). Recycled polycarbonate from electronic waste and its use in concrete: effect of irradiation. *Construction and Building Materials* 201: 778–785. https://doi.org/10.1016/j.conbuildmat.2018.12.147.

De la Colina-Martínez, A.L., Martínez-Barrera, G., Barrera-Díaz, C.E. et al. (2017). Characterization of recycled polycarbonate from electronic waste and its use in hydraulic concrete: improvement of compressive performance. *Advances in Concrete Construction* 5 (6): 563–573. https://doi.org/10.12989/acc.2017.5.6.563.

de Melo, N.S., Pondé Weber, R., and Miguez Suarez, J.C. (2007). Toughness behavior of gamma-irradiated polycarbonate. *Polymer Testing* 23 (3): 315–322. https://doi.org/10.1016/j.polymertesting.2006.10.014.

Dorn, P.B., Chou, C.S., and Gentempo, J.J. (1987). Degradation of bisphenol A in natural waters. *Chemosphere* 16 (7): 1501–1507. https://doi.org/10.1016/0045-6535(87)90090-7.

EPA (2015). *Migration of Bisphenol A from Polycarbonate Plastic of Different Qualities.* The Danish Environment Protection Agency.

European Parliament (2011). Directive 2011/65/EU of the European Parliament and of the Council of 8 June 2011 on the restriction of the use of certain hazardous substances in electrical and electronic equipment. *Official Journal of the European Union* 54 (L 174): 88. https://doi.org/10.1017/CBO9781107415324.004.

European Parliament (2012). Directive 2012/19/EU of the European Parliament and of the Council of 4 July 2012 on waste electrical and electronic equipment (WEEE). *Official Journal of the European Union* 55 (L 197): 38. https://doi.org/10.3000/19770677.L_2012.197.eng.

Garay-Rodríguez, L.F., Zermeñob, B., López de la Ob, K.A. et al. (2018). Photocatalytic degradation of bisphenol A: kinetic studies and determination of the reaction pathway. *Journal of Applied Research and Technology* 16: 334–245. https://doi.org/10.22201/icat.16656423.2018.16.5.737.

Gobierno de la Ciudad de México (2020). *Bases de datos—Red Automática de Monitoreo Atmosférico (RADIACIÓN).*

Güven, O., Alacakir, A., and Tan, E. (1997). An atomic force microscopic study of the surfaces of polyethylene and polycarbonate films irradiated with gamma rays. *Radiation Physics and Chemistry* 50 (2): 165–170. https://doi.org/10.1016/S0969-806X(97)00002-9.

Hacioğlu, F., Özdemir, T., Kinalır, K.G. and Usanmaz, A. (2016). Possible use of bisphenol-a polycarbonate in radioactive waste embedding. *Progress in Nuclear Energy* 90: 98–104. https://doi.org/10.1016/j.pnucene.2016.03.012.

Hokenek, S. (2009). Characterization of conductive polycarbonate films, M.S. dissertation. University of South Florida, USA.

Honsberg, C. and Bowde, S.G. (2020). *Photovoltaics Education Website*, www.pveducation.org.

IHS Markit (2018). *Polycarbonate Resins Chemical Economics Handbook.*

Jahantigh, F. and Nazirzadeh, M. (2017). Synthesis and characterization of TiO2 nanoparticles with polycarbonate and investigation of its mechanical properties. *International Journal of Nanoscience* 16 (1): 1–6. https://doi.org/10.1142/S0219581%D7;17500120.

Kalkar, A.K., Kundagol, S., Chand, S., and Chandra, S. (1992). Effect of gamma-irradiation on structural and electrical properties of poly (bisphenol-A carbonate) films. *International Journal of Radiation Applications and Instrumentation.* Part 39: 435. https://doi.org/10.1016/1359-0197(92)90056-L.

Katiyar, N. and Balasubramanian, K. (2014). Thermal modelling of hybrid composites of nano cenosphere and polycarbonate for a thermal protection system. *RSC Advances* 4: 47529–47535. https://doi.org/10.1039/c4ra07973f.

Krasnansky, V.J., Achhammer, B.G., and Parker, M.S. (1961). Effect of gamma radiation on chemical structure of plastics. *Polymer Engineering & Science* 1 (3): 133–138. https://doi.org/10.1002/pen.760010308.

Kumar, R., Singh, P., Gupta, S.K. et al. (2017). Radiation induced nano-scale free volume modifications in amorphous polymeric material: a study using positron annihilation lifetime spectroscopy. *Journal of Radioanalytical and Nuclear Chemistry* [Preprint]. https://doi.org/10.1007/s10967-017-5510-9.

Mayo, D.W., Miller, F.A., and Hannah, R.W. (2004). *Course Notes on the Interpretation of Infrared and Raman Spectra.* John Willey & Sons.

Méndez-Liñán, L., López-Garzón, F.J., Domingo-García, M., and Pérez-Mendoza, M. (2010). Carbon adsorbents from polycarbonate pyrolysis char residue: hydrogen and methane storage capacities. *Energy and Fuels* 24: 3394–3400. https://doi.org/10.1021/ef901525b.

Mercea, P. (2009). Physicochemical processes involved in migration of bisphenol a from polycarbonate. *Journal of Applied Polymer Science* 11: 579–593. https://doi.org/10.1002/app.29421.

Metz, C.M. (2016). Bisphenol a: understanding the controversy. *Workplace Health and Safety* 64 (1): 28–36. https://doi.org/10.1177/2165079915623790.

Mokhtari, S.A., Farzadkia, M., Esrafili, A. et al. (2016). Bisphenol A removal from aqueous solutions using novel UV/persulfate/H2O2/Cu system: optimization and modelling with central composite design and response surface methodology. *Journal of Environmental Health Science and Engineering* 14 (1): https://doi.org/10.1186/s40201-016-0255-x.

Mordor Intelligence (2018). *Polycarbonate Market—Segmented by Product Type, End-User Industry and Geography—Growth, Trends and Forecast (2018–2023).*

National Institute for Public Health and the Environment (2016). *Bisphenol A Part 2. Recommendations for Risk Management.*

Navarro-González, R. and Aliev, R. (2001). Mechanism of radiation-induced degradation of bisphenol-A polycarbonate. *Revista de la Sociedad Química de México* 45 (4).

Ohko, Y., Ando, I., Niwa, C. et al. (2001). Degradation of bisphenol A in water by TiO2 photocatalyst. *Environmental Science and Technology* 35 (11): 2365–2368. https://doi.org/10.1021/es001757t.

Parshin, A., Gunyakov, V.A., Zyryanov, V.Y., and Shabanov, V.F. (2013). Domain structures in nematic liquid crystals on a polycarbonate surface. *International Journal of Molecular Sciences* 14 (8): 16303–16320.

PerkinElmer (2010). *Thermogravimetric Analysis (TGA).*

Pondé Weber, R., Vecchio, K.S., and Miguez Suarez, J.C. (2010). Dynamic behavior of gamma-irradiated polycarbonate. *Matéria* 15 (2): 218–224.

Prospector (2016). *Lexan 143R.*

Ribeiro, A., Santos, V.D., and Brandalise, R.N. (2018). Assessment of a system for the prediction of water permeability coefficient in pervious concretes. *Ceramica* 67 (372): https://doi.org/10.1590/0366-69132018643722445.

Scientific Committee on Emerging and Newly Identified Health Risks (2015). *The safety of the use of bisphenol A in medical devices.* https://doi.org/10.2772/75546.

Shareef, A., Angove, M.J., Wells, J.D., and Johnson, B.B. (2006). Aqueous solubilities of estrone, 17β-estradiol,

17α- ethynylestradiol, and bisphenol A. *Journal of Chemical and Engineering Data* 51 (3): 879–881. https://doi.org/10.1021/je050318c.

Shintani, H. (2001). Determination of the endocrine disrupter bisphenol-A in the blood of uremia patients treated by dialysis. *Chromatographia* 53: 331–333. https://doi.org/10.1007/BF02490435.

Si, H.Y., Xiang, T.C., and Wang, R.T. (2014). Effects of pH and temperature on the degradation of polycarbonate in water. *Applied Mechanics and Materials*. https://doi.org/10.4028/www.scientific.net/AMM.522-524.346.

Sinha, D., Sahoo, K.L., Sinha, U.B. et al. (2004). Gamma-induced modifications of polycarbonate polymer. *Radiation Effects and Defects in Solids* 159 (10): 587–595. https://doi.org/10.1080/10420150412331330539.

Statista (2016). Production of polycarbonates worldwide by region in 2016 (in million metric tons). Global and European production of polycarbonates 2016. https://www.statista.com/statistics/650318/polycarbonates-production-worldwide-and-in-europe.

Sweileh, B.A., Al-Hiari, Y.M., Kailani, M.H., and Mohammad, H.A. (2010). Synthesis and characterization of polycarbonates by melt phase interchange reactions of alkylene and arylene diacetates with alkylene and arylene diphenyl dicarbonates. *Molecules* 15: 3661–3682.

Tayel, A., Zaki,M.F,El Basaty, A.B., and Hegazy, T.M. (2015). Modifications induced by gamma irradiation to Makrofol polymer nuclear track detector. *Journal of Advanced Research* 6 (2): 219–224. https://doi.org/10.1016/j.jare.2014.01.005.

Troiano, R. and Goodman, W. (2009). *UHPLC Separation and Detection of Bisphenol A (BPA) in Plastics. Application Note: Liquid Chromatography*. PerkinElmer.

Universidad de Barcelona (2019). *Materials: Policarbonato*. Available at: http://www.ub.edu/cmematerials/es/content/policarbonato (accessed 22 March 2019).

Vandenberg, L.N., Colborn, T., Hayes, T.B. et al. (2012). Hormones and endocrine-disrupting chemicals: low-dose effects and nonmonotonic dose responses. *Endocrine Reviews* 33 (3): 378–455. https://doi.org/10.1210/er.2011-1050.

Vom Saal, F.S., Nagel, S.C., Timms, B.G., and Welshons, W.V. (2005). Implications for human health of the extensive bisphenol a literature showing adverse effects at low doses: a response to attempts to mislead the public (multiple letters). *Toxicology* 212(2–3): 244–252. https://doi.org/10.1016/j.tox.2005.05.006.

Wang, R., Ren, D., Xia, S.et al.(2009). Photocatalytic degradation of Bisphenol A (BPA) using immobilized TiO2 and UV illumination in a horizontal circulating bed photocatalytic reactor (HCBPR).*Journal of Hazardous Materials*169(1–3): 926–932. https://doi.org/10.1016/j.jhazmat.2009.04.036.

Welshons, W.V., Nagel, S.C., and Vom Saal, F.S. (2006). Large effects from small exposures. III. Endocrine mechanisms mediating effects of bisphenol A at levels of human exposure. *Endocrinology* [Preprint]. https://doi.org/10.1210/en.2005-1159.

Wróbel, A.M., Kryszewski, M., Rakowski, W. et al. (1978). Effect of plasma treatment on surface structure and properties of polyester fabric. *Polymer* 19 (8): 908–912. https://doi.org/10.1016/0032-3861(78)90197-0.

Yadav, R., Naebe, M., Wang, X. et al. (2017). Structural and thermal stability of polycarbonate decorated fumed silica nanocomposite via thermomechanical analysis and in-situ temperature assisted SAXS. *Scientific Reports* 7 (7706). https://doi.org/10.1038/s41598-017-08122-7.

Yamamoto, T. and Yasuhara, A. (1999). Quantities of bisphenol A leached from plastic waste samples. *Chemosphere* 38 (11): 2569–2576. https://doi.org/10.1016/S0045-6535(98)00464-0.

Yang, C.W. (2015). Degradation of bisphenol A using electrochemical assistant Fe(II)-activated peroxydisulfate process. *Water Science and Engineering* 8 (2): 139–144. https://doi.org/10.1016/j.wse.2015.04.002.

18

Hydrometallurgical Processing of Electronic Waste

Muammer Kaya and Shokrullah Hussaini*

Eskisehir Osmangazi University, Mining Engineering Department, Eskisehir, Turkey
** Corresponding author*

18.1 Introduction

E-waste is considered one of the fastest-growing municipal solid waste streams in the world in terms of both volume and potentially hazardous environmental impact. In recent years, the increasing generation of waste electrical and electronic equipment (WEEE) and e-waste has been one of the most pressing environmental challenges. Figure 18.1 illustrates global e-waste creation in kilograms per inhabitant from 2014 to 2019, as well as e-waste generation estimates until 2030. Globally in 2019, 53.6 Mt of e-waste was created and only 17.4% of it was documented, collected, and legally recycled (Jadhao et al. 2022). On the other hand, illegal dumping, inadequate processing (open burning and uncontrolled chemical leaching), and landfilling account for the main disposal techniques used by the informal sector's recycling. The WPCBs (waste printed circuit boards), the most valuable part of e-waste, despite accounting for only about 4–7% by weight, represent a significant portion, about 40%, of the total e-waste metal recovery value (Kaya 2016; 2018a; 2019). E-waste comprises a wealth of valuable resources as well as a variety of hazardous elements (i.e. Cd, Pb, Hg, Cr, etc.), making it both a desirable secondary source and potential environmental contamination. Therefore, recovering noble metals (i.e. Pt, Au, Pd, Te, Ta, Se, Ge, etc.) from e-waste should be a top priority. WPCBs are diverse and extremely complex materials, the composition of which varies depending on their specific type, class, product, brand, and year of production. With time, the composition of PCBs is continuously changing, making it more difficult to obtain a stable material composition (Kaya 2016). The realization of recycling, on the other hand, should be based on maximum recovery and minimal negative environmental impact. The cycle from EEE to e-waste with reuse is demonstrated in Figure 18.2 (Islam et al. 2020; Kaya 2019).

In recent years, the formal recycling process of e-waste has been mainly executed throughout metallurgical routes. Chemical, physical, hydrometallurgical, pyrometallurgical, biometallurgical, and thermochemical processes, as well as combinations of these, can be used to recover metals from e-waste (Dolker and Pant 2019; Wang et al. 2017). The pyrometallurgical process involves the smelting of metals in a series of high-temperature furnaces. This procedure is well defined and allows the processing of e-waste on a large scale. Despite widely known drawbacks, such as high energy consumption, additional filtration system requirements due to the generation of harmful gases (such as dioxins and furans), and further necessity of purification of the products by hydrometallurgy, pyrometallurgy is still considered the conventional technique for e-waste recycling (Kaya 2019).

The hydrometallurgical method is more selective in terms of metal recovery from e-waste, has an easier to manage reaction, and is less hazardous to the environment than other approaches. Leaching is done in hydrometallurgical processes with a variety of solvents, including mineral/organic acids, alkali solutions, brines, and ionic liquids, and metals are leached off as soluble salts. Agents that cause leaching include cyanide, halide, thiosulfate, and thiourea, which are the most frequent for recovering metals from e-waste using this method (Chakraborty et al. 2022). Hydrometallurgical processes generate a huge amount of sludge and liquid waste, which must be carefully disposed of. Another disadvantage of the method is its slow leaching kinetics. Electro-oxidation can reduce acid usage while allowing important metals to be recovered without loss. The first phase in a hydrometallurgical process is leaching,

Electronic Waste Management: Policies, Processes, Technologies, and Impact, First Edition. Edited by Sunil Kumar and Vineet Kumar.

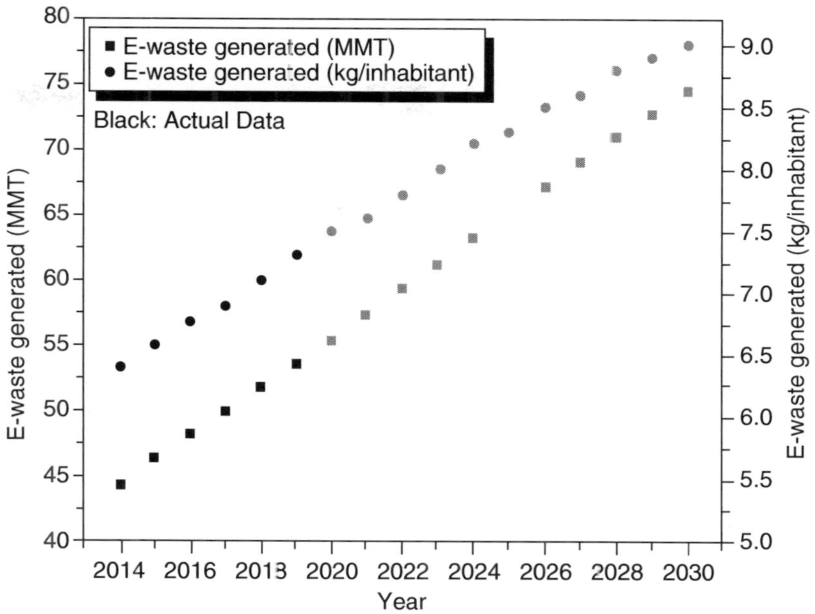

Figure 18.1 The total amount of e-waste produced over the world (Jadhao et al. 2022 / with permission of Elsevier).

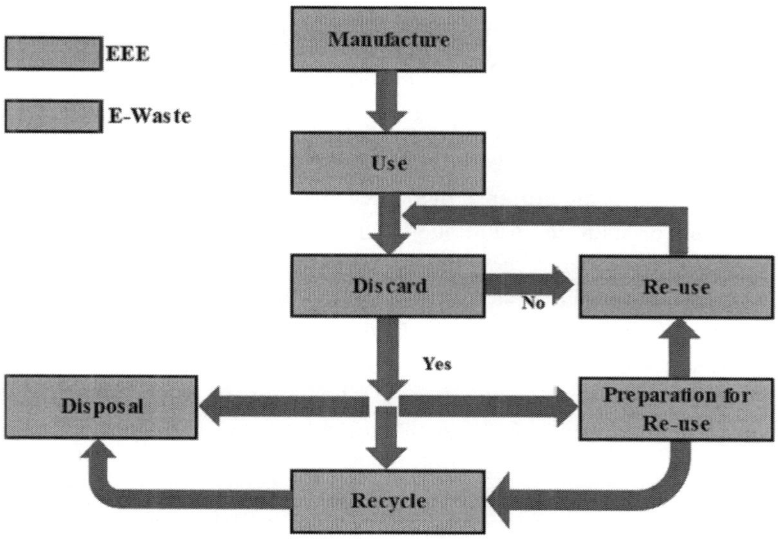

Figure 18.2 The recycling flow from EEE to e-waste.

and it is also the most crucial step in recovering base and precious metals from e-waste. The research on recovering metals from e-waste using the hydrometallurgical method began in the 1970s, with the initial intention of recovering precious metals like Au and/or Ag.

18.2 Recovery of Base Metal

Waste printed circuit boards (WPCBs) are rich secondary resources and important pieces of e-waste. PCBs are typically comprised of 30% ceramics (i.e. alumina, silica, mica, etc.), 30% organics (i.e. flame retardants, polymers, glass fiber, etc.), and 40% metals (i.e. Pb, Cd, Cu, Zn, Sn, and precious metals Au, Ag, etc.) (Feng et al. 2022). Previous base metal leach reagent media, conditions, and extraction results were summarized by Kaya (2019). In general, the first stage in hydrometallurgical processing is leaching, which involves dissolving the elements of the WPCBs in an appropriate lixiviant to generate a pregnant leach solution (PLS). For the first-step leaching of base metals (i.e. Cu, Pb, Sn, Zn, and Cd) from WPCBs, leaching with acid and an oxidant reagent is commonly utilized. Acids such as HNO_3, H_2SO_4, HCl, $HClO_4$, and aqua regia, together with different oxidants such as O_2, H_2O_2, Cl_2, and Fe^{3+}, are used to leach base metals, particularly Cu. The use of H_2O_2 is eco-friendly because it decomposes to water and oxygen gas. Organic EDTA and citric acid can also be used for BM extraction (Kaya 2016; 2016a; 2017, 2018, 2019a, 2019; 2021).

Recovery of Cu from WPCBs with both leaching and physical processing was investigated by Silvas et al. (2015). H_2SO_4 was used in the first step of leaching (1.0 M, 75°C, 1/10 S/L ratios in 4 h), and in the second step of leaching, oxidant (240 ml H_2O_2) was added to media under the same conditions. The pH was between 0 and 0.5 during the testing, while the ORP was between 220 and 280 mV (Silvas et al. 2015). Peroxysulfuric acid (H_2SO_5) functions as a significant oxidant for Cu in the presence of H_2SO_4 and H_2O_2:

$$H_2SO_4 + H_2O_2 \rightarrow H_2SO_5 + 2H_2O \quad (18.1)$$

$$Cu + H_2SO_5 \rightarrow Cu^{2+} + SO_4^{2-} + H_2O \quad (18.2)$$

Molecular oxygen is adsorbed onto the Cu surface as a result of the enhanced rate of H_2O_2 decomposition. The presence of oxygen on the surface of Cu increases the H_2SO_4 interaction (Yang et al. 2011).

Jadhao et al. (2021) recovered 100% Cu and 90% Ni from WPCB in alkali leaching (90 g/L ammonia, 180 g/L $(NH_4)_2SO_4$, 0.4 M H_2O_2, 20 ml/g L/S ratio, 4 h, 80°C and 700 rpm). They determined that H_2O_2 had a significant effect on dissolution. Additionally, the kinetic research was conducted using the shrinking core model, which demonstrated that the rate-controlling step for Ni and Cu extraction is interior diffusion. Finally, Cu with 98.38% purity was recovered on the Al cathode utilizing EW at a constant direct current voltage of 2.0 V (Jadhao et al. 2021). Recovery of Cu from WPCBs in alkaline media is sensitive to the pH, according to Li et al. (2021), and recovery of copper increases as the concentrations of Cu, initial pH, and temperature. They observed that Mextral 84H performed better than Mextral 54–100 in solvent extraction, although it requires a greater H_2SO_4 concentration for Cu stripping. Total Cu extraction was 91.6% with 15% Mextral 84H following 200 g/L H_2SO_4 stripping and 86.7% with 30% Mextral 54–100 following 80 g/L H_2SO_4 stripping (Li et al. 2021).

In the Cu–NH_3–H_2O system from the Eh-pH diagram, various soluble components such as $NH_4^{+,}$ NH_3, Cu^{2+}, H^+, $Cu(OH)_3^-$, Cu^+, $Cu(OH)_4^{2-}$, $CuNH_3^+$, $Cu_2(OH)_2^{2+}$, $Cu(NH_3)_2^{2+}$, $Cu(NH_3)^{2+}$, $Cu(NH_3)_4^{2+}$, and $Cu(NH_3)_3^{2+}$ coexist in the solution (Jadhao et al. 2021; Radmehr et al. 2013). Equations (18.3) and (18.4) can be used to express the total reaction for Cu and Ni extraction (Jadhao et al. 2021; Zuniga et al. 2010).

$$Cu + 1/2O_2 + 2NH_3 + 2NH_4^+ \rightarrow Cu(NH_3)_4^{2+} + H_2O \quad (18.3)$$

$$Ni + 1/2O_2 + 4NH_3 + 2NH_4^+ \rightarrow Ni(NH_3)_6^{2+} + H_2O \quad (18.4)$$

Lisinska et al. (2018) discovered that 5.0 M H_2SO_4 and 30% H_2O_2 considerably dissolved Cu in 6 hours and 70°C temperature of leaching; however, H_2O_2 had no effect on Al, Zn, or Fe dissolving (Lisinska et al. 2018). For Cu and Ni recovery from WPCBs, Kinoshita et al. (2003) performed two-stage leaching and SX (solvent extraction). In the first-stage leaching for 72 hours leaching time at 80°C of leaching temperature and using 0.1 M HNO_3 concentration, 98% Ni was effectively leached. In the second stage, the remaining Cu was recovered at 90°C leaching temperature for 6 hours leaching duration with a 1.0 M HNO_3 concentration. In solids, 99% purity Au was extracted in flakes. Cu was recovered by LIX984 solvent extraction and 4.0 M HNO_3 stripping (Lisinska et al. 2018).

When it comes to Cu ore, extraction by the hydrometallurgical method is well-established; nevertheless, there is no single flow chart that might be used to recycle Cu from WPCBs on a large scale. Cu is often dissolved in sulfuric acid media to form a Cu-PLS, which is then subjected to SX to improve Cu purity. Electrowinning is used to achieve pure Cu at the end. Qiu et al. (2020) and Choubey et al. (2021) investigated the recovery of Cu from WPCBs by electrolysis. Cu ion is generated from Cu at the anode and then proceeds to the cathode, where it receives electrons and is separated from Cu. In the electrolysis process, the lower the metal's standard electrode potential, the weaker the oxidized state's oxidation ability and the greater the reduced state's reduction ability. Figure 18.3 shows a schematic diagram of Cu separation by electrolysis (Choubey et al. 2021; Qiu et al. 2020). Zhang et al. (2018b) determined that without pretreatment, one-step $[BSO_3HP_y]HSO_4$ slurry electrolysis can effectively extract Cu metal powders from WPCBs. Current efficiency, recovery rate of Cu, particle size of Cu powders, and purity were 70.68%, 90.94%, 2.30 μm, and 81.69%, respectively, when 10% H_2SO_4 was replaced with $[BSO_3HP_y]HSO_4$.

In research on the dissolution of metals from WPCBs, it was discovered that using 2.0 M H_2SO_4 and 0.2 M H_2O_2 at 85°C, 100% of the Cu and Zn were recovered after 8 hours; nevertheless, by 12 hours, 95% of the Al, Ni, and Fe had been recovered. Whereas in the NH_4OH–$CuSO_4$–$(NH_4)_2S_2O_3$ and H_2SO_4–H_2O_2 systems, Pb dissolution was slow, therefore, the solid residue, primarily $PbSO_4$, was leached with NaCl to produce $PbCl_2$, which was then separated from the liquid (Oh et al. 2003). The effects of the beginning Cu concentration, the temperature, and particle size on the leaching of Cu were investigated by Yang et al. (2011). The results showed that small fragments (–1 mm) were treated effectively, and that beginning Cu concentration and temperature had no influence on leaching of Cu (Yang et al. 2011).

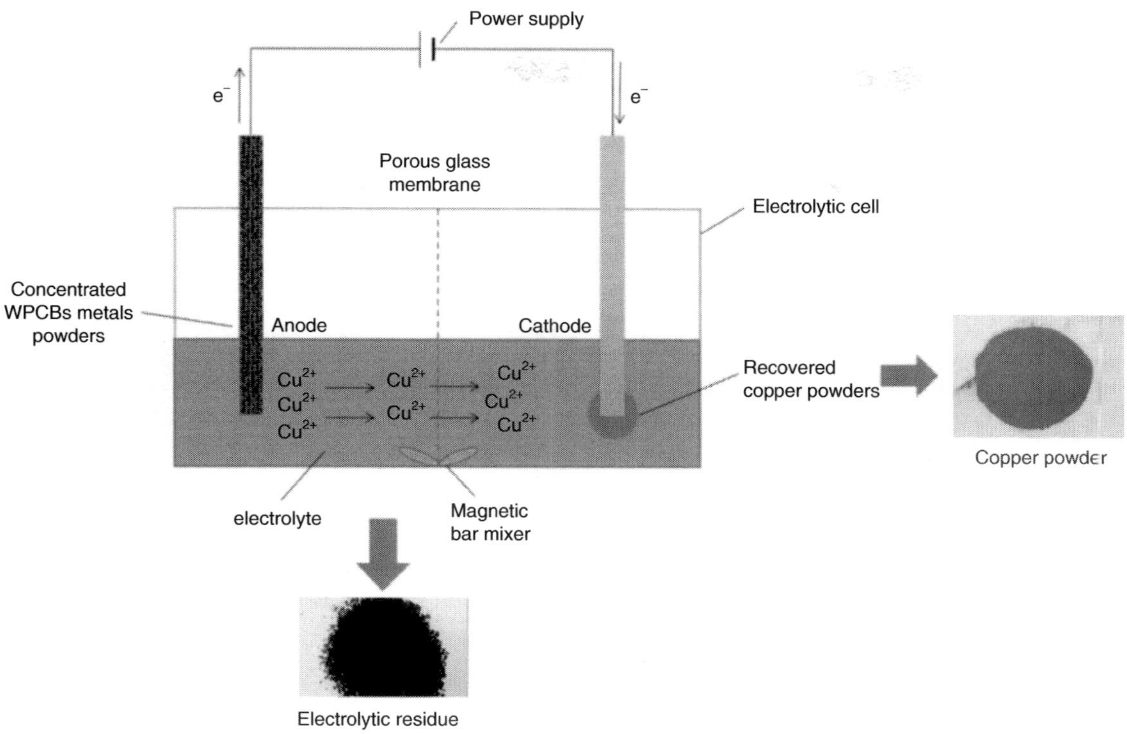

Figure 18.3 Schematic diagram of electrolytic recovery of Cu from the WPCB (Qiu et al. 2020 / with permission of Elsevier).

Furthermore, the high mixing rate had a detrimental influence on Cu leaching, which may be related to the fact that the increased mixing speed produced degradation of H_2O_2 (Yang et al. 2009).

18.3 Solder Leach

Sn, Pb, and Cu can be leached from PCB solder using inorganic acids (i.e. HNO_3, HCl, or HBF_4), organic acids (methanesulfonic acid (MSA)), and alkaline solution (NaOH). HNO_3 extracts both Pb and Cu, but could not extract Sn. HNO_3 and Sn form insoluble metastannic acid (H_2SnO_3), which can be easily dissolved by HCl at high temperature. However, the use of HCl presents hazardous gas evaluation (Kaya 2019). Almost full Pb and 98.7% Sn extraction at 90°C and L/S ratio of 20 in 120 minutes using 0.2M HNO_3 and 3.5M HCl was achieved by Jha et al. (2012). 3.5M MSA+0.5M H_2O_2 were used for selective desoldering of the electronic components from WPCBs in 45 minutes by Zhang et al. (2017b).

18.4 Recovery of Precious Metals

Precious metals (PMs) have a significant role in the world economy since they are used in a wide range of industrial products and activities. About 250 000 kgs of gold, among the most costly precious metals, are utilized in the electronics industry each year (almost 8% of global Au production) (Birloaga and Vegliò 2022). According to reports, the amount of silver, gold, and palladium in PCBs is 0.05–0.1%, 0.03%, and 0.01%, respectively, which is 10 times higher than the mass percentages of rich materials (Huang et al. 2022). The recycling of PMs from e-waste is critical. WPCBs, for example, have a 35–50 times greater Au concentration than natural gold ore, despite the fact that the Au content of PCBs is declining. Following the recovery of base metals, the extraction of precious metals from WPCBs includes leaching, purification, and recovery.

Due to the PM content, WPCBs have innate worth. PMs account for 80% of WPCB's intrinsic worth, which is less than 1% of its total value. Pretreatment, including chemical and physical pretreatments, is required before PM leaching from WPCBs. After dismantling the electric components (ECs), the next steps are to mechanically split and grind PCB scrap, and then enrich non-metals and metals using one of the methods, such as magnetic separation (MS), pneumatic separation (PS), ECS, electrostatic separation (ES), and screening depending on the differences in physical characteristics of non-metals and metals. Chemical pretreatment involves dissolving base metals like Pb, Cu, and Sn in solution, keeping precious metals in the residue, then isolating and purifying precious metals from the residue. H_2O_2 + H_2SO_4 pretreatment or HNO_3 pretreatment are two commonly utilized chemical pretreatments. A typical flow chart for recovering precious metals from WPCBs is shown in Figure 18.4.

Because of the strong metal complex generated, hydrometallurgical methods allow selective extraction of

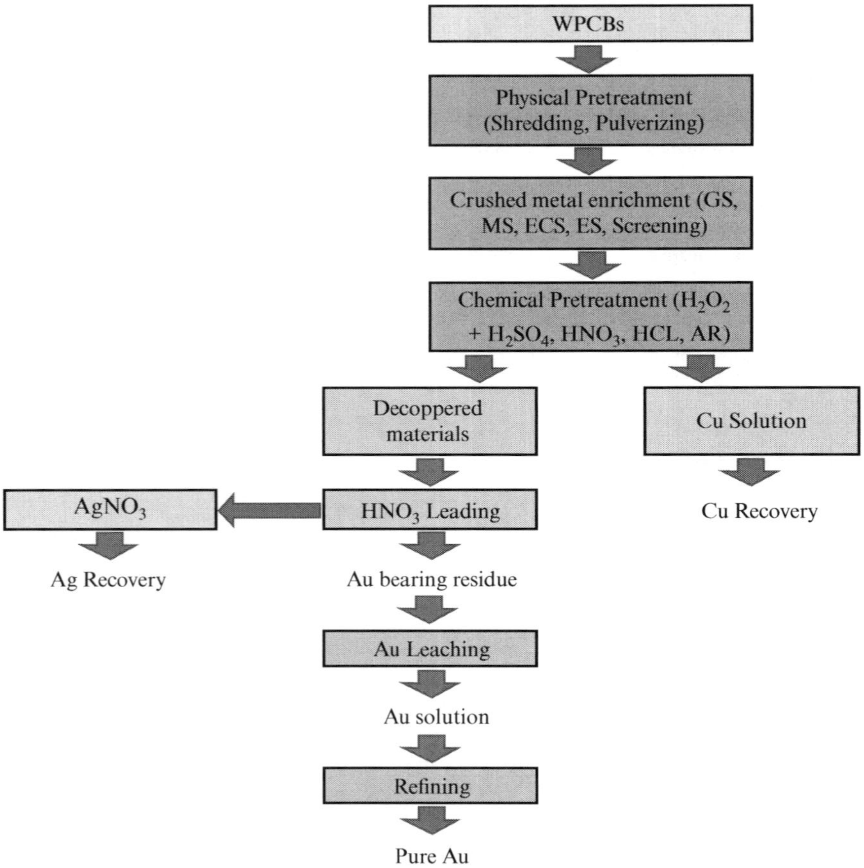

Figure 18.4 A typical precious metal recovery flow chart from WPCBs (GS: Gravity separation, MS: Magnetic separation, ECS: Eddy current separation, and ES: Electrostatic separation).

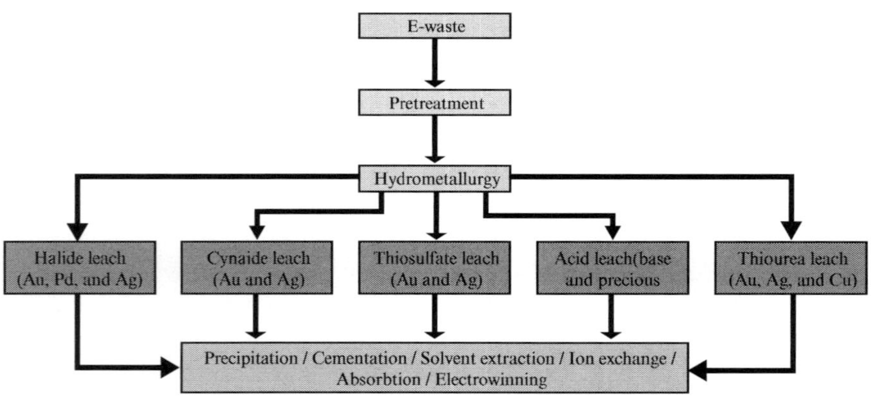

Figure 18.5 Recovery of precious metals by the different hydrometallurgical methods.

precious metals from resources utilizing a chemical solution based on a variety of reagents such as thiosulfate, AR, halogens, thiourea, and cyanide. Figure 18.5 illustrates the whole strategy for precious metal recovery utilizing hydrometallurgical techniques.

For more than a decade, cyanide as a gold lixiviant has controlled the mining industry because of its benefits, which include great efficiency, cheap cost, reduced doses, operation in an alkaline media, etc. Because the cyanide ion is steady at pH 10.2, most of the leaching processes of cyanic take place at this pH. Under pH 8.2, cyanide occurs as pretty unstable hydrogen cyanide (HCN), resulting in cyanide loss and danger to operators' safety. According to Islam et al. (2020), a sufficient amount of oxygen was required to complete the reaction of leaching with cyanide (Islam et al. 2020). The idea of cyanidation is to extract Au and Ag from the WPCB into an aqueous solution using alkali metal cyanide, then extract Au and Ag from the solution of cyanide through reduction. Leaching with cyanide is just effective for Au/Ag leaching of WPCBs. The gold/silver contained inside WPCBs is difficult to dissolve with cyanide. Although leaching with cyanide has still been widely used in the Au mining sector, the process produces a lot of cyanide-containing wastewater, which is hazardous to both the environment and operators. Some noncyanide leaching mechanisms, such as thiosulfate leaching, halide leaching, and thiourea leaching, have recently been receiving greater attention (Yang et al. 2012; Zhang et al. 2012).

There is a need for the advancement of an acceptable technique for recovering Au from secondary resources that have key characteristics such as selectivity, high recovery, economy, and environmental friendliness, even when used on a small scale. Hydrometallurgical, microbiological, electrochemical, pyrometallurgical, or mechanical methods, as well as their combinations, are currently accessible for Au recovery. Nevertheless, the majority of these technologies have environmental hazards and necessitate considerable material preprocessing (Huang et al. 2022). Although hydrometallurgical techniques have proven to be effective for leaching Au from WEEE, recovering Au from solution is frequently time-consuming and involves many steps.

18.4.1 Cyanide Leaching for Recovery of Precious Metals

Cyanide leaching is a century-old conventional process that is commonly used for Au and Ag recovery because of its simplicity and efficiency. However, because of its toxicity to humans and harmful effects on the environment, this technique is no longer widely employed. Recently, some researchers have begun to use nontoxic cyanide to recover Au, Pd, Ag, and Pt. For example, Li et al. (2022) used nonhazardous potassium ferricyanide, and potassium permanganate as an oxidant and glycine solution to recover Au, Pd, Ag and Pt. They discovered that increasing the concentration of oxidant (0.04–0.16 M) and decreasing the particle size of the sample considerably improved Au extraction. However, the concentration of glycine (0.5–1 M), increasing the temperature (55°C), as well as adding the oxidant in stages, did not improve the extraction of Au appreciably. Using this technique, 86.8% Au, 89.3%, Pd, 87.9% Cu, and 70.2% Ag could be extracted under the specified conditions (Li et al. 2022). Cyanide is widely utilized due to its great efficiency, low cost, and ease of usage. Leaching by cyanide accounts for over 75% of global Au output (Egan et al. 2016).

According to the process of cyanide leaching, throughout leaching, the noble metals generate complexes of cyanide in the aqueous solution. The reactions that occur when Au and cyanide come together are listed below (Jadhao et al. 2022). Equations (18.5) and (18.6) indicate the whole reaction:

$$4Au + 2H_2O + O_2 + 8(Na/K)CN \rightarrow 4(Na/K)OH + 4(Na/K)\left[Au(CN)_2\right] \quad (18.5)$$

$$\text{Overall: } 4Au + 2H_2O + O_2 + 8CN^- \rightarrow 4OH^- + 4Au(CN)_2^- \quad (18.6)$$

18.4.2 Halide Leaching for Recovery of Precious Metals

Leaching using iodide, bromide, and chloride are all instances of halide leaching. Depending on the chemistry of the solution, Au generates complexes of Au^{3+} and Au^{+} with iodide, bromide, and chloride. Just chloride/chlorine has been used on a large scale in the halides industry, and in this media effect of increasing reaction duration and temperature is significant in the recovery of Au. In addition, increased chloride content, which forms $AuCl_4^-$, with a low degree of hydration, was found to improve Au(III) recovery efficiency (Islam et al. 2020; Lu and Xu 2016; Quinet et al. 2005). Behnamfard et al. (2013) investigated the recovery of precious metal from WPCBs in three stages; the use of chloride reagents including the $HCl\text{-}NaClO\text{-}H_2O_2$ system resulted in 100% Pd, 100% Au, and 71.36% Ag leaching (Behnamfard et al. 2013). The iodide-iodine leaching system's main problems are its high reagent use and high expense. Wang et al. (2021) determined that in comparison to aqua regia, a DMF-$CuCl_2$–$CaCl_2$ solution demonstrated improved selectivity of Au from PCBs of mobile phones (Wang et al. 2021). The mechanism of halide leaching is given in Equation (18.7).

$$2Au + 2L^- + L_2 \rightarrow 2ML_2^- \quad (18.7)$$

(L^-: oxidant, L_2: halogen)

18.4.3 Aqua Regia Leaching for Recovery of Precious Metals

Aqua regia is regarded as an effective and favored leaching procedure due to its strong oxidation capabilities. Aqua regia is a combination of hydrochloric and nitric acids generated in a 3:1 volume ratio (Manigandan et al. 2020). This combination has a high dissolution ability and is used to dissolve Pt and Au selectively. Park and Fray (2009) suggested employing aqua regia to recycle high-purity precious metals (Au, Pd, and Ag) from WPCBs. The best duration time and S/L ratio were 3 hours and 1/20 g/mL, respectively. They recovered 97% Au, 98% Ag, and 93% Pd. Equations (18.8)–(18.10) show the reaction process for recovering gold from aqua regia (Jadhao et al. 2022).

$$6HCl + 2HNO_3 \rightarrow 3Cl_2 + 4H_2O + 2NO \quad (18.8)$$

$$2Au + 3HNO_3 + 9HCl \rightarrow 2AuCl_3 + 6\,H_2O + 3NOCl \quad (18.9)$$

$$HNO_3 + 3HCl \rightarrow 2H_2O + NOCl \quad (18.10)$$

Nitric acid functions as a strong oxidant in Equation (18.9), and HCl interacts with Au to form chloroaurate anions. Platinum also can be written with a same reaction mechanism, as shown in Equation (18.11) (Manigandan et al. 2020).

$$Pt + 8H^{+} + 4NO^{3-}\ Pt^{4+} + 4H_2O + 4NO_2 \quad (18.11)$$

According to Sheng and Etsell (2007), higher temperatures result in faster rates of Au leaching, probably because the creation of chlorine is faster at higher temperatures, allowing the Au to be oxidized more quickly. In addition, they determined that in the aqua regia solution, no stirring was required for the leaching of Au (Sheng and Etsell 2007).

18.4.4 Thiosulfate Leaching for Recovery of Precious Metals

Gold recovering with thiosulfate has been note in several research studies. There are two types of thiosulfate usually used for the Au leaching in the thiosulfate process. Cu^{2+} reacts with ammonia to generate the cupric tetramine complex, which aids in the stabilization of the gold-thiosulfate combination. Then for Au recovery with thiosulfate, ammonia and Cu^{+2} are essential. As demonstrated in Equation (18.12), the interaction between thiosulfate and Au produces a gold-thiosulfate $[Au\,(S_2O_3)_2{}^{3-}]$ complex. The reaction that converts thiosulfate to tetrathionate is shown in Equation (18.13) (Chu et al. 2003).

$$Au + 5S_2O_3^{2-} + Cu(NH_3)_4^{2+} \rightarrow Au(S_2O_3)_2^{3-} + 4NH_3 + Cu(S_2O_3)_3^{5-} \quad (18.12)$$

$$8S_2O_3^{2-} + 2Cu(NH_3)_4^{2+} \rightarrow S_4O_6^{2-} + 8NH_3 + 2Cu(S_2O_3)_3^{5-} \quad (18.13)$$

The two types of thiosulfate usually used are ammonium thiosulfate and sodium thiosulfate, respectively (Breuer and Jeffrey 2003; Chu et al. 2003; Ha et al. 2014, 2010; Jadhao et al. 2022; Tripathi et al. 2012). Thiosulfate is a chemically unstable compound with a low rate of metal extraction. Cu^{2+} can be employed as a catalyst to improve recovery rates.

Thiosulfate and thiourea leaching processes are well-known for being both environmentally friendly and cost-effective. While sodium or ammonium thiosulfate is alkaline, preventing the development of a destabilizing sulfate radical that would otherwise arise in an acidic media, this is not the case for thiourea. Furthermore, thiosulfate-based leaching has several advantages, including good selectivity, nontoxicity and noncorrosivity. Despite its advantages, the challenges associated with the reagent consumption rate and Au extraction render the process uneconomical. Thiosulfate is stable in an alkaline environment since tetrathionate, the oxidation product of thiosulfate, will revert to thiosulfate in about 60% of cases. These hydrometallurgical techniques can also be used to extract other precious and valued metals such as Pt, Pd, and Ag (Manigandan et al. 2020).

Ha et al. (2010) and Tripathi et al. (2012) researched the leaching of Au and found that increasing thiosulfate concentration up to 0.1–0.14 M increased the amount of Au recovery (Ha et al. 2010; Tripathi et al. 2012). However, when the concentration of Au increases, the efficiency of Au leaching reduces. This could occur because greater thiosulfate levels produce undesired compounds such as tetrathionate, trithionate, and sulfate (Breuer and Jeffrey 2003). Ha et al. (2014) enhanced Au leaching with thiosulfate from PCB of waste mobile phones. For 2.395 × 10–5 mol/(m^2.s) leaching of Au, the optimal conditions were 0.266 ammonia, 0.01 M Cu^{2+}, and 0.07 M thiosulfate (Ha et al. 2014). When using a solution containing 0.04 M $CuSO_4$ and 0.1 M ammonium thiosulfate at pH 10–10.5 for Au recovery, Tripathi et al. discovered 50% greater Au recovery from non-crushed and untreated PCB than from crushed samples. The temperature of the reaction was kept constant at 25°C with a S/L ratio of 40 g/L for 8 hours. The availability of a larger surface area in shredded PCB leads to more dissolution of Cu and the formation of undesired compounds such as polythionates and tetrathionate. As a result, thiosulfate is utilized in the production of undesired chemicals, lowering the recovery of Au from shredded PCB (Jadhao et al. 2022; Tripathi et al. 2012). Ficeriová et al. (2011) showed poorer Au and Ag leaching efficiency from the noncrushed and untreated PCB sample as compared to the processed WPCB by milling, which contradicts the aforementioned findings (Ficeriová et al. 2011). They recovered 93% Ag and 98% Au from pretreated samples after 48 hours of duration time with 1.0 M ammoniac, 0.2 M $CuSO_4$, and 0.5 M ammonium thiosulfate at 40°C. In addition, leaching in aqua regia solution for 2 hours resulted in up to 90% Pd extraction.

Gold was extracted from WPCBs using a 20% (w/v) potassium persulfate solution heated to boiling for 20 minutes. The sludge from the peeled Au is filtered, cleaned, dried, and then melted with potassium nitrate and borax (Syed 2006). One of the most powerful oxidizing agents is the persulfate ion. Persulfate bivalent ions produce unstable anion sulfate under typical conditions, which looks to be a particularly powerful oxidizing agent with good exchange and can recover all base metals (Ni, Cu, etc.). Sulfate salts and base metals are extracted after filtration. The purity of the Au recovered is 99.5%.

18.4.5 Thiourea Leaching for Recovery of Precious Metals

The theory of employing thiourea for Ag and Au extraction was first established in 1941, though many scientists began actively exploring this subject in 1977. Thiourea research has dominated headlines around the world in recent decades (Manigandan et al. 2020). Due to its remarkable extraction rates, excellent selectivity, and rapid kinetics, thiourea leaching has been examined for the recovery of precious metals. The shared electrons among sulfur and nitrogen atoms in thiourea give it an ability for forming a coordination bond with the Au and Ag. Because thiourea is soluble in acidic solutions, an acidic state should be provided during the leaching by thiourea process (Gurung et al. 2013; Jadhao et al. 2022). However, thiourea is cancerous (because of its limited stability in chemical solution) and sensitive to the base metals (As, Sb, Cu, and Pb) that are found in WPCBs. Owing to its large cost and consumption, thiourea is not a cost-effective method for Au recovery (Kaya 2019).

Many researchers have identified the kinetics of Au leaching in thiourea systems in the presence of various oxidants, including MnO_2, formamidine disulfide, $Fe_2(SO_4)_3$, H_2O_2, O_2, and Na_2O_2 (Batnasan et al. 2018). Batnasan et al. (2018) used $Fe_2(SO_4)_3$ as an oxidized agent for the recovery of precious metals, and they observed that Au and Pd leaching efficiency increases considerably. Increasing the concentration of Fe^{3+} to 5 g/L improved Au recovery (from 18% to 68%) and decreased Ag recovery (down to 1.6%); however, Pd dissolution increased a little (up to 7.1%). Fe^{+3} promotes leaching, according to Equations (18.14)–(18.16):

$$Au + Fe^{3+} + 2CS(NH_2)_2 \rightarrow Fe^{2+} + Au[CS(NH_2)_2]_2^+ \quad (18.14)$$

$$Ag + Fe^{3+} + 3CS(NH_2)_2 \rightarrow Fe^{2+} + Ag[CS(NH_2)_2]_3^+ \quad (18.15)$$

$$Pd + Fe^{3+} + 4CS(NH_2)_2 \rightarrow Fe^{2+} + Pd[CS(NH_2)_2]_4^+ \quad (18.16)$$

When compared to when they are not present, ferric ions can improve gold extraction by four times (Gurung et al. 2013). Increased oxidant has a negative effect on PMs recovery because thiourea oxidizes to generate formamidine disulfide and strong ferric sulfate complex, which quickly decomposes to sulfur, cyanimide, and thiourea, according to Equations (18.17)–(18.19). The sulfur produced forms a thin coating over the metal surface, limiting the mass transfer rate (Birloaga et al. 2014; Jadhao et al. 2022). Finally, this reduces Ag and Au recovery. Then for avoiding unwanted products, experimental control should be precise about which settings are required.

$$SC(NH_2)_2 + SO_4^{2-} + Fe^{3+} = (FeSO_4.SC(NH_2)_2)^+ \quad (18.17)$$

$$2Fe^{3+} + 2SC(NH_2)_2 = 2H^+ + 2Fe^{2+} + (SCN_2H_3)_2 \quad (18.18)$$

$$(SCN_2H_3)_2 = S + NH_2CN + SC(NH_2)_2 \quad (18.19)$$

The impact of thiourea content on the extraction of Au, Ag, and Pd is significant. With 12.5 g/L concentration of thiourea, the maximal extraction efficiencies of Au, Pd, and Ag were 91%, 11.9%, and 81%, respectively. For extracting precious metals, the presence of formamidine disulfide $(SCN_2H_3)_2$ is advantageous (Batnasan et al. 2018; Kaya 2019).

Birloaga et al. (2013) investigated the effect of Cu on Au recovery with thiourea extracted from WPCBs. They used leaching by oxidation to extract Cu from connection pins of the CPU and they were able to recover 90% of the Cu. Using 6, 10, and 20 g/L of ferric ions, sulfuric acid, and thiourea, respectively, with a solid/liquid ratio of 1/10, 84% Ag and 82% Au were extracted after oxidative leaching of Cu (Birloaga et al. 2013). Although the leaching experiment was done on WPCB without prior leaching by oxidation under the identical conditions of reaction, Au was not recovered. As a result, preparatory oxidative leaching is a crucial stage to improve the leaching of Au, and gold recovery after leaching by oxidation was almost 69%. Lee et al. (2018) investigated Au and Ag extraction using two types of leaching. In type one, the researchers used leaching by thiourea directly and found 100% Ag and 90% Au, respectively. In type two, to extract the base metals, H_2SO_4 leaching was used prior to leaching by thiourea. At the optimum conditions, 100% Ag and 98% Au were extracted in type two leaching (Lee et al. 2018). Table 18.1 summarizes the advantages and disadvantages of each leaching agent for PMs recovery.

Zhang et al. (2012) used the multi-criteria decision making, analytical hierarchy process (AHP) to compare various PMs leaching processes for the determination of leaching scores. The highest leaching score of 4.46 was obtained for cyanide leaching where economic feasibility and research level have the highest points, but the toxicity point is the lowest. Thiourea is the second-best leaching process for PMs from WPCBs.

Table 18.1 Disadvantages and advantages of various leaching lixiviants in hydrometallurgy.

Method	Disadvantages	Advantages
Cyanide leaching	Production of hazardous effluents	Successful gold leaching
	Extremely toxic	Low cost
Halide leaching	Need for specialist equipment	High rate of leaching
	Extremely corrosive	
	High cost	
Acid leaching	Production of hazardous effluents	High rate of metal leaching
	Corrosive	
	Toxic	
Thiosulfate leaching	pH sensitivity in the rate of leaching	Noncorrosive
		Low toxicity
	Slow in kinetics	
Thiourea leaching	Formation of by-product	Highly effective for extraction of precious metals
	High cost	
	High consumption of reagent	Less toxic
	Difficult monitoring of reaction parameters	

18.5 Supercritical Fluid Technology for Brominated Flame Retardant Leaching

Supercritical water or liquified CO_2 can be used as the solvent. Supercritical water (ScW) technology at a temperature above 374.3°C and pressure 22.1 MPa is generally regarded as a green/eco-friendly process. Because supercritical water has quite different physical properties (such as high diffusivity and solubility for organic compounds and has the decomposition power for brominated epoxy resin (BFR)) than normal water, ScW is an ideal pretreatment method for metals recovery from WPCBs. Li and Xu (2015) used ScW to decompose BFR and recover metals (99.80%) from WPCBs. However, PM separation needs further studies. Supercritical fluid extraction is faster, efficient, inexpensive, safe, environmentally friendly (non-toxic and non-flammable), does not emit hazardous substances, is readily available, and is recyclable. The critical point of CO_2 is 31.1°C and 7.38 MPa. CO_2 has a high diffusivity, low viscosity, high solubility, and no surface tension. Supercritical fluid extraction has high recovery rate of metals, recyclability of waste PCBs polymers, and production of gases of industrial interest (Liu et al. 2016b).

18.6 Purification and Separation of Metals from Leaching Solution

For purifying leaching solutions from metal pregnant leach solution (PLS) operations and achieving pure precious metal products, a variety of purification methods have been utilized, including cementation precipitation, adsorption, ion exchange, solvent extraction (SX), and electrowinning (EW) (Kaya 2019). Silver can be extracted using silver chloride precipitation, cementation with Al, and activated carbon adsorption. Alam et al. (2007) reported the removal of Ag by cementing it with Cu particles and then extracting it from an ammoniacal PLS with a solvent extraction method (Alam et al. 2007). Long Le et al. (2011) investigated cementation with Zn to remove impurities from thiourea leach solution toward Au and Ag extractions (Long Le et al. 2011).

The precious metals in the leach solution are concentrated and refined into an aqueous solution during the SX stage of the process. In the EW phase, the pure aqueous solution is employed to create a pure metal with high quality in physically and chemically characterization (Kaya 2019). Kinoshita et al. (2003) used LIX984 (200 g/L), LIX860IC (200 g/L), and LIX841 (200 g/L) as an organic solution for loading Cu from PLS. The S/L ratio was 100 g/L, and 11.18 g/L Cu was loaded to the organic phase in the process. They used 4.0 M HNO_3 for stripping Cu from the organic phase in 1 hour. In addition, the optimum pH was determined 1.0 and NaOH was used for arranging pH. They determined that 11.15 g/L Cu transferred to the aqueous phase for organic after the stripping experiment. 98% of the Cu was purified selectively, leaving the 4.3% Ni in the raffinate (Kinoshita et al. 2003). Lixiviant selection, pH, diluent, and organic extractant all have a significant impact on SX efficiency. For base metal (Cu/Ni) extraction, LIX 984 N or LIX 84 IC is commonly used as an SX lixiviant (Mir and Dhawan 2022).

Behnamfard et al. (2013) and Zhang and Zhang (2014) investigated the extraction of Pd from PLS with a 3.13–19.82 mg/L concentration of Pd, and 0.935 mg/L and 10.18 mg/L concentration of Au and Ag, respectively. They used sodium borohydrate and diisoamyl sulfide for precipitation and SX experiments (Behnamfard et al. 2013; Zhang and Zhang 2014). Kubota et al. (2019) used N-[N,N-di(2-ethylhexyl) amino-carbonylmethyl] glycine (D2EHAG) as both an extractant in the SX stage and a polymer inclusion membrane (PIM) integrated extractant to investigate the selective extraction of Au (96%) from aqua regia solution. In 48 hours, 1.0 M HCl and 0.1 M thiourea carried Au(III) PIM transport selectively and effectively (Kubota et al. 2019). The efficient loaded of Au(III) from chloride averment was successfully accomplished using membranes made up of 70% poly(vinylidene fluoride-co-hexafluoropropene) and 30% extractant with Cyphos IL 104 by Bonggotgetsakul et al. (2016).

Panda et al. (2020a) researched the recovery of Ag and Pd from multilayer ceramic capacitors (MLCCs). First, leaching in 2.0 M chloric acid was used to separate the nickel and concentrate the high purity Ag-Pd, which was then followed by leaching with 4.0 M HNO_3, potassium chloride precipitation, and finally, they used LIX 84IC in the SX step (Panda et al. 2020a). Cyganowski et al. (2017) investigated the extraction of Au from WPCBs by H_2SO_4 and HNO_3, followed by aqua regia and then used a kind of anion exchange (1-(2-aminoethyl) piperazine and ethylenediamine) for recovery of high purity Au. Pre-enrichment procedures, on the other hand, used up many inorganic acids in an unfavorable way (Cyganowski et al. 2017). The extraction of (>99%) Cu, Ni, and Ag from ICs was reported using HNO_3 leaching, KCl/NaCl precipitation, and a LIX 84IC / LIX 984 N liquid-liquid separation technique by Ajiboye et al. (2019) and Panda et al. (2020b).

18.7 Recovery of Rare Earth Elements from E-waste

Rare earth elements (REEs) are used in electrical equipment such as modern electronics, electrical motors, scanning equipment, medical diagnostics, and other applications due to their unique qualities of being extremely phosphorescent, magnetic, electrochemical conductive, and catalytic. REE recovery from e-waste is growing in popularity because these elements are required for modern and innovative renewable technologies (Gonzalez Baez et al. 2022; Ramprasad et al. 2022). REEs are mostly discovered as oxides in e-waste, like La_2O_3, Tb_2O_3, Ce_2O_3, Eu_2O_3, Y_2O_3, In_2O_3, and so on.

The very small fraction of REEs in each piece of equipment is one of the major hurdles in e-waste recycling. Because of this, there is difficulty recovering REEs from WEEE (Ambaye et al. 2020). Because REEs are more reactive than other metals, they will be removed in the waste phase if the pyrometallurgical method is used. Therefore, leaching should be used to recover REEs from e-waste.

Neodymium (Nd) is the most common and popular REE in most electrical devices. Neodymium alloys, which contain Fe and Br, are utilized as magnets in hard drives (HDs) of computers, automobiles, computers, wind turbines, audio devices, and loudspeakers. Magnets, often known as RE magnets, include about 27–31% (w/w) of REEs. In addition, Hard Disc Drives (HDDs) are made up of 2% magnets and an HDD contains 0.6% rare-earth materials (Kaya 2019; Lixandru et al. 2017; Ramprasad et al. 2022). Figure 18.6 shows some magnets from HDDs. The magnets contain a variety of additional REEs, including holmium (Ho), gadolinium (Gd), terbium (Tb), praseodymium (Pr), and dysprosium (Dy), based on the usage and composition of the alloy.

H_2SO_4 leaching at ambient temperature can be used to recycle Nd-based magnet scrap from HDD shredding. Selective leaching, oxidative roasting, and direct leaching are some of the methods that can be used. Equations (18.20)–(18.22) demonstrate reaction REEs with some acids (Ramprasad et al. 2022). They could be recovered as two types of salts after purification from the REE solution (Abrahami et al. 2015). Ueberschaar and Rotter (2015) investigated component compositions of HDDs. They discovered that for 544 g of HDD there are 2.6% NdFeB magnets and that 27.5 to 102.5 kg of NdFeB magnets can be recycled in a month (Ueberschaar and Rotter 2015).

$$nH+ + RE_2O_3 \rightarrow nH_2O + nRE^{3+} \tag{18.20}$$

$$nH_2O + nSO_4^{2-} + nRE^{3+} \rightarrow RE_2(SO_4)_n.nH_2O \tag{18.21}$$

$$nHCl + nRE^{3+} \rightarrow nRECl_3 \tag{18.22}$$

Liquid crystal displays (LCDs) and fluorescent lamps employ REE-containing phosphors, and demand is predicted to grow at an annual rate of 8% by 2025 (Awais et al.

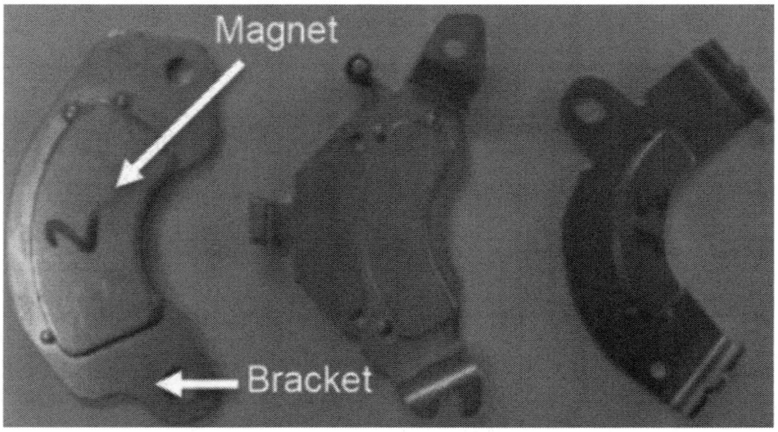

Figure 18.6 Some magnets from hard disc drives.

2017; Walton et al. 2015). Advanced fluorescent lights provide white light with phosphor components varying from 10 to 55% green, blue, and red, and these lamps are significant resources of Y, Tb, and Eu that are more important than light REEs. 10–20% REEs, 12% alumina, 20% fine glass, 45% halophosphate, and 5% others are generated from waste phosphor powder. Leaching and precipitation, as well as SX, can be employed to extract REEs from phosphor mixtures. Pavón et al. (2021) investigated the recovery of Eu and Y from fluorescent lamp waste. They found fluorescent lamp wastes were selectively recovered by solid-state chlorination followed by leaching with an organic solvent, obtaining 95.7% Y and 92.2% Eu (Pavón et al. 2021).

Technology alternatives for recovering materials from WEEE must consider macro and micro-perspectives of procedure efficiency and monetary benefit while maintaining sustainability. Recovering and recycling REEs has the potential to create value chains and jobs while reducing the environmental, public health, and mental problems connected with WEEE. NdFeB magnets, NiMH batteries, and lamp phosphors are the most interesting waste streams for REE recycling. Recycling REEs is difficult technically, but not impossible.

18.8 Other Technologies

In none of the traditional processes is it possible to achieve simultaneously high extraction efficiency, low environmental impact, recycling of waste streams, low processing cost, and high metal purity. Therefore, several technologies have been studied to replace, integrate, or improve the current processes of e-waste management, recycling, and recovery of metals, such as bioleaching (Esmaeili and Arshadi 2022; Kadivar et al. 2021), pyrolysis (Chen et al. 2021; Liu et al. 2021), electrolysis (Choubey et al. 2021; Wang et al. 2021a), hybrid technologies involving hydrometallurgical routes (Amato et al. 2020; Gande et al. 2021; Lyu et al. 2021), ionic liquids (Luo et al. 2019; Zhang et al. 2018a), supercritical extraction (Calgaro et al. 2015; Reisdörfer et al. 2020), and supercritical water technology.

18.9 Conclusions

The pros and cons of hydrometallurgical processing over conventional pyrometallurgical process are summarized in Table 18.2. Hydrometallurgical aqueous approaches offer more opportunities to eliminate material loss from e-waste recycling processes but have some potential environmental impacts in implementation. In e-waste recycling, both base and precious metals can be recovered by hydrometallurgical treatment. Small-scale applications and low-grade wastes can be suitable for leaching. After pretreatment for liberation and size reduction, suitable lixiviants leach easily and selectively dissolve metals from e-waste. After purification of PLS, metals can be recovered by precipitation, solvent extraction, electrowinning, etc. Inorganic/organic acids, alkaline solutions, brines, cyanide, thiourea, thiosulfate, and/or chelating agents can be used for metal extraction. Hydrometallurgy is easy to apply and simple to operate. Due to slow kinetics, long leaching times may be required. Weak organic acids require high temperatures and long residence times as compared to inorganic acids. For e-waste recycling, lixiviant concentration should be as low as possible with a high extraction rate.

A comparison of different types of metal recovery techniques is given in Table 18.3. Among hydro/pyro/biometallurgical methods, the cheapest process is hydrometallurgy, the lowest toxic process is biometallurgy, and the highest efficiency and industrial applicability and shortest process time requirement are obtained from pyrometallurgy.

Current industrial e-waste recycling methods include both pyrometallurgy and hydrometallurgy, which have a long history and wide applications. Biometallurgy is an emerging technology that also takes up a certain share of the WPCB recycling market. Many research and development studies have been performed to improve the performance

Table 18.2 Comparison of advantages and disadvantages of hydrometallurgy with pyrometallurgy.

Advantages of hydrometallurgical treatment	Disadvantages of hydrometallurgical treatment
Advanced treatment method	Strong hazardous inorganic acids/bases are used
Easy to apply	Generates acidic/alkaline wastewater
Easy and simple to operate	Corrosive reagents are used
Selective separation	Corrosion-resistant equipment used
Flexible separation	Generally, needs fine size reduction
Results are exact and expectable	High energy consumption for comminution
Single/multi-stage leaching is possible	Long process time due to slow kinetics
Less hazardous to the environment	Tedious
Low waste gas emission	Requires more steps
Low energy consumption	Risk of metal losses during subsequent steps
Low-temperature requirement	NMF cannot be recovered
Easily controlled	Free heavy metal ions present in the effluent
Low investment cost	Downstream processes are difficult
Small-scale operations	No one solution for all
Reagent recyclability	Easy working conditions
High recovery rates	
Cu, Ag, Au, Pd, Pt, Pb, Sn, Al, etc. can be recovered	
No slag generation	

Table 18.3 Comparison of different metal recovery techniques.

Recycling technique	Cost	Efficiency	Toxicity1	Recycling time	Industrial applicability
Hydrometallurgy	**Low**	Medium	Medium	Medium	Medium
Pyrometallurgy	High	**High**	High	**Short**	**High**
Biometallurgy	Medium	Low	**Low**	Long	Low

or cut the operating cost of these techniques. There are inherent environmental drawbacks and intensive energy consumption problems for these processes. Thus, it is extremely important to develop novel full-recovery methods for WPCB to improve the current methods of WPCB recycling with high safety and economic feasibility. Sustainable hydrometallurgical recycling technology should ensure that e-waste is processed in an ecological and economical-friendly manner, with high efficiency and lower carbon footprint, at a fraction of the costs involved with setting multibillion-dollar smelting facilities. Recycling offers a potent resource for our planet, transforming millions of tons of e-waste into reusable material assets.

E-waste recycling converts today's fastest-growing and hazardous solid waste problem into a challenge or an opportunity for economic (sustainable profitabity), social (generates new jobs and well-being for the society), and environmental (generates superior environmental performance and full utilization of depleted resources) point of view. Struggle against illegal imports of e-waste has become one of the major challenges for most developing countries where unregulated and rudimentary techniques are widely used. Formal recycling technology based on metal-centric and non-metal-centric hydrometallurgy should be developed for developing countries. Oxidative leaching, direct leaching, or reductive leaching usually can be used to extract critical/scarce/valuable metals of interest. In some cases, pressure leaching, mechanochemical-assisted leaching, chlorination-assisted leaching, or microwave-assisted leaching can be applied toward sustainability and a zero-waste scheme.

The hydrometallurgy process generally uses toxic, non-recyclable, and costly chemical reagents for the solubilization of the valuable metals present in e-waste, mainly H_2SO_4, HCl, HNO_3, and H_2O_2. Moreover, this technology has been used extensively for the separation and refining of pyrometallurgical products, gaining prominence as the traditional process for precious metal recovery through the use of cyanide, thiosulfate, and thiourea. Regardless of the low energy requirements in comparison to the pyrometallurgical process, the generation of highly toxic wastewater, longer reaction time, and operational complexity are disadvantages related to the hydrometallurgical route that cannot be neglected.

References

Abrahami, S.T., Xiao, Y., and Yang, Y. (2015). Rare-earth elements recovery from post-consumer hard-disc drives. *Mineral Processing and Extractive Metallurgy* 124: 106–115. https://doi.org/10.1179/1743285514Y.0000000084.

Ajiboye, E.A., Panda, P.K., Adebayo, A.O. et al. (2019). Leaching kinetics of Cu, Ni and Zn from waste silica rich integrated circuits using mild nitric acid. *Hydrometallurgy* 188: 161–168. https://doi.org/10.1016/j.hydromet.2019.06.016.

Alam, M.S., Tanaka, M., Koyama, K. et al. (2007). Electrolyte purification in energy-saving monovalent copper electrowinning processes. *Hydrometallurgy* 87: 36–44. https://doi.org/10.1016/j.hydromet.2006.12.001.

Amato, A., Becci, A., and Beolchini, F. (2020). Sustainable recovery of Cu, Fe and Zn from end-of-life printed circuit boards. *Resources, Conservation and Recycling* 158: 104792. https://doi.org/10.1016/j.resconrec.2020.104792.

Ambaye, T.G., Vaccari, M., Castro, F.D. et al. (2020). Emerging technologies for the recovery of rare earth elements (REEs) from the end-of-life electronic wastes: a review on progress, challenges, and perspectives. *Environmental Science and Pollution Research* 27: 36052–36074. https://doi.org/10.1007/s11356-020-09630-2.

Awais, M., Coelho, F., Degri, M. et al. (2017). Hydrocyclone separation of hydrogen decrepitated NdFeB. *Recycling* 2: 22. https://doi.org/10.3390/recycling2040022.

Batnasan, A., Haga, K., and Shibayama, A. (2018). Recovery of precious and base metals from waste printed circuit boards using a sequential leaching procedure. *JOM* 70: 124–128. https://doi.org/10.1007/s11837-017-2694-y.

Behnamfard, A., Salarirad, M.M., and Veglio, F. (2013). Process development for recovery of copper and precious metals from waste printed circuit boards with emphasize on palladium and gold leaching and precipitation. *Waste Management* 33: 2354–2363. https://doi.org/10.1016/j.wasman.2013.07.017.

Birloaga, I., Coman, V., Kopacek, B., and Vegliò, F. (2014). An advanced study on the hydrometallurgical processing of waste computer printed circuit boards to extract their valuable content of metals. *Waste Management* 34: 2581–2586. https://doi.org/10.1016/j.wasman.2014.08.028.

Birloaga, I., De Michelis, I., Ferella, F. et al. (2013). Study on the influence of various factors in the hydrometallurgical processing of waste printed circuit boards for copper and gold recovery. *Waste Management* 33: 935–941. https://doi.org/10.1016/j.wasman.2013.01.003.

Birloaga, I. and Vegliò, F. (2022). An innovative hybrid hydrometallurgical approach for precious metals recovery from secondary resources. *Journal of Environmental Management* 307: 114567. https://doi.org/10.1016/j.jenvman.2022.114567.

Bonggotgetsakul, Y.Y.N., Cattrall, R.W., and Kolev, S.D. (2016). Recovery of gold from aqua regia digested electronic scrap using a poly(vinylidene fluoride-co-hexafluoropropene) (PVDF-HFP) based polymer inclusion membrane (PIM) containing Cyphos® IL 104. *Journal of Membrane Science* 514: 274–281. https://doi.org/10.1016/j.memsci.2016.05.002.

Breuer, P.L. and Jeffrey, M.I. (2003). Copper catalysed oxidation of thiosulfate by oxygen in gold leach solutions. *Minerals Engineering* 16: 21–30. https://doi.org/10.1016/S0892-6875(02)00287-X.

Calgaro, C.O., Schlemmer, D.F., Da Silva, M.D.C.R. et al. (2015). Fast copper extraction from printed circuit boards using supercritical carbon dioxide. *Waste Management* 45: 289–297. https://doi.org/10.1016/j.wasman.2015.05.017.

Chakraborty, S.C., Zaman, M.W.U., Hoque, M. et al. (2022). Metals extraction processes from electronic waste: constraints and opportunities. *Environmental Science and Pollution Research*. https://doi.org/10.1007/s11356-022-19322-8.

Chen, Y., Liang, S., Xiao, K. et al. (2021). A cost-effective strategy for metal recovery from waste printed circuit boards via crushing pretreatment combined with pyrolysis: effects of particle size and pyrolysis temperature. *Journal of Cleaner Production* 280: 124505. https://doi.org/10.1016/j.jclepro.2020.124505.

Choubey, S., Goswami, P., and Gautam, S. (2021). Recovery of copper from waste PCB boards using electrolysis. *Materials Today: Proceedings* 42: 2656–2659. https://doi.org/10.1016/j.matpr.2020.12.596.

Chu, C.K., Breuer, P.L., and Jeffrey, M.I. (2003). The impact of thiosulfate oxidation products on the oxidation of gold in ammonia thiosulfate solutions. *Minerals Engineering* 16: 265–271. https://doi.org/10.1016/S0892-6875(02)00369-2.

Cyganowski, P., Garbera, K., Leśniewicz, A. et al. (2017). The recovery of gold from the aqua regia leachate of electronic parts using a core–shell type anion exchange resin. *Journal of Saudi Chemical Society* 21: 741–750. https://doi.org/10.1016/j.jscs.2017.03.007.

Dolker, T. and Pant, D. (2019). Chemical-biological hybrid systems for the metal recovery from waste lithium ion battery. *Journal of Environmental Management* 248: 109270. https://doi.org/10.1016/j.jenvman.2019.109270.

Egan, J., Bazin, C., and Hodouin, D. (2016). Effect of particle size and grinding time on gold dissolution in cyanide solution. *Minerals* 6: 68. https://doi.org/10.3390/min6030068.

Esmaeili, A. and Arshadi, M. (2022). Simultaneous leaching of Cu, Al, and Ni from computer printed circuit boards using penicillium simplicissimum. *Resources, Conservation and Recycling* 177: 105976. https://doi.org/10.1016/j.resconrec.2021.105976.

Feng, P., Wang, Z., Meng, L., and Guo, Z. (2022). Supergravity-enhanced liquation crystallization for metal recovery from waste printed circuit boards. *Chemical Engineering and Processing: Process Intensification* 173: 108813. https://doi.org/10.1016/j.cep.2022.108813.

Ficeriová, J., Baláž, B., and Gock, E. (2011). Leaching of gold, silver and accompanying metals from circuit boards (PCBs) waste, Acta Montan. *Slovaca* 16: 128–131.

Gande, V.V., Vats, S., Bhatt, N., and Pushpavanam, S. (2021). Sequential recovery of metals from waste printed circuit boards using a zero-discharge hydrometallurgical process. *Cleaner Engineering and Technology* 4: 100143. https://doi.org/10.1016/j.clet.2021.100143.

Gonzalez Baez, A., Pantoja Muñoz, L., Garelick, H., and Purchase, D. (2022). Characterization of industrially pre-treated waste printed circuit boards for the potential recovery of rare earth elements. *Environmental Technology and Innovation* 27: 102481. https://doi.org/10.1016/j.eti.2022.102481.

Gurung, M., Adhikari, B.B., Kawakita, H. et al. (2013). Recovery of gold and silver from spent mobile phones by means of acidothiourea leaching followed by adsorption using biosorbent prepared from persimmon tannin. *Hydrometallurgy* 133: 84–93. https://doi.org/10.1016/j.hydromet.2012.12.003.

Ha, V.H., Lee, J., Huynh, T.H. et al. (2014). Optimizing the thiosulfate leaching of gold from printed circuit boards of discarded mobile phone. *Hydrometallurgy* 149: 118–126. https://doi.org/10.1016/j.hydromet.2014.07.007.

Ha, V.H., Lee, J., Jeong, J. et al. (2010). Thiosulfate leaching of gold from waste mobile phones. *Journal of Hazardous Materials* 178: 1115–1119. https://doi.org/10.1016/j.jhazmat.2010.01.099.

Huang, T., Zhu, J., Huang, X. et al. (2022). Assessment of precious metals positioning in waste printed circuit boards and the economic benefits of recycling. *Waste Management* 139: 105–115. https://doi.org/10.1016/j.wasman.2021.12.030.

Islam, A., Ahmed, T., Awual, M.R. et al. (2020). Advances in sustainable approaches to recover metals from e-waste––A review. *Journal of Cleaner Production* 244: 118815. https://doi.org/10.1016/j.jclepro.2019.118815.

Jadhao, P.R., Ahmad, E., Pant, K.K., and Nigam, K.D.P. (2022). Advancements in the field of electronic waste recycling: critical assessment of chemical route for generation of energy and valuable products coupled with metal recovery. *Separation and Purification Technology* 289: 120773. https://doi.org/10.1016/j.seppur.2022.120773.

Jadhao, P.R., Pandey, A., Pant, K.K., and Nigam, K.D.P. (2021). Efficient recovery of Cu and Ni from WPCB via alkali leaching approach. *Journal of Environmental Management* 296: 113154. https://doi.org/10.1016/j.jenvman.2021.113154.

Jha, M.K., Kumari, A., Choubey, P.K. et al. (2012). Leaching of lead from solder material of waste printed circuit boards (PCBs). *Hydrometallurgy* 121–124: 28–34. https://doi.org/10.1016/j.hydromet.2012.04.010.

Kadivar, S., Pourhossein, F., and Mousavi, S.M. (2021). Recovery of valuable metals from spent mobile phone printed circuit boards using biochar in indirect bioleaching. *Journal of Environmental Management* 280: 111642. https://doi.org/10.1016/j.jenvman.2020.111642.

Kaya, M. (2016). Recovery of metals and nonmetals from electronic waste by physical and chemical recycling processes. *Waste Management* 57: 64–90. https://doi.org/10.1016/j.wasman.2016.08.004.

Kaya, M. (2016a). Recovery of metals from electronic waste by physical and chemical recycling processes. *International Journal of Chemical, Nuclear, Materials and Metallurgical Engineering* 10 (12): 232–243. scholar.waset.org/1307.6892/10003863.

Kaya, M. (2017). Recovery of metals and nonmetals from waste Printed Circuit Boards (PCBs) by physical recycling techniques. In: *Energy Technology 2017 Book* (ed. L. Zhang, J.W. Drelich, N.R. Neelameggham et al.). Springer. https://doi.org/10.1007/978-3-319-52192-3.

Kaya, M. (2018). Waste Printed Circuit Board (WPCB) recovery technology: disassembly and desoldering approach. In: *Reference Module in Materials Science and Materials Engineering/ Encyclopedia of Renewable and Sustainable Materials*. Elsevier. https://doi.org/10.1016/B978-0-12-803581-8.11246-9.

Kaya, M. (2018a). Chapter 3: Current WEEE recycling solutions. In: *Waste Electrical and Electronic Equipment Recycling––Aqueous Recovery Methods* (ed. F. Veglio and I. Birloaga), Woodhead Publishing Series in Electronic and Optical Materials. https://doi.org/10.1016/B978-0-08-102057-9.00003-2.

Kaya, M. (2019). *Electronic Waste and Printed Circuit Board Recycling Technologies, the Minerals, Metals & Materials Series*. Cham, Switzerland: Springer International Publishing. https://doi.org/10.1007/978-3-030-26593-9.

Kaya, M. (2019a). Waste Printed Circuit Board (WPCB) recycling: conventional and emerging technology approach. In: *Encyclopedia of Renewable and Sustainable Materials* (ed. S. Hashmi and I.A. Choudhury), 1–18. Elsevier Ltd. https://doi.org/10.1016/b978-0-12-803581-8.11296-2.

Kaya, M. (2021). Urban mining of rare earth elements from e-waste: progress, challenges, and perspectives. In: *Horizons in Earth Science Research Vol. 22* (ed. B. Veress and J. Szigethy), 33–116. New York, US: Nova Science Publications.

Kinoshita, T., Akita, S., Kobayashi, N. et al. (2003). Metal recovery from non-mounted printed wiring boards via hydrometallurgical processing. *Hydrometallurgy* 69: 73–79. https://doi.org/10.1016/S0304-386X(03)00031-8.

Kubota, F., Kono, R., Yoshida, W. et al. (2019). Recovery of gold ions from discarded mobile phone leachate by solvent extraction and polymer inclusion membrane (PIM) based separation using an amic acid extractant. *Separation and Purification Technology* 214: 156–161. https://doi.org/10.1016/j.seppur.2018.04.031.

Lee, H., Molstad, E., and Mishra, B. (2018). Recovery of gold and silver from secondary sources of electronic waste processing by thiourea leaching. *JOM* 70: 1616–1621. https://doi.org/10.1007/s11837-018-2965-2.

Li, H., Oraby, E., and Eksteen, J. (2021). Recovery of copper and the deportment of other base metals from alkaline glycine leachates derived from waste printed circuit boards (WPCBs). *Hydrometallurgy* 199: 105540. https://doi.org/10.1016/j.hydromet.2020.105540.

Li, H., Oraby, E., and Eksteen, J. (2022). Extraction of precious metals from waste printed circuit boards using cyanide-free alkaline glycine solution in the presence of an oxidant. *Minerals Engineering* 181: 107501. https://doi.org/10.1016/j.mineng.2022.107501.

Li, K. and Xu, Z.M. (2015). Application of supercritical water to decompose brominated epoxy resin and environmental friendly recovery of metals from waste memory module. *Environmental Science and Technology* 49 (3): 1761–1767. https://doi.org/10.1021/es504644b.

Lisinska, M., Saternus, M., and Willner, J. (2018). Research of leaching of the printed circuit boards coming from waste mobile phones. *Archives of Metallurgy and Materials* 63: 143–147. https://doi.org/10.24425/118921.

Liu, K., Zhang, Z., and Zhang, F.S. (2016b). Advanced degradation of brominated epoxy resin and simultaneous transformation of glass fiber from waste printed circuit boards by improved supercritical water oxidation processes. *Waste Management* 56: 423–430. https://doi.org/10.1016/j.wasman.2016.05.022.

Liu, Y., Zhang, J., Yang, X. et al. (2021). Efficient recovery of valuable metals from waste printed circuit boards by microwave pyrolysis. *Chinese Journal of Chemical Engineering* 40: 262–268. https://doi.org/10.1016/j.cjche.2020.11.008.

Lixandru, A., Venkatesan, P., Jönsson, C. et al. (2017). Identification and recovery of rare-earth permanent magnets from waste electrical and electronic equipment. *Waste Management* 68: 482–489. https://doi.org/10.1016/j.wasman.2017.07.028.

Long Le, H., Jeong, J., Lee, J.-C. et al. (2011). Hydrometallurgical process for copper recovery from waste Printed Circuit Boards (PCBs). *Mineral Processing and Extractive Metallurgy Review* 32: 90–104. https://doi.org/10.1080/08827508.2010.530720.

Lu, Y. and Xu, Z. (2016). Precious metals recovery from waste printed circuit boards: a review for current status and perspective. *Resources, Conservation and Recycling* 113: 28–39. https://doi.org/10.1016/j.resconrec.2016.05.007.

Luo, D., Zhu, N., Li, Y. et al. (2019). Simultaneous leaching and extraction of indium from waste LCDs with acidic ionic liquids. *Hydrometallurgy* 189: 105146. https://doi.org/10.1016/j.hydromet.2019.105146.

Lyu, J., Liu, Y., Lyu, X. et al. (2021). Efficient bromine removal and metal recovery from waste printed circuit boards smelting flue dust by a two stage leaching process. *Journal of Cleaner Production* 322: 129054. https://doi.org/10.1016/j.jclepro.2021.129054.

Manigandan, S., Rajmohan, K.S., and Varjani, S. (2020). Current trends in gold recovery from electronic wastes. In: *Current Developments in Biotechnology and Bioengineering* (ed. A. Pandey, S. Negi, C.R. Soccol, et al.), 307–325. Elsevier. https://doi.org/10.1016/B978-0-444-64321-6. 00016-1.

Mir, S. and Dhawan, N. (2022). A comprehensive review on the recycling of discarded printed circuit boards for resource recovery. *Resources, Conservation and Recycling* 178: 106027. https://doi.org/10.1016/j.resconrec.2021.106027.

Oh, C.J., Lee, S.O., Yang, H.S. et al. (2003). Selective leaching of valuable metals from waste printed circuit boards. *Journal of the Air and Waste Management Association* 53: 897–902. https://doi.org/10.1080/10473289.2003.10466230.

Panda, R., Dinkar, O.S., Jha, M.K., and Pathak, D.D. (2020a). Hydrometallurgical processing of waste multilayer ceramic capacitors (MLCCs) to recover silver and palladium. *Hydrometallurgy* 197: 105476. https://doi.org/10.1016/j.hydromet.2020.105476.

Panda, R., Jha, M.K., Pathak, D.D., and Gupta, R. (2020b). Recovery of Ag, Cu, Ni and Fe from the nitrate leach liquor of waste ICs. *Minerals Engineering* 158: 106584. https://doi.org/10.1016/j.mineng.2020.106584.

Park, Y.J. and Fray, D.J. (2009). Separation of zinc and nickel ions in a strong acid through liquid-liquid extraction. *Journal of Hazardous Materials* 163: 259–265. https://doi.org/10.1016/j.jhazmat.2008.06.085.

Pavón, S., Lorenz, T., Fortuny, A. et al. (2021). Rare earth elements recovery from secondary wastes by solid-state chlorination and selective organic leaching. *Waste Management* 122: 55–63. https://doi.org/10.1016/j.wasman.2020.12.039.

Qiu, R., Lin, M., Ruan, J. et al. (2020). Recovering full metallic resources from waste printed circuit boards: a refined review. *Journal of Cleaner Production* 244: 118690. https://doi.org/10.1016/j.jclepro.2019.118690.

Quinet, P., Proost, J., and Van Lierde, A. (2005). Recovery of precious metals from electronic scrap by hydrometallurgical processing routes. *Mining, Metallurgy and Exploration* 22: 17–22. https://doi.org/10.1007/BF03403191.

Radmehr, V., Koleini, S.M.J., Khalesi, M.R., and Tavakoli Mohammadi, M.R. (2013). Ammonia leaching: a new approach of copper industry in hydrometallurgical processes. *Journal of the Institution of Engineers (India): Series D* 94: 95–104. https://doi.org/10.1007/s40033-013-0029-x.

Ramprasad, C., Gwenzi, W., Chaukura, N. et al. (2022). Strategies and options for the sustainable recovery of rare earth elements from electrical and electronic waste. *Chemical Engineering Journal* 442: 135992. https://doi.org/10.1016/j.cej.2022.135992.

Reisdörfer, G., Bertuol, D.A., and Tanabe, E.H. (2020). Extraction of neodymium from hard disk drives using supercritical CO_2 with organic acids solutions acting as cosolvents. *Journal of CO2 Utilization* 35: 277–287. 834. https://doi.org/10.1016/j.jcou.2019.10.008.

Sheng, P.P. and Etsell, T.H. (2007). Recovery of gold from computer circuit board scrap using aqua regia. *Waste Management & Research: The Journal for a Sustainable Circular Economy* 25: 380–383. https://doi.org/10.1177/0734242×07076946.

Silvas, F.P.C., Jiménez Correa, M.M., Caldas, M.P.K. et al. (2015). Printed circuit board recycling: physical processing and copper extraction by selective leaching. *Waste Management* 46: 503–510. https://doi.org/10.1016/j.wasman.2015.08.030.

Syed, S. (2006). A green technology for recovery of gold from non-metallic secondary sources. *Hydrometallurgy* 82: 48–53. https://doi.org/10.1016/j.hydromet.2006.01.004.

Tripathi, A., Kumar, M., Sau, D.C. et al. (2012). Leaching of gold from the waste mobile phone printed circuit boards (PCBs) with ammonium thiosulphate. *International Journal of Metallurgical Engineering* 1: 17–21. https://doi.org/10.5923/j.ijmee.20120102.02.

Ueberschaar, M. and Rotter, V.S. (2015). Enabling the recycling of rare earth elements through product design and trend analyses of hard disk drives. *Journal of Material Cycles and Waste Management* 17: 266–281. https://doi.org/10.1007/s10163-014-0347-6.

Walton, A., Yi, H., Rowson, N.A. et al. (2015). The use of hydrogen to separate and recycle neodymium–iron–boron-type magnets from electronic waste. *Journal of Cleaner Production* 104: 236–241. https://doi.org/10.1016/j.jclepro.2015.05.033.

Wang, H., Zhang, S., Li, B. et al. (2017). Recovery of waste printed circuit boards through pyrometallurgical processing—A review. *Resources, Conservation and Recycling* 126: 209–218. https://doi.org/10.1016/j.resconrec.2017.08.001.

Wang, J., Huang, Z., Yang, D. et al. (2021a). A semi-scaled experiment for metals separating and recovering from waste printed circuit boards by slurry electrolysis. *Process Safety and Environmental Protection* 147: 37–44. https://doi.org/10.1016/j.psep.2020.09.030.

Wang, R., Zhang, C., Zhao, Y. et al. (2021). Recycling gold from printed circuit boards gold-plated layer of waste mobile phones in mild aqua regia *system. Journal of Cleaner Production* 278: 123597. https://doi.org/10.1016/j.jclepro.2020.123597.

Yang, J., Wu, Y., and Li, J. (2012). Recovery of ultrafine copper particles from metal components of waste printed circuit boards. *Hydrometallurgy* 121–124: 1–6. https://doi.org/10.1016/j.hydromet.2012.04.015.

Yang, J.G., Tang, C.B., He, J. et al. (2011). A method of extracting valuable metals from electronic wastes, State Intellectual Property Office of the People's Republic of China, ZL 200910303503.5.

Yang, T., Xu, Z., Wen, J., and Yang, L. (2009). Factors influencing bioleaching copper from waste printed circuit boards by Acidithiobacillus ferrooxidans. *Hydrometallurgy* 97: 29–32. https://doi.org/10.1016/j.hydromet.2008.12.011.

Zhang, D.J., Dong, L., Li, Y.T. et al. (2018a). Copper leaching from waste printed circuit boards using typical acidic ionic liquids recovery of e-wastes' surplus value. *Waste Management* 78: 191–197. https://doi.org/10.1016/j.wasman.2018.05.036.

Zhang, X., Guan, J., Gua, Y. et al. (2017b). Effective dismantling of waste PCB assembly with methanesulfonic acid containing hydrogen peroxide, AIChE. *Environmental Progress & Sustainable Energy* 36 (3). https://doi.org/10.1002/ep.

Zhang, Y., Chen, M., Tan, Q. et al. (2018b). Recovery of copper from WPCBs using slurry electrolysis with ionic liquid $[BSO_3HPy]\cdot HSO_4$. *Hydrometallurgy* 175: 150–154. https://doi.org/10.1016/j.hydromet.2017.11.004.

Zhang, Y., Liu, S., Xie, H. et al. (2012). Current status on leaching precious metals from waste printed circuit boards. *Procedia Environmental Sciences* 16: 560–568. https://doi.org/10.1016/j.proenv.2012.10.077.

Zhang, Z. and Zhang, F.-S. (2014). Selective recovery of palladium from waste printed circuit boards by a novel non-acid process. *Journal of Hazardous Materials* 279: 46–51. https://doi.org/10.1016/j.jhazmat.2014.06.045.

Zuniga, M., Parada, L.F., and Asselin, E. (2010). Leaching of a limonitic laterite in ammoniacal solutions with metallic iron. *Hydrometallurgy* 104: 260–267. https://doi.org/10.1016/j.hydromet.2010.06.014.

19

Influence of Recycled e-Polycarbonate on Mortar Composites Mechanical and Thermal Study

Ana Laura De la Colina Martínez[1,*] *and David Joaquín Delgado Hernández*[2]

[1] *Doctora en Ciencia de Materiales por la Facultad de Química, Universidad Autónoma del Estado de México, Paseo Colón esquina Paseo Tollocan S/N, 50000, Toluca, México, México*
[2] *Facultad de Ingeniería, Universidad Autónoma del Estado de México, Avenida Universidad S/N, Cerro de Coatepec, Ciudad Universitaria, 50110, Toluca, México*
* *Corresponding author*

19.1 Introduction

The global demand for polycarbonate reached around 4.5 million metric tons (Mt) by the end of 2016 (Greenfield 2018; IHS Chemical 2012). The polycarbonate market is expected to register a compound annual growth rate (CAGR) of 5.5% from 2018 to 2023 (Mordor Intelligence 2018). To date, polycarbonate is extensively used in electrical and electronic devices. However, their rapid development, combined with a short life cycle, has resulted in a continuous increase of waste from electrical and electronic equipment (e-waste or WEEE) (Achilias and Antonakou 2015). In 2016, 44.7 Mt of e-waste were generated globally (an increase of 3.3 Mt from 2014) (Baldé et al. 2017). Experts foresee e-waste increasing to 52.2 Mt by 2021. In 2016, only 8.9 Mt (20%) of all this waste was recycled. In 2016 the top producer of WEEE in the Americas was the US, with 6.3 Mt. The second was Brazil, with 1.5 Mt, and the third was Mexico, with 1 Mt; in Europe, the total e-waste generated was 12.3 Mt (Baldé et al. 2017).

Using plastic waste in cement mortar could be one of the best ecological solutions, due to its economic and environmental advantages. In this regard, several works have been performed or are under way to evaluate the properties of cement-composites containing various types of plastic waste as aggregate, filler, or fiber (Saikia and De Brito 2012). Polyethylene terephthalate (PET) plastic waste has been the most probed (Choi et al. 2009; Marzouk et al. 2007; Safi et al. 2013). Other waste plastics have also been studied, such as high-impact polystyrene (HIPS) (Wang and Meyer 2012), polyvinyl chloride (PVC) (Aciu et al. 2018), and polycarbonate (Hannawi et al. 2010; Kaur and Pavia 2020).

However, recycled and extruded polycarbonate from electronic waste (recycled e-polycarbonate) as aggregate in cement mortars has not yet been studied. To show the effects of it in cement mortar, the aims of this work are to characterize sands and recycled e-polycarbonate particles utilized and to obtain mechanical, thermal, and physical properties of cement mortars prepared with recycled e-polycarbonate, Portland cement, sand, and water.

19.2 Materials and Methods

The following section contains the materials, methods, and standards employed to carry out this research.

19.2.1 Materials

Two different granulometric distributions of sand size 0/4 mm were used and are shown in Figure 19.1. They were F and G sands. The F sand (Figure 19.1a) has a major number of fine particles compared to the G sand (Figure 19.1b). Sands were provided by the local company MONCA, located in Gijón, Asturias, Spain.

Recycled e-polycarbonate particles 1 mm diameter and 3 mm length in size (1 × 3 mm), were obtained by extrusion from electrical outlets wall switch waste. Recycled e-polycarbonate (Figure 19.1c) was provided by the Polyvima Company (located at San Lorenzo Tepaltitlán, Toluca, México). The physical and mechanical properties of the recycled e-polycarbonate were provided for the company according to the Resin Pellets and/or Makrolon® polycarbonate sheet, which are in accordance with Covestro and Professional Plastics (Covestro 2015; Professional Plastics 2017).

Electronic Waste Management: Policies, Processes, Technologies, and Impact, First Edition. Edited by Sunil Kumar and Vineet Kumar.

Figure 19.1 (a) F sand. (b) G sand. (c) Recycled e-polycarbonate.

19.2.2 Samples Preparation

Cement mortars were prepared by mixing Portland cement Type I (sulfate resistant, with a resistance medium-high of 42.5 MPa to 28 days), sand, water, and recycled e-polycarbonate. Recycled e-polycarbonate was used as partial replacement for sand in cement mortar as is shown in Table 19.1. The idea was to examine how recycled e-polycarbonate affects the mechanical, thermal, and physical properties of mortars when it substitutes sand-size particles of two different sands. In the first stage, both sands (F and G), used in FA and GA mortars respectively, as "acquired," were replaced by recycled e-polycarbonate in the percentage indicated in the table. In the second stage, both sands (F and G), used in FG and GG mortars respectively, were sieved, and only the sand particles that were "retained" in the 2 mm mesh size were replaced by recycled e-polycarbonate in the percentage indicated in Table 19.1. Six different kinds of mortars were elaborated and tested: (a) cement mortar with cement, sand, and water, which was denominated as control cement mortar (F0 and G0); (b) cement mortar, namely FA and FG, elaborated with F sand and recycled e-polycarbonate particles at concentrations of 0, 3, and 6% by volume; and (c) cement mortar, namely GA and GG, elaborated with G sand and recycled e-polycarbonate particles at concentrations of 0, 3, 6, 15, and 30% by volume. For all mortars the water/cement ratio (w/c) was 0.5 (236 g of water and 472 g of cement).

The mix of mortar components was made in an automatic mixer. Mortar prismatic specimens (40 × 40 × 160 mm) were elaborated by compaction in a steel mold according to UNE-EN 196–1:2018 standard (AENOR 2018). Additionally, the UNE-EN 1015–11:2020 standard (AENOR 2020) was also applied. Three test specimens for each type of cement mortar were obtained. The test specimens were cured in climatic chambers according to the UNE-EN 1015–11:2020 standard (AENOR 2020). For the first 7 days, the specimens were cured at 20°C ± 2°C and a relative humidity (RH) of 95% ± 5%. For the last 21 days, the specimens were cured at 20°C ± 2°C and a 65% ± 5% RH. With the intention of knowing the behavior of the cement mortar over time, the test specimens were maintained at 20°C ± 2°C and a 65% ± 5% RH prior to the mechanical and physical tests, which were performed to 36 and 56 days.

19.2.3 Test Methods

19.2.3.1 Granulometric Characterization of Sand and Recycled e-Polycarbonate

Sand and recycled e-polycarbonate granulometric characterizations were made according to the UNE-EN 933–1:2012 standard (AENOR 2012). On the other hand, the sieve analysis was made with the mesh series according to the UNE-EN 933–2:2022 standard (AENOR 2022).

Table 19.1 Constituents of control cement concrete and concrete with recycled e-polycarbonate.

Mortar with fine sand at 36 days		Mortar with fine sand at 56 days		Mortar with grainy sand at 36 days		Mortar with grainy sand at 56 days		Sand replacement by recycled e-polycarbonate (vol%)	Recycled e-poly carbonate (g)	Sand (g)
FA-36	FG-36	FA-56	FG-56	GA-36	GG-36	GA-56	GG-56			
✓	✓	✓	✓	✓	✓	✓	✓	0	0	1416
✓	✓	✓	✓	✓	✓	✓	✓	3	19	1373
✓	✓	✓	✓	✓	✓	✓	✓	6	38	1331
				✓	✓	✓	✓	15	96	1204
				✓	✓	✓	✓	30	192	990

19.2.3.2 Determination of Compressive Strengths

Compressive strengths of mortar specimens were measured in a Universal Testing Machine brand MEM 101, according to the UNE-EN 1015–11:2020 standard (AENOR 2020), located at Polytechnic School of Engineering of Gijón, University of Oviedo, Asturias, Spain. For the compressive strength determination, three specimens of each trial formulation were tested at 36 and 56 days.

19.2.3.3 Determination of Flexural Strengths

Flexural strengths of mortar specimens were measured in a Universal Testing Machine brand MEM 101, according to the UNE-EN 1015–11:2020 standard (AENOR 2020), located at Polytechnic School of Engineering of Gijón, University of Oviedo, Asturias, Spain. For the flexural strength determination, three specimens of each trial formulation were tested at 36 days.

19.2.3.4 Determination of Thermal Conductivity of Hardened Mortar

The determination of the thermal conductivity of hardened mortar was made with a non-destructive thermal testing instrument, namely C-Therm TCi Thermal Conductivity Analyzer TH130041. This instrument is located at the Departmental Building of Viesques N°7, Gijón, University of Oviedo, Asturias, Spain. For the thermal conductivity determination, two specimens (40 × 40 × 40 mm) of each trial formulation were tested at 36 days.

19.2.3.5 Determination of Water Absorption Coefficient Due to Capillary Action of Hardened Mortar

The determination of the water absorption coefficient due to capillary action was measured according to UNE-EN 1015–18:2003 standard (AENOR 2003). For the determination of the water absorption coefficient, three specimens of each trial formulation were tested at 36 days.

19.2.3.6 Determination of Density of Hardened Mortar

The density of the mortar specimens was estimated using a digital balance scale 10 kg, 0.1 g precision accurate. The test specimens were cured according to the UNE-EN 1015–11:2020 standard (AENOR 2020). The test specimens were maintained at 20°C ± 2°C and a 65% ± 5% RH before testing. For each trial formulation, three specimens (40 × 40 × 160 mm) were tested at 28 and 36 days.

19.3 Results and Discussion

Having established the standards used to characterize the samples, in the next section the main results of the research are presented.

19.3.1 Sieve Analysis of Sand 0/4 mm and Recycled e-Polycarbonate

The particle-size distributions of F and G sands are shown in Figure 19.2a. The sieve analysis reveals a distinctive Fineness Modulus (FM) for each sand. F sand exhibits a FM of 3.8 while G sand shows a FM of 4.4. These FM values indicate that the F sand contains more little particles than the G sand.

The particle-size distribution of recycled e-polycarbonate is shown in Figure 19.2b. The sieve analysis revealed a concentration of 99% of recycled e-polycarbonate retained in the 2 mm mesh size.

19.3.2 Compressive Strengths of Mortar

Mortars with recycled e-polycarbonate were elaborated. Different recycled e-polycarbonate concentrations were added to the cement mortar.

In Figure 19.3a the compressive strength values of FA and FG mortars are shown. Two clear behaviors are observed: FG-36 and FG-56 mortars show greater values than FA-36 and FA-56 mortars. FG-56 mortars show that recycled e-polycarbonate in 3% and 6%, with sand 3.8 FM, generates important improvements on the mortar resistance (10% and 3%, respectively). Such behaviors can be attributed to the presence of more fine sand particles in FG mortars than in FA mortars, since the sand particles in FG mortars (size 2 mm) were replaced by recycled e-polycarbonate.

These improvements in compressive strength contrast completely with the values reported by Hannawi et al. (2010), where considerable decreases (compared with 52.5 MPa from standard mortar) of 6.8% and 27.2% were reported in compressive strength for mortars containing, respectively, 3% and 10% of polycarbonate sharpened particles size 3.15 mm.

As can be seen, the recycled e-polycarbonate (size 1 × 3 mm), in contrast with sharpened particles (size 3.15 mm), can be utilized in mortars (with F sand) in concentrations up to 6% without affecting the compressive strengths, even if the cement type is 42.5 MPa and not 52.5 MPa.

In Figure 19.3b compressive strength values of GA and GG mortars are shown. In general terms, as the recycled e-polycarbonate concentration increases the compressive strength decreases. Such behaviors can be explained since this sand shows a FM of 4.4, and the reduced quantity of fine-sand particles has an impact on the resistances. Without these fine-sand particles, compressive strength of mortar goes down (at 36 and 56 days), even when the sand particles (2 mm) are replaced by recycled e-polycarbonate (1 × 3 mm) in low dosages (3% or 6%).

However, this decrease in compressive strength contrasts with the values reported by Hannawi et al. (2010), showing that the recycled e-polycarbonate (size 1 × 3 mm),

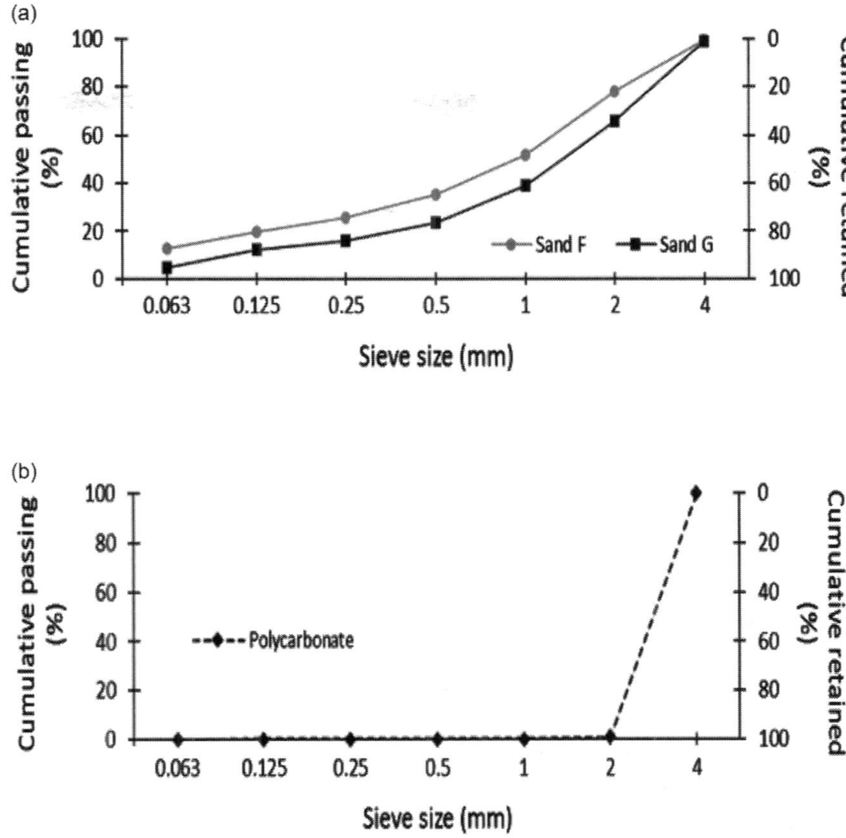

Figure 19.2 Granulometric distribution. (a) F and G sands. (b) Recycled e-polycarbonate.

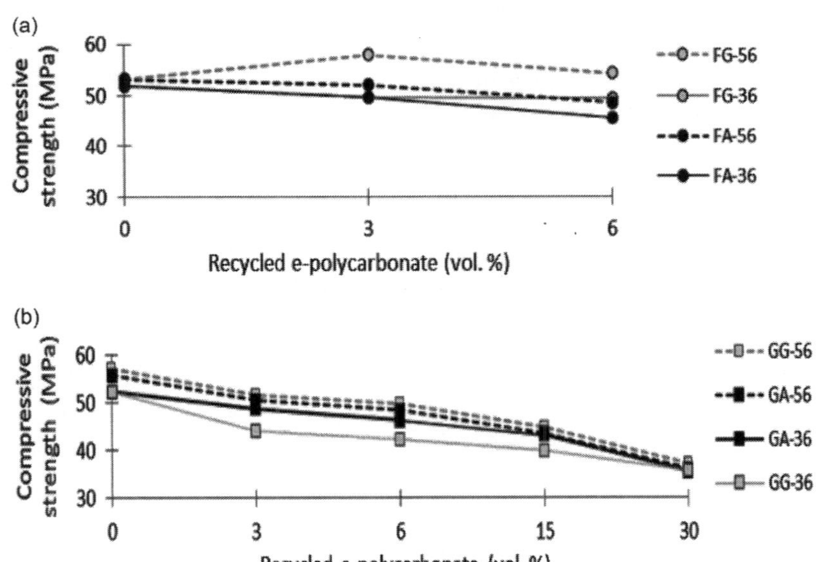

Figure 19.3 Compressive strengths. (a) FA and FG mortars. (b) GA and GG mortars.

in contrast with sharpened particles (size 3.15 mm), can be utilized in mortars (with G sand) in concentrations up to 15%, without affecting the compressive strengths.

19.3.3 Flexural Strength of Mortar

Mortars with recycled e-polycarbonate particles were added to the mortar. In Figure 19.4a flexural strength values of FA and FG mortars are shown. In this case, as the recycled e-polycarbonate concentration increases the flexural strength decreases.

With F sand, the mortar shows better flexural strength than with G sand. For instance, these values decreased from 10 to 7.5 MPa (Figure 19.4a), and from 8.5 to 6.5 MPa (Figure 19.4b) when F and G sand were replaced (respectively) 6% by recycled e-polycarbonate particles.

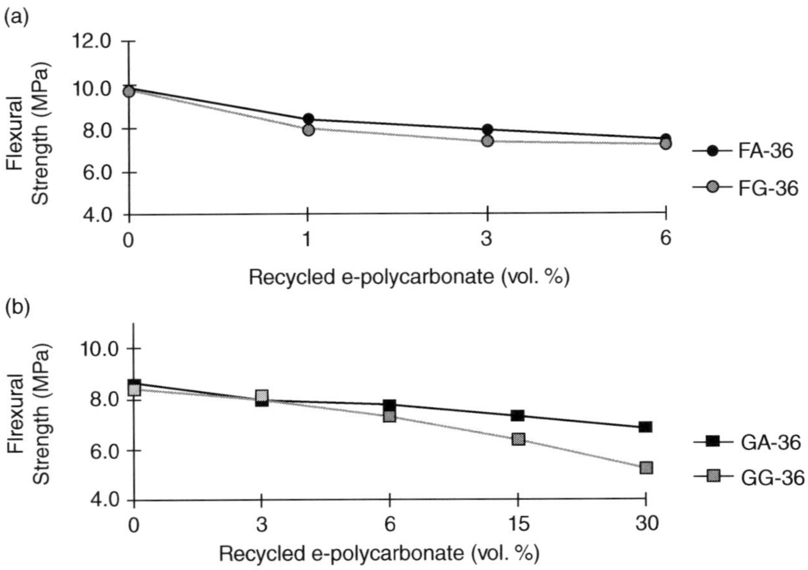

Figure 19.4 Flexural strengths. (a) FA and FG mortars. (b) GA and GG mortars.

The flexural strength values reported in the present study contrast fully with the values reported by Hannawi et al. (2010), where no significant changes (close to 5 MPa) were observed for mortars containing up to 20% of polycarbonate. These findings could be explained by the different shape and texture of each plastic aggregate. For example, the recycled e-polycarbonate particles utilized here are plus rounded and smoothed (since they were extruded) in contrast to the sharpened particles of polycarbonate (since they were cut) used by Hannawi et al. (2010).

The recycled e-polycarbonate (size 1 × 3 mm), in contrast with sharpened particles (size 3.15 mm), can be utilized in mortars (with F and G sands) in concentrations up to 6%, with better flexural strengths than in the Hannawi et al. (2010) study. All these observations suggest that the shape of the recycled e-polycarbonate (pellets) could offer an advantage over the sharpened particles of polycarbonate.

19.3.4 Determination of Thermal Conductivity of Hardened Mortar

Recycled e-polycarbonate particles were added to the cement mortar. In Figure 19.5a the thermal conductivities of FA, FG, GA, and GG mortars are shown. In general, as the recycled e-polycarbonate concentration increases the thermal conductivity decreases. In addition, GA-36 and GG-36 show a more pronounced decrease in values. These results contrast significantly with those of Hannawi and Prince-Agbodjan (2015). They show that 6% of recycled e-polycarbonate is sufficient to obtain a thermal conductivity below 1 W/mK. However, Hannawi and Prince-Agbodjan (2015) show that 50% of sharp polycarbonate particles are required to obtain a thermal conductivity close to 1 W/mK.

On the other hand, to explain the GA-36 and GG-36 behavior it is necessary to consider several factors. For instance, the polycarbonate thermal conductivity is 0.19 W/mK and the mortar thermal conductivity without aggregates, according to the American Concrete Institute (ACI) (2002), goes from 0.79 to 1.45 W/mK when the mortar density goes from 1.92 to 2.40 g/cm^3. This is based on the formula: $k_c = 0.079e^{1.2089d}$, where k_c is the thermal conductivity (W/mK) and d is the density (g/cm^3). This mortar was named ACI mortar.

In Figure 19.5b the relationships between the thermal conductivities and densities of ACI, FA-36, and GG-36 mortars are shown. As can be seen, the density values of the three mortars (ACI, FA-36, and GG-36) are very similar. Nevertheless, the thermal conductivity differs in each case. For example, when ACI (mortar without aggregates) and FA-36 densities diminish, the thermal conductivities gradually decrease. However, when the density diminishes for GG-36, the thermal conductivity sharply diminishes up to 30% (0.87 W/mK) when recycled e-polycarbonate concentration scarcely reaches 6%. The decrease in thermal conductivity improves the mortar thermal insulation (Lafarge Tarmac 2015).

In this context, Wang and Meyer (2012) and Aciu et al. (2018) have shown a decrease in thermal conductivity when sand was replaced by plastic waste particles of HIPS and PVC, respectively.

19.3.5 Water Absorption Coefficient by Capillary Action of Hardened Mortar

Recycled e-polycarbonate particles were added to the mortar. In Figure 19.6 the water absorption coefficients of FA, FG, GA, and GG mortars are shown. The values are constant for all mortars (0.1 mg/cm^{2}s$^{1/2}$). This behavior shows that the recycled e-polycarbonate addition has no influence on the mortar's water absorption, even when the samples contain 30% of it. These water absorption values

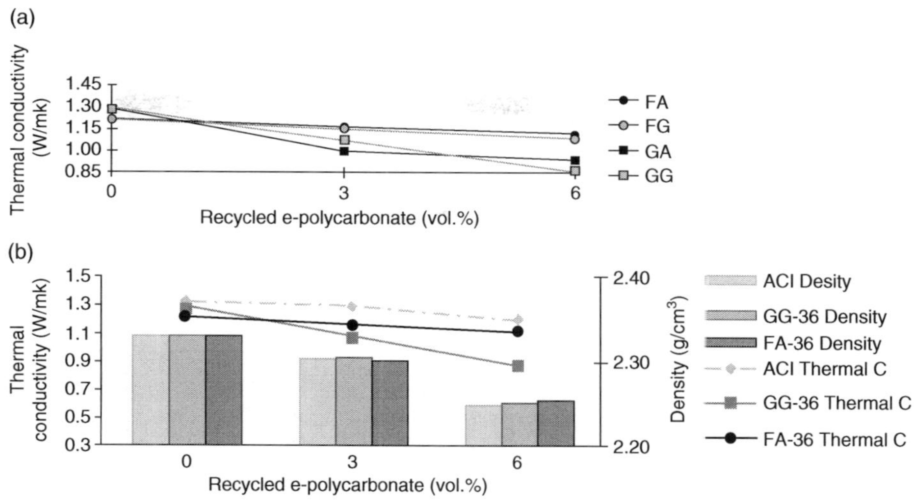

Figure 19.5 (a) Thermal conductivity of FA-36, FG-36, GA-36, and GG-36 mortars. (b) Correlation between thermal conductivities and densities of ACI, FA-36, and GG-36 mortars.

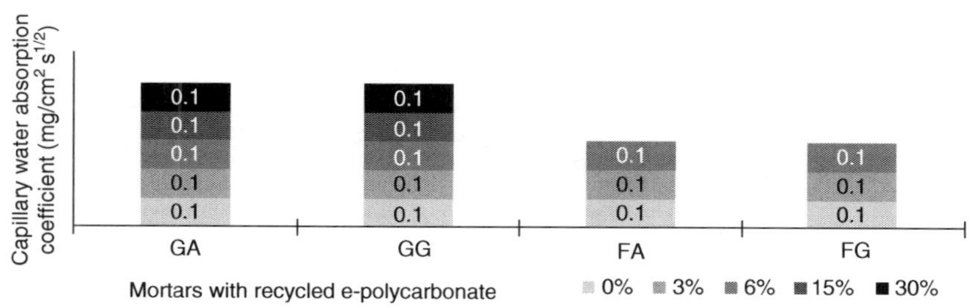

Figure 19.6 Water absorption coefficients of GA, GG, FA, and FG mortars.

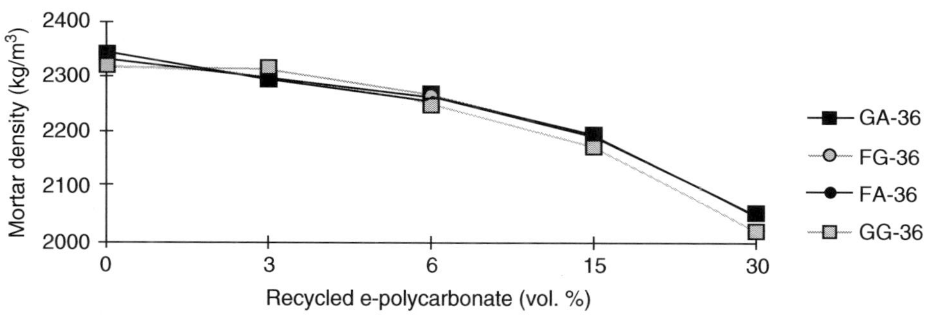

Figure 19.7 Densities of GA, GG, FA, and FG mortars.

do not affect the mortar durability to sodium chloride and sulfates, because they do not increase. These results contrast with the values reported by Hannawi et al. (2010) and Kaur and Pavia (2020), where the water absorption values decrease if the plastic aggregates increase.

19.3.6 Determination of Density

Recycled e-polycarbonate particles were added to the cement mortar. The densities at 36 days from FA, FG, GA, and GG mortars are shown in Figure 19.7. In general, as the recycled e-polycarbonate concentration increases the density decreases, since the polycarbonate density is 1.2 g/cm^3.

This decrease in the mortar density has been reported by various authors (Aciu et al. 2018; Choi et al. 2009; Hannawi et al. 2010; Marzouk et al. 2007; Safi et al. 2013; Wang and Meyer 2012), who used waste particles such as PET, HIPS, PVC, and polycarbonate as sand-substitution aggregates within the composite material.

19.4 Conclusions

This experimental work, to characterize new mortar composites containing extruded (size 1 × 3 mm) and recycled polycarbonate from electronic waste (recycled e-polycarbonate), registers the following conclusions. In general, mortar shows better compressive strengths when Fine (F) sand is used. The recycled e-polycarbonate can be utilized in mortars, elaborated with F and Grainy (G) sands, in concentrations up to 6% and 15%, respectively, without affecting the compressive strengths. The mortar shows better flexural strengths when it is elaborated with F sand than with G sand. The flexural values decreased from 10 to 7.5 MPa (for F sand), and from 8.5 to 6.5 MPa (for G sand), when sand was replaced 6% by recycled e-polycarbonate particles. The increments in recycled e-polycarbonate concentration provoke an important diminution in thermal conductivity, which can reach 0.87 W/mK when recycled e-polycarbonate reaches 6%. Recycled e-polycarbonate concentration increments do not affect the water absorption coefficient ($0.1\ mg/cm^2s^{1/2}$) even if it achieves a 30% level of G sand. The increments in recycled e-polycarbonate concentration provoke diminutions in mortar densities (up to 25% once recycled e-polycarbonate arrives at 30% of G sand).

Consequently, the environmental contamination caused by polycarbonate from electronic waste could be reduced if recycled e-polycarbonate were used as aggregate in cement mortar, and the recycled e-polycarbonate can be seen as an additive that benefits numerous mortar characteristics.

References

Achilias, D.S. and Antonakou, E.V. (2015). Chemical and thermochemical recycling of polymers from waste electrical and electronic equipment. In: *Recycling Materials Based on Environmentally Friendly Techniques* (ed. D.S. Achilias), 37–61. London, United Kingdom: IntechOpen.

Aciu, C., Ilutiu-Varvara, D.-A., Manea, D.-L. et al. (2018). Recycling of plastic waste materials in the composition of ecological mortar. *Procedia Manufacturing* 22: 274–279.

AENOR (2003). UNE-EN 1015–18:2003 Methods of test for mortar for masonry—Part 18: determination of water absorption coefficient due to capillary action of hardened mortar (definitive 2023).

AENOR (2012). UNE-EN 933–1:2012 Tests for geometrical properties of aggregates—Part 1: determination of particle size distribution—sieving method (definitive 2023).

AENOR (2018). UNE-EN 196–1:2018 Methods of testing cement—Part 1: determination of strength (definitive 2023).

AENOR (2020). UNE-EN 1015–11:2020 Methods of test for mortar for masonry—Part 11: determination of flexural and compressive strength of hardened mortar.

AENOR (2022). UNE-EN 933–2:2022 tests for geometrical properties of aggregates—Part 2: determination of size distribution—test sieves, nominal size of apertures.

American Concrete Institute (2002). *ACI 122R-02 Guide to Thermal Properties of Concrete and Masonry Systems, Concrete.*

Baldé, C.P., Forti, V., Gray, V. et al. (2017). *The Global E-waste Monitor 2017—Quantities, Flows, and Resources.* Bonn/Geneva/Vienna: United Nations University (UNU), International Telecommunication Union (ITU), and International Solid Waste Association (ISWA).

Choi, Y.W., Moon, D.J., Kim, Y.J., and Mohamed Lachemi, M. (2009). Characteristics of mortar and concrete containing fine aggregate manufactured from recycled waste polyethylene terephthalate bottles. *Construction and Building Materials* 23 (8): 2829–2835. https://doi.org/10.1016/j.conbuildmat.2009.02.036.

Covestro (2015). *Makrolon® Polycarbonate Sheet. Fabrication Guide, Research Report BRO0000915.*

Greenfield, M. (2018). *Global plastic production from 1950 to 2016 (in million metric tons).* The Statistics Portal.

Hannawi, K., Kamali-Bernard, S., and Prince, W. (2010). Physical and mechanical properties of mortars containing PET and PC waste aggregates. *Waste Management* 30 (11): 2312–2320.

Hannawi, K. and Prince-Agbodjan, W. (2015). Transfer behaviour and durability of cementitious mortars containing polycarbonate plastic wastes. *European Journal of Environmental and Civil Engineering* 19 (4): 467–481.

IHS Chemical (2012). After Major Downturn, Global Demand for Polycarbonate Growing Again, Says IHS Chemical Report.

Kaur, G. and Pavia, S. (2020). Physical properties and microstructure of plastic aggregate mortars made with acrylonitrile-butadiene-styrene (ABS), polycarbonate (PC), polyoxymethylene (POM) and ABS/PC blend waste. *Journal of Building Engineering* 31: 101341. https://doi.org/10.1016/j.jobe.2020.101341.

Lafarge Tarmac (2015). *Low Thermal Conductivity Concrete Solution Guide.* Marston Green, Birmingham, UK: Lafarge Tarmac.

Marzouk, O.Y., Dheilly, R.M., and Queneudec, M. (2007). Valorization of post-consumer waste plastic in cementitious concrete composites. *Waste Management* 27 (2): 310–318. https://doi.org/10.1016/j.wasman.2006.03.012.

Mordor Intelligence (2018). *Polycarbonate Market—Segmented by Product Type, End-user Industry and Geography—Growth, Trends and Forecast (2018–2023).*

Professional Plastics (2017). *Polycarbonate Material Data Sheet General Purpose, Glazing Grade.*

Safi, B., Saidi, M., Aboutaleb, D., and Madani Maallem, M. (2013). The use of plastic waste as fine aggregate in the self-compacting mortars: effect on physical and mechanical properties. *Construction and Building Materials* 43: 436–442. https://doi.org/10.1016/j.conbuildmat.2013.02.049.

Saikia, N. and De Brito, J. (2012). Use of plastic waste as aggregate in cement mortar and concrete preparation: a review. *Construction and Building Materials* 34: 385–401.

Wang, R. and Meyer, C. (2012). Performance of cement mortar made with recycled high impact polystyrene. *Cement and Concrete Composites* 34 (9): 975–981. https://doi.org/10.1016/j.cemconcomp.2012.06.014.

20

Generation, Collection, and Recycling Policies of E-waste Across the Asian Region

Rajwinder Singh[1,], Arti Thanki[2], Akhilesh Kumar Yadav[3], Anmol Kaur[4], and Karanvir Singh Sohal[5]*

[1] *Research Scholar, Department of Civil Engineering, DR BR Ambedkar National Institute of Technology, Jalandhar, Punjab, India*
[2] *Senior Scientist, UniStar Environment & Research Labs Pvt. Ltd., Dahej, Gujarat, India*
[3] *Associate Professor, Department of Mining Engineering, Indian Institute of Technology Bharti Hindu University, Varanasi, Uttar Pradesh, India*
[4] *Postgraduate Student, Satish Chandar Dhawan Govt. College, Ludhiana, Punjab, India*
[5] *Structural Designer, Geo Media Engineering and Consultancy Services Pvt. Ltd., Bathinda, Punjab, India*
* *Corresponding author*

20.1 Introduction

In the current market, there is a constant influx of new products and other electronic goods, while at the same time, older products are rapidly falling out of use and becoming outmoded. As a result, the amount of electronic waste that is produced is continuously climbing at a rapid rate year after year (Shahabuddin et al. 2022). Electronic waste or E-waste is mainly defined as the byproduct of all of the components of electronic and electrical equipment (EEE) that have been thrown away without the intent of being recycled or reused (Baldé et al. 2016; Nnorom and Osibanjo 2008). More than two billion metric tons of waste are generated every year all over the world, with E-waste making up 53.6 million tons of that total (Statista 2019). The amount of E-waste generated on a global scale has been reported to be steadily increasing. In terms of numbers, the total generation of E-waste in the year 2015 was around 42.3 Mt, which increased to 53.6 Mt in the year 2019, representing an overall 21% increase (Tiseo 2021). In addition, it has also been noticed that approximately 83% of the total E-waste produced in the year 2019 was not accounted for appropriately, and as a result, there are greater chances of its open burning or illegal dumping, both of which pose a significant risk to both human health and the environment (Statista 2019). However, in 2019, the residual 17% of E-waste was subjected to the collection and recycling processes in a suitable manner globally (Forti et al. 2020). When looking at the amount of E-waste generated in each region in 2019, it is seen that Asia produced the most (46.4% of the total), followed by the United States (24.4%), Europe (22.4%), Africa (5.4%), and Oceania (1.3%). Due to the high population density in Asia (4.40 billion people), the continent produced the most overall E-waste, but its residents produced significantly less E-waste per person than those living in Europe (16.2 kg/capita), Oceania (16.1 kg/capita), or the Americas (13.3 kg/capita). In the year 2019, the Asian region was responsible for the generation of E-waste of around 5.37 kg/capita (Ian Tiseo 2023; Pyramid 2019). However, Africa had the world's lowest rate of E-waste generation at 2.5 kg/ capita (Forti et al. 2020).

The recycling of E-waste is important because it is composed of a diverse variety of constituents, including plastics, metals, and glasses. Due to the difficulty involved in the collection and recycling processes, only a fraction (17.4%) of the world's E-waste is currently being reused or recycled (Forti et al. 2020; Sthiannopkao and Wong 2013). Recycling E-waste is a challenging task due to the fact that it is composed of toxic metals that vary in their physicochemical properties. It is possible for these potentially harmful compounds to pose a threat to the environment if they are not disposed of in an appropriate manner (Chen et al. 2015; Widmer et al. 2005).To lessen the dangers to human health and the deterioration of the environment caused by landfilling, it has become an international priority to develop effective methods of waste management and to find environmentally responsible ways to dispose of E-waste. The management of E-waste could be hampered by insufficient infrastructure and ineffective scientific technologies, despite the fact that the recovery of resources through E-waste could bring employment and commercial opportunities (Ikhlayel 2018).

Electronic Waste Management: Policies, Processes, Technologies, and Impact, First Edition. Edited by Sunil Kumar and Vineet Kumar.

In the current situation, the appropriate management of E-waste is the main problem since it has been neglected in the first place. As found observed in the literature, several experimental and numerical studies have been carried out while providing information on the generation rate, prediction, and collection systems involved in the management of E-waste in major continents of the world (Andeobu et al. 2021; Fuller 2020; Kahhat et al. 2008). However, the data on the current environmental concerns, policies, and possible solutions to mitigate these issues either have not been discussed in detail or are outdated, particularly in the context of the Asian region (Golev et al. 2016; Herat and Agamuthu 2012; Widmer et al. 2005). Therefore, considering all of the problems mentioned above, the aim of the current article is to provide the latest information regarding the generation rate, processes involved in the recycling and recovery of E-waste such as collection, recovery methods, problems, and opportunities involved in the management of E-waste across the Asian region. In addition to this, a summary of the management policies and legislation and other aspects of E-waste in Asia and the rest of the continents have also been articulated. While collecting the data to achieve the proposed objectives, only those relevant and recent studies were taken into account, which covers the wide spectrum of E-waste management and its attributes. Based on the best knowledge of the authors, no such study has been carried out that focuses and highlights the important information regarding the aforementioned aspects of E-waste with a particular attention to the Asian continent.

20.2 Types and Typical Configurations of Ingredients of E-waste

From household items, lighting equipment, and toys to medical devices, a variety of items fall under the class of E-waste. The adoption of more modern lifestyles, the acceleration of technological innovation, and the acceleration of global economic growth have all contributed to the growing production of this waste, which has resulted in its becoming an environmental and health concern. The normal configuration of components existing in E-waste is illustrated in Figure 20.1.

E-waste is usually categorized on the basis of its origin and application. Some of the appliances are presented in Figure 20.2. However, the typical percentage of E-waste in each category might vary based on consumer behavior, population size, the needs of inhabitants, and dependency of the people on the electronic and electrical appliances (Forti et al. 2020). An approximate percentage of 42.1% of E-waste has been reported to be produced from household items, followed by the IT sector (33.9%), inhabitants' devices (13.7%), and other smaller appliances from household items (4.7%) (European Union 2012).

20.3 Processes Involved in the Recycling and Recovery of E-waste

The economic, environmental, and health benefits of properly managing E-waste include recycling and recovering it. Only when E-waste is recycled and valuable elements are recovered can it be considered a resource. In the following sections, the collection, recycling, and treatment facilities of E-waste are discussed in detail.

20.3.1 Collection

The first thing that needs to be done in order to properly manage electronic trash is to collect the garbage that is produced by companies and homes. The collection of E-waste can be carried out by adopting these approaches (Awasthi and Li 2017; Hicks et al. 2005): (i) the authorized

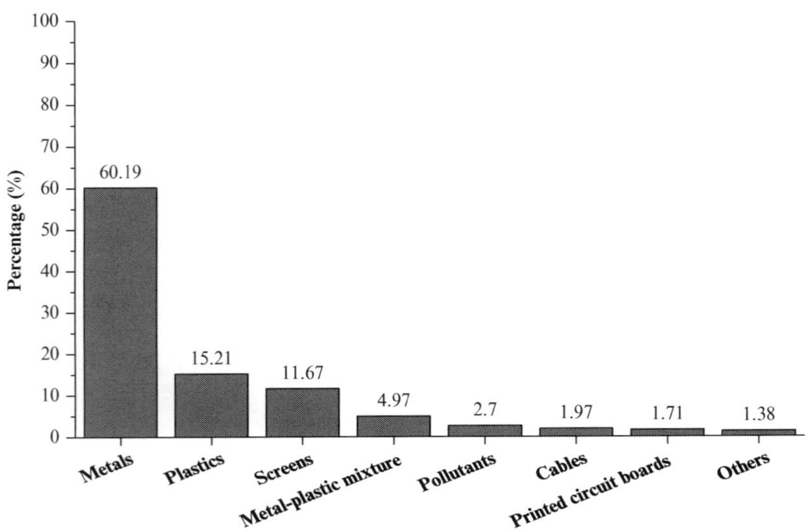

Figure 20.1 Structure of various components such as metals, plastic, and others present in E-waste (Ilankoon et al. 2018 / with permission of Elsevier).

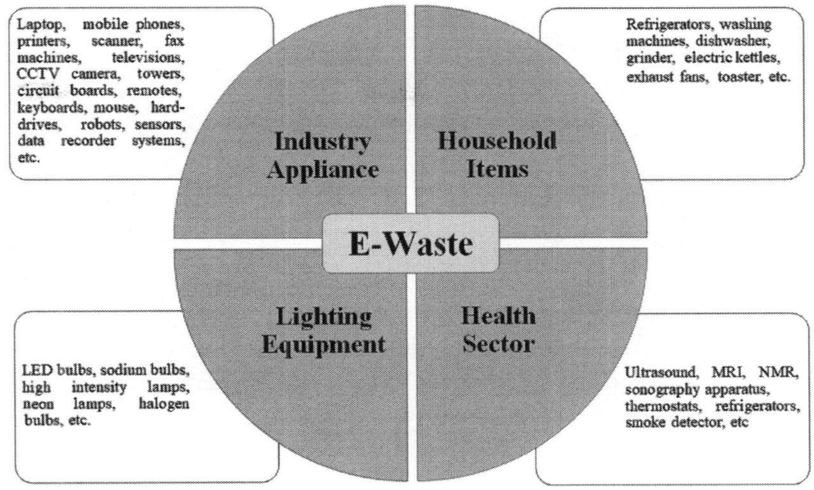

Figure 20.2 Major categories of E-waste based on their generation and list of corresponding items (Babu et al. 2007; Ongondo et al. 2011).

withdrawal scheme; (ii) separation of the remaining waste; (iii) collection apart from the authorized take-back scheme for countries with advanced waste management; and (iv) collection apart from the authorized withdraw scheme for nations with and without advanced waste management. Most collection sites will have distinct containers or boxes for different types of E-waste at this stage, since doing so according to best practices is essential. This is especially true for battery-containing E-waste, which necessitates specific handling and poses a serious risk if it is dumped in with regular trash. According to the latest report, with a tentative value of US$5.48 billion, Europe was the biggest collector of E-waste in 2019 (Forti et al. 2020). This region collected 5.1 million tons or 42.3% of the world's total E-waste. Asia was the second-best collector, with a total of 2.9 million tons (11.7%), followed by the Americas with 1.2 million tons (9.4%), Oceania with 0.06 million tons (8.8%), and Africa with 0.03 million tons (0.9%). There is currently US$57.0 billion worth of electronic garbage being produced all over the world. Out of this, only $10 billion worth of E-waste is recovered through appropriate techniques (Forti et al. 2020). The overall worth of collected E-waste versus the amount collected in the various regions is presented in Figure 20.3. The collection systems in each region are determined by their policies, available funds, and public awareness (Golev et al. 2016; Sthiannopkao and Wong 2013).

20.3.2 Storage

Despite the fact that safe storage of E-waste is a risky task, proper channels are acquired to store it while at the same time focusing on some sort of segregation, if possible. Items having large size and consisting of circuits such as CRT, TVs, and monitors are usually heavily polluted with lead and are separated out from the whole E-waste. The screens having film and glass are sent back to the other processing units for

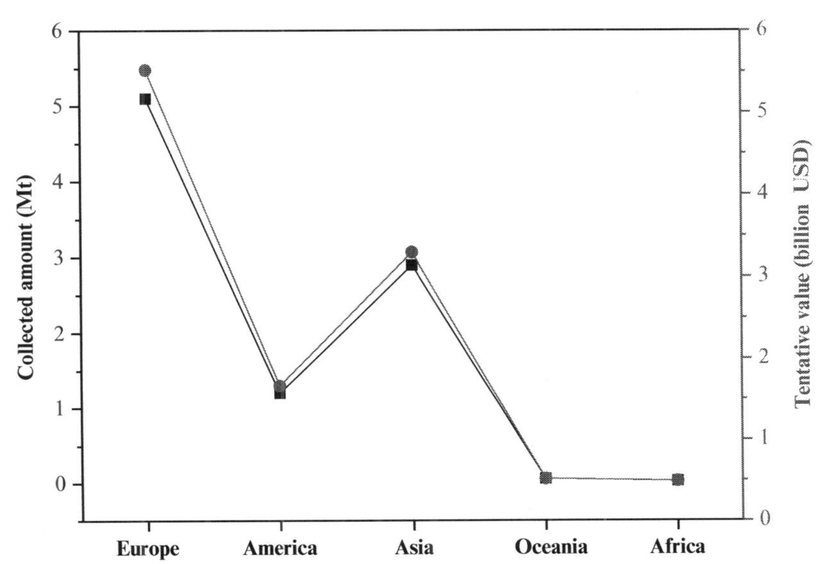

Figure 20.3 Collection of E-waste versus its corresponding value in the different continents of the world. (Adapted from Forti et al. 2020).

reuse purposes, while the rest of the items follow for their dismantling (Babu et al. 2007; Cui and Forssberg 2003).

20.3.3 Pretreatment and Recovery Methods

Followed by the storage of the E-waste, the distinct materials present in it are recovered through various types of facilities (Shaikh et al. 2020).

20.3.3.1 Manual Sorting, Dismantling, Shredding

Manual sorting of E-waste begins with the removal of various materials (such as batteries and bulbs) from the trash stream for further processing. Dismantling of components for reuse or the recovery of valuable materials may also occur at this stage (Babu et al. 2007). Once the E-waste has been shredded into little pieces, the materials may be sorted more accurately. In underdeveloped countries, manual separation and small-scale metal recovery are widespread. Low-intensity magnet drums and electric conductivity-based equipment are employed to isolate the ferrous and non-ferrous metals from the whole E-waste. During hand disassembly, dangerous and precious elements are often separated (Golev et al. 2016).

20.3.3.2 Mechanical Separation

The mechanical separation of the various materials is a series of sequential processes (Cui and Forssberg 2003; Kang and Schoenung 2005; Ravi 2012). Mechanical pretreatment includes sizing reduction, crushing and shredding, and adopting heavy machines. During normal and peak times, mechanical separation is undertaken so as to hasten the pace of the process. Developed countries generally use semi-automatic (labor-intensive and power-driven process) separation for metal recovery. The two most important phases of separation are employed through magnetic and water-based systems.

20.3.3.3 Recovery

The materials that have been segregated from the clusters of E-waste can be sold or reused in local or national recovery facilities. Plastic and steel are two examples of items that are diverted into another recycling stream, while others may be treated on-site and sold alongside usable components that have been isolated at the beginning of the production process (Garlapati 2016). After the separation of the useful metals from the respective items, they are shifted to the ultimate recovery step through pyrometallurgy and hydrometallurgy (Duan et al. 2016; Khaliq et al. 2014). Using high temperatures, pyrometallurgy purifies and extracts ferrous metal through following refining, smelting, combustion, and burning processes, while hydrometallurgy is utilized to recover non-ferrous metals, such as Cu, Pb, and Zn.

20.4 Generation, Problems, and Opportunities Associated with E-waste

It is difficult to get accurate statistics on global E-waste creation since the quantity of utilized WEEE that is nearing the end of its useful life cannot be quantified directly. Based on manufacturing or sales statistics, and the expected life duration of WEEE, most estimations are based on best guesses (Herat and Agamuthu 2012; Jaiswal et al. 2015). Many governments have implemented E-waste inventories to figure out the quality and types of E-waste that are being generated. Usually, consumers are more likely to keep using WEEE at home and in their workplaces, making assessment more difficult. Information about the overall generation of E-waste in major Asian countries and other continents is summarized in Table 20.1.

20.4.1 Description of Problems in the Context of the Asian Region

In the present scenario, the scientific literature has started to make people aware of the problems that are related to E-waste. As mentioned in the previous sections, E-waste consists of a variety of distinct materials that have extremely high toxicity toward human health and the environment. The items, such as central processing units (CPU) of computers, mobile phones, laptops, and cathode ray or radon filaments, are used in the daily routine and consist of a variety of potentially harmful chemicals and heavy metals such as cadmium, lead, mercury, silver, chromium, zinc, lead, tin, and copper, which can have a serious impact on human health as well as on the environment. Therefore, their practice in daily chores should be restricted or eliminated (Barontini and Cozzani 2006; Birnbaum and Staskal 2004). However, it has also been found that in the Asian region, the people living and working close to the E-waste disposal sites are affected the most. Because of illegal recycling, the problems are governed by the employees working without personal safety equipment, at low salary levels, using labor-intensive technology, with negligible regulation and registration, and in poor socioeconomic position (Annamalai 2015). In addition to this, several toxic pollutants have also been detected in the air, bottom ash, soil, and water streams (Csi and Jiang 2006; Han et al. 2009; Smith et al. 2006). For example, phthalate esters are synthetic substances present in products such as electronics as well as cables and wires and are readily released into the environment during the dismantling/destruction process of electronic equipment, as these compounds are not chemically bound to the items (Ma et al. 2013; Wang et al. 2015; Zhang et al. 2019). Studies have reported

Table 20.1 Production rate of E-waste in Asia and other regions in the respective year.

Country	Overall generation	Source
India	7.71 lakh tonnes (2018–19), 10.14 lakh tonnes (2019–20)	(Athrady 2022)
China	10.1 million tons (2020)	(Neha 2020)
Japan	860 000 (2005)	(Herat and Agamuthu 2012)
Malaysia	688 000 metric tons 2008	(Soo et al. 2013)
Pakistan	1790 kilotons (2018–2019)	(Shaikh 2021)
UK	1.6 million tons (2019)	(Malloy 2021)
Switzerland	66 042 (2003)	(Herat and Agamuthu 2012)
Germany	853 124 tons (2018)	(Kotschik 2021)
Africa	2.9 Mt of (2019)	(Lebbie et al. 2021)
Australia	465 818 tons (2016–17)	(Spencer 2020)
USA	1.8 million tons (2012)	(Schumacher 2016)

elevated lead concentrations in human beings involved in informal recycling practices, which were caused by the burning of items containing lead (Haryanto 2016; Liu et al. 2009; Parween et al. 2018; Pascale et al. 2016). For instance, in China, e-waste is considered a commodity, and as a result of lower labor costs, informal laborers are capable of paying greater prices through curbside collection services. Consequently, informal employees collect greater proportions of E-waste (30–90% of WEEE created) than formal workers (7.8–24%) (Salhofer et al. 2016). Almost 90% of e-waste is processed in unsafe, uncontrolled conditions by the informal sector. Unemployment, a lack of education, and a poor degree of economic growth are some of the factors that force individuals to sacrifice their health for financial gain (Awasthi and Li 2017; Townsend 2011). Unlike in Europe and the United States, e-waste is valued in China, resulting in informal and formal industrial marketplaces (Castro et al. 2023; Garg 2021).

20.5 Legislation and Policies in Asia and Other Countries

Brief information regarding the legislation and policies imposed in Asia and other countries is highlighted in Table 20.2.

Based on Table 20.2, it can be clearly noticed that most of the regions do have rules and regulations that were invented almost a decade ago to mitigate the problems linked with the appropriate organization of E-waste, but still, there is the requirement to adopt a proper channel to handle E-waste.

20.5.1 Other Initiatives Regarding the Management of E-waste in the World

20.5.1.1 Basel Convention

The Basel Convention was enacted in 1989 with the title "The Control of Transboundary Movements of Hazardous Wastes and Their Disposal." It is the most extensive worldwide environmental agreement on hazardous waste and other waste that has ever been negotiated. The parties involved in the convention were supposed to ensure that the waste was handled and disposed of in a manner that was safe for the environment. Waste that is toxic, poisonous, explosive, corrosive, flammable, ecotoxic, and contagious are all included in the scope of the convention. In addition, parties were obligated to reduce the quantities of trash that are transported, to process and dispose of waste as close as is practicable to the location where the waste was generated, and to prevent or reduce the amount of waste that is produced at the point of origin (Kumar and Singh 2013).

20.5.1.2 Solving the E-waste Problem (StEP)

Founded in 2007, the United Nation's Solving the E-waste Problem (StEP) initiative brings together enterprises, academic institutions, governmental agencies, international organizations, and NGOs to develop ecologically, economically, and properly sound E-waste recovery, reuse, and prevention strategies. This initiative worked on the basis of five task forces, including Policy, ReDesign, ReUse, and ReCycle, along with Capacity Building (Sthiannopkao and Wong 2013).

20.5.1.3 3Rs

3R is an environmental system and plan for waste minimization that stands for Reduce, Reuse, and Recycle. 3R was

Table 20.2 Regulations, recycling techniques, and rates of E-waste in Asia and other countries (Awasthi, Cucchiella et al. 2018; Herat and Agamuthu 2012; Shahabuddin et al. 2022).

Countries	Regulations	Main recycling approach	Recycling rate and source
India	Management of E-waste, 2016 and 2022	Normal mechanical techniques	21% (Jadhav 2022)
Pakistan	National Environment Policy, 2005	Manual sorting, normal mechanical techniques	6% (Imran et al. 2017)
Japan	Specified Home Appliance Recycling Law (SHARL), 1998	Advanced mechanical processing, manual sorting	55–70% (Ternald 2019)
China	Technical Policy on Pollution Prevention and Control of WEEE, (2006); Management of the Recycling and Disposal of Waste Electrical and Electronic Equipment (2011)	Advanced pyrometallurgical processes	40% (Werf 2020)
Bangladesh	Electrical and Electronic Waste Rules, 2011	Manual collection, sorting techniques	20–30% (Alam and Bahauddin 2015)
Malaysia	Environmental Quality Regulations, 2005	Mechanical techniques	25% (Lim 2019)
Vietnam	The Law on Environmental Protection, 2014	Mechanical techniques	20% (Hai 2014)
United States	Resource, Conservations and Recovery act (RCRA), 1976; Environmental Protection Agency (EPA): Cathode ray tube rules	Advance mechanical processing, pyrometallurgical, hydrometallurgical processes	40% (Awasthi and Wang et al. 2018)
European Union	Waste Electrical and Electronic Equipment Directive 2012/2019/EU; Restriction of Hazardous Substances Directive (RoHS) 2011/65/EU	Advanced mechanical techniques	33–40% (Awasthi and Wang et al. 2018)
African countries	Electronic Waste Management Policy, 2012	Manual sorting, mechanical techniques	17.4% (Fuller 2020)

announced in Japan in the year 2006. It works on the hierarchy principle, and its purpose is to provide as much product benefit as possible with as little trash as possible. People and organizations apply the 3R principles in order to assess purchase possibilities that are both viable and environmentally sound (Sthiannopkao and Wong 2013).

20.5.2 Extended Producer Responsibility (EPR)

Extended Producer Responsibility (EPR) is a policy idea that has led to successful waste management laws in a number of developed countries. The policy was made so that local governments would not have to spend as much money to collect and treat waste, especially waste that is hard and expensive to treat or recycle in a safe way. EPR is based on a simple principle: producers who design and promote items should be responsible for their disposal at the end of their useful lifespan. Although EPR systems may differ, a few fundamentals remain constant. On the basis of the experiences of industrialized economies, certain fundamental lessons can be derived as prerequisites for effective EPR systems. Over the last two decades, several industrialized countries have implemented EPR systems for a wide range of products, including packaging, cars, e-products, home hazardous waste, pharmaceuticals, batteries, and accumulators. China and India have recently created EPR laws after seeing the rapid acceptance of EPR in advanced countries.

Several studies have examined the adoption of EPR for E-waste management in developing Asian nations (Herat 2021; Kojima et al. 2009; Pariatamby and Victor 2013; Widmer et al. 2005). These prior research studies focused on the mismatch between the specific issues and conditions in emerging Asia and the overall EPR policy approach that countries in the region are striving to apply.

20.5.2.1 Difficulties with the Implementation of EPR

The challenge of identifying the producer is one of the greatest obstacles to the effective operation of EPR systems in emerging economies. Due to its cheap operating expenses, the informal sector is able to give consumers relatively higher reimbursements for end-of-life. In contrast to established, regulated recycling enterprises, informal ones typically require less cash to launch.

In the majority of countries where EPR law has been enacted, a waste collection infrastructure was already in existence, typically run or commissioned by municipalities. The introduction of an EPR system primarily included a financial transfer from the local government to producers. In industrialized economies, environmental awareness of E-waste is typically strong, which aids in its appropriate disposal. In developing economies, understanding of these issues is often poor; this, combined with poverty and a dearth of alternative work possibilities, makes it simple to disregard environmental protection in E-waste handling. The lack of effective monitoring methods to guarantee that these commodities are handled appropriately in importing nations is an essential issue. Therefore, EPR systems for both industrialized and developing nations would need to be constructed with global resource circulation in mind.

Multiple subcomponents addressing distinct concerns, physical infrastructure and compliant behavior, and enough institutions and capacity to impose laws at all levels, including municipal, are required for an EPR system. Therefore, Asian nations must analyze their capacity relative to the EPR system's resource requirements and intended aims.

20.6 Discussion

In spite of the regulations and mechanisms that have been described in the above sections, more than half of the E-waste that is collected for recycling in rich countries is shipped to the poorer parts of the world to be processed or disposed of there. Over the course of the past few years, China, India, Pakistan, and Nigeria have been recognized as the top four importers of E-waste (Duan et al. 2016; Imran et al. 2017; Sthiannopkao and Wong 2013; Yu et al. 2010). The people dealing with the shipment and processing of E-waste find it easy to import this waste by labeling it as used products, which the Basel Convention does not cover, yet, these four countries intend to address the importation of hazardous waste. Local law enforcement officials typically do not place a high priority on the task of enforcing regulations concerning the import or processing of E-waste, and they are frequently persuaded to look the other way when violations occur. Apart from these aspects, other management strategies could involve the application of the by-products generated after segregating the valuable materials from E-waste in industrial (Ottoni et al. 2020; Realff et al. 2004), agricultural (Islam et al. 2021; Singh and Patel 2022), construction (Sanito et al. 2021; Sodhi et al. 2021; Dixit and Singh 2022; Singh and Patel 2022a), pharmaceutical (Jayapradha 2015; Sanito et al. 2021), metallurgical (Priya and Hait 2017; Thakur and Kumar 2020), and other important sectors as raw materials have been reported, but extensive research in these sectors is still required.

20.7 Conclusions and Recommendations

The pace at which electronic garbage is produced around the world is far higher than that of the vast majority of other types of waste. Even though several programs have been started in order to achieve sustainable management of E-waste, there are still a substantial number of problems and difficulties that need to be solved. Developing nations, particularly those in the Asian region, are facing a significant challenge since their governments do not have the strict regulations or the infrastructure in place to deal with this problem in a sustainable way. The resolutions to mitigate the problems and difficulties associated with E-waste are required to be implemented strictly in the Asian region, just as in developed countries such as the United States and the United Kingdom. The customers and stakeholders are required to work together cooperatively. E-waste presents a challenge because of the presence of potentially harmful components, but it also offers a solution to the problem of the natural resources upon which electronic and electrical equipment producers rely. It is of the utmost importance to recycle E-waste in an appropriate manner to successfully harvest the secondary sources. By imposing rigorous recycling policies for the management of E-waste, the health and environmental problems can also be lowered. These policies can be broken down into three basic stages: appropriate collection, sorting/disassembly, and pre-processing and end-processing. In developing nations, the informal E-waste recycling sector is responsible for the majority of these three operations. Even if the first two of these stems can be carried out properly in Asian countries, their adverse impact can be lowered. To efficiently recover valuable materials, it will be necessary to have end-processing facilities that are up to date with the latest technological advances, such as integrated smelters. With the exception of strong growing nations within the Asian region, the majority of developing countries do not have the financial resources necessary to make the substantial investments required by such technologies. It is recommended that developing nations be provided with the right technology for the management of their E-waste problem while taking into account the countries' social, environmental, and economic constraints.

Acknowledgment

The authors acknowledge the support provided by their colleagues to improve the quality of the manuscript.

References

Alam, M. and Bahauddin, K.M. (2015). Electronic waste in Bangladesh: evaluating the situation, legislation and policy and way forward with strategy and approach. *Present Environment and Sustainable Development* 9 (1): 81–101. https://doi.org/10.1515/pesd-2015-0005.

Andeobu, L., Wibowo, S., and Grandhi, S. (2021). An assessment of e-waste generation and environmental management of selected countries in Africa, Europe and North America: a systematic review. *Science of the Total Environment* 792: 148078. https://doi.org/10.1016/j.scitotenv.2021.148078.

Annamalai, J. (2015). Occupational health hazards related to informal recycling of E-waste in India: an overview. *Indian Journal of Occupational and Environmental Medicine* 19 (1): 61. https://doi.org/10.4103/0019-5278.157013.

Athrady, A. (2022). Generation of e-waste in India sees 31% annual growth. *Deccan Herald*.

Awasthi, A.K., Cucchiella, F., D'Adamo, I. et al. (2018). Modelling the correlations of e-waste quantity with economic increase. *Science of the Total Environment* 613–614: 46–53. https://doi.org/10.1016/j.scitotenv.2017.08.288.

Awasthi, A.K. and Li, J. (2017). Management of electrical and electronic waste: a comparative evaluation of China and India. *Renewable and Sustainable Energy Reviews* 76 (February): 434–447. https://doi.org/10.1016/j.rser.2017.02.067.

Awasthi, A.K., Wang, M., Wang, Z. et al. (2018). E-waste management in India: a mini-review. *Waste Management and Research* 36 (5): 408–414. https://doi.org/10.1177/0734242×18767038.

Babu, B.R., Parande, A.K., and Basha, C.A. (2007). Electrical and electronic waste: a global environmental problem. *Waste Management and Research* 25 (4): 307–318. https://doi.org/10.1177/0734242×07076941.

Baldé, C.P., Wang, F., and Kuehr, R. (2016). *Transboundary movements of used and waste electronic and electrical equipment: Estimates from the European Union using trade statistics*. Available at: http://www.step-initiative.org/files/_documents/other_publications/UNU-Transboundary-Movement-of-Used-EEE.pdf%0AA; http://www.step-initiative.org/files/step-2014/Publications/GreenandWhitePapers/UNU-Transboundary-Movement-of-Used-EEE.pdf (accessed 27 August 2023).

Barontini, F. and Cozzani, V. (2006). Formation of hydrogen bromide and organobrominated compounds in the thermal degradation of electronic boards. *Journal of Analytical and Applied Pyrolysis* 77 (1): 41–55. https://doi.org/10.1016/j.jaap.2006.01.003.

Birnbaum, L.S. and Staskal, D.F. (2004). Brominated flame retardants: cause for concern? *Environmental Health Perspectives* 112 (1): 9–17. https://doi.org/10.1289/ehp.6559.

Castro, F.D., Botelho Júnior, A.B., Bassin, J.P. et al. (2023). E-waste policies and implementation: a global perspective. In: *Waste Management and Resource Recycling in the Developing World* (ed. P. Singh, P. Verma, and A.C.S. Batalhão), 271–307. Elsevier. https://doi.org/10.1016/B978-0-323-90463-6.00016-6.

Chen, M., Zhang, S., Huang, J., and Chen, H. (2015). Lead during the leaching process of copper from waste printed circuit boards by five typical ionic liquid acids. *Journal of Cleaner Production* 95: 142–147. https://doi.org/10.1016/j.jclepro.2015.02.045.

Csi, Z. and Jiang, G. (2006). Determination of polybrominated diphenyl ethers in soil from e-waste recycling site. *Talanta* 70 (1): 88–90. https://doi.org/10.1016/j.talanta.2006.01.016.

Cui, J. and Forssberg, E. (2003). Mechanical recycling of waste electric and electronic equipment: a review. *Journal of Hazardous Materials* 99 (3): 243–263. https://doi.org/10.1016/S0304-3894(03)00061-X.

Dixit, S. and Singh, P. (2022). Investigating the disposal of E-waste as in architectural engineering and construction industry. *Materials Today: Proceedings* 56: 1891–1895. https://doi.org/10.1016/j.matpr.2021.11.163.

Duan, H., Hu, J., Tan, Q. et al. (2016). Systematic characterization of generation and management of e-waste in China. *Environmental Science and Pollution Research* 23 (2): 1929–1943. https://doi.org/10.1007/s11356-015-5428-0.

European Union (2012). *Directive 2012/19/EU of the European Parliament and of the Council on Waste Electrical and Electronic Equipment (WEEE)*. European Parliment. Available at: https://eur-lex.europa.eu/LexUriServ/LexUriServ.do?uri=OJ:L:2012:197:0038:0071:en:PDF (accessed 20 June 2022).

Forti, V., Baldé, C.P., Kuehr, R., and Bel, G. (2020). *The Global E-waste Monitor 2020—ITU*. United Nations University.

Fuller, L. (2020). Africa recycles less than 0.1% e-waste. E-Waste Expo. Available at: https://www.ewaste-expo.com/africa-recycles-less-than-0-1-e-waste/#:~:text=TheITUnotesthatwhile,declinedto17.4%25in2019.%5C (accessed 20 June 2022).

Garg, C.P. (2021). Modeling the e-waste mitigation strategies using grey-theory and DEMATEL framework. *Journal of Cleaner Production* 281: 124035. https://doi.org/10.1016/j.jclepro.2020.124035.

Garlapati, V.K. (2016). E-waste in India and developed countries: management, recycling, business and biotechnological initiatives. *Renewable and Sustainable Energy Reviews* 54: 874–881. https://doi.org/10.1016/j.rser.2015.10.106.

Golev, A., Schmeda-Lopez, D.R., Smart, S.K. et al. (2016). Where next on e-waste in Australia? *Waste Management* 58: 348–358. https://doi.org/10.1016/j.wasman.2016.09.025.

Hai, H.T. (2014). *Electric and Electronic Waste Recycling in Vietnam*. Available at: https://www.epa.gov/sites/production/files/2014-08/documents/vietnam_country_presentation_2-_prof._hai.pdf (accessed 14 August 2023).

Han, W., Feng, J., Gu, Z. et al. (2009). Polybrominated diphenyl ethers in the atmosphere of Taizhou, a major E-waste dismantling area in China. *Bulletin of Environmental Contamination and Toxicology* 83 (6): 783–788. https://doi.org/10.1007/s00128-009-9855-9.

Haryanto, B. (2016). Lead exposure from battery recycling in Indonesia. *Reviews on Environmental Health* 31 (1): 13–16. https://doi.org/10.1515/reveh-2015-0035.

Herat, S. (2021). E-waste management in Asia Pacific region: review of issues, challenges and solutions. *Nature Environment and Pollution Technology* 20 (1): 45–53. https://doi.org/10.46488/NEPT.2021.v20i01.005.

Herat, S. and Agamuthu, P. (2012). E-waste: a problem or an opportunity? Review of issues, challenges and solutions in Asian countries. *Waste Management & Research: The Journal for a Sustainable Circular Economy* 30 (11): 1113–1129. https://doi.org/10.1177/0734242×12453378.

Hicks, C., Dietmar, R., and Eugster, M. (2005). The recycling and disposal of electrical and electronic waste in China—legislative and market responses. *Environmental Impact Assessment Review* 25 (5 SPEC. ISS.): 459–471. https://doi.org/10.1016/j.eiar.2005.04.007.

Ikhlayel, M. (2018). An integrated approach to establish e-waste management systems for developing countries. *Journal of Cleaner Production* 170: 119–130. https://doi.org/10.1016/j.jclepro.2017.09.137.

Ilankoon, I.M.S.K., Ghorbani, Y., Chong, M.N. et al. (2018). E-waste in the international context—a review of trade flows, regulations, hazards, waste management strategies and technologies for value recovery. *Waste Management* 82: 258–275. https://doi.org/10.1016/j.wasman.2018.10.018.

Imran, M., Haydar, S., Kim, J. et al. (2017). E-waste flows, resource recovery and improvement of legal framework in Pakistan. *Resources, Conservation and Recycling* 125 (January): 131–138. https://doi.org/10.1016/j.resconrec.2017.06.015.

Islam, M.K., Khatun, M.S., Arefin, M.A. et al. (2021). Waste to energy: an experimental study of utilizing the agricultural residue, MSW, and e-waste available in Bangladesh for pyrolysis conversion. *Heliyon* 7 (12): e08530. https://doi.org/10.1016/j.heliyon.2021.e08530.

Jadhav, R. (2022). Around 78% of India's e-waste is not being collected or disposed by the government. *Business Line*. Available at: thehindubusinessline.com/data-stories/data-focus/around-78-of-indias-e-waste-is-not-being-collected-or-disposed-by-the-government/article65406820.ece (accessed 20 June 2022).

Jaiswal, A., Samuel, C., Patel, B.S., and Kumar, M. (2015). Go green with WEEE: eco-friendly approach for handling e-waste. *Procedia Computer Science* 46 (Icict 2014): 1317–1324. https://doi.org/10.1016/j.procs.2015.01.059.

Jayapradha, A. (2015). Scenario of E-waste in India and application of new recycling approaches for E-waste management. *Journal of Chemical and Pharmaceutical Research* 7 (3): 232–238.

Kahhat, R., Kim, J., Xu, M. et al. (2008). Exploring e-waste management systems in the United States. *Resources, Conservation and Recycling* 52 (7): 955–964. https://doi.org/10.1016/j.resconrec.2008.03.002.

Kang, H.Y. and Schoenung, J.M. (2005). Electronic waste recycling: a review of U.S. infrastructure and technology options. *Resources, Conservation and Recycling* 45 (4): 368–400. https://doi.org/10.1016/j.resconrec.2005.06.001.

Khaliq, A., Rhamdhani, M.A., Brooks, G., and Masood, S. (2014). Metal extraction processes for electronic waste and existing industrial routes: a review and Australian perspective. *Resources* 3 (1): 152–179. https://doi.org/10.3390/resources3010152.

Kojima, M., Yoshida, A., and Sasaki, S. (2009). Difficulties in applying extended producer responsibility policies in developing countries: case studies in e-waste recycling in China and Thailand. *Journal of Material Cycles and Waste Management* 11 (3): 263–269. https://doi.org/10.1007/s10163-009-0240-x.

Kotschik, G. (2021). Electrical and electronic waste in Germany, umweltbundesamt. Available at: https://www.umweltbundesamt.de/en/topics/waste-resources/product-stewardship-waste-management/electrical-electronic-waste#electrical-and-electronic-waste-in-germany (accessed 20 June 2022).

Kumar, U. and Singh, D.D.N. (2013). E-waste management through regulations. *International Journal of Engineering Inventions* 3 (2): 6–14. Available at. www.ijeijournal.com.

Lebbie, T.S., Moyebi, O.D., Asante, K.A. et al. (2021). E-waste in Africa: a serious threat to the health of children. *International Journal of Environmental Research and Public Health* 18 (16): https://doi.org/10.3390/ijerph18168488.

Lim, B. (2019). Reducing the impact of e-waste. *Straits Times*.

Liu, Q., Cao, J., Li, K.Q. et al. (2009). Chromosomal aberrations and DNA damage in human populations exposed to the processing of electronics waste. *Environmental Science and Pollution Research* 16 (3): 329–338. https://doi.org/10.1007/s11356-008-0087-z.

Ma, T.T., Christie, P., Luo, Y.M., and Teng, Y. (2013). Phthalate esters contamination in soil and plants on agricultural land near an electronic waste recycling site.

Environmental Geochemistry and Health 35 (4): 465–476. https://doi.org/10.1007/s10653-012-9508-5.

Malloy, K. (2021). UK on track to become Europe's biggest E-waste contributor. *Resource*. Available at: https://resource.co/article/uk-track-become-europe-s-biggest-e-waste-contributor (accessed 20 June 2022).

Neha, A. (2020). World's e-waste 'unsustainable', says UN report citing China, India and U.S. *Reuters*. Available at: https://www.reuters.com/article/us-global-waste-un-report-idUSKBN243255 (accessed 20 June 2022).

Nnorom, I.C. and Osibanjo, O. (2008). Electronic waste (e-waste): material flows and management practices in Nigeria. *Waste Management* 28 (8): 1472–1479. https://doi.org/10.1016/j.wasman.2007.06.012.

Ongondo, F.O., Williams, I.D., and Cherrett, T.J. (2011). How are WEEE doing? A global review of the management of electrical and electronic wastes. *Waste Management* 31 (4): 714–730. https://doi.org/10.1016/j.wasman.2010.10.023.

Ottoni, M., Dias, P., and Xavier, L.H. (2020). A circular approach to the e-waste valorization through urban mining in Rio de Janeiro, Brazil. *Journal of Cleaner Production* 261: 120990. https://doi.org/10.1016/j.jclepro.2020.120990.

Pariatamby, A. and Victor, D. (2013). Policy trends of e-waste management in Asia. *Journal of Material Cycles and Waste Management* 15 (4): 411–419. https://doi.org/10.1007/s10163-013-0136-7.

Parween, A., Khan, M.M., Upadhyay, T., and Tripathi, R.V. (2018). Prevalence of elevated blood lead level in children of India. *Nature Environment and Pollution Technology* 17 (3): 703–710. Available at. https://neptjournal.com/upload-images/NL-65-6-(4)B-3521.pdf.

Pascale, A., Sosa, A., Bares, C. et al. (2016). E-waste informal recycling: an emerging source of lead exposure in South America. *Annals of Global Health* 82 (1): 197. https://doi.org/10.1016/j.aogh.2016.01.016.

Priya, A. and Hait, S. (2017). Comparative assessment of metallurgical recovery of metals from electronic waste with special emphasis on bioleaching. *Environmental Science and Pollution Research* 24 (8): 6989–7008. https://doi.org/10.1007/s11356-016-8313-6.

Pyramid, P. (2019). *Population Pyramids of the World from 1950 to 2100, Population Pyramid*. Available at: https://www.populationpyramid.net/asia/2019 (accessed 10 February 2023).

Ravi, V. (2012). Evaluating overall quality of recycling of e-waste from end-of-life computers. *Journal of Cleaner Production* 20 (1): 145–151. https://doi.org/10.1016/j.jclepro.2011.08.003.

Realff, M.J., Raymond, M., and Ammons, J.C. (2004). E-waste: an opportunity. *Materials Today* 7 (1): 40–45. https://doi.org/10.1016/S1369-7021(04)00054-9.

Salhofer, S., Steuer, B., Ramusch, R., and Beigl, P. (2016). WEEE management in Europe and China—a comparison. *Waste Management* 57: 27–35. https://doi.org/10.1016/j.wasman.2015.11.014.

Sanito, R.C., You, S.-J., and Wang, Y.-F. (2021). Application of plasma technology for treating e-waste: a review. *Journal of Environmental Management* 288: 112380. https://doi.org/10.1016/j.jenvman.2021.112380.

Schumacher, K.A. (2016). Electronic waste management in the U.S.: practice and policy. Doctoral dissertation. University of Delaware. Available at: http://udspace.udel.edu/handle/19716/20333 (accessed 15 August 2023).

Shahabuddin, M., Uddin, M.N., Chowdhury, J.I. et al. (2022). A review of the recent development, challenges, and opportunities of electronic waste (e-waste). *International Journal of Environmental Science and Technology* https://doi.org/10.1007/s13762-022-04274-w.

Shaikh, S. (2021). A critical analysis of e-waste management and recycling in Pakistan: a life cycle assessment. PhD thesis. Victoria University. Available at: https://vuir.vu.edu.au/42515 (accessed 12 July 2023).

Shaikh, S., Thomas, K., and Zuhair, S. (2020). An exploratory study of e-waste creation and disposal: upstream considerations. *Resources, Conservation and Recycling* 155 (December 2019): 104662. https://doi.org/10.1016/j.resconrec.2019.104662.

Singh, R. and Patel, M. (2022). Strength and durability performance of rice straw ash-based concrete: an approach for the valorization of agriculture waste. *International Journal of Environmental Science and Technology* https://doi.org/10.1007/s13762-022-04554-5.

Singh, R. and Patel, M. (2022a). Effective utilization of rice straw in value-added by-products: a systematic review of state of art and future perspectives. *Biomass and Bioenergy* 159 (March): 106411. https://doi.org/10.1016/j.biombioe.2022.106411.

Smith, P., Bond, M., and Kağitçibaşi, Ç. (2006). *Understanding Social Psychology across Cultures: Living and Working in a Changing World*. London, United Kingdom: SAGE Publications Ltd. https://doi.org/10.4135/9781446212028.

Sodhi, A.K., Bhanot, N., Singh, R., and Alkahtani, M. (2021). Effect of integrating industrial and agricultural wastes on concrete performance with and without microbial activity. *Environmental Science and Pollution Research* 29 (57): 1–17.

Soo, V.K., Featherston, C., and Doolan, M. (2013). E-waste assessment in Malaysia. In: *Re-engineering Manufacturing for Sustainability* (ed. A.Y.C. Nee, B. Song, and S.-K. Ong). Singapore: Springer. 389–395. https://doi.org/10.1007/978-981-4451-48-2_64.

Spencer, L. (2020). Can Australia do better on e-waste? *Arnnet*. Available at: https://www.arnnet.com.au/

article/680299/can-australia-do-better-e-waste/#:~:text=However%2Cthemostrecentfigures,ofe-wasteinAustralia (accessed 20 June 2022).

Statista (2019). Electronic waste generated worldwide from 2010 to 2019 (in million metric tons). Statista. Available at: https://www.statista.com/statistics/499891/projection-ewaste-generation-worldwide/#:~:text=Since2010%2Cthevolumeof,tonsinjustfiveyears (accessed 20 July 2022).

Sthiannopkao, S. and Wong, M.H. (2013). Handling e-waste in developed and developing countries: initiatives, practices, and consequences. *Science of the Total Environment* 463–464: 1147–1153. https://doi.org/10.1016/j.scitotenv.2012.06.088.

Ternald, D. (2019). *E-waste management in Japan Best Practices*. Available at: https://www.itu.int/en/ITU-D/Regional-Presence/AsiaPacific/SiteAssets/Pages/Events/2019/Policy-awareness-workshop-on-E-waste/E-wastemanagementinJapan.pdf (accessed 12 June 2023).

Thakur, P. and Kumar, S. (2020). Metallurgical processes unveil the unexplored "sleeping mines" e-waste: a review. *Environmental Science and Pollution Research* 27 (26): 32359–32370. https://doi.org/10.1007/s11356-020-09405-9.

Tiseo, I. (2023). Share of e-waste in Asia documented to be collected and properly recycled in 2019, by subregion. *Statista*. Available at: https://www.statista.com/statistics/1154682/ewaste-documented-recycling-asia (accessed 14 February 2023).

Tiseo, I. (2021). *Global E-Waste—Statistics & Facts. Statista*. Available at: https://www.statista.com/topics/3409/electronic-waste-world%0Awide/#dossierKeyfgures.%0A (accessed 20 June 2022).

Townsend, T.G. (2011). Environmental issues and management strategies for waste electronic and electrical equipment. *Journal of the Air and Waste Management Association* 61 (6): 587–610. https://doi.org/10.3155/1047-3289.61.6.587.

Wang, B., Wang, H., Zhou, W. et al. (2015). Urinary excretion of phthalate metabolites in school children of China: implication for cumulative risk assessment of phthalate exposure. *Environmental Science & Technology* 49 (2): 1120–1129. https://doi.org/10.1021/es504455a.

Werf, M.V.D. (2020). e-Waste, a valuable circular mine field for CN–EU collaboration. Brussels, Belgium: Dr2Consultants. Available at: https://dr2consultants.cn/e-waste-a-valuable-circular-mine-field-for-cn-eu-collaboration/#:~:text=AccordingtotheChinesegovernment,theinternationale-wasteclassification (accessed 20 June 2022).

Widmer, R., Oswald-Krapf, H., Sinha-Khetriwal, D. et al. (2005). Global perspectives on e-waste. *Environmental Impact Assessment Review* 25 (5 SPEC. ISS.): 436–458. https://doi.org/10.1016/j.eiar.2005.04.001.

Yu, J., Williams, E., Ju, M., and Shao, C. (2010). Managing e-waste in China: policies, pilot projects and alternative approaches. *Resources, Conservation and Recycling* 54 (11): 991–999. https://doi.org/10.1016/j.resconrec.2010.02.006.

Zhang, B., Zhang, T., Duan, Y. et al. (2019). Human exposure to phthalate esters associated with e-waste dismantling: exposure levels, sources, and risk assessment. *Environment International* 124: 1–9. https://doi.org/10.1016/j.envint.2018.12.035.

21

Recent Innovations in Green Recycling Technologies of E-waste Management

Ruchi Bharti, Monika Verma, Ajay Thakur, Renu Sharma, and Annu Pandey*

Department of Chemistry, University Institute of Sciences, Chandigarh University, Mohali, Punjab, India
** Corresponding author*

21.1 Introduction

The past few decades have shown tremendous growth in science and technology. The evolution of new gadgets and equipment has simplified the lifestyle of individuals. However, due to the careless and uncontrolled disposal of toxic wastes into ecosystems that may pose dangers, environmental management of this waste has gathered significant attention worldwide. The amount of waste produced each year globally, which is currently close to 1.3 billion tons, is predicted to rise to 4.3 billion tons by 2025. Electronic-waste (E-waste) output is accelerating rapidly worldwide by outpacing its growth three times faster compared to other waste types. It is a major contributor to environmental contamination. Hence, it is imperative to create affordable and efficient recycling ways for its reduction.

E-waste contains precious and toxic metals that call for special processing and recycling techniques to reduce environmental pollution and any harmful impacts. E-waste components can be effectively recycled if their compositions are known and understood. E-waste can be managed using various recycling techniques and procedures, but choosing and implementing the right one requires careful consideration and depends on the material composition, making it crucial to understand the E-waste's constituent parts. Nearly 60 different components, including precious and dangerous substances, are found in advanced electronic equipment. The most costly and complex components have been utilized in printed circuit boards (PCBs) (Rankin 2011).

PCBs are the essential component on which electrical gadgets are assembled. They use sheets of copper that have conductive pathways to connect and support electrical components. The most commercially valuable part of E-waste is thought to be PCBs. However, because such a combination of valuable and potentially hazardous components is restricted in such small quantities, recycling of essential materials is greatly hindered. Processing PCBs is challenging due to the heterogeneous mixture of organics, metals, fiberglass, and hazardous chemicals, including heavy metals and polymers. Recycling waste PCBs (WPCBs) is vital for managing trash and retrieving priceless metals. Typically, 30% or less of WPCBs are metals like aluminum, zinc, lead, nickel, tin, iron, copper, and valuable metals (Duan et al. 2011).

Along with metals, hazardous compounds including brominated flame retardants (BFR) and polyvinyl chloride (PVC) are also present (Cui and Zhang 2008). Due to the wide range of components and complexity in isolating them, their recycling is time-consuming and expensive. Metals are typically extracted and recovered from WPCBs using Marques et al. techniques (Canal Marques et al. 2013). Marques et al. techniques refer to the methods developed by Marques and his colleagues for the extraction and recovery of metals from WPCBs. These techniques are based on pyrometallurgy and hydrometallurgy, which are two traditional methods for metal extraction and purification. Pyrometallurgy involves the use of high temperatures to extract metals from WPCBs, while hydrometallurgy involves the use of aqueous solutions to dissolve metals and then recover them through various chemical processes. E-waste recycling in the industry is primarily accomplished by pretreatment and pyrometallurgical procedures. One of the world's largest producers of copper, Aurubis Global, effectively handles E-waste along with other copper materials and valuable metals. In a submerged furnace, two reduction processes take place, accompanied by an oxidation at an anode furnace. Electrowinning is used to

Electronic Waste Management: Policies, Processes, Technologies, and Impact, First Edition. Edited by Sunil Kumar and Vineet Kumar.

purify the copper, and the resulting 99% copper is poured into the anode (Abdelbasir et al. 2018). Other foreign businesses have used various industrial techniques to recover metals from E-waste, including the Kosaka plant in Japan, Ronesker Smelter in Sweden, Noranda Process in Quebec, and Eumicor Integrated Smelting and Refining Facility (Khaliq et al. 2014).

Thermal processing can only partially extract metals, which results in a minimal amount of metal being recovered. Also, additional leaching or electrochemical processing is necessary. Recent studies on the energy recovery from computing waste demonstrate the utilization of plastics in electronic waste. They suggest that if an effective emission control system is established, thermal treatment of this waste offers a potential method for energy recovery from E-waste (Cui and Zhang 2008).

In recent years, there have been many studies on hydrometallurgical processes (HMPs) for metal recycling of E-waste (Ashiq et al. 2019; Chakraborty et al. 2022; Huy Do et al. 2023), probably due to cheaper costs, less environmental effect, and greater ease of handling than using pyrometallurgy. Recycling precious and ordinary metals is the research subject of the hydrometallurgical process. Due to its high availability in PCB waste, copper is the primary metal sought for recycling in hydrometallurgy. Leaching, purification, and extraction are the primary phases of hydrometallurgical treatment. Waste PCBs can be treated by leaching using four methods: with acid, ammonia, ammonium salt, or chloride. Some of these processes are used in industry (Zeng et al. 2012). Typically, copper is dissolved in leachants, improved copper purity is achieved via solvent extraction, and pure copper is achieved by electrowinning (Sum 1991). However, because of their sluggishness, need for much time, and adverse environmental effects, several E-waste recycling methods have drawbacks that prevent their usage on an industrial scale.

Some of the leachants used in HMPs are harmful and provide a significant health threat to humans; as a result, they should be produced or utilized in line with strict safety regulations. In addition, concerns exist over the economic differences between pyrometallurgical and hydrometallurgical methods of extracting metals from E-waste. Other drawbacks of pyrometallurgical processes include producing much slag, losing valuable metals, and difficulty recovering Al, Fe, and other metals (Cui and Zhang 2008; Ning et al. 2017).

Therefore, it has become essential to describe and go into great detail about the reprocessing of E-waste, particularly WPCBs, while keeping in mind the idea of the "green process." Many green integrated recycling techniques with excellent efficacy for treating E-waste are given and analyzed, focusing on ecological preservation and resource recycling usage for the metals.

E-waste has become a growing concern globally due to the rapid advancement of technology and the increasing consumption of electronic devices. The improper disposal of E-waste has led to significant environmental and health hazards, including soil and water pollution, the release of toxic chemicals, and the spread of hazardous waste in communities. This chapter provides a comprehensive examination of E-waste and its impact on the environment. The issue of E-waste is thoroughly discussed, including its generation, composition, and disposal methods. The chapter also highlights the need for effective E-waste management strategies to mitigate the negative effects on the environment and human health. By examining the current state of E-waste and its impact, this chapter aims to raise awareness and promote the development of sustainable solutions to address this pressing issue.

21.2 Categories of E-waste

All outdated electronic and electrical equipment is referred to as "E-waste." It is divided into many kinds following the European WEEE Directives 2002/96/EC and 2012/19/EU (Directive 2002/95/EC, Directive 2012/19/EU). As shown in Figure 21.1, these divisions include toys, small and large household appliances, automatic dispensers, monitors, medical devices, sports equipment, lighting systems, consumer goods, telecom and information technology equipment, etc. (Sharma et al. 2012).

21.3 Composition and Characteristics of E-waste

One type of E-waste is E-structural waste, which refers to electronic waste that is generated from large structures, such as buildings and infrastructure, rather than from small household appliances and personal electronics. E-structural waste's makeup is influenced by a variety of factors, including the type of E-waste and the type of the electronic product, its maker, the time it was produced, and the age of the trash. Included are items such as air-conditioning units, elevators, lighting systems, and other large electronic components that are commonly found in buildings and infrastructure. This type of waste is generated due to the aging and replacement of these systems, and it can pose significant environmental and health hazards if not properly managed. The proper disposal of E-structural waste is important to mitigate the negative impact on the environment and to ensure the safe and sustainable recovery of valuable materials.

E-waste categories	
Category 1	Medical devices
Category 2	Consumer electronics
Category 3	Household appliances
Category 4	Toy, leisure, and sports equipment
Category 5	Automatic dispensers
Category 6	IT and telecommunication equipment
Category 7	Electrical and electronic tools
Category 8	Monitoring and controlling equipment

Figure 21.1 Different categories of electronic waste.

The amount of valuable metals in trash from information technology and telecom systems is higher than that of scrap from domestic appliances. For example, a cell phone has about 40 different elements, including precious metals like palladium (Pd), gold (Au), silver (Ag), and primary metals like tin (Sn) and copper (Cu) (Geyer and Doctori Blass 2009; Li et al. 2008; Liu et al. 2009; Schmidt 2002). Hazardous substances like zinc (Zn), lead (Pb), Cu, barium (Ba), mercury (Hg), lead (Pb), chromium (Cr), and various rare earth metals may be present on the circuit boards of the majority of electronic equipment. Cathode ray tubes (CRTs) used in televisions and computer monitors may also contain these substances. As a result, landfill disposal of CRTs is prohibited in most nations. It has been challenging to design rules and adapt them in controlling E-waste because of the changing chemical composition brought on by technological advancement. E-waste composition is influenced by a number of aspects, such as the budget, the resale business, the recycling sector, trash treatment initiatives, and control implementation. The distinctive material fractions of the components of E-waste are displayed in Figure 21.2 (Widmer et al. 2005).

21.4 E-waste Management Methods

As part of waste management, waste items are gathered and processed, transported, disposed of, or managed and recycled to decrease their adverse impact on the well-being of humans and the ecosystem. It is done in order to retrieve assets from waste components as well. The following are a few of the main procedures used to deal with E-waste.

21.4.1 Landfill Disposal

Burying electronic garbage is one of the common ways to dispose of it. Landfilling can make use of mining spaces or underground depths. Over time, E-waste dumped in a landfill may naturally leak contaminants into the environment. For example, battery leaching can result in the release of heavy metals (such as Hg, cadmium, and nickel) and acids. Additionally, landfills for E-waste may contaminate groundwater (Schmidt 2002; Yang 1993). After soaking in the soil, the E-waste can reach groundwater and contaminate it. Contaminated water mixes with many other water bodies, such as streams and rivers. Also, if it is consumed, the contaminated water can be dangerous for

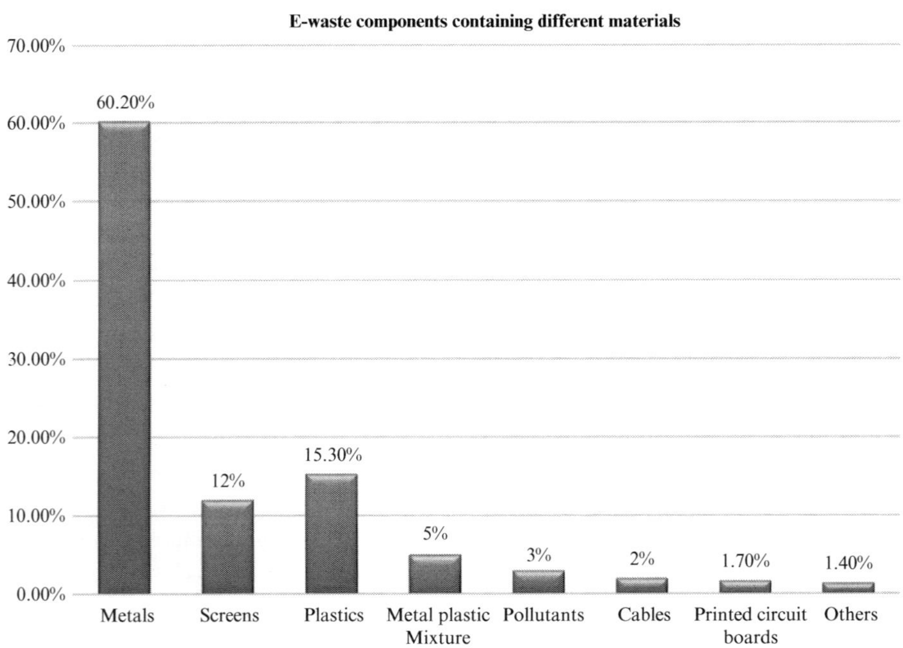

Figure 21.2 Materials fraction in different types of electronic waste.

humans and other animals. According to the nature of waste and the extent of decomposition, decomposing substances in landfills dissolve and leak through the ground as trash leachate with high levels of dangerous chemicals.

21.4.2 Thermal Treatment

Pyrolysis or incinerations are the two thermal processes used to remediate E-waste. Waste can be disposed of by burning it through incineration (Lee et al. 2007). Typically, it is a substitute for other disposal techniques, particularly landfilling. The amount of waste and its combustible components' energy level can be reduced through incineration. Because of this, the amount of the waste materials is reduced during burning, and the energy contained in the waste materials can be used. Pyrolysis, which involves heating a substance without oxygen to create fumes, oils, and charcoal, is another form of thermal treatment.

Burning the PVC or plastic circuit board causes the carcinogenic polycyclic aromatics and dioxins in the vapors, polychlorinated dibenzofurans, and gases like carbon-sulfur and nitrogen oxides are also released. Heavy metal oxides in trace amounts may be found in the haze. Despite being a straightforward, affordable solution, burning, due to its major disadvantages, has been outlawed because of environmental contamination (Bi et al. 2010; Luo et al. 2011; Owens et al. 2007.) The burning of E-waste plants makes up a large portion of the annual emissions of mercury and cadmium (Funcke et al. 1997; Stewart and Lemieux 2003).

21.4.3 Reuse Method

In the reuse method, used or repurposed original hardware has undergone modest alterations. The volume of created E-waste can be reduced with this strategy. However, the incentives provided by the retailers to monetize the old appliances by trading them in for new ones are simply marketing ploys to increase sales volume. However, when used in a business setting, the benefits to the consumer of the new-for-old exchange practice are still hypothetical (Vidyadhar 2016).

21.4.3.1 Recycling Method

Recycling is the practice of reusing discarded products for the same application they were originally designed for or for a completely different one. E-waste recycling involves disassembling and destroying outdated technology to recover its constituent parts. It is a crucial step in treating trash and recovering valuable resources. In comparison to disposal, it is also a good alternative. When waste materials are used instead of virgin resources, the US Environmental Protection Agency (EPA) has identified distinct real benefits, such as energy savings and a decrease in pollution. E-waste recycling's primary goal is to maximize material recovery while minimizing hazardous items' adverse environmental effects. To attain these goals, a thorough understanding of the E-waste constituents is required (Jianzhi Li et al. 2004).

21.4.3.2 Waste Printed Circuit Boards (WPCB)

Reuse of PCBs is regarded as among the most difficult undertakings when having to deal with the growing amount of E-waste due to their complicated composition and the resulting complex mixture of materials. Additionally, PCBs are crucial for a variety of devices, including smart phones, televisions, computers, laptops, and electrical and electronic components. PCBs are made of three distinctive materials attributable to components installed on the substrate, laminate printed conducting tracks, and a non-conducting substrate. Typically, the substrate is made of paper fortified with phenolic resin or glass fiber fortified with epoxy resin, both of which include BFR (Jianzhi Li et al. 2004). Other significant PCB elements include polymers and industrial plastics made of polyester, epoxies, polyethylene, and polypropylene.

WPCBs have drawn a lot of interest from researchers and workers in the industry due to their abundance of resources and possible risks to the environment and human health when recycled poorly and without sufficient controls. Financial viability, recovery efficiency, and environmental impact influence metal extraction. Pretreatment, size reduction, and metallurgical treatment are the typical three stages of the WPCB recycling process for the best possible metal recovery. Pretreatment entails thoroughly disassembling harmful pieces through thermal or chemical desoldering and compositional analysis (Jianzhi Li et al. 2004; Wäger et al. 2011; Yang et al. 2017). Then, to make the materials smaller, they are shredded, crushed, and screened (Guo et al. 2009). Finally, to recover and purify the metals, the metallurgical treatment uses heat treatment (Yang et al. 2013), leaching (Ghosh et al. 2015), electrolysis (Cristina et al. 2017), and biological processes (Zhou et al. 2007).

21.4.3.3 E-waste Recycling and Treatments

Due to the tight regulations that are upheld in industrialized nations like Japan, Australia, United States (US), and the European Union (EU), incinerating, landfilling, and exporting E-waste are now prohibited (Herat and Agamuthu 2012). Due to growing environmental concerns and the possibility of reusing metals or components, it is therefore vital to collect harmful and significant metals from E-waste before discharging them into the surroundings. However, the main obstacles to their widespread usage are the adverse environmental effects and high energy requirements. In addition, pyrometallurgical techniques could be expensive and energy-intensive.

By using adequate off-gas treatment, they can prevent the creation of mixed halogenated dioxins, furans, and vast amounts of slag. Thiosulfate, thiourea, halide, and cyanide are widely used leaching agents in the HM treatment of E-waste to recover the most valuable components (Ashiq et al. 2019). Precious metals have also been extracted from PCBs using bio-hydrometallurgy (Ghosh et al. 2015; Sahni et al. 2016). Useful metals can also be obtained from leaching solutions through adsorption employing biomass waste (including microbial biomass) (Gurung et al. 2013).

Many studies have employed chelating compounds to effectively and efficiently remove metals like Pb, Cu, and Zn from E-waste. Chelating agents can create stable and soluble metal complexes, making them helpful in extracting metals (Neale et al. 1997). It has been discovered that Cu, Pb, and Zn may be extracted up to 86% using EDTA (ethylenediaminetetraacetic acid) and nitrilotriacetic acid (NTA). Even if there are several conventional ways available, sometimes they only move the contaminants from one location to another. Therefore, scientists are working to create more eco-friendly procedures to address this issue effectively. Most recycling techniques try to physically and chemically extract valuable metals from E-waste.

In waste management, waste items are frequently gathered, moved, dumped, or treated and repurposed to decrease their negative impacts on environment or human health. There are various recycling methods, and the best methodology depends on the type of material, the amount of metal in it, and the volume. However, E-waste must undergo an essential process, as shown in Figure 21.3, which includes gathering, shattering and separating, processing, and final metal retrieval.

Developing a new method for recycling WPCBs must consider several factors, including novelties, societal consequences, environmental impact, a uniform waste treatment strategy, as well as the operation's economics. In terms of size, composition, components, and configuration, the WPCBs are diverse and complex. It becomes more challenging to obtain a steady material composition due to the ongoing change in PCB composition. WPCB recycling methods can generally be divided into two categories: chemical recycling methods and physical recycling methods.

Metals and non-metals are commonly separated from WPCBs via physical methods, and these methods depend on a number of factors, such as form separation, thickness, and electrostatic and electric conductivity properties (Cui and Zhang 2008). However, gasification, burning, and pyrolysis are all included in chemical recycling forms. The metallic fraction can be processed using biotechnological, hydrometallurgical, or pyrometallurgical methods, although the repurposing procedure becomes more difficult when the constituents are only present in trace amounts.

The fundamental reason for recycling is to recover metal values, which make up over 30% of the total weight of WPCBs, whereas the analyzed elements (around 70%) have considerably lower economic value. The main objective of the majority of recycling methods is to retrieve the most precious metals from WPCBs utilizing techniques that are often not exceptionally environmentally friendly.

21.4.4 Physical Treatment Processes

Physical treatment processes are applied when different metals and non-metals contained in E-waste are freed and come apart by some shredding and crushing processes, during the upgrading step. With the help of extraction techniques like physical, chemical, and combination pyro/hydro leaching separation pathways, precious metals, in particular Cu, Pd, Ag, and Au, can be extracted. Physical procedures involve pulverizing and dry crushing followed by high-voltage electrostatic separation to get a variety of

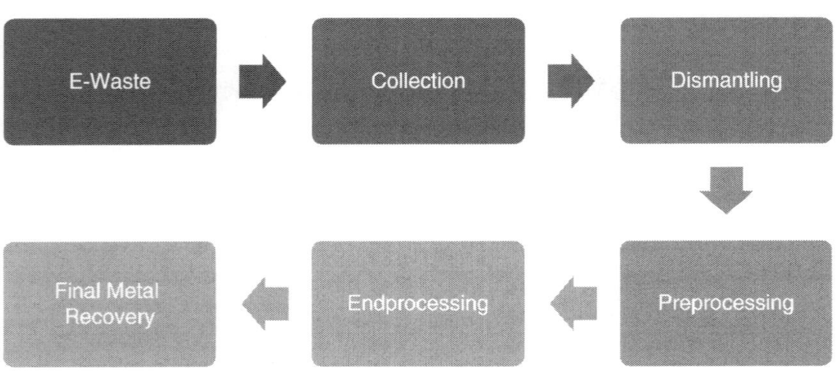

Figure 21.3 Some basic steps involved in the electronic waste recycling process.

conductive metal powders (such as Ag, Au, Sn, Al, Zn, Pb, Cu, etc.) and non-conductive, non-metallic resin powder. The consequence of a recycling process can be measured from two perspectives: material recovery proficiency and the impact on the environment.

Since no secondary pollutants were produced throughout the process, the mechanical approach has been recognized as the best alternative for recovering E-waste. In order to separate magnetic minerals from other materials, magnetic separation is typically utilized in the mineral processing or solid waste business (Li et al. 2007). On the other hand, Eddy's current separation has been utilized to recover non-ferrous metals, including Zn, Pb, Al, and Cu, from solid E-waste. Various physical methods for recycling E-waste are listed in Table 21.1. These approaches offer several benefits. They are quick, appropriate, and eco-friendly procedures. They require inexpensive machinery and energy, and their products have a wide range of potential uses. Some performance flaws include metal loss and dust generation during grinding and shredding. Today, certain industrialized nations typically physically process E-waste to recover basic materials. They use electrostatic Eddy current separations, selective hand disassembly, shredding, and magnetics in their process flow charts.

Table 21.1 Different physical electronic waste recycling methods (Adapted from Abdelbasir et al. 2018).

Method	Materials recovered or separated	Advantages	Disadvantages
Mechano-chemical activation	Soluble metallic Compounds were separated (Cu, Pd, and Ag).	Highly efficient and quick leaching speed. Avoids the use of corrosive mineral acids, consequently preventing the emission of NO_2 or NO pollutants.	Care must be taken in management of diluted acidic wastewaters and a small amount of ammonia-containing wastewater.
Inverse flotation process	Separation of the non-metallic and metallic fractions.	Does not require any harsh chemicals, simply air and water, thus eliminating environmental pollution.	Lowers pollution connected with E-waste by preventing the generation of more E-waste.
Centrifugal separation and vacuum pyrolysis	Solder and organic materials from WPCBs.	Solder can be recovered in an efficient manner that is both clean and economical.	Needs further treatment for recovering various metals contained in the pyrolysis residue.
Vibration separator	Separates the non-metallic and metallic PCB components (up to 95% segregation).	The expense of dewatering and sludge disposal is limited by dry treatment in the absence of water. Presents an opportunity for simpler process integration.	Higher metal liberation requires cutting and milling in successive stages.
Electrostatic separation	Semiconductors and conductors are separated in the initial stage. The weaved glass is extracted from the mixture of particles. Copper is segregated from silicon on circuits in the second stage.	No wastewater discharge, simple design, and minimal energy use.	Higher roll speed and voltage level.
Rotation Disassembling cylinder (RDC) and vibration segregating screen	Separation of electric components from WPCBs.	There is no detrimental effect on the environment or laborers' health. There is no leaking of gas, and it is suitable for most types of WPCBs.	Since they operate at low temperatures, several harmful chemicals are emitted throughout this process.
Multi-step separation methods including dense medium separation, magnetic separation, and sieving	Disassembly of electric and electronic components from WPCBs.	Increases the percentage of recycled metal in WPCBs and, after recycling, reduces emissions. In addition, the dismantled EECs again can be reused as repair parts in PCB production.	The maximal separation effectiveness is less than 60%. As a result, other ways for improving separation efficiency and exploiting these EEC should be presented.
Super gravity separation	Pb–Sn alloy.	A certain environmental benefit in the asset-saving society.	Sn recuperation is minimal due to the development of some substantial Sn solutions.

21.4.5 Chemical Treatment Processes

Several studies investigated the chemical processes used to extract valuable metals, copper, lead, and zinc from E-waste. These techniques are based on the traditional hydrometallurgical process of extracting metal from its ores. The precious metals in E-waste are dissolved using acids or alkalis, then segregated and refined to increase their metal content and eliminate contaminants. Next, solvent extraction, adsorption, or ion exchange procedures are used to isolate the desired metal. Finally, metals are extracted from solutions using chemical or electrorefining procedures. Cui and Zhang (2008) reviewed and detailed the methods used to extract metals from E-waste. Compared to pyrometallurgical processes, hydrometallurgical chemical processes have additional advantages since they are more specific, predictable, and manageable (Paretsky et al. 2004). Table 21.2 provides a brief explanation of the green hydrometallurgical processes.

Chemical techniques have been employed to recover metals from E-waste effectively, but they have several drawbacks that prevent their usage on an industrial scale. The economy of recycling is impacted by the laborious, time-consuming, and chemical processes. However, it takes much time to mechanically handle the E-waste and lower the size needed for effective dissolution. During the liberation process, about 20% of the metals are mechanically lost, significantly lowering the overall metal recovery. Due to the solid corrosive acids and oxidizing circumstances, halide leaching is very challenging to implement; specialized rubber or stainless steel equipment is also needed for the leaching process. Additionally, there is a chance that metal will be lost during later phases, which could harm the overall recovery of metals.

21.4.6 Bioleaching Treatment Processes

One of the most promising methods in metallurgical processing is the bioleaching process, which has been well thought out. It is a green technology because it uses less energy and has lower operating costs than chemical processes. It works pretty well for removing metals from low-grade ores. It is technically possible to recover base metals like Cu, Ni, Zn, and Cr and valuable metals like Au and Ag from electronic waste via bioleaching. Many academics have recently investigated the extraction of metals from electronic scrap debris. When acidophilic bacteria were utilized, heavy metals were excellently leached.

Table 21.2 Different green hydrometallurgical processes used to recover different materials with their advantages and disadvantages (Adapted from Abdelbasir et al. 2018).

Method	Materials recovered or separated	Advantages	Disadvantages
Supercritical methanol process	Ultrafine copper from PCBs	Simple, efficient process, no need to use any reducing agents or surfactants.	Involves a longer time and consumes a higher energy.
Ionic liquid acids	Copper from PCBs	No disassembling of the electronic components from the PCBs is needed.	Effective only for copper.
A green process for cuprous chloride synthesis	Copper from printed circuit boards	No corrosive acids or strong oxidants were used. High-quality cuprous chloride with a purity of 98.7% was obtained.	Effective only for copper, and needs pretreatment.
Chelation technology	Copper	Simple to operate. Is a low-cost process (use of inexpensive reagents and a resin of low commercial price). Fast (the global process involves a short duration of time).	Mechanical Pretreatment required.
Pyro-hydro hybrid process	Base metals (copper, iron, and nickel)	The process is cleaner and environmentally friendly with zero waste concepts and enables the subsequent physical beneficiation and hydrometallurgical treatment to be easier.	The process has potential to be commercialized after some validation on pilot-scale studies.
Chemical desoldering and leaching (patented green chemistry and engineering methodologies)	>99% copper recovery	No shredding or grinding, cost-effective. No need for using cyanide or aqua regia. Low volumes of chemicals required. Near room temperature operation.	It needs a site for location near sources of E-waste.
Chelation technology	Copper (83.8% recovery)	Zero discharge to the environment.	Entails high cost for grinding.

For instance, Wang et al. (2009) used *Acidobacillus* bacteria to mobilize metals from WPCBs. In order to solubilize metals, the bacteria *Acidithiobacillus ferrooxidans* (*A. ferrooxidans*) and *Acidithiobacillus thiooxidans* (*A. thiooxidans*) were raised and adapted in WPCBs. Unfortunately, very few studies mention using fungi to recover metals from E-waste.

The utilization of fungi is constrained because of a steady supply of nutrients for fungal development, turnover handling, and lengthy processing times. In addition to these restrictions, fungal bioleaching has several advantages over bacterial bioleaching. For example, since fungi can grow at high pH levels, they are effective for bioleaching alkaline materials. In addition, they are helpful in the metal-leaching process because they may quickly leach metals and hide organic acids that bind metal ions.

Despite these benefits, there is still a knowledge gap regarding using fungi to leach metals from E-waste. An external source of nutrients is necessary to leach metals from E-waste because the E-waste lacks the energy source needed to develop the bacteria. The most common bacteria and microorganisms for leaking E-waste are displayed in Table 21.3. The commercial performance of the bioleaching process is still in its infancy, despite the technique's many benefits. This can be attributed to the procedure's plodding pace. Many bioleaching techniques take a long time, anywhere from 48 to 245 hours, to recover metals, but they do not always recover all the metals found in E-waste. Therefore, it is essential to create a quick and affordable bioleaching method that may be used in industry (Yaashikaa et al. 2022).

21.4.7 Vacuum Recycling Processes

Vacuum methods do not pollute the wastewater and can be used to recover metals from E-waste. In that application, vacuum pyrolysis can be used to remove the components from the E-waste so that the metals can be easily recovered. Using a two-step vacuum pyrolysis procedure, the Zhou research group (Zhou et al. 2007) could extract metals from WPCBs. At temperatures between 400 and 600°C, the main procedure separated and recovered the solder alloy. The next step involved pyrolyzing WPCBs and heating the leftover material under a vacuum to separate the solder by centrifugal force. The resultant residues still require different treatments for the glass fiber and other inorganic metals and minerals. The vacuum technique separated the cadmium (Cd) and zinc (Zn) metals from the WPCBs. The researchers used a vacuum furnace, which helped them to separate Zn and Cd successfully because of their vapor pressure differences. Because the Pb-Bi alloy was created with low pressure, it was more challenging to separate the lead that was present in the solders of WPCBs.

Using coke powder as a reducing agent, He et al. successfully recovered metallic indium from waste liquid crystal display (LCD) panels, a particular type of E-waste. Under conditions of high temperature, a reducing environment converted indium oxide (In_2O_3) to metallic indium. The recovery rate of indium reached 90 wt% at the processing conditions of 1223 K and 1 Pa with 30 wt% carbon additions for 30 minutes. Indium was also recovered from scrap LCD panels using a chlorinated vacuum-separation process. Ammonium chloride (NH_4Cl), a chlorinating agent, was utilized at a temperature of 400°C. The employed NH_4Cl-to-glass powder had a weight ratio of 1:2, and the ideal particle size was smaller than 0.13 mm. The recovered indium chloride was 99.50% pure, with a 98.02% recovery rate.

Vacuum recycling techniques for recycling E-waste have been proven to have financial and environmental advantages.

Table 21.3 Various microorganisms used for bioleaching processes (Adapted from Abdelbasir et al. 2018).

Microorganisms	Leached metal	Recovery (%)
Acidithiobacillus ferrooxidans, *A. thiooxidans*, and mixture	Cu, Pb, Zn	99, 74.9, 88.9
Aspergillus niger	Cu, Sn	65
Penicillium simplicissimum	Al, Ni, Pb, Zn	>95
Aspergillus niger, *Acidithiobacillus thiooxidans*	Cu	83
Thermosulfidooxidans sulfobacilllus + *Thermoplasma acidophilum*	Cu, Zn, Al, Ni	86, 80, 64, 74
A. ferrooxidans	Ni, Cd	54
Lactobacillus acidophilus	Au	85
A. ferrooxidans + *A. thiooxidans*	Cu, Ni, Al, Zn	> 90
A. ferrooxidans	Cu	99
A. thiooxidans	Cu	74.9
Sulfobacillus thermosulfidooxidans	Ni, Cu, Al, Zn	81, 89, 79, 83
Chromobacterium violaceum	Au	68.5

Additionally, there will not be any wastewater or dust emissions because gas flow can be efficiently controlled. This method of recycling metal from E-waste is effective and environmentally benign, but it still needs to be improved to move forward. On the one hand, vacuum methods are very advantageous for extracting metals like Zn, Pb, and Cd with high-saturation vapor pressure and low boiling points. Contrarily, the vacuum condensation method is not as effective in separating expensive and rare metals with low-saturation vapor pressure. Therefore, before its modern industrial application, many potential difficulties need to be cleared up (Zhan and Xu 2014).

21.4.8 Green Computing Practices

The study and practice of using computing resources wisely are known as "green computing." This methodology aims to reduce the use of hazardous materials, increase energy efficiency throughout the product's lifetime, and promote recycling or biodegradability of used items and manufacturing waste (Anam and Syed 2013). Green information technology (IT) is another name for green computing. The goals of green computing, also known as green IT, are to make computers more economically viable and to enhance their use. Green IT techniques include creating environmentally friendly production methods, computers that use less energy, and enhanced disposal and recycling techniques. The following four complementing strategies are used to promote green computing concepts at all appropriate levels:

Green Use: Using computers and their accessories in an energy-efficient and environmentally responsible manner.

Green Disposal: Recycling unwanted electronic equipment or repurposing an old computer are examples of environmentally friendly disposal.

Green Design: Creating digital products such as computers, servers, printers, and projectors that use less energy.

Green Production: Using green manufacturing techniques to produce computers and other components can lessen their harmful effects on the environment.

There are various ways in which green computing will have a good effect and many advantages thanks to the ever rapidly developing technology and developments. The advantages of green computing are substantial, not just from the perspective of the customer, industry, or nation but also from a worldwide perspective. In addition, using technology has a good impact on the environment and our costs when we consume less energy, waste, and money. However, it is not well established and readily available, and its startup cost is also high. Overall, though, the advantages of green computing will lead to cost savings, energy conservation, and environmental protection. Some other initiatives implemented to minimize the E-waste problems are shown in Table 21.4.

Table 21.4 Initiatives implemented to minimize E-waste problems.

Initiative	Description
Seco/Empa E-waste program	In collaboration with multiple authorities and local partners, Empa implemented the project set up by Seco to improve WEEE (Waste from Electrical and Electronic Equipment) recycling methodologies. In this, all the WEEE recycling systems in the world are studied and analyzed, and the recycling technology-based knowledge, framework, and information are exchanged.
European Recycling Platform (ERP)	Founded by Electrolux, Braun, Sony, and Hewlett Packard in late 2002 to enable product manufacturers to comply with WEEE guidelines. Its objectives include assessing, organizing, and putting into action a pan-European system for waste disposal and recycling solutions.
Electronic Product Stewardship Canada (EPS Canada)	Sixteen top electronics firms form this industry-led organization. It is a collaboration between government and business corporate that was created to build and adapt solutions that reduce hurdles in Canada's waste management.
National Electronics Product Stewardship Initiative (NEPSI)	A step was taken to target the WEEE management framework at the national level in the USA with the help of a multi-stakeholder dialogue. This dialogue involved representatives from environment groups, local and state governments, the recycling industry, retailers, manufacturing industries, etc.
WEEE forum	Established in 2002 with a group of representatives who take care of take-back systems and ensure that each producer takes care of their responsiveness when their products are end-cycled.
The Electronics TakeBack Coalition (ETBC)	According to the Extended Producer Responsibility (EPR), some non-governmental groups in the US are working to save the environment from E-waste by urging the electronics industry to launch a take-back program for their products at their end-life.

Table 21.4 (Continued)

Initiative	Description
Silicon Valley Toxic Coalition (SVTC) and computer take-back campaign	A diverse organization engaged in research, backing, and works at ground level to improve human health as well as protecting environment in response to the rapidly expanding tech industry.
The StEP initiative (Solving the E-waste Problem)	In the conference held in Berlin (Electronic Goes Green), the United Nations took an initiative in which it decided to create an international platform to develop and provide WEEE systems in different countries so that the supply chain can be reversed at the global level.
Partnership for Action on Computing Equipment (PACE)	This multi-stakeholder partnership, which took place under the Basel Convention, aims to manage computing equipment at the end of its term in such a way as to reduce its impact on the environment.
The Nairobi Declaration	An agreement by the parties to the Basel Convention that accelerates the efforts to solve the global E-waste problem.
The Mobile Phone Partnership Initiative (MPPI)	End-of-life mobile phones are a very important issue from an environmental point of view and to reduce the waste, many mobile phone manufacturers and associated companies have taken an initiative to develop and encourage such ways and methods to solve this hurdle and reduce the environmental problem.
The Basel Ban Amendment	This Basel Convention amendment outright restricts the shipment of hazardous materials from the established (mainly Organization for Economic Co-operation and Development) nations to underdeveloped nations. Unfortunately, the Basel Prohibition Amendment has still not taken effect.
The Basel Convention on the control of trans-boundary movements of hazardous wastes and their disposal	The most important global environmental treaty, which plays a great role in controlling the exchange of hazardous wastes between nations. Especially, this treaty limits the provision of dumping of harmful waste from the developed countries to those countries that are less developed.

21.5 Conclusion

The issue of E-waste management has become increasingly relevant in today's rapidly growing technological world. The need for sustainable solutions for managing E-waste has led to the development of various green recycling technologies aimed at reducing the environmental impact of E-waste and maximizing the recovery of valuable materials. The innovations in green recycling technologies, such as the development of new mechanical recycling processes and the use of biotechnology for the extraction of valuable metals, have shown promising results in terms of both environmental protection and resource recovery.

This chapter provided a comprehensive overview of the current state of E-waste management, focusing on the different categories of E-waste, their compositional and characteristic features, and the various methods of E-waste management. A critical evaluation of the available E-waste management practices was conducted, including the traditional landfill disposal and thermal treatment methods, as well as the more innovative approaches such as reuse and recycling. The recycling methods were further divided into physical, chemical, bioleaching, and vacuum processes, and the benefits and drawbacks of each were analyzed. Additionally, the potential role of green computing practices in reducing E-waste generation was discussed as a promising avenue for future research and implementation.

This chapter aimed to provide valuable insights into the ongoing efforts to promote sustainable E-waste management and to contribute to the development of more environmentally friendly and economically feasible methods for the management of E-waste. The future of E-waste management is promising, with a growing emphasis on sustainable, environmentally-friendly methods. The development of innovative technologies, such as bioleaching and vacuum recycling, shows promise in improving the efficiency and reducing the environmental impact of E-waste management. It is imperative that governments, industry, and individuals continue to work together to promote the responsible disposal and recycling of E-waste to protect the environment and human health.

Abbreviations

E-waste	Electronic waste
PCBs	Printed circuit boards
WPCBs	Waste printed circuit boards
BFR	Brominated flame retardants
PVC	Polyvinyl chloride
HMPs	Hydrometallurgical processes
Pd	Palladium
Au	Gold
Ag	Silver
Sn	Tin
Cu	Copper
Zn	Zinc
Pb	Lead
Ba	Barium
Hg	Mercury
Cr	Chromium
CRTs	Cathode ray tubes
EPA	Environmental Protection Agency
US	United States
EU	European Union
EDTA	Ethylenediaminetetraacetic acid
NLA	Nitrilotriacetic acid
A. ferrooxidans	*Acidithiobacillus ferrooxidans*
A. thiooxidans	*Acidithiobacillus thiooxidans*
LCD	Liquid crystal display
In_2O_3	Indium oxide
NH_4Cl	Ammonium chloride
WEEE	Waste from electrical and electronic equipment
ERP	European recycling platform
NEPSI	National Electronics Product Stewardship Initiative
EPR	Extended producer responsibility
ETBC	Electronics Take-back Coalition
SVTC	Silicon Valley Toxic Coalition
PACE	Partnership for Action on Computing Equipment
MPPI	Mobile Phone Partnership Initiative

References

Abdelbasir, S.M., El-Sheltawy, C.T., and Abdo, D. (2018). Green processes for electronic waste recycling: a review. *Journal of Sustainable Metallurgy* [online]. https://doi.org/10.1007/s40831-018-0175-3.

Anam, A. and Syed, A. (2013). Green computing: e-waste management through recycling. *International Journal of Scientific & Engineering Research* [online] 4 (5). Available at: https://www.ijser.org/researchpaper/Green-Computing-E-waste-management-through-recycling.pdf (accessed 1 November 2020).

Ashiq, A., Kulkarni, J., and Vithanage, M. (2019). Chapter 10—hydrometallurgical recovery of metals from E-waste. In: *ScienceDirect* (eds. M.N.V. Prasad and M. Vithanage), 225–246. Butterworth-Heinemann.

Bi, X., Simoneit, B.R.T., Wang, Z. et al. (2010). The major components of particles emitted during recycling of waste printed circuit boards in a typical e-waste workshop of South China. *Atmospheric Environment* 44 (35): 4440–4445. https://doi.org/10.1016/j.atmosenv.2010.07.040.

Canal Marques, A., Cabrera Marrero J.-M., and de Fraga Malfatti, C. (2013). A review of the recycling of non-metallic fractions of printed circuit boards. *Springerplus* 2: 521. https://doi.org/10.1186/2193-1801-2-521.

Chakraborty, S.C., Zaman, W.U. Md., Hoque, M. et al. (2022). Metals extraction processes from electronic waste: constraints and opportunities. *Environmental Science and Pollution Research* 29 (22): 32651–32669. https://doi.org/10.1007/s11356-022-19322-8.

Cristina, C., Salvatore, P., Carmelo, S., and Rosalinda, I. (2017). Study of a novel electrochemical method for copper recovery from waste printed circuit boards. *Chemical Engineering Transactions* 57 (57): 1621–1626. https://doi.org/10.3303/CET1757271.

Cui, J. and Zhang, L. (2008). Metallurgical recovery of metals from electronic waste: a review. *Journal of Hazardous Materials* 158 (2–3): 228–256. https://doi.org/10.1016/j.jhazmat.2008.02.001.

Directive 2002/95/EC of the European Parliament and of the Council of 27 January 2003 on the restriction of the use of certain hazardous substances in electrical and electronic equipment (2003). *Official Journal of the European Union* [online]. Available at: https://cir.nii.ac.jp/crid/1570009750918886656 (accessed 30 June 2022).

Directive 2012/19/EU of the European Parliament and of the Council of 4 July 2012 on waste electrical and electronic equipment (WEEE) (recast) (Text with EEA relevance) (n.d.) [online]. Available at: https://eur-lex.europa.eu/legal-content/EN/TXT/PDF/?uri=CELEX:32012L0019.

Duan, H., Hou, K., Li, J., and Zhu, X. (2011). Examining the technology acceptance for dismantling of waste printed circuit boards in light of recycling and environmental concerns. *Journal of Environmental Management* 92 (3): 392–399. https://doi.org/10.1016/j.jenvman.2010.10.057.

Funcke, W., Hemminghaus, H.J., Mark, F.E., and Vehlow, J. (1997). PXDF/D in flue gas from an incinerator charging wastes containing Cl and Br and a statistical description of the resulting PXDF/D combustion profiles. *Organohalogen Compounds* 31: 93–98.

Geyer, R. and Doctori Blass, V. (2009). The economics of cell phone reuse and recycling. *The International Journal of Advanced Manufacturing Technology* 47 (5–8): 515–525. https://doi.org/10.1007/s00170-009-2228-z.

Ghosh, B., Ghosh, M.K., Parhi, P. et al. (2015). Waste printed circuit boards recycling: an extensive assessment of current status. *Journal of Cleaner Production* 94: 5–19. https://doi.org/10.1016/j.jclepro.2015.02.024.

Guo, J., Guo, J., and Xu, Z. (2009). Recycling of non-metallic fractions from waste printed circuit boards: a review. *Journal of Hazardous Materials* 168 (2–3): 567–590. https://doi.org/10.1016/j.jhazmat.2009.02.104.

Gurung, M., Adhikari, B.B., Kawakita, H. et al. (2013). Recovery of gold and silver from spent mobile phones by means of acidothiourea leaching followed by adsorption using biosorbent prepared from persimmon tannin. *Hydrometallurgy* 133: 84–93. https://doi.org/10.1016/j.hydromet.2012.12.003.

Herat, S. and Agamuthu, P. (2012). E-waste: a problem or an opportunity? Review of issues, challenges and solutions in Asian countries. *Waste Management & Research* 30 (11): 1113–1129. https://doi.org/10.1177/0734242×12453378.

Huy Do, M., Tien Nguyen, G., Dong Thach, U. et al. (2023). Advances in hydrometallurgical approaches for gold recovery from E-waste: a comprehensive review and perspectives. *Minerals Engineering* 191: 107977. https://doi.org/10.1016/j.mineng.2022.107977.

Khaliq, A., Rhamdhani, M., Brooks, G., and Masood, S. (2014). Metal extraction processes for electronic waste and existing industrial routes: a review and Australian perspective. *Resources* 3 (1): 152–179. https://doi.org/10.3390/resources3010152.

Lee, J., Song, H.T., and Yoo, J.-M. (2007). Present status of the recycling of waste electrical and electronic equipment in Korea. *Resources, Conservation and Recycling* 50 (4): 380–397. https://doi.org/10.1016/j.resconrec.2007.01.010.

Li, J., Lu, H., Guo, J. et al. (2007). Recycle technology for recovering resources and products from waste printed circuit boards. *Environmental Science & Technology* 41 (6): 1995–2000. https://doi.org/10.1021/es0618245.

Li, J., Shrivastava, P., Gao, Z., and Zhang, H.-C. (2004). Printed circuit board recycling: a state-of-the-art survey. *IEEE Transactions on Electronics Packaging Manufacturing* 27 (1): 33–42. https://doi.org/10.1109/TEPM.2004.830501.

Li, Y., Xu, X., Liu, J. et al. (2008). The hazard of chromium exposure to neonates in Guiyu of China. *Science of the Total Environment* 403 (1–3): 99–104. https://doi.org/10.1016/j.scitotenv.2008.05.033.

Liu, Q., Li, K.Q., Zhao, H. et al. (2009). The global challenge of electronic waste management. *Environmental Science and Pollution Research* 16 (3): 248–249. https://doi.org/10.1007/s11356-009-0131-7.

Luo, C., Liu, C., Wang, Y. et al. (2011). Heavy metal contamination in soils and vegetables near an e-waste processing site, south China. *Journal of Hazardous Materials* 186 (1): 481–490. https://doi.org/10.1016/j.jhazmat.2010.11.024.

Neale, C.N., Bricka, R.Y., and Chao, A.C. (1997). Evaluating acids and chelating agents for removing heavy metals from contaminated soils. *Environmental Progress* 16 (4): 274–280. https://doi.org/10.1002/ep.3300160416.

Ning, C., Lin, C.S.K., Hui, D.C.W., and McKay, G. (2017). Waste printed circuit board (PCB) recycling techniques. *Topics in Current Chemistry* 375 (2). https://doi.org/10.1007/s41061-017-0118-7.

Owens, C.V., Lambright, C., Bobseine, K. et al. (2007). Identification of estrogenic compounds emitted from the combustion of computer printed circuit boards in electronic waste. *Environmental Science & Technology* 41 (24): 8506–8511. https://doi.org/10.1021/es071425p.

Paretsky, V.M., Antipov, N.I., and Tarasov, A.V. (2004). Hydrometallurgical method for treating special alloys, jewelry, electronic and electrotechnical scrap. In: *Proceedings of the TMS Fall Extraction and Processing Conference*, 713–721.

Rankin, W.J. (2011). *Minerals, Metals and Sustainability: Meeting Future Material Needs*. CSIRO Publishing.

Sahni, A., Kumar, A., and Kumar, S. (2016). Chemo-biohydrometallurgy—a hybrid technology to recover metals from obsolete mobile SIM cards. *Environmental Nanotechnology, Monitoring & Management* 6: 130–133. https://doi.org/10.1016/j.enmm.2016.09.003.

Schmidt, C.W. (2002). e-Junk explosion. *Environmental Health Perspectives* 110 (4): A188–A194. https://doi.org/10.1289/ehp.110-a188.

Sharma, P., Fulekar, M.H., and Pathak, B. (2012). E-waste—a challenge for tomorrow. *Research Journal of Recent Sciences* 1 (3): 86–93. Available at: http://www.isca.in/rjrs/archive/v1/i3/14.ISCA-RJRS-2012-041_Done.php (accessed 30 June 2022).

Stewart, E.S. and Lemieux, P.M. (2003). Emissions from the incineration of electronics industry waste. *IEEE Xplore*. https://doi.org/10.1109/ISEE.2003.1208088.

Sum, E.Y.L. (1991). The recovery of metals from electronic scrap. *JOM* 43 (4): 53–61. https://doi.org/10.1007/bf03220549.

Vidyadhar, A. (2016). A review of technology of metal recovery from electronic waste. In: *E-waste in Transition—from Pollution to Resource* (ed. F.-C. Mihai). https://doi.org/10.5772/61569.

Wäger, P.A., Hischier, R., and Eugster, M. (2011). Environmental impacts of the Swiss collection and recovery systems for waste electrical and electronic equipment (WEEE): a follow-up. *Science of the Total Environment* 409 (10): 1746–1756. https://doi.org/10.1016/j.scitotenv.2011.01.050.

Wang, J., Bai, J., Xu, J., and Liang, B. (2009). Bioleaching of metals from printed wire boards by Acidithiobacillus ferrooxidans and Acidithiobacillus thiooxidans and their mixture. *Journal of Hazardous Materials* 172: 1100–1105.

Widmer, R., Oswald-Krapf, H., Sinha-Khetriwal, D., Schnellmann, M., and Böni, H. (2005). Global perspectives on e-waste. *Environmental Impact Assessment Review* 25 (5): 436–458.

Yaashikaa, P.R., Priyanka, B., Senthil Kumar, P. et al. (2022). A review on recent advancements in recovery of valuable and toxic metals from e-waste using bioleaching approach. *Chemosphere* 287: 132230. https://doi.org/10.1016/j.chemosphere.2021.132230.

Yang, G.C.C. (1993). Environmental threats of discarded picture tubes and printed circuit boards. *Journal of Hazardous Materials* 34 (2): 235–243. https://doi.org/10.1016/0304-3894(93)85008-3.

Yang, T., Zhu, P., Liu, W. et al. (2017). Recovery of tin from metal powders of waste printed circuit boards. *Waste Management* 68: 449–457. https://doi.org/10.1016/j.wasman.2017.06.019.

Yang, X., Sun, L., Xiang, J. et al. (2013). Pyrolysis and dehalogenation of plastics from waste electrical and electronic equipment (WEEE): a review. *Waste Management* 33 (2): 462–473. https://doi.org/10.1016/j.wasman.2012.07.025.

Zeng, X., Zheng, L., Xie, H. et al. (2012). Current status and future perspective of waste printed circuit boards recycling. *Procedia Environmental Sciences* 16: 590–597. https://doi.org/10.1016/j.proenv.2012.10.081.

Zhan, L. and Xu, Z. (2014). State-of-the-art of recycling E-wastes by vacuum metallurgy separation. *Environmental Science & Technology* 48 (24): 14092–14102. https://doi.org/10.1021/es5030383.

Zhou, P.G., Zheng, Z., Peng, X.C., and Wang, Y. (2007). Leaching of copper from printed circuit board by Thiobacillus ferrooxidans. *Environmental Pollution Control* 29 (2): 119–122.

22

Electronic Waste Recycling in Maintaining a Circular Economy

Muammer Kaya and Angela Manka Tita*

Eskisehir Osmangazi University, Mining Engineering Department, Eskisehir, Turkey
** Corresponding author*

List of Abbreviations and Definitions

BFR	Brominated Flame Retardants
CE	Circular economy
CFC	Chlorofluorocarbon
COP21	Conference of Parties (UN climate change conference)
CRT	Cathode ray tube
EC	Electronic component
EEE	Electrical and electronic equipment
EEE POM	Electrical and electronic equipment placed on the market
EPR	Extended producer responsibility
EoL	End-of-life
E.U.	European Union
e-waste	Electronic waste
IW	Incinerated waste
IT	Information technology
GDP	Gross domestic product
GHG	Greenhouse gas
GPS	Global Positioning Service
HCFC	Hydrochlorofluorocarbons
LCD	Liquid crystal display
LE	Linear economy
LED	Light emitting diode
LIB	Lithium-ion battery
LoW	List of wastes
WSW	Municipal solid waste
NmF	Non-metal fraction
NWR	National waste and recycling
MSW	Municipal solid waste
PCB	Polychlorinated biphenyls
PGM	Platinum group metal
PM	Precious metals
PVC	Polyvinylchloride
REE	Rare earth element
RoHS	Restriction of hazardous substances
SDG	Sustainable development goals of the United Nations
SM	Strategic metals
SteP	Solving the e-waste problem
TEE	Temperature exchanging equipment
OECD	Organization for Economic Cooperation and Development
UN	United Nations
UNU-	A categorization of 54 electronic/electrical products that make up a
KEYS	classification system for WEEE statistics
WEEE	Waste electrical and electronic equipment
WSR	Waste shipment regulation

22.1 Introduction

Waste electrical and electronic equipment (WEEE) and e-waste are the fastest-growing types of municipal solid waste (MSW) and represent the most hazardous stream worldwide. This growth results from several trends, including the increased number of information technologies (ITs) users and the progressive obsolescence of these electronic devices. Despite being composed of many hazardous elements, the e-waste also contains some base, strategic, and critical raw materials, and therefore, this e-waste

Electronic Waste Management: Policies, Processes, Technologies, and Impact, First Edition. Edited by Sunil Kumar and Vineet Kumar.

emerges as a significant supply of secondary resources. Thus, the e-waste requires proper processing to be used as a secondary raw material source, to mitigate the mining impacts from virgin mines, and to avoid the negative effects on the environment and human health. Less than 18% of the e-waste generated globally was officially collected and recycled in 2019. The developed countries have a well-established system for e-waste collection and automatic processing technology but it is insufficient to handle the amount of e-waste. The developing countries have a significant contingent of collectors (or waste pickers) who have begun to understand the value of technological waste but have not yet realized the hazardous risks involved in handling these products.

Globally, developing and underdeveloped countries can also be determined as one of the greatest e-waste generators. Regarding waste quantities, these countries are one of the biggest e-waste "urban mines" in the world, and therefore, the Electrical and Electronic Equipment (EEE)'s reverse supply chain has become a growing business in these countries.

The electronics industry is considered one of the most polluting, especially due to the hazardous components and manufacturing processes, the material extraction from ores, and energy requirements for production, use, and the reverse chain, besides the shortening of the lifespan of such products. E-waste is, then, a heterogeneous mixture composed of ferrous, non-ferrous, polymeric, and ceramic materials (Kaya 2016, 2017). The main concerns about e-waste are related to its hazardous potential since these electronic devices are often composed of toxic metals and brominated flame retardants (BFR), which might generate detrimental environmental impacts, in addition to health diseases to individuals when exposed to such inadequately disposed of or handled e-waste (Kaya 2018, 2019).

A large portion of e-waste contains precious metals (PM) that are especially concentrated in the printed circuit boards (PCB) found in practically all WEEE. The PCBs are laminated boards with circuit traces and connected to electrical components (ECs), which are usually composed of high-added value elements, such as Au, Cu, Pd, Ag, etc. PM and strategic critical minerals (rare earth elements (REEs)) account for about 80% of the e-waste's intrinsic value, even though they do correspond to less than 1% of the total equipment weight. The value of PM in waste PCB (WPCB) makes it more profitable to urban mine from e-waste than to mine the ores, which emphasizes the economic and environmental importance of recycling (Kaya 2019a, 2021).

In developing and underdeveloped countries, even though most e-waste components can be recycled, there is no large-scale state-of-the-art technology installed to recover valuable metals from WPCB. As a result, the formal WPCB streams end up being exported to large European and American plants overseas, thus transferring to other countries the highest aggregated value generated in the segment. The literature indicates the urgent need for developing eco-friendly feasible technologies for the recycling of most e-waste components, especially PCBs, REE-containing permanent magnets, lithium-ion batteries (LIBs), Au, Ag, Ni Dy, Li, Co, Ni, etc. containing parts. Recycling and other value recovery options contribute to decreasing dependency on a permanent supply of essential resources, encouraging recycling companies and reducing environmental hazards derived from inadequate e-waste disposal.

This responsibility becomes even more important when considering climate change and the goals for reducing greenhouse gas (GHG) emissions, especially CO_2. The decarbonization of human activities is necessary to succeed in the long-term goal of the COP21 climate change agreement in Paris: to keep the increase of the global average temperature to 1.5°C maximum compared to pre-industrial levels by the end of the twenty-first century. The current production, consumption, and business models need to be completely rethought to align with the changes toward more circular, sustainable, and carbon-neutral patterns, especially for sectors that generate high impacts, such as in the case of electronics.

The growing amounts of electronic devices will make it increasingly difficult to achieve the United Nations Sustainable Development Goals (SDGs) for 2030, especially the SDGs related to environmental protection and public health safety. Previous studies emphasized that e-waste management is closely related to many of the United Nation (UN)'s 17 SDGs, such as SDG 3 on good health and well-being, SDG 6 on clean waste and sanitation, SDG 8 on decent work and economic growth, SGD 12 on responsible production and consumption, SDG 13 on climate action, and SDG 14 on life below water. The circular economy (CE) may encourage the achievement of the SDGs.

22.2 Definitions

22.2.1 E-waste

Electronic waste is fundamentally any electronic waste that has been discarded at the end of its useful life. To give a more holistic definition, we could consider quoting from a known international organization to give some uniformity. The StEP initiative (Solving the E-waste Problem) is an international organization focused on coming up with solutions to e-waste problems all over the world. StEP defines e-waste as follows: "E-waste is a term used to cover items of all types of EEE and its parts that have been discarded by the user as a solid waste without the intention of re-use." (Kumar et al. 2017)

The Basel Convention, in Article 2, defines waste as "substances or objects, which are disposed of or are intended to be disposed of or are required to be disposed of by the provisions of national law" (Forti et al. 2018).

22.2.2 Recycling

The term recycling was not documented until 1924 when it was used in the oil refinery field and similar fields. "Re-" stands for "again" or "back" or "anew," while "-cycle" means a complete set or series, usually repeated in the same order. According to article 3(17) of the EU directive from 2008, recycling is described as "any recovery operation by which waste materials are reprocessed into new products, substances, or products whether for the original use or other purposes. It includes the reprocessing of organic polymeric material but does not include energy recovery and the reprocessing into materials that are to be used as fuels or for backfilling operations."

Recycling can be categorized into three types: feedstock recycling, product recycling, and material recycling (Bartl 2014). Feedstock/raw material/chemical recycling refers to the process in which the waste is brought back into its original components. A common example is a de-polymerization process. Product recycling, on the other hand, is a process in which the recycled product is no longer the original product but is now used for a different purpose, even though the physical and chemical components of the material did not change. A common example is the use of broken glass bottles as building materials. As for material recycling, which happens to be the most applied category, it is a process in which the physical components of the material are destroyed, but the chemical part stays intact. An example is the melting and reusing of metals.

Besides these recycling categories, the recycling process can also be divided into closed-loop and open-loop allocations: Closed-loop as the name goes, applies to product systems in which there is no change in the fundamental properties of the material being recycled. An example is magnet-to-magnet recycling. An open-loop procedure exercises in product systems where there is a change in the fundamental properties of the material, and different product systems emerge when the material is recycled (Bartl 2014).

22.2.3 The Circular Economy

The term circular economy is becoming more and more commonly used, especially in the domain of recycling. The CE cannot be mentioned without the linear economy (LE), as they are antagonistic terms: In the LE, raw materials are used to make new products, which are directly discarded at the end of their useful life. A circular economy, on the contrary, seeks to close the gap between the natural ecosystem's cycles and production.

Sometimes, CE and recycling are used interchangeably, but other theories say that they are different. The Ellen MacArthur Foundation, which is an organization working on making circular economies around the world, says that in a CE, there is no waste production at all, whereas recycling entails the transformation of already existing waste into useful products. Their definition of a CE is: "In our conventional LE, we borrow materials from the Earth's crust, manufacture products from them, and finally throw them away as waste. In contrast, in a CE, we stop waste production in the first place" (https://ellenmacarthurfoundation.org/topics/circular-economy-introduction/overview).

The World Economic Forum, which is an organization also working actively toward a CE, defines a CE as follows: "A circular economy is a restorative or regenerative industrial system that is by intention and design. It replaces EoL concept with restoration, shifts towards renewable energy use, eliminates the use of hazardous chemicals, which impair reuse and return to the biosphere, and aims for the elimination of waste through the novel design of systems, products, materials, and business models" (Circular Economy—Definition, Principles, Benefits and Barriers 2020). Figure 22.1 compares the LE with a CE.

22.3 Categories/Classification of E-waste

The range of varying EEE products is so vast that classifying them into reasonable and practical groups is very useful, especially for easy recycling operations. There are so many classifications for WEEE such as the WEEE Directive categories, the UNU-KEYS, the NWR classification, etc. Forti et al. (2020) divided the 54 WEEE product-centric categories into six general categories:

- Temperature exchange equipment
- Monitors and screens
- Light lamps
- Large equipment (excluding and including photovoltaics)
- Small equipment
- Small information technology and telecommunication equipment

They performed this categorization based on the UNU-KEYS classification, which is a product categorization made up of 54 products and constructed in a manner that each category shares similar average weights, end-of-life properties, material contents, and lifetime ranges across. The EU uses these UNU-KEYS too, to carry out some e-waste generated calculations, as well as Electrical and

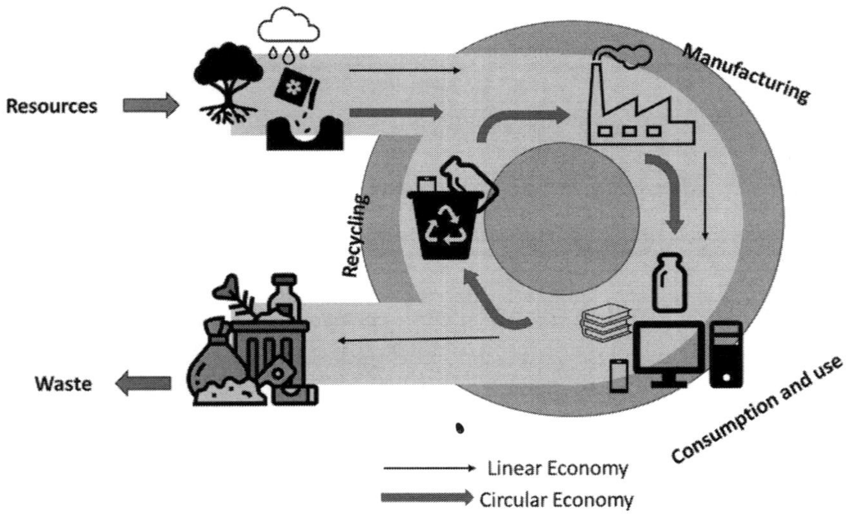

Figure 22.1 Circular economy versus linear economy.

Electronic Equipment Placed on the Market (EEE POM) calculations, making the UNU-KEYS classification a perfect link between the existing classifications and the EU classification. Table 22.1 presents the classification, definition, and typical equipment of e-waste. Figure 22.2 shows the decision tree for the categorization of WEEE.

Electronic waste can be classified also according to the presence of toxic material in the waste. This classification

Table 22.1 Classification of E-waste. (Adapted from Forti et al. 2020).

Category	Definition	Typical equipment
1) **Temperature exchange equipment (TEE)**	EEE with internal circuits where substances besides water (e.g. oil, gas, refrigerant, or a secondary fluid) are employed for cooling purposes and/or heating and/or dehumidifying.	Refrigerators, deep freezers, air conditioners (ACs), heat pumps, and dehumidifying equipment.
2) **Monitors and screens**	EEE is intended to supply information and images on an electronic display— regardless of its dimensions—such as cathode ray tubes (CRT), light-emitting diode (LED) displays, liquid crystal displays (LCD), or other kinds of electronic displays.	TVs, monitors, screens, PCs, notebooks, tablets, e-readers, LCD photo frames, etc.
3) **Light lamps**	EEE for the creation of light, which means electrical light sources (including LED lamps) that can be installed or exchanged by the end-user in specific electric and electronic equipment such as luminaires, projectors, solariums, etc.	Fluorescent lamps, high-intensity discharge lamps, LED lamps, LED straight fluorescent lamps, fluorescent lamps, compact fluorescent lamps, high-intensity discharge lamps, and low-pressure Na lamps.
4) **Large equipment**	EEE is not allocated to categories 1, 2, or 3. The external dimension is less than 50 cm.	Dishwashing/washing machines, clothes dryers, electric stoves, large medical devices, large printing machines, photocopying equipment, ACs, and photovoltaic panels.
5) **Small equipment**	EEE that is not allocated to categories 1, 2, 3, 4, or 6. The external dimension is less than 50 cm.	Microwave ovens, vacuum cleaners, ventilation equipment, electric kettles, toasters, scales, electric shavers, calculators, radios, video cameras, electrical and electronic toys, small electrical and electronic tools, small medical devices, small monitoring and control instruments.
6) **Small information technology and telecommunication equipment**	EEE that is not allocated to categories 1, 2, 3, 4, or 6. The external dimension is less than 50 cm.	Smartphones, GPS devices, calculators, routers, PCs, printers, and telephones.

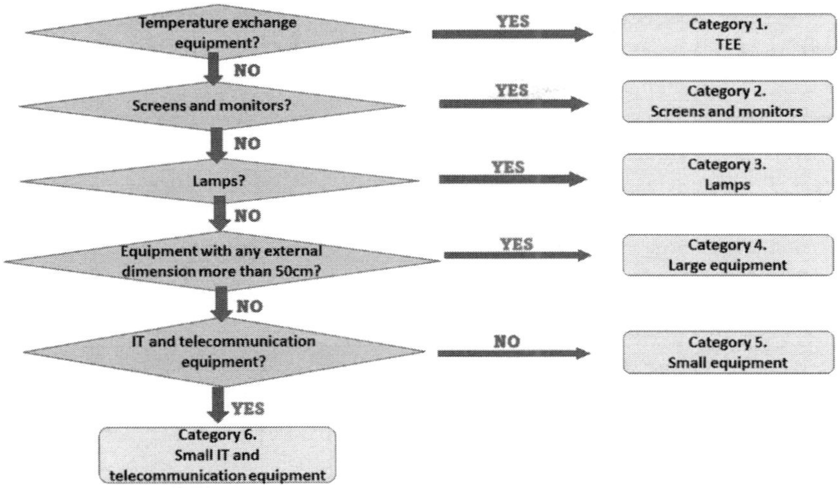

Figure 22.2 Decision tree for categorization. (*Source:* Adopted from EWRN 2018).

of waste falls into hazardous/toxic and non-hazardous waste. There is a strong debate globally when it comes to this classification (Ottoni 2021). The hazardous elements/chemical substances contained in WEEE, which make them termed "hazardous WEEE," include lead (Pb), mercury (Hg), chromium VI (Cr^{6+}), strontium (Sr), barium (Ba), cadmium (Cd), nickel (Ni), arsenic (As), antimony (Sb), gallium (Ga), americium (Am), beryllium (Be), selenium (Se), indium (In), bromide (Br), barium (Ba), zinc (Zn), lithium (Li), cobalt (Co), phosphorus (P), chlorofluorocarbon (CFC), polyvinylchloride (PVC), polychlorinated biphenyls (PCB), ethanediol, methyl cellosolve, etc.

On the other hand, the harmless materials contained in this WEEE include Sn, Cu, Al, Mg, Fe, Si, Ag, Au, PGM-Pd, Ru, PGM-P, Ta, Te, Ge, REE-Nd, REE-cerium oxide, REE-Y, Eu, Tb, REE-La, REE-Gd, C, S, Bi, Nb, W, Ni, etc. (Kaya 2019).

The European List of Wastes (LoW) classifies waste in the EU for administration, that is to say, for the supervision and the permits in waste generation and management. The LoW is used by many European, Central Asian, and Caucasian countries to collect data regarding waste statistics. The LoW defines 839 waste types, organized into 20 chapters, with each type of waste given a 6-digit code. The LoW codes that refer to e-waste are 13, which are subgrouped into hazardous and non-hazardous waste as seen in Table 22.2 (Forti et al. 2018).

22.4 Global (International) E-waste Statistics and E-waste Value

In 2019, global e-waste studies reported 53.6 Mt of e-waste, which demonstrated a startling increase compared to the previous five years: a 21% increase. Projecting these results into 2030 (less than a decade), approximate 74 Mt waste worldwide is predicted to be generated. In 2014, just about half of this amount was generated, implying that global e-waste generation is fast-paced, owing to the frequent

Table 22.2 The European list of wastes under hazardous and non-hazardous components. (Adapted from Forti et al. 2018).

Hazardous	Non-hazardous
Single-use cameras with batteries	Single-use cameras without batteries
Transformers and capacitors containing PCBs	Single-use cameras with batteries
Discarded equipment containing or contaminated by PCBs	Discarded equipment
Discarded equipment containing chlorofluorocarbons (HCF or HFCH)	Discarded EEE
Discarded equipment with free asbestos	
Discarded equipment with hazardous components	
Fluorescent tubes and other Hg-containing waste	
Discarded equipment with CFCs	
Discarded EEE with hazardous components	

updates in technology, short life cycles and short life spans, difficulty to repair, and alteration in the way consumers behave toward the electronics.

Figure 22.3 shows the top ten global e-waste producers. China (22.1 Mt), the USA (6.9 Mt), and India (3.2 Mt) are the first three e-waste-producing countries in the world. The reports of e-waste generation in the year 2019 according to continents, in decreasing order as follows: Asia, with the highest e-waste generation at 24.9 Mt, followed by America at 13.1 Mt, then Europe at 12 Mt, Africa at 2.9 Mt, and Oceania at 0.7 Mt. Of all this waste, just about 17.4% were accounted for, while the remaining percentage, its use, and environmental and health impact remained uncertain. It is irrefutable that e-waste contains massive quantities of resources that could be recycled, such as plastics, precious metals, base metals, and even non-metals. Unfortunately, rather than processing this waste to extend the resource lifespan, through secondary recovery, the waste was either dumped or incinerated. The metals in the e-waste are approximately valued at $65 billion, which is a fortune and higher than the GDP of many countries (Murthy and Ramakrishna 2022).

From the statistics of Kumar et al. (2017), they suggested that higher GDP countries are more likely to generate more e-waste, while high-population countries do not necessarily generate significantly high amounts of e-waste if they have a lower GDP or lower purchasing power.

22.5 Electrical and Electronic Equipment Sales

The sales data between the years 2013 and 2016 for various electronics in the world are shown in Table 22.3.

Table 22.3 Quantity of electronics sold (Kumar et al. 2017 / with permission of Elsevier).

Items	Units (in millions)	Year
Total mobile phones	12 445	2015
Desktops and laptops	239	2016
Total televisions	7.1	2015
Printers	106 000	2014
E-book reader	20.2	2015
Home appliances	583	2015
Electric ovens	0.73 (the USA)	2015
Refrigerator	11.1 (the USA)	2013
Automatic washers	9.7 (the USA)	2013

22.6 State-of-the-Art WPCB/WEEE Recycling Systems

There are four steps involved in the e-waste recycling value chain (Kaya 2018):

- Collection and sorting/classification
- Desoldering and disassembly/dismantling
- Pre-processing (communition and separation)
- End-processing (smelting, leaching, and refining)

22.6.1 Collection and Sorting

The collection step of WEEE is the first and foremost step that will eventually determine how much material is available for the recycling process. From region to region, these collection policies vary, with the collection rates being more affected by social factors than by the collection methods. The

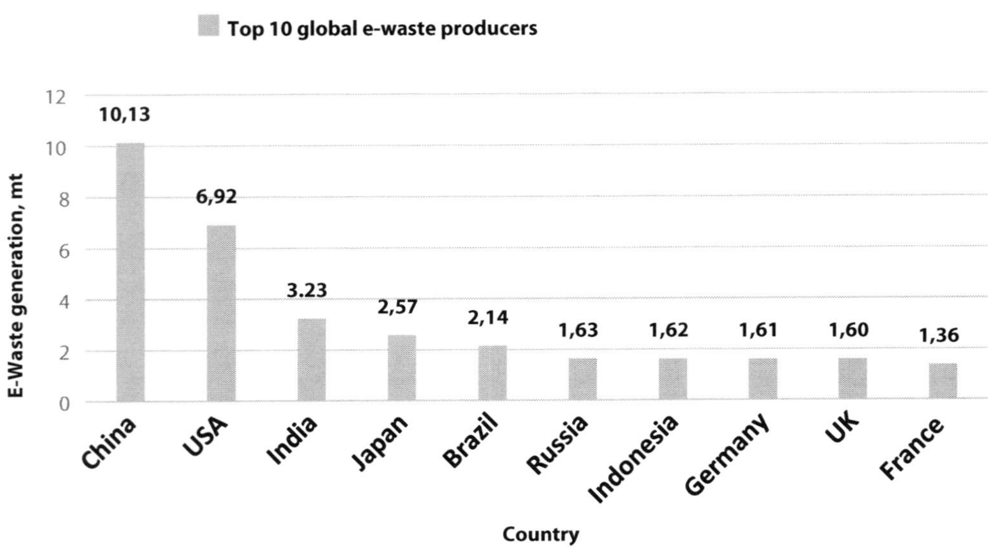

Figure 22.3 Top ten global e-waste producers (Murthy and Ramakrishna 2022 / MDPI / Public Domain CC BY 4.0).

economic value of WEEE increases as the recycling chain progresses to very low economic value at the start of the chain and very high economic value after processing. The collection, dismantling, and pre-processing techniques may differ concerning the material types of the waste, as well as the recycling technologies that are available to use (Kaya 2018).

22.6.2 Dismantling/Disassembly

Disassembly could be defined as "A systematic approach that enables the detachment of electronic components (ECs) from a PCB/WEEE or the separation of a product into all its subparts for a particular purpose" (Vermesan et al. 2019). This is a vital step in the WEEE recycling process. Dismantling aims to free the ECs and channel them to their appropriate following treatment processes. It is in this step that the hazardous components need to be separated from the valuable ones, and their fates treated differently: the useful components depending on the state will be reused or channelled to efficient recycling processes, while the hazardous components will be removed and destroyed, or stored, or treated safely. Dismantling is the most costly and time-consuming operation in the recycling chain (Kaya 2018). Also, it is a necessary process, as the reuse of components is its priority, followed by the hazardous and valuable components' separation. The valuable parts of e-waste generally could be removed by manual dismantling, mechanical treatment, or a combination. Most recycling plants use manual dismantling (Rousis et al. 2008), and different sorts of tools are used in this process for the removal of hazardous components and to recover valuable components.

The materials and components of WEEE render them complex and non-homogeneous, which makes recycling a challenge. It is always necessary to identify and quantify both the hazardous and the useful materials, as well as to look out for the physical properties. WEEE comprises a mix of different ECs of different shapes and sizes, with some of the components being hazardous, and needing their removal for a separate treatment process. Table 22.4 shows components of WEEE that should be treated selectively due to their hazardous content.

Manual dismantling of PCBs before the shredding stage will greatly prevent PM and SM losses. Added to this are other advantages such as its cost-effectiveness, the use of simple unsophisticated electric/pneumatic tools, and the labor force is very cheap because special skills are not required to carry out the manual dismantling. The use of manual or automated dismantling could be selective or simultaneous depending on the objective. Selective dismantling targets either the value or the hazardous components for special storage or treatment.

However, regarding the use of automated disassembly, Boks and Tempelman (1998) researched the potential of future disassembly in the automotive industry. The authors selected a panel consisting of about 70 specialists whose opinions stood as the data for this research. These panel specialists' task was to give their opinion on not only the technical feasibility, but also the economic attractiveness of limited, partial, and full automation of all disassembly operations. Limited automation represented 20–50%, partial automation represented 50–90%, and full automation represented 90–100% of all dismantling operations.

Table 22.4 Major hazardous components in waste electrical and electronic equipment. (Adapted from Rousis et al. 2008).

Components	Description
PCBs	Cd occurs in certain components.
Batteries	Pb, Cd, Hg are present.
Asbestos waste	This waste should be treated selectively.
LCDs	Surfaces >220 cm should be removed.
CRTs	Pb is present in the cone glass and fluorescent coating.
Polychlorinated biphenyl (PCB)-containing capacitors	These capacitors should be removed and destroyed safely.
Hg-containing components	In thermostats, relays, sensors, and switches of PCBs, measuring equipment, discharge lamps, medical equipment, telecommunication, data transmission, and smartphones.
CFC, HCFC or HFC-containing equipment	CFCs or HCFCs present in the circuit have to be suitably extracted and destroyed or recycled if possible.
Gas discharge lamps	Hg should be removed.
Toner cartridges, liquid and pasty, color toner	Should be removed and separated from e-waste.
Plastics containing halogenated flame retardants (HFR)	These HFRs are capable of producing toxic components during incineration.

Concerning technical feasibility for full automation, 65% of these specialists believed in a breakthrough in automated disassembly by 2010. Just slightly above half of the panelists thought it would happen in Germany, while about one-third of them thought that by 2020 full automation disassembly of white and brown goods would not be economically viable. What they considered as the main hindrances against a commercially buoyant automated disassembly activity included the following: too many variations in the types of products available, very limited amounts of similar or same products, disassembly-unfriendliness of the product designs, logistic issues, and an unstable waste amount to be disassembled (Rousis et al. 2008).

22.6.3 Pre-processing/Upgrading

This step ensures the liberation and upgrading of the desired materials by physical separation and size reduction. Figure 22.4 shows the diagram demonstrating the flow of the shredding and separation process. This is the step that prepares the materials for the purification and refining end-processing steps such as leaching and smelting. The processes that are mechanical include shredding, pulverizing, sorting, classification, gravity separation, magnetic separation, etc. (Kaya 2018, 2019). Shredding or comminution involves particle size reduction, and equipment such as knife mills, metal shredders, and hammer crushers are used for size reduction of e-waste. Mechanical separation serves as the means of separation of various streams from the shredded material. Magnetic separators separate ferromagnetic iron, steel, and rare earth metals from other materials, while gravity separators recover heavy base and precious metals such as Ag, Au, and Cu from light non-metal materials. These density/gravity separators include air tables, cyclones, and centrifugal separators. Sensors are also employed nowadays for separation purposes, such as infrared sensors for varying plastic separation and optical sensors for glass separation (Kaya 2019; Kumar et al. 2017).

22.6.4 End-Processing/Purification and Refining

End-processing is the step in which valuable metals are recovered from the WEEE that has passed through pre-processing. At this step, there is purification, refining, and recovery. The most widely applied processes are hydrometallurgy, pyrometallurgy, and bio-hydrometallurgy (Kumar et al. 2017). Sometimes pyrometallurgy and hydrometallurgy can be used together in a hybrid process.

Hydrometallurgy involves leaching using specific types of reagents depending on the metals of interest, which will put the valuable metals into their ionic states in solution. Hydro- or electro-metallurgical refining involves processes such as purification using precipitation or cementation, solvent extraction (SX), and electrowinning (EW). Hydrometallurgical precipitation is a chemical reaction in which a solid substance is formed from a solution through the addition of a reagent, change of pH, evaporation, or temperature change. In other words, precipitation will occur when the limit of solubility is exceeded. Cementation is another form of precipitation in which the added reagent to the solution is a metal of higher electro-positivity than that of the metal in the solution. SX is a purification and separation method to separate metals using two immiscible liquids, generally water, which will be the aqueous phase, and an organic solvent, which will form the organic phase. The relative solubilities of the metals in these solvents are the parameter that will bring about the separation, as a net transfer of metals from one phase to another occurs. EW and electrorefining processes involve the passing of direct electrical current in an electrolyte solution from an anode to

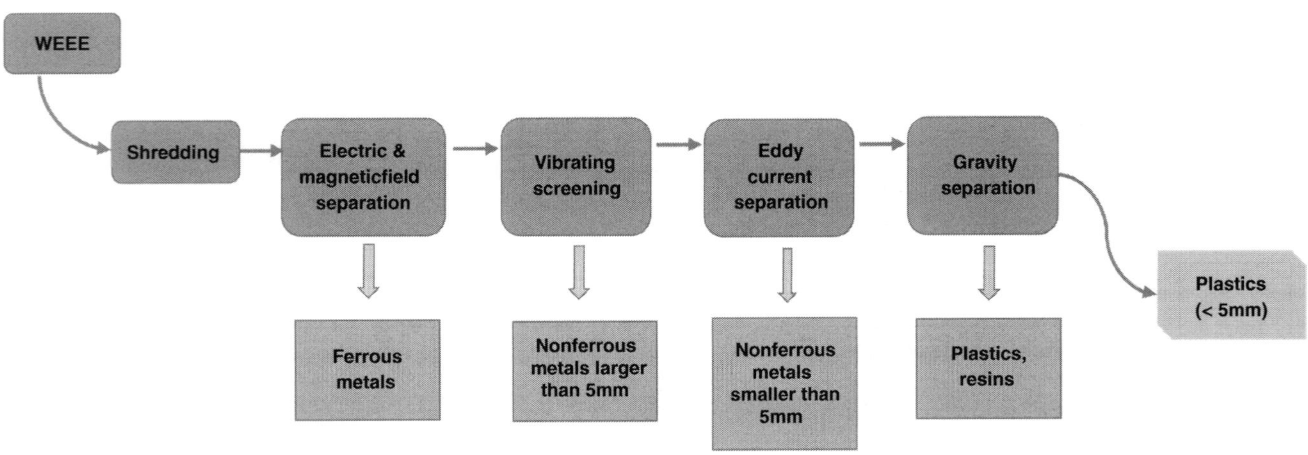

Figure 22.4 Diagram demonstrating the flow of the shredding and separation process (Vermesan et al. 2019 / MDPI / Public Domain CC BY 4.0).

a cathode where the solid metal is deposited. In the EW process, non-ferrous metals such as Zn, Cu, Ag, and Au can be recovered and >99% pure metal is obtainable by EW, while the ferrous fraction can be sent to steel plants for Al and Fe recovery (Kaya 2019).

Pyrometallurgy is another conventional extractive method that involves introducing the materials under very high temperatures to obtain a mixture of the desired metals. Pyrometallurgy is mostly applied to metals such as Cu, Pd, Ag, and Au. Bio-hydrometallurgy is a treatment process in which microorganisms (i.e. fungi and bacteria) are used to enhance the leaching activity of the metals in the solution. This process is continuously gaining popularity, especially for Cu and Au leaching, due to its environmental friendliness, low cost, and simple mode of operation. The most widely used bacterium in bioleaching is the acidophilic bacterium called *Thiobacillus ferroxidans* (Cui et al. 2021).

22.7 E-waste Legislation Across the Globe

From region to region across the globe, different e-waste policies have been put in place to ensure the application of sustainable ways of managing waste. More than 70% of the world's population has at least incorporated e-waste policies in their systems, but within this fraction, illegal e-waste management practices are still performed. Table 22.5 presents the status of e-waste in the different continental regions worldwide.

22.7.1 Asia

The economic situation of a country greatly influences its e-waste management. The necessity of e-waste management in Asia has been recognized in Japan, China, Korea, India, and Singapore such that they have enacted laws for e-waste management. Asian countries have been using the EU's directives as a model for the creation of e-waste legal systems. So far, many regulations regarding e-waste have been established, which encourage concepts like CE, as well as the granting of licenses and permits for conscientious e-waste handling.

The world's highest e-waste producer is China, in Eastern Asia. China alone generated 10.1 Mt of e-waste in 2019, and its pivotal position in the global EEE industry is due to its very high population (most populous in the world), implying a high demand for electronics, as well as its advancement in the manufacturing industry of EEE. China has national legislation for regulating the collection of e-waste divided into 14 types: television, washing machine, refrigerator, PC, air conditioner, water heater, range hood, fax machine, gas water heater, telephone, mobile phone, photocopier, monitor/screen, and printer/plotter. Besides China in East Asia, countries like Japan and South Korea are also advancing considerably in regulating e-waste. Japan regulates e-waste using two acts: the Act on Recycling of Specified Kinds of Home Appliances and the Act on Promotion of Recycling of Small Waste Electrical and Electronic Equipment. Japan also was among the pioneer countries to apply an extended producer responsibility (EPR)-based system on e-waste. Regarding Central and Western Asia, the advancement of e-waste legislation and management is still very slow (Forti et al. 2020).

22.7.2 The Americas

The Americas consist of the Caribbean, Northern America, Central America, and South America. This region has been producing massive quantities of e-waste for decades. The USA and Canada, followed by Brazil and Mexico, are the biggest producers of e-waste in this area, but they have no strict national laws that regulate e-waste management. The USA has not done any ratification on the Basel Convention Ban,

Table 22.5 E-waste status in the different continents across the globe (Forti et al. 2020; Murthy and Ramakrishna 2022).

	Asia	The Americas	Europe	Africa	Oceania
E-waste generated (Mt; kg per capita)	24.9 Mt; 5.6 kg per capita	13.1 Mt; 13.3 kg per capita	12.0 Mt; 16.2 kg per capita	2.9 Mt; 2.5 kg per capita	0.7 Mt; 16.1 kg per capita
E-waste is legally collected and recycled properly (%; Mt)	11.7%; 2.9 Mt	9.4%; 1.2 Mt	42.5%; 5.1 Mt	0.9; 0.03 Mt	8.8%; 0.06 Mt
Number of countries having e-waste policy	17	10	37	13	1
Economic value of raw materials in e-waste (billion dollars)	26.4	14.2	12.9	3.2	0.7
Countries with the highest e-waste generation	China, India, and Japan	USA, Brazil, and Mexico	Russia, Germany, and the UK	Egypt, Nigeria, and South Africa	Australia and New Zealand
Population covered (millions)	4445	984	740	1152	42

and even though some of the states have applied separate e-waste laws in their zones (25 states, as well as the District of Columbia, have implemented some policies), there is still no federal policy that governs e-waste management. All the states that have enacted laws in the USA apply the EPR approach, except for California and Utah (Forti et al. 2020). The export of e-waste could be considered legal in the USA as a result of the absence of any jurisdiction over foreign trade for export prevention. In the absence of this federal legislation, huge quantities of e-waste are exported to countries such as Mexico, Africa, and China, where informal processing of these materials takes place, leading to pollution and environmental footprints. Nevertheless, some generic measures have been put in place for domestic e-waste regulation and management through some acts and frameworks (Patil and Ramakrishna 2020).

Canada, just like the USA, lacks national e-waste legislation for regulation and management, though all the provinces and territories have their local policies in place, except for Nunavut (the least populated territory in Canada), which does not yet have e-waste regulations in its territory. Compared to the USA, Canada has a wider product scope.

Major parts of Latin America also are not bound by e-waste national policies. However, countries like Mexico, Peru, Costa Rica, Chile, Colombia, Ecuador, and Bolivia have legislation for e-waste management, and these countries have all applied the EPR system. There has been considerable progress in e-waste regulation in Latin America over the past half to one decade, even though this growth is limited to just a few countries such as Mexico, Costa Rica, Colombia, and Peru, where they are working toward improving their already established e-waste legislation systems. Brazil is creating the bases to implement a formal e-waste regulatory system.

The absence of uniformity in e-waste regulations, because there is no general legislation to lean toward, is an issue in the Americas. The individual states each have their policies that have differences at so many levels:

- At the level of approach (EPR versus shared responsibility versus public sector programs)
- At the jurisdiction level (federal versus state versus city)
- How fundamental principles are defined
- How roles and responsibilities are allocated
- The stakeholders involved, etc.

22.7.3 Africa

Africa produces about 5.4% of global e-waste (Murthy and Ramakrishna 2022) but besides that, a huge amount of WEEE is (illegally) imported from developed countries into Africa, where mostly crude and informal recycling processes are employed (Grant and Oteng-Ababio 2012), with high pollution consequences. All this is happening because of the absence of policies governing e-waste management and recycling in most African countries. Just a few countries like Cameroon, Nigeria, Ghana, Egypt, South Africa, Rwanda, Ivory Coast, Kenya, and Madagascar have published e-waste legislation, even though the application and enforcement of these laws seem to be far-fetched. Some African countries, however, have come together to adopt a practical and sustainable regional e-waste strategy; for example, the East African Community comprising Tanzania, Kenya, Burundi, Rwanda, Sudan, and Uganda has formed a local e-waste management framework called EACO.

In summary, e-waste regulation issues are similar for the different African countries. These issues include:

- The recycling sector is dominated by an informal system, which leads to a high pollution risk.
- Adequate recycling facilities are limited.
- There is a lack of an EPR system and collection system.
- Public awareness is limited.
- There are limited funds for managing hazardous waste.

22.7.4 Europe

The European Union commission passed the Waste Shipment Regulation (WSR) in 1993, immediately after the signing of the Basel Treaty. This resulted in the banning of the exportation of hazardous e-waste to any country. Ten years later, the Restriction of Hazardous Substances (RoHS) Directive (2012/95/EC) was introduced, which encouraged the use of eco-friendly materials in production and packaging to reduce hazardous chemicals contained in the e-waste. Additionally, this directive helped to improve the recycling rate of home-generated WEEE.

The greater part of the e-waste in Europe is regulated under the WEEE Directive (2012/19/EU). The WEEE Directive works hand-in-glove with the EU and Norway, and it is an extensive management regulation of e-waste that regulates the collection, recycling, reuse, and recovery of WEEE, targeting all six categories of e-waste. According to the Directive, the information regarding how much e-waste has been processed and recycled must also be reported to the National Enforcement Authority. Member states must support that EEE is designed and produced in a way that disassembly and recycling after EoL are facilitated. In pursuit of eco-friendly processing, the WEEE Directive sets procedure standards for treating different materials and ECs of electronic waste, together with where the e-waste is stored. EPR is also included in this Directive, where producers take the responsibility to recycle the EoL products. Some of the new developments in the WEEE Directive are the initiation of the open scope as well as specified reporting guidelines (Murthy and Ramakrishna 2022; Forti et al. 2020).

22.7.5 Oceania

Oceania is made up of Australia, New Zealand, and the Pacific Island subregion. Regarding the implementation of e-waste legislation, only Australia has adopted e-waste management laws. The National Television and Computer Recycling Scheme (NTCRS), which was implemented in Australia under the Product Stewardship Act 2011, is a regulation that gives householders and small enterprises a means to connect to industry-funded services for the collection and recycling of computers. It has stood as a paradigm for the other countries in Oceania in the enforcement of systems like EPR. New Zealand as well as other Pacific Island countries are also beginning to emulate these laws by introducing product stewardship frameworks. Collaboration with the EU is also another way in which some smaller countries on the Pacific Island manage their hazardous waste (Patil and Ramakrishna 2020).

22.8 Recycling Reasons–The Potential of E-waste in a Circular Economy

The CE promotes the recirculation of secondary base, precious, rare, and critical materials contained in WEEE for pollution prevention and materials management. E-waste generation/accumulation acts as a strict threat to the environment and also becomes an urban mine for future supply chain-related risks. Globally, EPR policy provides access to the WEEE urban mining. EPR can be implemented either voluntarily or mandatorily. Without collection targets, the implementation of EPR is called voluntary; while under national legislation collection targets are called mandatory implementation. The EPR-authorized producers must channel their e-waste to authorized recyclers or producer responsibility organizations. Therefore, authorized recyclers or producer responsibility organizations have access to the secondary raw materials embedded in e-waste (Kaya 2021).

E-waste is considered a vital secondary source of materials, especially within the model of a circular economy. Problems that are associated with primary mining, material scarcity, fluctuations in market price, availability of resources, access to the resources, overfilled landfills, etc. could be greatly mitigated if there is an improvement in the mining of secondary resources. The benefits of recycling can be divided into four groups: economic benefits, environmental benefits, energy-saving benefits, and public health and safety reasons.

A circular economy tries to minimize generated waste and to use resources continuously. The value of raw materials, products, and natural resources is maintained in the CE for a longer time, and waste creation should be eliminated as much as possible. CE models use reuse, sharing, refurbishment, remanufacturing, repair, and recycling to generate a closed-loop system, minimizing resource use, eliminating waste pollution, and reducing the carbon footprint. The CE goal is to hold products, equipment, and infrastructure in use for a much longer time, therefore improving the productivity of these natural resources. All waste produced becomes a by-product of another industrial process. This sustainable CE approach is opposite to the traditional LE, which has a "take, make, use, dispose, and pollute" model of production. In the nineteenth century, LE gained importance during the industrial revolution and now leads to serious pollution and over-exploitation of scarce natural resources. CE is described in the three basic principles of designing to avoid waste and pollution, keeping materials and products in use as long as possible, and regenerating natural systems.

Recycling is one of the major ingredients of the CE because of its catalytic role in transforming post-consumer materials into valuable products, feeding used materials back into the value chain, and therefore achieving the "waste-to-resource" model. This converts waste into fewer pure materials with minimum energy/water consumption and fewer environmental influences. There are different ways to close the material loop in CE (Binnemans et al. 2013):

- Direct recycling of clean and homogeneous pre-consumer scrap and manufacturing industrial residues (i.e. swarf, slags, etc.).
- Urban/technospheric mining of EoL post-consumer goods and other products (used, heterogeneous, and complex).
- Landfill mining of historically landfilled pre/post-consumer Municipal Solid Waste (MSW), or Incinerated Waste (IW) streams.

Recycling is both a labor- and capital-intensive growing industry that requires a substantial amount of financial funding and different skills for the collection, separation, and extraction operations. Despite the financial limitations, recycling can provide a low-cost and eco-friendly solution for e-waste management between the mid-term and long-term. A profitable regional/national/international industry can be developed from the recovery of precious, critical, strategic, and base metals embedded in e-waste (www.bir.org).

22.8.1 Economic Benefits

EEE is very complex concerning its designs and production, and it may contain up to 69 elements from the periodic table. These elements include PMs (Ag, Au, Pt, Cu, Ru, Ir,

Pd, Os, Rh, etc.), non-critical metals (Fe and Al), and critical metals (e.g., Co, In, Pd, Sb, Bi, Ge, etc.). As of 2019, the value of selected raw materials that 25 Mt of e-waste contained was estimated at around US$57 billion, and most of the total weight of the raw waste material came from Al, Fe, and Cu. In that same year, the demand for Al, Fe, and Cu was estimated at around 39 Mt to produce new electronics. It can be seen that there is a great gap between the secondary metals found in e-waste and their demand for the production of EEE, which is a consequence of the fast and continuously growing technology (Forti et al. 2020; Kaya 2021).

The most expensive part of e-waste is represented by waste PCBs, which account for more than 40% of the total metal value of e-waste. As of 2012, it was reported that yearly, the electronic industry uses up 7 500 tons of Ag and 320 tons of Au and that US$21 billion could be generated from the urban mining of e-waste. In 2016, it was concluded that over 70% of Ag and 80% of Au and PGMs were in monitors, screens, and small information technology equipment. The possible revenue from waste PCBs was estimated at US$21 200/ton (Kaya 2021; Kumar et al. 2017).

Comparing conventional mining with urban mining, the metal concentration found in the latter is considerably higher. As time goes by, the global ore grade also diminishes, leading to the excavation of more complex ore to meet global demand. In 2014, the Electronics Takeback Coalition estimated that for one million units of mobile phones, which weigh about 148.4 tons, 24 kg of Au, 250 kg of Ag, 9 kg of Pd, and 9 tons of Cu are contained. For this same quantity of Au, Ag, Pd, and Cu a run of mine ore will need 23 762 tons, 1160 tons, 3333 tons, and 1500 tons, respectively. E-waste prevents the extinction of certain critical minerals (i.e. Ga and In).

E-waste recycling is a newly developing industrial sector, a potential job creator, and a facilitator of the CE, acting as a catalyst and channel of the materials that are fed back into the circular system. Therefore, the recycling process presents a win-win-win opportunity as—if harnessed correctly and efficiently—it offers the opportunity to address economic, environmental, and social challenges. Recycling industries globally employ approximately 1.6 million people worldwide and handle yearly more than 600 Mt of recyclables. With annual revenue of more than $200 billion, the recycling sector has now become a major driver for future development. (www.bip.org). In 2015, close to 100 000 workers were employed as informal e-waste recyclers. It is also estimated that recycling computers has the potential of creating 296 more jobs yearly for every 10 000 tons of computer waste that are processed (Unhelkar 2010).

22.8.2 Environmental Benefits

The major environmental benefit of recycling is that it permits the separation of hazardous waste from useful waste, which in turn keeps the former out of landfills, therefore decreasing environmental risks associated with the disposal of this waste. Toxic materials including lead, cadmium, mercury, chromium, chlorofluorocarbons (CFCs), hydrochlorofluorocarbons (HCFCs), and poly/brominated flame retardants (BFR) are contained in e-waste streams and, if disposed of in the landfills or incinerated, will cause adverse effects to the environment.

Also, the total global demand for new metal production is greatly reduced when e-waste is recycled. Reduction in new metal production means a reduction in the emission of greenhouse gases, which is valuable to the planet. For one computer with a monitor, 0.24 tons of fossil fuel, 21.8 kg of chemicals, and 1.5 tons of water are required (Kyle 2016), which results in a high amount of greenhouse gas emissions. Just looking at Al, Fe, and Cu, recycling them in 2019 helped save up to 15 Mt of CO_2 equivalent emissions compared to using them as virgin raw materials (Forti et al. 2020). Also, some countries have landmass scarcity, such that more and more disposal of waste in landfills without recycling will require more land that is not available. Recycling this waste will provide space for other activities. Countries like Japan and regions like Europe, which suffer landmass scarcity, will be needing viable solutions to save more space.

22.8.3 Energy-Saving Benefits

A significant amount of energy is saved when metals are recycled, compared to virgin materials. This energy-saving and energy consumption is connected to greenhouse gas emissions, as the metals are produced. Energy saving of recycled materials over virgin materials is about 60% for Zn, 65% for Pb, 74% for Fe and steel, 85% for Cu, 95% for Al, 80% for plastic, and 64% for paper (Kumar et al. 2017).

Taking aluminum as an example, the recycling of 10 kg of Al saves about 95% of the energy that is necessary to mine virgin Al. At the same time, 13 kg of bauxite waste residue, 20 kg of CO_2, and 0.11 kg of SO_2 and many other emissions and impacts are prevented (Kaya 2021; Kyle 2016). Likewise, iron and steel recycling saves considerable energy, as up to 74% of energy is saved when recycled material is used rather than virgin material. Approximately 80% of water and air pollution is reduced, up to 97% of mining wastes are reduced, 40% of water usage is reduced, and 90% is saved in the use of virgin materials. Desktop recycling saves 80% of energy and resources, while laptops save 87% of energy and resources (Kumar et al. 2017).

The CE leverages a green design deal that leads to the design of products that are easy to disassemble, reuse, or recycle in a raw material-efficient manner to save energy and resources. The CE is a great opportunity to significantly reduce environmental problems and create economic gains at the same time. "84 Bt of natural resources are extracted from earth crust (minerals, ores, fossil fuels, biomass) which can be used as physical products, construction, mobility, pharmaceuticals, consumables, food; however, less than 10% of it has been recycled back into our systems" (UNIDO 2018).

22.8.4 Public Health and Safety Reasons

E-waste contains some hazardous materials, as shown previously in Table 22.2. These hazardous materials have been recently found to affect public health adversely, which can cause kidney and liver damage and mental impairment. Over the last few years, many studies on adverse health effects from e-waste have increased, and recent research has found that informal mining of e-waste harms health such as in alteration of neurodevelopment, birth outcomes, DNA damage, cardiovascular effects, respiratory effects, effects on the immune system, hearing loss, skin diseases, and cancer (Forti et al. 2020). Table 22.6 shows some hazardous materials in e-waste and their effect on humankind.

Table 22.6 Some toxic materials and their effects on human health (Kumar et al. 2017 / with permission of Elsevier).

Material	Impact on human health
Cadmium	Kidney and bone structure damage
Lead	Irreversible effects on the nervous system
Antimony	Serious skin problems
Mercury	Central nervous system and kidney damage
Polychlorinated biphenyls	Immune system suppression, nervous system and liver damage, reproductive system damage, cancer vulnerability
Polybrominated diphenyl ether	Growth hormone interference, brain development interference in animals
Polychlorinated naphthalene	Effects on liver, skin, reproductive system, and nervous system
Nonylphenol	Intersex in fish, DNA damage, sperm function damage in humans

22.9 E-waste Concerns, Opportunities, and Challenges

Based on the data from the *Global E-waste Monitor 2020* (Forti et al. 2020), less than 18% of e-waste is being legally collected and recycled. This implies that a larger proportion of e-waste is recycled informally by scrap dealers who use rudimentary methods, resulting in adverse health and environmental impacts. These informal scrap dealers employ mostly vulnerable social groups like children, prisoners, immigrants, women, etc., to carry out these high-risk operations (Kaya 2018). There is a high need for e-waste collection and recycling rates to be improved, to curb these issues that arise from informal mining.

The recycling sector also experiences high costs in recycling, and the recycling challenges involved especially in separating the hazardous from the useful materials, as well as the safe storage/disposal of these hazardous materials. For example, some materials like germanium and indium are so dispersedly used in products that their recovery becomes challenging. In addition, product designs and assembly of components are not recycle-friendly, making disassembly and separation difficult.

Despite these challenges, recycling e-waste offers numerous advantages, as indicated in the previous section. Devices such as mobile phones and computers contain relatively high concentrations of base and PMs such as Au, for instance, 280 grams per ton of e-waste. In addition, the methods of separation and recycling can be economically viable if carried out manually, because material losses will be <5%. Also, separate collection and recycling will in turn be economically viable, especially when they contain high levels of concentrations of precious metals.

22.10 Recycling of Polymers/Plastics

Light, cheap, and versatile polymeric plastics are the principal materials of present-day CE. Their production and consumption are expected to multiply over the next 10 years. WEEE contains about 15% of plastics without BFR and 5% with BFR. Electronic component populated PCBs have a share of 30% polymeric plastics. PCBs' non-metal fraction (NmF) includes approximately 70% epoxy resins and 28% glass fiber, and 2% unseparated metals. The most well-known plastic polymers in waste PCBs are: 10–16% PE, 4.8% PS, 4.8% PP, 4.8% epoxy, 2.4% PTPE, 2.4% PVC, and 0.9% nylon (UNIDO 2018; Kaya 2020a). When plastics are open-air burned without control, toxic dioxins and furans occur. However, globally less than 15% of all plastic is collected for recycling after use, and huge quantities are left in the environment. Plastic recycling saves about 80%

of energy over virgin materials (Kaya 2016). Light plastics are diamagnetic and can be separated from heavy metals by either density separation (such as air current separation, fluidized bed separators, shaking tables, jigs, and heavy medium separation) or Corona electrostatic separation (to separate non-metallic plastics from non-ferrous Al-Cu metals) (Kaya 2017, 2018, 2018a). Non-metals (plastics and glass fiber) can be used as fillers for modified asphalt, phenolic molding product, and wood-plastic composite products. In most industrial recycling plants, organic plastics are used as reducers and energy sources in the metal smelting process (Kaya 2018, 2020c).

The non-metal fraction (NmF) of e-waste can be directly and chemically recycled. Combustion, incineration, pyrolysis, depolymerization with supercritical fluids, plasma treatment, gasification, hydrothermal methods, hydrogenolysis degradation, and chemical leaching are some of the processes used. Chemical recycling converts polymer materials (resin/plastic) into smaller monomer molecules (such as liquid, oil, or gaseous products), which are suitable for the production of new petrochemical and plastic feedstock (Kaya 2021). Chemical recycling products can be used as fuel for energy production. A comparison of incineration and pyrolysis shows that pyrolysis reduces toxic gas emissions (Hg, Pb, or Cd); syngas, oil, and solid residue are produced, and HBr/Br_2 evolves in part of the oil/gas. The disadvantages of pyrolysis are the high cost of operation, Ar/N_2 gas requirement, and less heat recovery as compared to incineration (Kaya 2020d). Currently, pyrolysis has the following restrictions to limit industrial-scale operations:

- Low gas and liquid production yields decrease economic benefits.
- A mixture of pyrolysis residues and metals should be further hydrometallurgically separated and purified, and hazardous components should be properly disposed of.

Energy in the form of heat, steam, and electricity can be obtained from incineration, which reduces the volume, stabilizes waste, and recovers energy and sterilizes waste. During energy recovery, CFCs and other BFRs create harmful emissions (like CO_2, NO_x, SO_x, VOCs, PMs, PAHs, dioxins, furans, etc.).

22.11 Recycling of Metals

Metals are the heart and centerpiece of the global CE. Metals are valuable materials with properties of toughness, thermal and electrical conductivity, and high-performance temperatures. However, metals are also part of the challenge society is facing in its transformation to a low-carbon, resource-efficient Green Economy. Metals are a scarce natural resource, and their management, consumption, and production need a suitable recycling economy. Additionally, metals are different from other materials in that they are intrinsically recyclable and can be used over and over again, reducing the need for mining and treating new raw materials and thus protecting substantial amounts of energy and water while minimizing environmental degradation in the process (Kaya 2021).

In developing/underdeveloped countries, >90% of metals are recovered informally, but recycling by the informal sector poses serious health risks and threats. In these countries, the informal sector uses women, children, prisoners, and immigrant labor, which increases environmental pollution and health hazards. Despite this, e-waste recycling offers an important economic opportunity, so the numerous challenges associated with recycling and the informal sector need to be investigated (Kaya 2020a).

Populated waste PCBs contain 40% metals. The most frequent metals in PCBs are Cu, Fe, Al, Sn, Pb, Zn, Ni, Sb, Au, Ag, Pd, Pt, and Co. Weight share distribution of PCBs changes Plastic > Cu > Fe > Al > Ag > Au > Pd and value share distribution changes Cu > Al > Au > Pd > Ag (Kaya 2016). Based on the UNDP report, globally less than one-third of 60 economically significant metals are recycled at rates of more than 50%. More than half of the critical and REE metals are recycled at rates less than 1%. Kaya (2020c) gives a summary table about industrial-scale e-waste or WPCB recycling lines along with processes and products in the world.

22.12 Conclusions

E-waste has potential wealth and also poses a hazardous menace to the environment. It must be exploited feasibly for the CE and environmental concerns. Reuse, recycling, and materials recovery through cradle-to-cradle waste management from e-waste on an industrial scale can resolve the environmental problems and furthermore add value to the national CE. Sustainable CE is a newly developing economic model targeted at minimizing the waste amount and the continuous use of natural resources. Products, materials, and natural resources should be kept in the economy as long as possible, and waste formation should be eliminated as much as possible. The linear economy (LE) or cradle-to-grave approach is no longer relevant due to severe effects like environmental deterioration, resource depletion, crises, rigorous environmental policy, and landfilling or disposal into nature or ocean/sea. E-waste is the fastest-escalating municipal solid waste stream and a significant toxicity contributor. The CE is a strategy for pollution prevention and e-waste management. The concept of the CE offers sustainable growth, healthy life, and decent jobs for new employment while seriously protecting the environment and scarce natural resources on the earth. The CE significantly contributes to material value recovery from e-waste, not only from the material value recovery prospects but also from different rigid environmental directives and regulations perspectives.

The CE leverages the green design concept for easy dismantling, reusing, or recycling in a secondary resource-efficient manner to conserve energy, water, and raw materials. The CE concept becomes a new sustainable and state-of-the-art model for flexible growth for the economies of both emerging and developed countries. The CE opens many new lucrative economic opportunities for emerging countries especially.

The new concept of urban mining from e-waste is an important potential source of secondary raw materials (i.e. metals and polymers). Recycling e-waste generates significant economic potential and reduces the import dependency of raw materials for emerging countries. Post-consumer wastes are transformed into valuable substances, feeding waste materials back into the value chain and achieving the waste-to-resource model.

Currently, about 80% of e-waste is undocumented for proper collection and recycling. Yearly consumed e-waste amount changes from 5–30 kg/inhabitant. Europe has the highest rate of collection and the most developed waste legislation and directives. Recycling and reuse of metals are a necessity for a resource-efficient sustainable economy. However, cost-effective and eco-friendly recycling industrial flow charts have already been developed only for some base and precious metals. EoL recycling of REE and LIB components is still negligible. Informal recycling leads to the destruction of the environment, loss of valuable material, and causes serious health hazards for its residents. The global "from waste to resources" call can be achieved by integrated and eco-friendly waste policy and management. The product-centric approach automatically dismantles ECs from WPCBs and recycles them separately according to their metal contents. E-waste recycling has three perspectives—namely, economic, environmental, and marketability. Future perspectives of WEEE recycling should concentrate on both maximum material recovery rates and more types of substances (especially PMs, SMs, REEs and other critical metals) recovery. Developments of novel separation and extraction processes should consider a significant improvement in recovery rates and separation efficiencies; better automatic/autonomous control; short process duration; lower labor intensity, CAPEX, and APEX; and minimum environmental impacts simultaneously. Recent cutting-edge technologies should take into consideration industrial-scale applicability, automation in pre-treatment, reduction in process steps, and consumed reagent concentrations/costs and pay more attention to recovering cheap NmF. Finally, e-waste is regarded as a significant secondary source for achieving higher sustainability under the CE.

References

Bartl, A. (2014). Moving from recycling to waste prevention: a review of barriers and enables. *Waste Management Research* 32: 3–18. https://doi.org/10.1177/0734242×14541986.

Binnemans, K., Jones, P.T., Blanpain, B. et al. (2013). Recycling of rare earths: a critical review. *Journal of Cleaner Production* 51: 1–22.

Boks, C. and Tempelman, E. (1998). Future disassembly and recycling technology: results of a Delphi study. *Futures* 30 (5): 425–442.

Bureau of International Recycling. www.bir.org.

Circular Economy—Definition, Principles, Benefits and Barriers (2020). Youmatter. https://youmatter.world/en/definition/definitions-circular-economy-meaning-definition-benefits-barriers (accessed 10 March 22).

Circular Economy Introduction (2022). https://ellenmacarthurfoundation.org/topics/circular-economy-introduction/overview (accessed 14 March 22).

Cui, X., Yuan, X., Li, H. et al. (2021). Copper and zinc recovery from sulfide concentrate by novel artificial microbial community. *Metals* 12: 45. https://doi.org/10.3390/met12010045.

EWRN (2018). WEEE2—definition and understanding of the 6 categories.

Forti, V., Baldé, C.P., Kuehr, R., and Bel, G. (2020). *The Global E-waste Monitor*. 120. Bonn/Geneva/Vienna: United Nations University.

Forti, V., Baldé, K., and Kuehr, R. (2018). *E-waste Statistics: Guidelines on Classifications, Reporting and Indicators*. United Nations University, Sycle Books.

Grant, R. and Oteng-Ababio, M. (2012). Mapping the invisible and real "African" economy: urban E-waste circuitry. *Urban Geography* 33: 1–21. https://doi.org/10.2747/0272-3638.33.1.1.

Kaya, M. (2016). Recovery of metals and nonmetals from electronic waste by physical and chemical recycling processes. *Waste Management* 57: 64–90. https://doi.org/10.1016/j.wasman.2016.08.004.

Kaya, M. (2017). Recovery of metals and nonmetals from waste printed circuit boards (PCBs) by physical recycling techniques. In: *Energy Technology 2017 Book* (eds. L.

Zhang, J.W. Drelich, N.R. Neelameggham et al.), 433–451. Springer. https://doi.org/10.1007/978-3-319-52192-3_43.

Kaya, M. (2018). Waste printed circuit board (WPCB) recovery technology: disassembly and desoldering approach. *Reference Module in Materials Science and Materials Engineering/ Encyclopedia of Renewable and Sustainable Materials* 4: 1–19. Elsevier. https://doi.org/10.1016/B978-0-12-803581-8.11246-9.

Kaya, M. (2018a). Chapter 3: Current WEEE recycling solutions, in waste electrical and electronic equipment recycling—aqueous recovery methods. In: *Woodhead Publishing Series in Electronic and Optical Materials* (eds. F. Veglio and I. Birloaga), 33–93. Elsevier. https://doi.org/10.1016/B978-0-08-102057-9.00003-2.

Kaya, M. (2019). Waste printed circuit board (WPCB) recycling: conventional and emerging technology approach. *Encyclopedia of Renewable and Sustainable Materials* Elsevier Ltd., 1–18. https://doi.org/10.1016/b978-0-12803581-8.11296-2.

Kaya, M. (2019a). *Electronic Waste and Printed Circuit Board Recycling Technologies, the Minerals, Metals & Materials Series*. Cham, Switzerland: Springer International Publishing. https://doi.org/10.1007/978-3-030-26593-9.

Kaya, M. (2020a). Chapter 1: E-waste and E-waste recycling. In: *Electronic Waste and Printed Circuit Board Recycling Technology* (ed. M. Kaya), 1–30. Springer. https://doi.org/10.1007/978-3-030-26593-9_1.

Kaya, M. (2020b). Chapter 2: Printed circuit boards (PCBs). In: *Electronic Waste and Printed Circuit Board Recycling Technology* (ed. M. Kaya), 33–55. Springer. https://doi.org/10.1007/978-3-030-26593-9_2.

Kaya, M. (2020c). Chapter 8: Industrial-scale E-waste/WPCB recycling lines. In: *Electronic Waste and Printed Circuit Board Recycling Technology* (ed. M. Kaya), 177–209. Springer. https://doi.org/10.1007/978-3-030-26593-9_8.

Kaya, M. (2020d). Chapter 9: Recycling of NMF from WPCBs. In: *Electronic Waste and Printed Circuit Board Recycling Technology* (ed. M. Kaya), 211–218. Springer. https://doi.org/10.1007/978-3-030-26593-9_9.

Kaya, M. (2021). Industrial applications of the circular economy for e-waste recycling. In: *Horizons in Earth Research* (ed. B. Veress and J. Szigethy), vol. 22, ch. 3. 117–184.

Kumar, A., Holuszko, M., and Espinosa, D.C.R. (2017). E-waste: an overview on generation, collection, legislation and recycling practices. *Resources, Conservation and Recycling* 122: 32–42. https://doi.org/10.1016/j.resconrec.2017.01.018.

Kyle, B. (2016). Facts and figures on E-waste and recycling.

Murthy, V. and Ramakrishna, S. (2022). A review on global E-waste management: urban mining towards a sustainable future and circular economy. *Sustainability* 14: 647. https://doi.org/10.3390/su14020647.

Ottoni, M.D.S.O. (2021). Assessment of a reverse logistics framework for waste printed circuit boards (WPCB) in Brazil, 173.

Patil, R.A. and Ramakrishna, S. (2020). A comprehensive analysis of e-waste legislation worldwide. *Environmental Science and Pollution Research* 27: 14412–14431. https://doi.org/10.1007/s11356-020-07992-1.

Rousis, K., Moustakas, K., Stylianou, M. et al. (2008). Management of waste from electrical and electronic equipment: the case of television sets and refrigerators. *Journal of Environmental Engineering and Science* 7: 105–114. https://doi.org/10.1139/S07-035.

Unhelkar, B. (2010). *Handbook of Research on Green ICT: Technology, Business and Social Perspectives: Technology, Business and Social Perspectives*. IGI Global.

UNIDO (2018). Circular economy: development of recycling industries. *Meeting Proceedings*, 14–15, November 2018, Vienna, Austria.

Vermesan, H., Tiuc, A.-E., and Purcar, M. (2019). Advanced recovery techniques for waste materials from IT and telecommunication equipment printed circuit boards. *Sustainability* 12: 74. https://doi.org/10.3390/su12010074.

23

The E-waste Scenario

Analytical Techniques for Effective Management

Hema Diwan, Nitha T.S., and Juan Mandy*

National Institute of Industrial Engineering, Powai, India 400087
[*] Corresponding author

23.1 Introduction

Global waste is forecasted to reach 74 million metric tons by 2030, and e-waste is becoming one of the fastest-growing waste streams worldwide (Ikhlayel 2017). This can be inferred from increased electrical and electronic hardware utilization, life cycle limits, and repair decision limits. The United Nations *Global E-waste Monitor 2020* estimated production of around 53.6 million tons of electronics in 2019, hinting at an increase of approximately 20% in five years. Global waste is forecasted to reach 74 million metric tons by 2030, with new research showing a nearly significant increase in just 16 years. According to a survey, Asia provided the largest amount of e-waste in 2019, followed by the USA and Europe. Of this, the maximum quantum of the e-waste generated was assembled and repurposed, showcasing an untapped opportunity to embrace a circular economy for reaping environmental and economic benefits. The opportunity is untapped because an estimated $57 billion in valuable, recoverable components (more than most countries' GDP) ends up in landfills rather than adequately disposed of. E-waste sees minimal disassembly, even though this sector has immense potential for reuse, recycling, and refurbishment. E-waste is a critical issue because of the nature of hazards and toxic properties associated with it (Milovantseva and Saphores 2012; Roy et al. 2022; UNEP 2015).

23.1.1 E-waste Nomenclature

E-waste, mainly waste of electrical and electronic equipment, also known as WEEE, is defined by the European Union (EU) as electrical or electronic hardware that is garbage, including all parts, subassemblies, and disposables that are essential for the item at the hour it is disposed of. In India, e-waste is characterized as electrical and electronic hardware (EEE) disposed of as waste by customers (exclusively or in mass), as well as rejected from assembling, restoration, and fixing processes, as characterized by the Central Pollution Control Board (CPCB) Waste (Management) Rules of 2016. The term "e-waste" alludes to all end-of-life hardware we wish to dispose of, to destroy, and reuse.

Electronic waste (e-waste) primarily (Grant et al. 2013; Man et al. 2013) includes computing devices, televisions, monitors, and electronic devices (Frazzoli et al. 2010; Grant et al. 2013). Expansions in industrialization have led to a proliferation of electrical and electronic (EEE) products, eventually leading to an increase in e-waste once these products achieve their life cycle. After China and the US, India is the world's third biggest creator of e-waste per year, with 3.2 million tons. Consumer electronics is one of the fastest-growing market segments, with over $200 billion in revenue in the US alone (Consumer Technology Association 2017). Metals, plastics, and glass constitute most of the e-waste. Important metals include palladium, iron, lead, zinc, silver, gold, tin, and nickel. Computer circuit boards are a rich source of rare metals popularly known as platinum group metals (Ikhlayel et al. 2017; Forti et al. 2022). Expansions in industrialization have led to an expansion of electrical and electronic gear (EEE) available, eventually leading to an increase in e-waste. As per the Worldwide E-waste Screen 2020, 53.6 million tons of electronic trash were created internationally in 2019, with 17.4% reused. This environmental liability can be seen as an opportunity, as an estimated $57 billion in valuable, recoverable components (which is more than most countries' GDP) ends up in landfills rather than being scientifically disposed of (Frazzoli et al. 2010; Grant et al. 2013; Rajesh et al. 2022).

Electronic Waste Management: Policies, Processes, Technologies, and Impact, First Edition. Edited by Sunil Kumar and Vineet Kumar.

23.2 Research Developments in the Field

Elsevier's Scopus database has been explored to understand the e-waste scenario globally with particular reference to analytical tools like LCA. Scopus is a rich database for scientific research and includes titles from Elsevier, Springer, Wiley, Taylor & Francis, and Emerald. The Biblioshiny interface of R studio software has been used to retrieve data and analyze the evolving field of study.

A descriptive analysis of the data retrieved showed a total of 1865 research papers published in the area of interfacing e-waste and analytical tools.

The search string used *E-waste*, *LCA*, *MFA. as the significant keywords, identifying 1865 research papers. The majority 1438 documents published were research articles; the rest were reviews, book chapters, and editorials. Of this, 131 research papers were single-authored, whereas the collaboration index was 3.65. The annual scientific publication count is also observed, as shown in Table 23.1

Table 23.1 Descriptive analysis.

Descriptive statistics	
Description	**Results**
Main information about data	
Time span	1961–2022
Sources (journals, books, etc.)	745
Documents	1865
Average years from publication	7.84
Average citations per document	29.68
Average citations per year per document	4.028
References	1
Document types	
Article	1438
Book	2
Book chapter	68
Conference paper	204
Conference review	8
Data paper	2
Editorial	2
Erratum	2
Letter	1
Note	5
Review	133
Document contents	
Keywords plus (ID)	0
Author's keywords (DE)	5278
Authors	
Authors	6448
Author appearances	7940
Authors of single-authored documents	118
Authors of multi-authored documents	6330
Authors collaboration	
Single-authored documents	131
Documents per author	0.289
Authors per document	3.46
Co-authors per document	4.26
Collaboration index	3.65

23.2.1 Keyword Analysis

The trending searches in the domain of e-waste management research are e-waste, life cycle assessment, LCA, sustainability, waste management, recycling, circular economy, and carbon footprint mapping. The search pattern indicates an emerging trend in quantitative research hinting at the role of data analytics in assessing, predicting, and forecasting e-waste generation, which is critical for its scientific management. Also evident is the interfacing of technological innovations in waste management technologies, be it chemical, thermal, or biological technologies like pyrolysis, incineration, or landfilling.

Another stream of research visible was energy recovery options concerning e-waste. This trend indicates the rising integration of circularity in the waste sector, particularly electronic waste (Table 23.1, Figures 23.1a, 23.1b).

23.2.2 Most Influential Documents

The H index was used to understand the most influential research in interfacing e-waste and analytics. It was observed that the top research papers were by Rebitzer et al. (2004) and (Figure 23.2).

23.2.3 Most Relevant Affiliations

The data was also analyzed to know the significant academic institutions publishing in the domain of study. It was observed that the Technical University of Denmark, Ghent University, Purdue University, University of California, Delft University of Technology, and University of Tehran (Figures 23.3a, 23.3b, 23.3c) appeared most prominently.

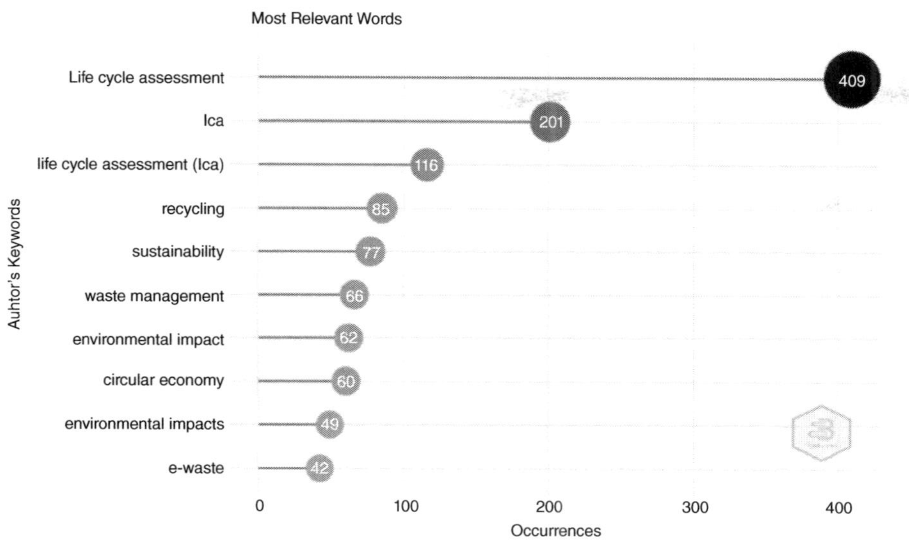

Figure 23.1a Most relevant words.

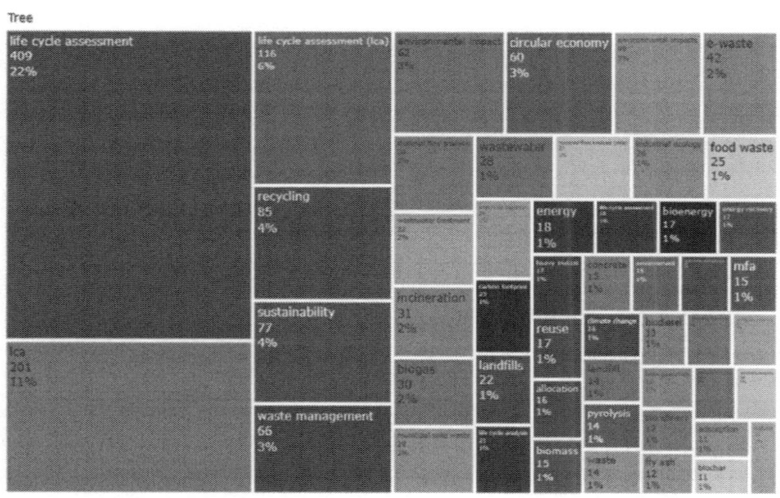

Figure 23.1b Most relevant keywords.

23.2.4 E-waste Generation: Indian Scenario

India is one of the world's significant importers of e-waste, representing around 3% of the world's overall equipment gadgets creation. Homegrown hardware yield represents 2.3% of India's GDP. There has been a recent increase in the production of portable handsets, LCD TVs, and LED items like TVs, as interest in these devices has continued to grow. The volume of cell phone creation reached 225 million dollars (22.5 billion rupees) units in 2017–18 from 60 million dollars (6 million) units in 2014–15. Over the past three or four years, 268 cell phone creation units have been set up in the country.

Expected annual changes in waste generation (13%) are complex to prevent, specifically in terms of unit and volume of e-waste and recycled products, considering only domestic generation and usage patterns. With the addition of import information, the expected year-over-year change in disposal age may increase overall. Given the ideal situation of 13% growth, with or without extrapolation, India will generate 7 million tons of waste annually by 2025 to about 160 million tons by 2050.

23.3 E-waste and Circular Systems

To induce circularity in the waste management space, maximum thrust is needed in realigning the forward and reverse supply chains toward closed resource optimization. In this goal, reverse supply chains can be considered one of the most potent tools for e-waste management to introduce circularity in the waste sector. Reverse

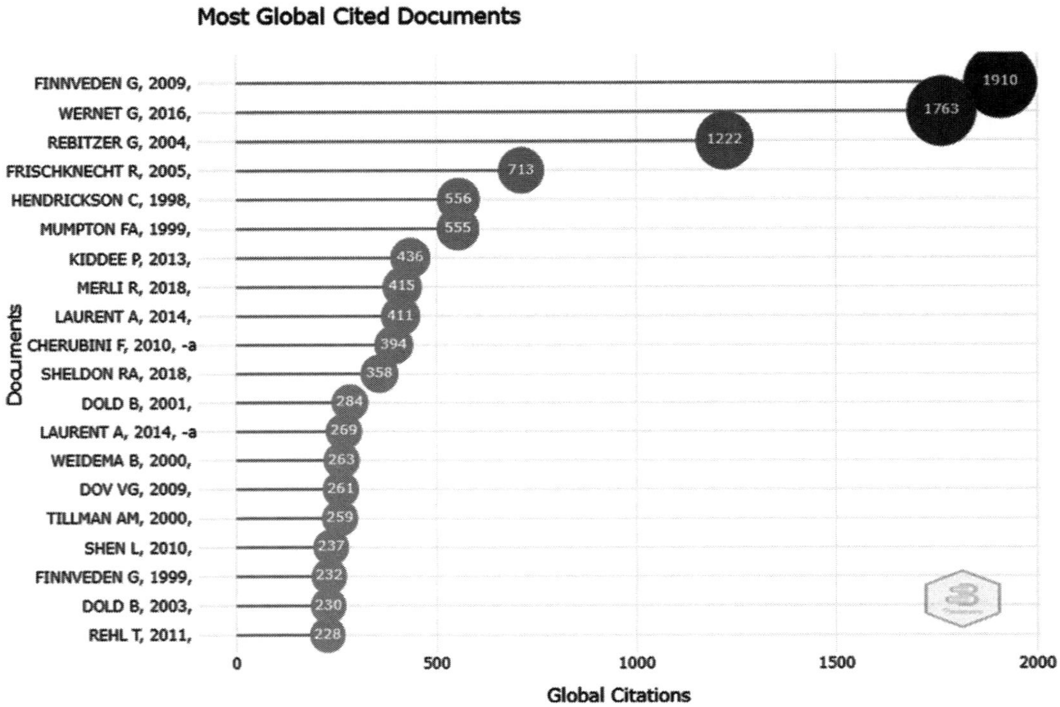

Figure 23.2 Most influential documents.

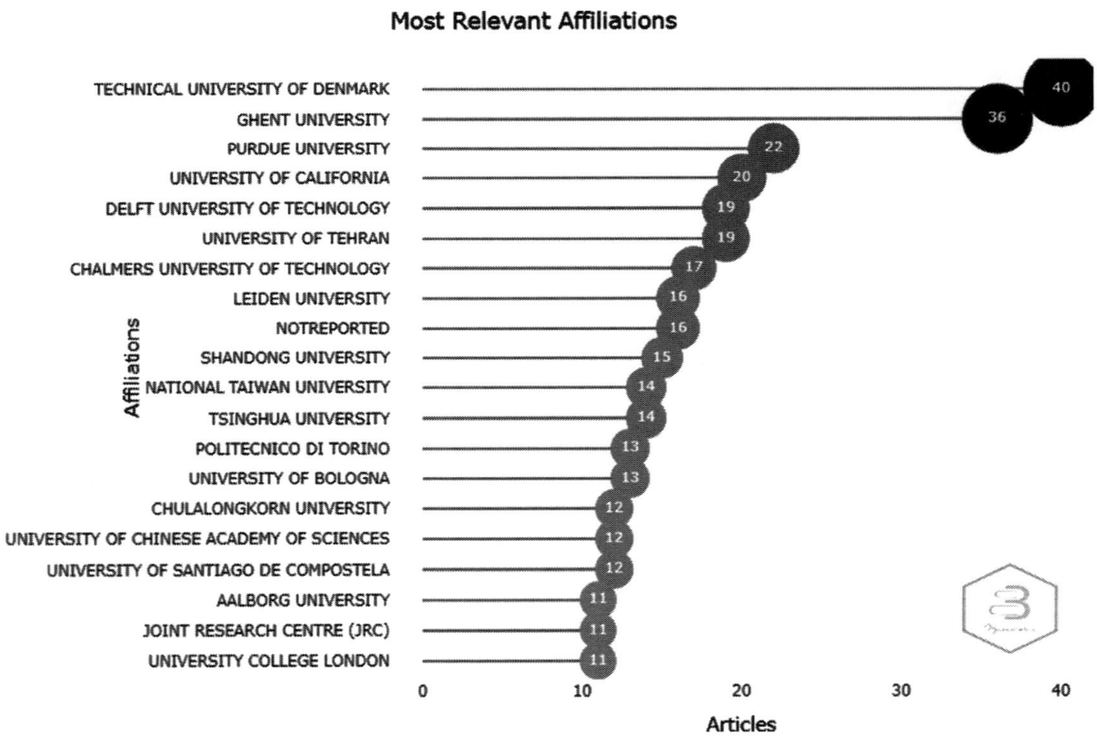

Figure 23.3a Most relevant affiliations.

supply chains are a subset of a closed-loop supply chain that leads only not to e-waste management but also to resource conservation and economic value creation (Tian et al. 2022).

23.3.1 Process Flow of E-waste Recycling

23.3.1.1 Product Acquisition

The business of collecting and collating returned products from consumers for waste treatment is known as product

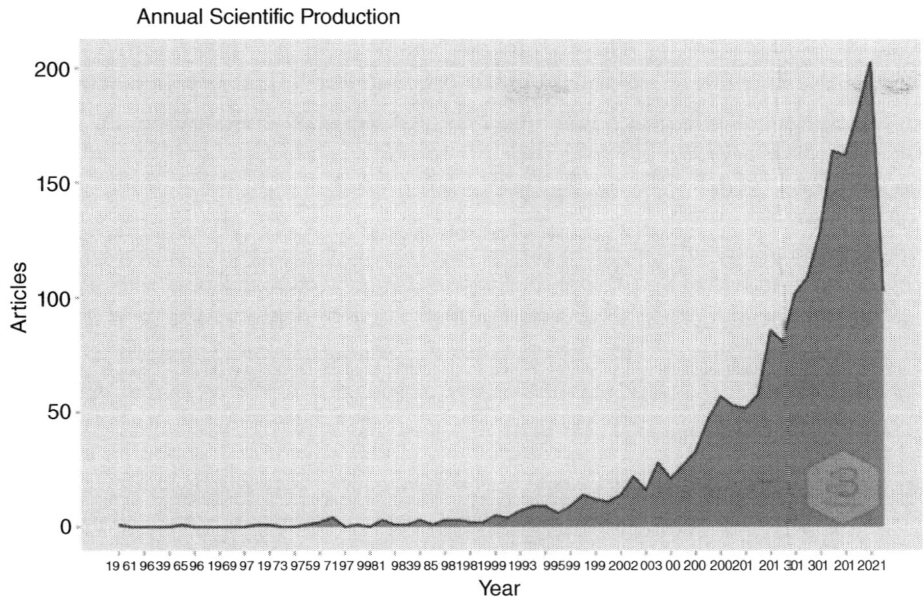

Figure 23.3b Annual scientific production.

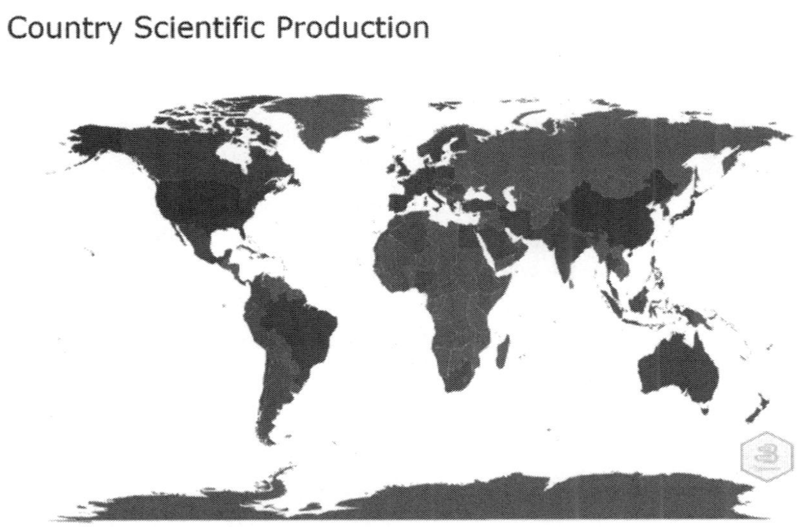

Figure 23.3c Scientific production by country.

acquisition. Return products are collected in three ways: through the original supply chain, through an existing reverse supply chain, or through the unorganized waste sector. Product returns or recalls are typical examples of forward chain collection.

23.3.1.2 Reverse Logistics

The WEEE stream is managed at the next level by employing reverse logistics functions to recover used products. Returned products are shipped to facilities for inspection, sorting, and arrangement in this process. Logistic costs often make up most of the entire expense of the reverse system. The reverse supply chain is sometimes used synonymously with reverse logistics (RL); however, there exists a difference in the connotation of both as reverse logistics is a subset of reverse supply chain (Prahinski and Kocabasoglu 2006). Reverse logistics is involved in transportation, warehouse, and inventory management (Kang et al. 2020).

Manual sorting, filing, separation, and dismantling are the initial phases of waste processing and reuse, followed by mechanical processing, including destruction, shredding, and other strategies. The recovered material is transferred to the appropriate entity to process and recover the assets and material. Waste recycling can be a broad term that includes many different stakeholders.

Activities like recycling and reuse, as well as remanufacturing and refurbishment, can be a part of a successful reverse logistics plan. Companies may switch to profitable closed-loop supply chains by incorporating reverse logistics and sustainability into daily operations. The key components are collection, categorization, and materialization. For practical disposal or value recovery, materials and products must return to their original locations as part of the supply chain (Ikhlayel 2017, 2018; Doan et al. 2019; Garg and Adhana 2019).

Reduce:

- Reduce the generation of e-waste through smart procurement and maintenance of goods.
- In innovative procurement, tools like artificial intelligence (AI) can be used for

1) Spending classification
2) Anomaly detection
3) Vendor matching

Reuse:

- Reuse the electronic equipment that is still functioning by donating and selling it to someone who can still use it.
- Reuse the working parts that are left after disassembling.

Recycle:

- Coupling recycling initiatives by applying assessment tools, artificial learning, and machine learning can save electronics manufacturers billions of dollars while also putting an end to e-waste pollution. AI, ML, and robotics electronics can be recycled quickly, and valuable parts can be saved for reuse.

23.3.1.3 Inspection and Disposition

The examination and disposal procedures aim to spot the standard of used products and, therefore, the appropriate strategy for recycling each product within the reverse supply chain system.

23.3.1.4 Refurbishment

If a product has to be upgraded or fixed after the inspection and disposition step, it will be delivered to a refurbishment practice such as a renovating or refashioning center.

23.3.1.5 Redistribution and Sales

The reverse supply chain relocation and trading methodology are very similar to the forward inventory network methodology. The previous option offers recovered or recycled materials or parts on the market, and the last option manages new items. Individual customers who belong to the conventional market, or new customers in various business areas who do not require or can handle the cost of new hardware, are targeted as buyers of these parts or materials (Figures 23.4a, 23.4b, 23.4c).

23.4 E-waste Management

Policy Instrument: *Extended Producer Responsibility*

Extended Producer Responsibility (EPR) is emerging as the tool for creating formal waste management setups across countries. Extended producer responsibility is an environmentally friendly arrangement in which producer responsibility applies to items at a post-sales stage, including final removal.

As the waste sector is highly unorganized and informal, EPR is a policy instrument that can formalize the role of

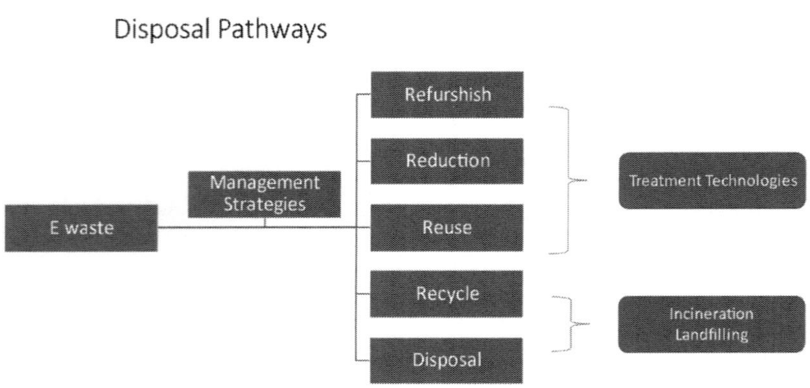

Figure 23.4a Disposal strategies.

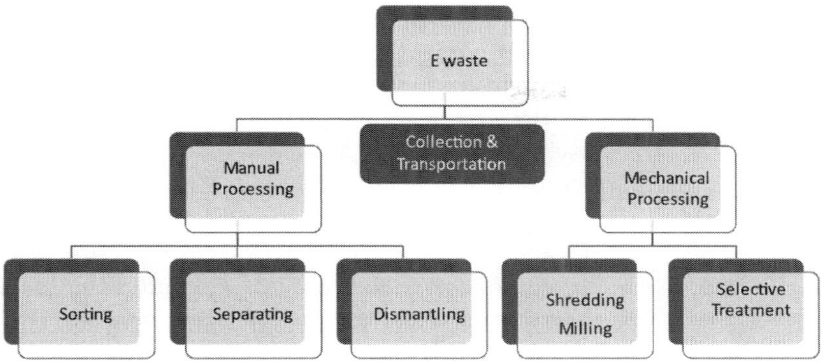

Figure 23.4b E-waste management strategies.

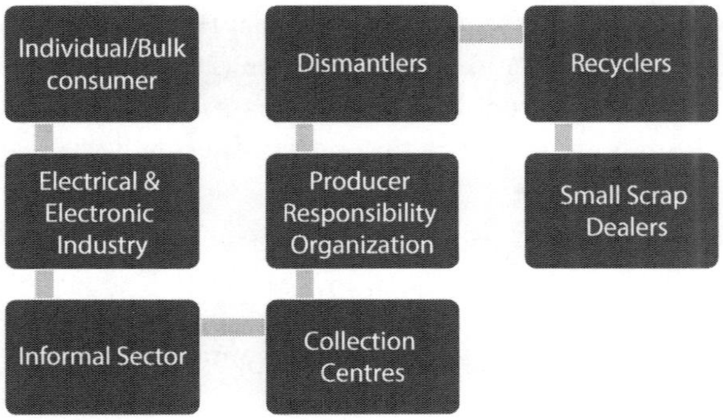

Figure 23.4c Value chain partners.

critical stakeholders like producers in the waste value chain (E-waste Rules 2016).

The commitment of makers at the post-purchaser phase of an item's anticipation, which includes distribution, transportation, capacity, dismantling, reusing, and removal, is alluded to as natural assurance. EPR advocates a push and pull mechanism, which creates a demand and supply loop to formalize the e-waste market. The mechanism is extended from product to component, consumables, spares, and parts, entailing a systems-level approach to e-waste management (Dhote 2017). Besides this, a target-based approach similar to those prevalent in countries like Japan, the UK, the Netherlands, and South Korea has been adopted (Table 23.2).

23.4.1 Analytical Tools

The scientific literature and e-waste figures identify vital emerging areas in waste management, reverse logistics, reverse supply chains, recycling, material recovery, and material flow analysis. However, with the rapid progress of information technology, analytical methods based on waste and disposal data are expanding. These approaches aim to accomplish adequate waste management by reducing the environmental footprint and corresponding expenses and increasing the societal acceptance of waste management. Wise waste management could also be established by combining environmentally friendly and economically viable alternatives. The industry 4.0 architecture facilitates

Table 23.2 E-waste categories.

Category I: Information technology and telecommunication equipment	Category II: Consumer electrical and electronics
Centralized data processing: Mainframes, minicomputers	Television sets (including sets based on liquid crystal display and light emitting diode technology)
Personal computing: Personal computers	Refrigerators
Laptop computers	Washing machines
Notebook computers	Air conditioners excluding centralized air-conditioning plants
Notepad computers	Fluorescent and other mercury-containing lamps
Printers including cartridges	
Copying equipment Electrical and electronic typewriters User terminals and systems	
Telex and telephones	
Cellular telephones and answering systems	

the design of data-driven approaches to address environmental problems. "Big data" is an ideal realization technology for considering all kinds of data and gaining insights from waste data through horizontal and vertical analysis.

Material Flow Analysis (MFA) and Life Cycle Analysis (LCA) are two tools commonly used in waste management and linked to WEEE management. Both strategies are frequently mentioned in the literature as critical for making WEEE management decisions. Many experts worldwide see MFA as a great way to study, research, and direct complex waste systems. This tool is well known for identifying challenges and breaks in the value chain and designing management strategies to address them. This is chiefly useful for complex and diverse waste flows. LCA is a popular and successful environmental management tool for gauging a product's or service's environmental impact and identifying hotspots and opportunities for improvement. Some scholars have proposed LCA as a critical tool to measure the ecological effects of waste management technologies and to evaluate the ecologically sound execution of waste management strategies, which includes WEEE management (Hong et al. 2014; Hong et al. 2015).

23.4.2 Material Flow Analysis (MFA)

Material flow analysis (MFA) is a widely recognized and committed device in modern science that examines input materials and studies the trajectory and movement of each material flow within a frame. The MFA method can be applied to analyze the flows of resources in a system of study. For example, it can be used to study the flows of recycled resources at a material level in WEEE. MFA studies can help make inventories of the amounts/concentrations of unused or in-use products/materials/components. This can be used to set the strategies for their management, e.g., reuse, recycling, or recovery (Lin 2008; Murakami et al. 2009).

MFA is considered to be a vital decision support tool in resource and management because it can explore the path of material inflow to a recycling plant as a space and time material store intricacy. The tool is used to determine the material stocks at a regional level. It can be used for identifying challenges and breaks in the value chain and designing management strategies to address them. The tool's most significant hurdles are data quantity and quality, both of which can be solved; however, with the rapid progress of information technology, analytical methods based on waste and disposal data are expanding.

Various research provides approaches to address the growing scenarios of e-waste management (Lin et al. 2010). Studies have calculated environmental impacts by combining MFA and LCA. Future research on complex MFA-LCA strategies features a loss of potential because it is extremely useful for deciding in terms of efficient management of finite resources (Lau et al. 2013; Ismail and Marlia (2021). As long as the circular economy is a relatively young research area that is constantly evolving, further research is required on the utilization of the notion of the circular economy in waste management. Research studies of PVC waste in Thailand (Nakem et al. 2016) have attempted to use MFA and LCA toward sustainable management. Similar studies have been conducted by Thushari et al. (2020a, 2020b); Withanage and Habib (2021); and Yahom et al. (2016). Used MFA to identify underlying socio-economic causes of WEEE flows in Switzerland. Similar work was done by Lau et al. (2013), who performed a MFA of WEEE flows in Hong Kong. Van Eygen et al. (2016) and Niero and Olsen (2016) used MFA to conduct simulation studies on the mass balance of recycling aluminum cans in several loops by using mathematical modeling.

23.4.3 Life Cycle Assessment (LCA)

LCA is a popular and successful environmental management tool for gauging a product's or service's environmental impact and identifying hotspots and opportunities for improvement (Hong et al. 2014; Mishra et al. 2019; Tefa et al. 2022). Some scholars have proposed LCA to evaluate the ecologically sound execution of WEEE management (Liu et al, 2017; Ardolino et al. 2022; Sakthivel et al. 2022). LCA is used to create ecologically sound electronic products and reduce e-waste complexities. It identifies potential environmental consequences so that eco-design goods, processes, or services can be developed. Carcinogens, global climate change, ozone, toxicity, land usage, and acidification are all samples of environmental effect categories defined by LCA. All those LCA applications are utilized to enhance product environmental performance and generate environmentally friendly products. The performance can be measured across the subsystems like collection, transportation, treatment, and disposal stages of the value chain. LCA can also be performed upstream and downstream of the supply chain dealing with electronic products. LCA applications enhance product environmental performance and generate environmentally friendly products. The LCA can be applied at material recovery, energy recovery, and disposal stages and can help forecast the damage potential beforehand; accordingly, the waste strategies can be devised. LCA studies to understand the impacts of different WEEE treatment options have been conducted by (Hong et al. 2014; Hong et al. 2015; Lau et al. 2013; Van Eygen et al. 2016; Liu et al. 2020, 2021a, 2021b).

Table 23.3 Impact assessment categories.

Impact assessment methods	Impact assessment categories	Metrics
CML	Acidification	Acidification potential
Eco Indicator	Climate change	GWP 100 years/20/500
ILCD 2011	Resource depletion	Depletion of abiotic resources
ReCipe	Ecotoxicity	Marine ecotoxicity, terrestrial ecotoxicity, marine sedimental ecotoxicity
Traci	Energy use	
USEcoTOx	Eutrophication	PO_4^{2-}, NO_3^-
	Human toxicity	HTP 100, HTP 20, HTP 500
	Land use	Land use, land competition
	Ozone layer depletion	ODP steady-state NMVOC; ODP 10, 15, 20, 25, 30, 40
	Photochemical oxidation	EBIR (low NOx); High NOx (including NMVOC); low NOx, MIR (very high NOx); MOIR (high NOx)

Scenarios viz. conventional e-waste management can be simulated to understand the environmental impacts. Allesch and Brunner (2016) used scenario analysis to plan and design waste management interventions through MFA.

Research studies by Ismail and Hanafish (2020), Shumon et al. (2014), and Yong et al. (2019) highlight the policy-related intervention in the e-waste management domain (Table 23.3).

23.4.4 Multi-Criteria Analysis (MCA)

For better planning, development, and implementation of waste management programs, it is imperative to predict and forecast waste generation at regional levels.

Multi-Criteria Analysis (MCA) is a dynamic methodology for making key decisions and solving tedious and time-consuming multi-measurement problems with quantitative or potentially subjective parts. This can be a multi-step cycle to address issues related to the design and decision of innovative construction of waste management frameworks. MCA-related techniques have been applied to analyze multiple criteria and prioritizing them. Hula et al. (2003) worked on the Pareto analysis for identifying EOL strategies to meet economic feasibility. Also, the impact of the European Union Directive on WEEE on EOL was studied to understand the trade-offs between energy recovery and cost.

Multiple linear regression, artificial neural networks, and Bayesian models have been attracting attention in the accurate and reliable forecasting of waste generation.

The ARIMA model, which stands for Auto Regressive Integrated Moving Average, is one of the popular tools that uses time series data to predict future waste loads, helping devise better planning strategies. These models can be used to plan for resource needs, waste collection, waste transportation, disposal at dumpsites, and so on (Dodo et al. 2022). Other time series methods include the Box–Jenkins ARIMA. The grey wave forecasting model can be employed in e-waste management by predicting and forecasting future waste generation loads and designing strategies at regional levels.

23.5 Conclusion

E-waste has been a point of traction for reorienting waste management planning as effective management of WEEE remains a serious challenge in developing countries. Government, industry, policymakers, and consumers are emphasizing integrated waste management principles as the current pollution control approach is becoming redundant, and they are looking for more preventive strategies. The sector today stands unorganized, and there is a need for stakeholders from industry and government to put systems and mechanisms in place and operationalize reuse, recycling, and reduction in production. Reverse logistics and reverse supply chains have to be aligned to induce circularity in the e-waste sector. The use of technology for data gathering and analysis using smart data techniques is emerging as a way forward for the waste sector.

References

Ardolino, A., Palladini, A.R., and Arena, U. (2022). A preliminary social assessment of innovative management options for mixed and hazardous plastics waste. In: *Wastelca_3: Life Cycle Sustainability Assessment for Waste Management and Resource Optimization*. Italy: ECI Symposium Series.https://dc.engconfintl.org/lca_waste_3/45.

Consumer Technology Association (2017). 19th Annual consumer technology ownership and market potential study. April 2017.

de Meester, S., Nachtergaele, P., Debaveye, S. (2019). Using material flow analysis and life cycle assessment in decision support: a case study on WEEE valorization in Belgium. *Resources, Conservation and Recycling* 142: 1–9.

Dhote, J. (2017). E-waste (management) rule 2016: need to gear-up awareness by environmentalists for its effective implementation. I.J.R.I.T.C.C. Impact Factor: 5.837. 5.

Doan, L.T.T., Amer, Y., and Lee, S.-H. (2019). E-waste reverse supply chain: a review and future perspectives, electronic waste: management and recovery technologies. *Applied Science* 9 (23): 5195.

Dodo, U.A., Ashigwuike E.C., and Nwachukwu, J.E. (2022). Forecast using an A.R.I.M.A. model: a focus on Abuja City. Nigeria IEEE NIGERCON. IEEE Municipal Solid Waste Generation.

Forti, V., Baldé, C.P., Kuehr, R., and Bel, G. (2022). *The Global E-waste Monitor 2020: Quantities, Flows and the Circular Economy Potential*. Bonn/Geneva/Rotterdam: United Nations University (U.N.U.)/United Nations Institute for Training and Research (U.N.I.T.A.R.), co-hosted S.C.Y.C.L.E. Programme, International Telecommunication Union (I.T.U.) and International Solid Waste Association (I.S.W.A.).

Frazzoli, C., Orisakwe, O.E., Dragone, R., and Mantovani, A. (2010). Diagnostic health risk assessment of electronic waste on the general population in developing countries' scenarios. *Environmental Impact Assessment Review* 30: 388–399.

Garg, N. and Adhana, D. (2019). E waste management in India: a study of current scenario (31 January 2019). *International Journal of Management, Technology and Engineering* IX (I).

Grant, K., Goldizen, F.C., Sly, P.D. et al. (2013). Health consequences of exposure to e-waste: a systematic review. *Lancet Global Health* 1: e350-e361.

Hong, I.-H., Lee, Y.-T., and Chang, P.-Y. (2014). Socially optimal and fund-balanced advanced recycling fees and subsidies in a competitive forward and reverse supply chain. *Resources, Conservation and Recycling* 82: 75–85.

Hong, J., Shi, W., Wang, Y. et al. (2015). Life cycle assessment of electronic waste treatment. *Waste Management* 38: 357–365.

Hula, A., Jalali, K., Hamza, K. et al. (2003). Multi-criteria decisionmaking for optimization of product disassembly under multiple situations. *Environmental Science & Technology* 37: 5303–5313.

Ikhlayel, M. (2017). Environmental impacts and benefits of state-of-the-art technologies for E-waste management. *Waste Management* 68: 458–474. https://doi.org/10.1016/j.wasman.2017.06.038. Epub 2017 Jun 26. P.M.I.D.: 28662843.

Ikhlayel, M. (2018). An integrated approach to establish e-waste management systems for developing countries. *Journal of Cleaner Production* 170: 119–130.

Ismail, H. and Hanafiah, M.M. (2020). A review of sustainable e-waste generation and management: present and future perspectives. *Journal of Environmental Management* 264: Article 110495.

Ismail, H. and Marlia, M.H. (2021). Evaluation of e-waste management systems in Malaysia using life cycle assessment and material flow analysis. *Journal of Cleaner Production* 308: 127358. ISSN 0959–6526.

Kang, K.D., Kang, H., Ilankoon, I.M.S.K., and Chong, C.Y. (2020). Electronic waste collection systems using internet of things (IoT): household electronic waste management in Malaysia. *Journal of Cleaner Production* 252: 119801. ISSN 0959–6526.

Lau, W.K., Chung, S.S., and Zhang, C. (2013). A material flow analysis on current electrical and electronic waste disposal from Hong Kong households. *Waste Management* 33: 714–721.

Lau, W.K.Y., Liang, P., Man, Y.B. et al. (2014). Human health risk assessment based on trace metals in suspended air particulates, surface dust, and floor dust from e-waste recycling workshops in Hong Kong, China. *Environmental Science and Pollution Research* 21: 3813–3825.

Lin, C. (2008). A model using home appliance ownership data to evaluate recycling policy performance. *Resources, Conservation and Recycling* 52 (1322–1328).

Lin, C.H., Wen, L., and Tsai, Y.M. (2010). Applying decision-making tools to national e-waste recycling policy: an example of analytic hierarchy process. *Waste Management* 30: 863–869.

Liu, G., Xu, Y., Tian, T. et al. (2020). The impacts of China's fund policy on waste electrical and electronic equipment utilisation. *Journal of Cleaner Production* 251: 119582.

Liu, X., Gu, S., Yang, S. et al. (2021a). Heavy metals in soil-vegetable system around E-waste site and the health risk assessment. *The Science of the Total Environment* 779: Article 146438. https://doi.org/10.1016/j.scitotenv.2021.146438.

Liu, X., Lin, K., and Wang, L. (2021b). Stochastic evolutionary game analysis of e-waste recycling in environmental regulation from the perspective of dual governance system. *Journal of Cleaner Production* 319: 128685.

Liu, Z., Tang, J., Li, B.-y., and Wang, Z. (2017). Trade-off between remanufacturing and recycling of WEEE and the environmental implication under the Chinese fund policy. *Journal of Cleaner Production* 167: 97–109.

Man, M., Naidu, R., and Wong, M.H. (2013). Persistent toxic substances released from uncontrolled e-waste recycling and actions for the future. *Science of the Total Environment* 463–464: 1133–1137.

Milovantseva, N. and Saphores, J.D. (2012). Time bomb or hidden treasure? Characteristics of junk TVs and of the U.S. households who store them. *Waste Management* 33 (3): 519–529.

Mishra, S., Roy, M., and Mohanty, K. (2019). Microalgal bioenergy production under zero-waste biorefinery approach: recent advances and future perspectives. *Bioresource Technology* 292: Article 122008.

Murakami, S., Oshugi, H., Murakami-Suzuki, R. et al. (2009). Average lifespan of mobile phones in in-use and hibernating stocks in Japan. *Journal of Life Cycle Assessment, Japan* 5 (1): 139–145.

Nakem, S., Pipatanatornkul, J., and Papong, S. (2016). Material flow analysis (M.F.A.) and life cycle assessment (L.C.A.) study for sustainable management of P.V.C. wastes in Thailand. In: *Computer Aided Chemical Engineering* (ed. Z. Kravanja and M. Bogataj), 1689–1694. ScienceDirect, Elsevier.

Niero, M. and Olsen, S.I. (2016). Circular economy: to be or not to be in a closed product loop? A life cycle assessment of aluminium cans with inclusion of alloying elements. *Resources, Conservation and Recycling* 114: 18–31. https://doi.org/10.1016/j.resconrec.2016.06.023.

Prahinski, C. and Kocabasoglu, C. (2006). Empirical research opportunities in reverse supply chains. *Omega* 34 (519–532).

Rajesh, R., Kanakadhurga, D., and Prabaharan, N. (2022). Electronic waste: a critical assessment on the unimaginable growing pollutant, legislations and environmental impacts. *Environmental Challenges* 7: 100507. ISSN 2667–0100.

Rebitzer, E.T., Frischknecht, R., Hunkeler, D. et al. (2004). Life cycle assessment, Part 1: framework, goal and scope, inventory analysis and applications. *Environment International* 30: 701–720.

Roy, H., Rahman, T.U., Suhan, M.B.K. et al. (2022). A comprehensive review on hazardous aspects and management strategies of electronic waste. *Bangladesh Perspectives, Heliyon* 8 (7): e09802.

Sakthivel, S., Tesfamicael, S.A., Mekonnen, S. et al. (2022). Recycled cotton/polyester and polypropylene nonwoven hybrid composite materials for household applications. *The Journal of the Textile Institute* 113 1 (45–53): https://doi.org/10.1080/00405000.2021.1913862.

Shumon, H.S., Ahmed, S., and Islam, T. (2014). Electronic waste: present status and future perspectives of sustainable management practices in Malaysia. *Environmental Earth Science* https://doi.org/10.1007/s12665-014-3129-5.

Tefa, L., Bianco, I., Blengini, G.A., and Bassani, M. (2022). Integrated and comparative structural-LCA analysis of unbound and cement-stabilized construction and demolition waste aggregate for subbase road pavement layers formation. *Journal of Cleaner Production* 352: 131599. ISSN 0959–6526.

Thushari, G.G.N. and Senevirathna, J.D.M. (2020a). Plastic pollution in the marine environment. *Heliyon* 6 (8): e04709. ISSN 2405–8440.

Thushari, I., Vicheanteab, J., and Janjaroen, D. (2020b). Material flow analysis and life cycle assessment of solid waste management in urban green areas, Thailand. *Sustainable Environmental Research* (30): 21. https://doi.org/10.1186/s42834-020-00057-5.

Tian, T., Liu, G., and Yasemi, H. (2022). Managing e-waste from a closed-loop lifecycle perspective: China's challenges and fund policy redesign. *Environmental Science Pollution Research* 29: 47713–47724. https://doi.org/10.1007/s11356-022-19227-6.

UNEP, United Nations Environment Program (2015). Illegally traded and dumped e-waste worth up to $19 billion annually poses risks to health, deprives countries of resources. https://www.unep.org/news-and-stories/press-release/illegally-traded-and-dumped-e-waste-worth-19-billion-annually-poses (accessed 29 May 2015).

Van Eygen, E., De Meester, S., Tran, H.P., and Dewulf, J. (2016). Resource savings by urban mining: the case of desktop and laptop computers in Belgium. *Resources, Conservation & Recycling* 107 (C): 53–64.

Wager, P.A., Hischier, R., and Eugster, M. (2011). Environmental impacts of the Swiss collection and recovery systems for waste electrical and electronic equipment (WEEE): a follow-up. *Sciences Total Environmental* 409: 1746–1756.

Withanage, S.V. and Habib, K. (2021). Life cycle assessment and material flow analysis: two under-utilized tools for informing e-waste management. *Sustainability* 13 (14): 1–21. July.

Xue, M. and Xu, Z. (2017). Application of life cycle assessment on electronic waste management: a review. *Environmental Management* 59: 693–707. https://doi.org/10.1007/s00267-016-0812-1.

Yahom, O., Malakul, P., and Charoensaeng, A. (2016). Material flow analysis and life cycle assessment study for sustainable management of wastes of metropolitan electricity authority (M.E.A). *Computer Aided Chemical Engineering* 38: 1695–1700.

Yong, Y.S., Lim, Y.A., and Ilankoon, I. (2019). An analysis of electronic waste management strategies and recycling operations in Malaysia: challenges and future prospects. *Journal of Cleaner Production* 224: 151–166.

Index

Electronic Waste Management: Policies, Processes, Technologies, and Impact, First Edition. Edited by Sunil Kumar and Vineet Kumar.

d

e

f

i

n

q

r

S